2nd International Conference of Indonesian Society for Remote Sensing (ICOIRS 2016)

IOP Conference Series: Earth and Environmental Science
Volume 47

Yogyakarta, Indonesia
17 - 19 October 2016

ISBN: 978-1-5108-3521-4
ISSN: 1755-1307

Printed from e-media with permission by:

Curran Associates, Inc.
57 Morehouse Lane
Red Hook, NY 12571

Some format issues inherent in the e-media version may also appear in this print version.

Printed by Curran Associates, Inc. (2017)

Institute of Physics
Dirac House, Temple Back
Bristol BS1 6BE UK

Phone: 44 1 17 929 7481
Fax: 44 1 17 920 0979

techtracking@iop.org

Additional copies of this publication are available from:

Curran Associates, Inc.
57 Morehouse Lane
Red Hook, NY 12571 USA
Phone: 845-758-0400
Fax: 845-758-2633
Email: curran@proceedings.com
Web: www.proceedings.com

Table of contents

Volume 47

2nd International Conference of Indonesian Society for Remote Sensing (ICOIRS)

17–19 October 2016, Yogyakarta, Indonesia

Accepted papers received: 10 November 2016
Published online: 14 December 2016

Preface

011001
OPEN ACCESS
2nd International Conference of Indonesian Society for Remote Sensing (ICOIRS) 2016

011002
OPEN ACCESS
Peer review statement

Papers

GIS and Cartography

012001
OPEN ACCESS
Map Analysis and Spatial Statistic: Assessment of Spatial Variability of Agriculture Land Conversion at Urban Fringe Area of Yogyakarta

Bowo Susilo.....1

012002
OPEN ACCESS
Image Mining in Remote Sensing for Coastal Wetlands Mapping: from Pixel Based to Object Based Approach

N M Farda, P Danoedoro, Hartono and A Harjoko.....14

012003
OPEN ACCESS
Application of OpenStreetMap (OSM) to Support the Mapping Village in Indonesia

Nurin Swasti Kanthi and Taufik Hery Purwanto.....26

Image and Spectral Data Acquisition and Processing

012004
OPEN ACCESS
Combined aerial and terrestrial images for complete 3D documentation of Singosari Temple based on Structure from Motion algorithm

Husnul Hidayat and A B Cahyono.....33

012005
OPEN ACCESS
The Low-Cost UAV-Based Remote Sensing System Capabilities for Large Scale Cadaster Mapping

Catur Aries Rokhmana and Susilo Utomo.....44

012006
OPEN ACCESS
Aerial Photographs of Landslide on Clapar, Madukara District of Banjarnegara as a Spatial Geodatabase

Anggi Permata, Anwar Juniansyah, Eka Nurcahyati, Mousafi Dimas Afrizal, Muhammad Adnan Shafry Untoro, Na'ima Arifatha, Raden Ramadhani Yudha Adiwijaya, Barandi Sapta Widartono and Taufik Hery Purwanto.....50

012007
OPEN ACCESS
Orthorectification of Sentinel-1 SAR (Synthetic Aperture Radar) Data in Some Parts Of South-eastern Sulawesi Using Sentinel-1 Toolbox

Athar Abdurahman Bayanudin and Retnadi Heru Jatmiko.....58

012008
OPEN ACCESS
Identifying Sugarcane Plantation using LANDSAT-8 Images with Support Vector Machines

Sidik Mulyono and Nadirah.....71

012009
OPEN ACCESS
GEOBIA For Land Use Mapping Using Worldview2 Image In Bengkak Village Coastal, Banyuwangi Regency, East Java

Fitzastri Alrassi, Emil Salim, Anastasia Nina, Luthfi Alwi, Projo Danoedoro and Muhammad Kamal.....79

Remote Sensing and GIS Applications in Coastal and Marine Environment

012010
OPEN ACCESS
Comparison of Chlorophyll-a Measurement Using Multi Spatial Imagery and Numerical Model in Bali Strait

Rizki Hanintyo and Eko susilo.....90

012011
OPEN ACCESS
Coral reef destruction of Small island in 44 years and destructive fishing in Spermonde Archipelago, Indonesia

Nurjannah Nurdin, Teruhisa Komatsu, Chair Rani, Supriadi, Sitti Fakhriyyah and Agus.....105

012012
OPEN ACCESS
Overview of National Thematic Data Integration (An Experience on One Map Mangrove Sulawesi)

A W Rudiastuti, D M Yuwono, Niendyawati, G H Pramono and B D Rahmanto.....114

012013
OPEN ACCESS
Remote sensing and GIS-based site suitability analysis for tourism development in Gili Indah, East Lombok

Debrina Agnes, Akbar Nandatama, Bagus Andi Isdyantoko, Fajri Aditya Nugraha, Giusti Ghivarry, Perwira Putra Aghni, Renaldi ChandraWijaya and Prima Widayani.....129

012014
OPEN ACCESS
Above Ground Carbon Stock Estimates of Mangrove Forest Using Worldview-2 Imagery in Teluk Benoa, Bali

E D Candra, Hartono and P Wicaksono.....133

012015
OPEN ACCESS
Plankton Biomass Models Based on GIS and Remote Sensing Technique for Predicting Marine Megafauna Hotspots in the Solor Waters

MIH Putra, SA Lewis, EM Kurniasih, D Prabuning and E Faiqoh.....144

012016
OPEN ACCESS
Distribution of Glass Eel by the Water Surface Salinity Using Landsat TM at Pelabuhan Ratu Bay, West Java

D S Irianto, Supriatna and TjiongGiok Pin.....163

012017
OPEN ACCESS
Monitoring The Land Accretion Development at Coastal Area of Blanakan, Subang Indonesia

Nandi, Ginna Meriana and Lili Somantri.....171

012018
OPEN ACCESS
Evaluating the performance of Lyzenga's water column correction in case-1 coral reef water using a simulated Wolrdview-2 imagery

Masita Dwi Mandini Manessa, Muhammad Haidar, Syarif Budhiman, Gatot Winarso, Ariyo Kanno, Tatsuyuki Sagawa and Masahiko Sekine.....181

012019
OPEN ACCESS
The Relationship between the Health of Coral Reefs to The Damages of Jelateng Sub Watersheds in the Northern Coast of Sekotong, West Nusa Tenggara

A Zaenuri, A N Saputra, D A Kudiastuti, I Fajria, L AWijayanti and P A Aryaguna.....194

Remote Sensing and GIS Applications in Disaster Risk Reduction and Management

012020
OPEN ACCESS
The Mapping of Temporary Evacuation Site (TES) and Tsunami Evacuation Route in North Pagai Island, Mentawai Islands Regency - Indonesia

Aprizon Putra and Herdiana Mutmainah.....204

012021
OPEN ACCESS
Observing Deformation at Mt. Raung East Java Based on PALSAR-2 Imagery by Using Interferometric SAR

Arliandy P. Arbad, Achmad Ardy and Ridwan A. Ashari.....220

012022
OPEN ACCESS
Measuring Deformation in Jakarta through Long Term Synthetic Aperture Radar (SAR) Data Analysis

Agustan, Albertus Sulaiman and Takeo Ito.....227

012023
OPEN ACCESS
Drought Vulnerability Mapping with Geomorphological Approach in Yogyakarta Special Region (DIY) and Central Java

Sudaryatno.....234

012024
OPEN ACCESS
Modelling tropical cyclone hazards under climate change scenario using geospatial techniques

M A Hoque, S Phinn, C Roelfsema and I Childs.....242

Remote Sensing and GIS Applications in: Geomorphology and Land Resources

012025
OPEN ACCESS
Modeling of Erosion on Jelateng Watershed Using USLE Method, Associated with an Illegal Mining Activities (PETI)

I N Ananda, F V Aswari, D A Narmaningrum, A S A Nugraha, M A A Asidiqi and Y Setiawan.....249

012026
OPEN ACCESS
A comparison of lineament and fracture trace extraction from LANDSAT ETM+ panchromatic band and panchromatic aerial photograph in Gunungsewu karst area, Java-Indonesia

E Haryono, B S Widartono, H Lukito and S B Kusumayuda.....259

Remote Sensing and GIS Applications in: Hydrology and Water Resources

012027
OPEN ACCESS
Determination of The Water Catchment Area in Semarang City Using a Combination of Object Based Image Analysis (OBIA) Classification, InSAR and Geographic Information System (GIS) Methods Based On a High-Resolution SPOT 6 Image and Radar Imagery

Yudo Prasetyo, Setyo Ardi Gunawan and Zia Ul Maksum.....269

012028
OPEN ACCESS
The Aplication of Landsat 8 OLI for Total Suspended Solid (TSS) Mapping in Gajahmungkur Reservoir Wonogiri Regency 2016

Apriwida Yanti, Bowo Susilo and Pramaditya Wicaksono.....281

012029
OPEN ACCESS
Total Suspended Solid (TSS) Mapping of Wadaslintang Reservoir Using Landsat 8 OLI

Muhammad Fauzi and Pramaditya Wicaksono.....293

012030
OPEN ACCESS
The effect of choosing three different C factor formulae derived from NDVI on a fully raster-based erosion modelling

Bambang Sulistyo.....301

Remote Sensing and GIS Applications in: Land Use Mapping

012031
OPEN ACCESS
Spatial Temporal Land Use Change Detection Using Google Earth Data

Adi Wibowo, Khairulmaini Osman Salleh, F.Th.R Sitanala Frans and Jarot Mulyo Semedi.....311

012032
OPEN ACCESS
Coupling of Markov chains and cellular automata spatial models to predict land cover changes (case study: upper Ci Leungsi catchment area)

K. Marko, F. Zulkarnain and E. Kusratmoko.....322

Remote Sensing and GIS Applications in: Precision Agriculture

012033
OPEN ACCESS
Effectiveness of Vegetation Index Transformation for Land Use Identifying and Mapping in the Area of Oil palm Plantation based on SPOT-6 Imagery (Case Study: PT.Tunggal Perkasa Plantations, Air Molek, Indragiri Hulu)

H A Setyowati, S H Murti B S and E S Sukentyas.....332

012034
OPEN ACCESS
Linear Spectral Mixture Analysis of SPOT-7 for Tea Yield Estimation in Pagilaran Estate, Batang Central Java

F Fauziana, P Danoedoro and S Heru Murti.....344

012035
OPEN ACCESS
Applications of Remote Sensing and Geographic Information System to Identify Rice Planting Season During El Nino Years: Case Study in the Pringsewu District, Province of Lampung

S D Ratri, E Kusratmoko and F S Hardiyanti.....354

012036
OPEN ACCESS
Protection of Paddy Field and Recommendation of Regional Planning in Cianjur Regency, West Java, Indonesia

Khursatul Munibah, Rani Yudarwati, Enni Dwi Wahyunie and Hermanu Wijaya.....362

Remote Sensing and GIS Applications in: Urban and Regional Planning

012037
OPEN ACCESS
Role of Remote Sensing and Geographyc Information System Mapping for Protected Areas Land Rice Field Subak, Buffer Zones, and Area Conversion (Case Studies In Gianyar Regency, Bali Province)

Indayati Lanya and N. Netera Subadiyasa.....372

012038
OPEN ACCESS
Urban Spatial Pattern and Interaction based on Analysis of Nighttime Remote Sensing Data and Geo-social Media Information

Nila Ratnasari, Erika Dwi Candra, Defa Herdianta Saputra and Aji Putra Perdana.....380

012039
OPEN ACCESS
Spatial temporal analysis of urban heat hazard in Tangerang City

Adi Wibowo, Kuswantoro, Ardiansyah, Andry Rustanto and Iqbal Putut Ash Shidiq.....390

012040
OPEN ACCESS
Using Visible Infrared Imaging Radiometer Suite (VIIRS) Imagery to identify and analyze light pollution

Wahyu Nurbandi, Febrina Ramadhani Yusuf, Ruwanda Prasetya and Mousafi Dimas Afrizal.....399

Remote Sensing and GIS Applications in: Vegetation and Forest Monitoring

012041
OPEN ACCESS
Impact of human interventions on mangrove ecosystem in spatial perspective

Abd. Rasyid, M. Akbar AS, Nurjannah Nurdin, Ilham Jaya and Ibrahim.....409

012042
OPEN ACCESS
Comparison Effectiveness of Pixel Based Classification and Object Based Classification Using High Resolution Image In Floristic Composition Mapping (Study Case: Gunung Tidar Magelang City)

Prama Ardha Aryaguna and Projo Danoedoro.....415

012043
OPEN ACCESS
The Utilization of Landsat 8 Multitemporal Imagery and Forest Canopy Density (FCD) Model for Forest Reclamation Priority of Natural Disaster Areas at Kelud Mountain, East Java

S Himayah, Hartono and P Danoedoro.....424

012044
OPEN ACCESS
Species Distribution Model using Hyperspectral Data Application in Peatland, Central Kalimantan

Muhammad Dayuf Jusuf.....434

Integration of Remote Sensing and GIS

012045
OPEN ACCESS
Learning Methods of Remote Sensing In the 2013 Curriculum of Secondary School

Nandi Lili Somantri.....439

012046
OPEN ACCESS
The Use of Hotspot Spatial Clustering and Multitemporal Satellite Imagery to Facilitate Peat Land Degradation in West Kalimantan, Indonesia (Case Study in Mensiku Miniwatershed of Kapuas River)

I Yanuarsyah, Y Suwarno and S. Hudjimartsu.....444

012047
OPEN ACCESS
Land Cover Mapping using GEOBIA to Estimate Loss of *Salacca zalacca* Trees in Landslide Area of Clapar, Madukara District of Banjarnegara

Anggi Permata, Anwar Juniansah, Eka Nurcahyati, Mousafi Dimas Afrizal, Muhammad Adnan Shafry Untoro, Na'ima Arifatha, Raden Ramadhani Yudha Adiwijaya and Nur Mohammad Farda.....454

FOREWORD

On behalf of the local organizing committee, I would like to welcome you all to the 2nd International Conference of Indonesian Society for Remote Sensing (ICOIRS) to be organized from 17 to 20 October 2016 in Yogyakarta, Indonesia. This year's ICOIRS is also the 22nd of the Yearly Scientific Meeting of Indonesian Society for Remote Sensing (PIT MAPIN XXII). The main theme of the 2nd ICOIRS is "Remote Sensing for a Better Governance". As we all know that remote sensing has become a part of our daily life; from simply viewing the streets from mobile phones to selecting the evacuation route and shelters when a disaster strikes or monitoring the extend of forest fire. These examples corroborate the need to discuss the strategic position of remote sensing in today's national and global governance issues. Therefore, this conference aims to elaborate the advantages and challenges of remote sensing as an essential element in the current governance through sharing the knowledge among the participants. Over 100 papers and 300 participants are expected in this conference.

ICOIRS 2016 (PIT MAPIN XXII) would not have been possible without the many people who have worked with us in planning and organizing the technical program, exhibition, and social events. We thanks to MAPIN for its tremendous guidance and support on organizing the event; special thanks to the local organizing committee members who have all work extremely hard to make this conference happen.

Yogyakarta is one of the foremost cultural centers of Java. This region is located at the foot of the active Merapi volcano and borders with Indian Ocean at the south lowest part. The city itself has a special charm, which seldom fails to captivate the visitor. It is renowned as a center of education (*Kota Pelajar*), making it the right place for exchanging ideas, engaging discussions and enriching knowledge. While you're here, take time to have fun and find memories with the various attractions that Yogyakarta has to offer.

Finally, I would like to ask all of you to attend this exciting conference by presenting papers, discussing technical issues, exchanging ideas, and enjoying the city of Yogyakarta together. We are looking forward to see you all at ICOIRS 2016 in Yogyakarta.

Drs. Projo Danoedoro, M.Sc., Ph.D.
ICOIRS 2016(PIT MAPIN XXII) Chairman

IOP Conf. Series: Earth and Environmental Science **47** (2016) 011001 doi:10.1088/1755-1315/47/1/011001

HELD BY

PUSPICS
FACULTY OF GEOGRAPHY
UNIVERSITAS GADJAH MADA

MAPIN/ISRS
YOGYAKARTA
CHAPTER

BADAN INFORMASI
GEOSPASIAL

IKATAN GEOGRAF INDONESIA
INDONESIAN GEOGRAPHICAL SOCIETY

SPONSORS

DigitalGlobe

ametisgeo

RIEGL
LASER MEASUREMENT SYSTEMS

GPS LANDS
(M) SDN BHD

SKYMAP GLOBAL

BHUMI
PRASAJA

HEXAGON
GEOSPATIAL

AIRBUS
DEFENCE & SPACE

PT. CITRA GAMA SAKTI

COMPNET
HELPFUL & RELIABLE

DI YOGYAKARTA

The Special Region of Yogyakarta is widely known as the centre of Javanese culture as well as the centre of learning. Its designation as a center of learning is marked by the existence of 120 state and private education institutions, with a student population of over 300,000. The Special Region of Yogyakarta has 3,400,000 inhabitants, 511,000 of whom reside in the city of **Yogyakarta**. The city is the second tourist destination of Indonesia. It is renowned as a centre of classical Javanese fine art and culture such as batik, ballet, drama, music, poetry, and puppet shows. Most tourists come to Yogyakarta for its strong Javanese culture and tradition. This makes it prominent among other Javanese cities. Yogyakarta was the Indonesian capital during the Indonesian National Revolution from 1945 to 1949. There are two seasons in Yogyakarta; wet and dry season. The weather in rainy season is often unpredictable, the sun still shines during the day and it will rain in the afternoon. The city features a lengthy wet season running from October until June and a short dry season that only covers the months of July, August and September. The city averages roughly 2200 mm of precipitation annually. Yogyakarta experiences particularly heavy rainfall from November through April. In the dry season, the weather is quite comfortable and sometimes could be very hot. However, temperatures remain relatively constant throughout the course of the year, with average high temperatures at around 30 degrees Celsius and average lows at around 22 degrees Celsius. For further information, please visit http://www.visitingjogja.com/ or http://www.indonesia-tourism.com/yogyakarta/ .

UNIVERSITAS GADJAH MADA

Universitas Gadjah Mada (internationally known as Gadjah Mada University; abbreviation: UGM) is an Indonesian public research university located in Yogyakarta, Indonesia, founded on December 19, 1949. UGM is the oldest and largest institution of higher learning in Indonesia. The 360-acre university comprises of 18 Faculties, 1 Graduate School, 1 Vocational School, 68 undergraduate study programs, 23 diploma study programs, 104 master and specialist study programs, and 43 doctorate study programs. It has approximately 55,000 students, 1,187 foreign students, 2,240 employees, and 2,273 lecturers. For further information, please visit http://ugm.ac.id/en.

The Faculty of Geography was established in 1963. Currently it has three departments: Geographic Information Science, Environmental Geography and Regional Development Department. The Faculty of Geography offers undergraduate and post graduate programs. The undergraduate program consists of three bachelor degree programs i.e. Bachelor in Geography and Environmental Science, Bachelor in Cartography and Remote Sensing, and Bachelor in Regional Development. The post graduate program offers 7 study programs with master degree, and a PhD program, including Master Degree programs in Geography, Remote Sensing, Coastal and Watershed Management and Planning, Environmental Science, Population Studies, Environmental Management, and Geo-Information for Spatial Planning and Risk Management, along with a Ph.D. program in Geography. For further information, please visit http://geo.ugm.ac.id/.

IOP Conf. Series: Earth and Environmental Science **47** (2016) 011001 doi:10.1088/1755-1315/47/1/011001

PUSPICS

PUSPICS is a national and regional center for remote sensing and integrated surveys. Initially it provided training and educational program for image interpretation and mapping for government officers, lecturers, private sectors or individuals. The program was developed in 1976 through cooperation between BAKOSURTANAL (now BIG) and the Faculty of Geography UGM. In the last 20 years, PUSPICS – in cooperation with BIG – held regular Training Cooperation for Developing Countries (TCDC) focusing on surveys and mapping. The participants came from Asian and Pacific countries such as Azerbaijan, Mongolia, Maldives, Afghanistan, Nepal, Cambodia, Myanmar, Laos, Philippines, Timor Leste, and Fiji. PUSPICS has strong research collaboration with national institutions such as BIG, LAPAN, Indonesian ministries and local governments, as well as international institutions such as World Bank, ITC, ACIAR, UNSW and UQ. These experiences and collaborations affirm the contribution of PUSPICS to the rapid development of geographic information sciences in Indonesia and Asia. For further information, please visit http://puspics.ugm.ac.id/.

KEYNOTE SPEAKER

Prof. Dr. Ir. Dewayany, M.AppSc.
President of Indonesian Society for Remote Sensing (MAPIN), Indonesia

Prof. Dr. Hartono, DEA., DESS.
Professor, Department of Geographic Information Science, Faculty of Geography, Universitas Gadjah Mada, Indonesia

Dr. Priyadi Kardono, M.Sc.
Head of Indonesian Geospatial Information Agency (BIG), Indonesia

Prof. Alfredo Huete
Professor, Department of Plant Functional Biology & Climate Change, University of Technology Sydney, Australia

Dr. Matthew Blackett
Senior Lecturer Geography, Environment and Disaster Management, Faculty of Business, Environment, and Society, Coventry University, United Kingdom

Dr. Wataru Takeuchi
Associate Professor, Graduate Program of Environmental Sciences, University of Tokyo, Japan

Prof. Chao-Hung Lin
Professor, Department of Geomatics, National Cheng-Kung University, Taiwan

Dr. Rahman Syaifoel

Indonesian Diaspora Network, The Netherlands

IOP Conf. Series: Earth and Environmental Science **47** (2016) 011001 doi:10.1088/1755-1315/47/1/011001

LOCATION MAP

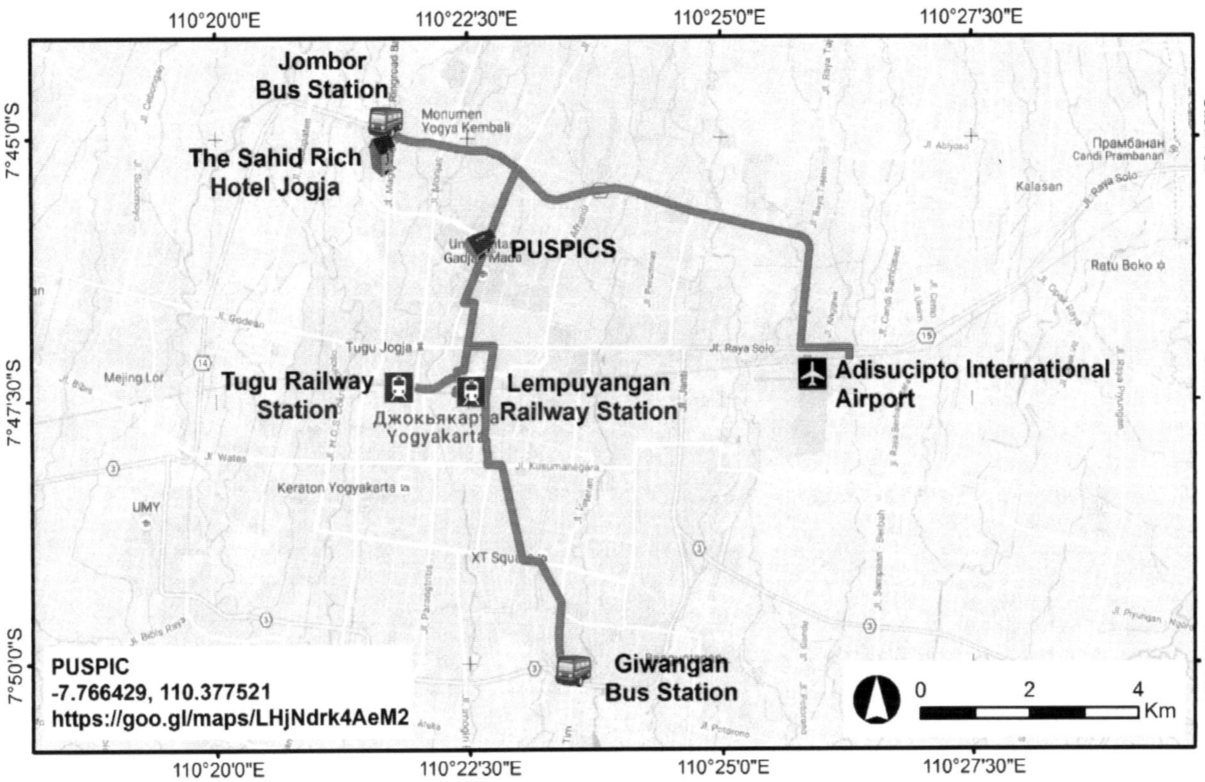

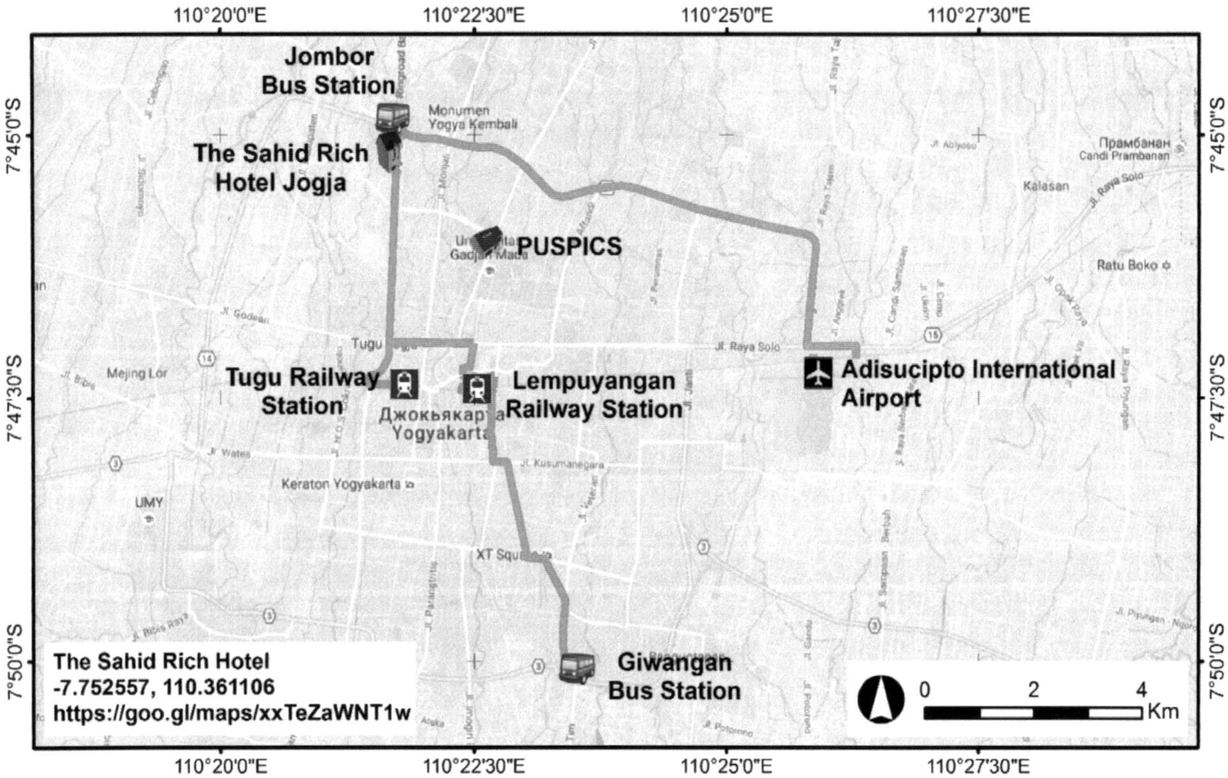
110°20'0"E
110°22'30"E
110°25'0"E
110°27'30"E
7°45'0"S
7°47'30"S
7°50'0"S
Jombor Bus Station
The Sahid Rich Hotel Jogja
PUSPICS
Tugu Railway Station
Lempuyangan Railway Station
Adisucipto International Airport
Giwangan Bus Station
Keraton Yogyakarta
UMY
Kalasan
Ratu Boko
Mejing Lor
XT Square
The Sahid Rich Hotel
-7.752557, 110.361106
https://goo.gl/maps/xxTeZaWNT1w
0 2 4 Km

IOP Conf. Series: Earth and Environmental Science **47** (2016) 011001 doi:10.1088/1755-1315/47/1/011001

THE VENUE

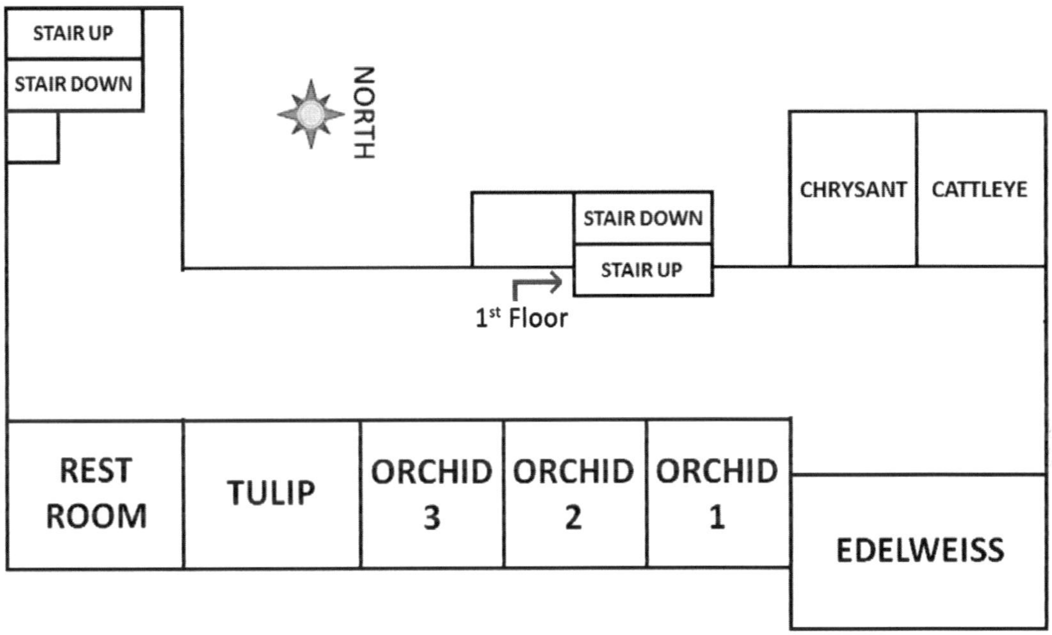

1st Floor

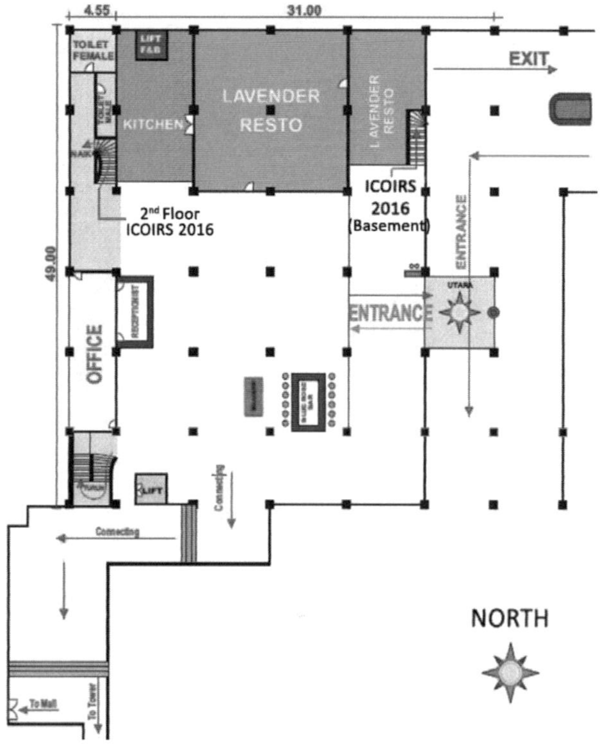

2nd Floor

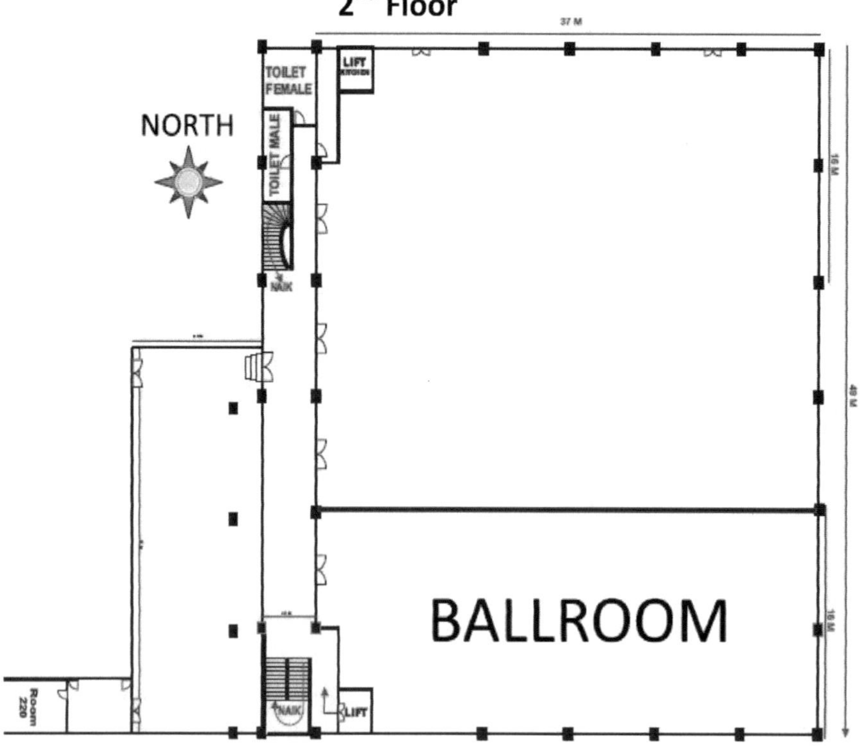

IOP Conf. Series: Earth and Environmental Science **47** (2016) 011001 doi:10.1088/1755-1315/47/1/011001

SCHEDULE OF ICOIRS 2016 / PIT MAPIN XXII

Time	Monday, 17 October 2016				Tuesday, 18 October 2016				Wednesday, 19 October 2016				Thursday, 20 October 2016
07.00 - 08.30	Registration												Cultural or Natural Excursion
08.30 – 10.30	Welcome & Opening Cer. Keynote 1 & 2				Keynote 3,4 & 5				Keynote 6, 7, & 8				
10.30 - 11.00	Coffee Break												
11.00 – 12.15	M-APS1	M-APS2	M-APS3	M-APS4	Poster Session				W-APS1	W-APS2	W-APS3	W-APS4	
12.15 – 13.15	Lunch Break												
13.15 – 14.30	M-BPS1	M-BPS2	M-BPS3	M-BPS4	T-BPS1	T-BPS2	T-BPS3	T-BPS4	W-BPS1	W-BPS2	W-BPS3	W-BPS4	
14.30 – 15.00	Coffee Break												
15.00 – 16.15	M-CPS1	M-CPS2	M-CPS3	M-CPS4	T-CPS1	T-CPS2	T-CPS3	T-CPS4	Closing Ceremony				

M-APS1: day-**Monday**; time-**A (11-12.15)**; session-**Parallel Session**; location – 1 **(Ball Room)**

Day: **M-** (Monday); **T-**(Tuesday); **W-**(Wednesday)

Time: **A-**(11.00-12.15); **B-**(13.15-14.30); **C-**(15.00-16.15)

Location: **1-**(Ball room), **2-** (Orchid1); **3-** (Orchid2); **4-** (Orchid3)

IOP Conf. Series: Earth and Environmental Science **47** (2016) 011002 doi:10.1088/1755-1315/47/1/011002

Peer review statement

All papers published in this volume of *IOP Conference Series: Earth and Environmental Science* have been peer reviewed through processes administered by the proceedings Editors. Reviews were conducted by expert referees to the professional and scientific standards expected of a proceedings journal published by IOP Publishing.

Map Analysis and Spatial Statistic: Assessment of Spatial Variability of Agriculture Land Conversion at Urban Fringe Area of Yogyakarta

Bowo Susilo

Department of Geographic Information Science, Faculty of Geography, Universitas Gadjah Mada

bowosusilo@ugm.ac.id

Abstract. Urban development has brought various effects, one of which was the marginalization of the agricultural sector. Agricultural land is gradually converted to other type of land uses which considered more profitable. Conversion of agricultural land cannot be avoided but it should be controlled. Early identification on spatial distribution and intensity of agricultural land conversion as well as its related factor is necessary. Objective of the research were (1) to assess the spatial variability of agricultural land conversion, (2) to identify factors that affecting the spatial variability of agricultural land conversion. Research was conducted at urban fringe area of Yogyakarta. Spatial variability of agricultural land conversion was analysed using an index called Relative Conversion Index (RCI). Combined of map analysis and spatial statistical were used to determine the center of agricultural land conversion. Simple regression analysis was used to determine the factors associated with the conversion of agricultural land. The result shows that intensity of agricultural land conversion in the study area varies spatially as well as temporally. Intensity of agricultural land conversion in the period 1993-2000, involves three categories which are high, moderate and low. In the period of 2000-2007, the intensity of agricultural land conversion involves two categories which are high and low. Spatial variability of agricultural land conversion in the study area has a significant correlation with three factors: population growth, fragmentation of agricultural land and distance of agricultural land to the city

1. Introduction

Urban development has brought various effects, one of which was the marginalization of the agricultural sector (Mazzocchi et al, 2013). The main cause of the marginalization is inequality of profit between sectors. Agricultural sector is deemed unable to give more profit, compared with other economic sectors. The marginalization process occurs gradually, especially in urban fringe or suburban areas. Agricultural land gradually converted to other land uses, which are considered more profitable.

The intensity of agricultural land conversion varies from one location to another. Such variation is thought to have a relationship with several factors. Mazzocchi et al (2013), mentions three factors related to the conversion of agricultural land, particularly in the suburb area. The first factor is the growth of the urban population. According to population projections by ONU Population Division (2009, in Mazzocchi et al, 2013), by 2015 half of the population in developing countries will live in urban areas. The second factor is the spatial factors, especially the distance of agriculture lands to town or urban area. The closer the agricultural land to the city, the higher the pressure on the land and so the greater the possibility of land conversion. The third factor is the economic mechanism that controls the value of agricultural land. Agricultural land located in suburban areas, more economically valuable for the development of housing and commercial buildings.

Yogyakarta is widely known as a city with distinct characteristics, so it attracts people from various regions in Indonesia and even foreign countries to come. They stayed for a while, or in a considerable period of time, even permanently. Educational activities and tourism, directly or indirectly, will trigger demand of land for development of facilities and supporting infrastructure. Conversion of agricultural land, particularly in urban fringe area of Yogyakarta, is inevitable. However, the policy makers need to control it. To do this, information on the distribution and spatial patterns of agricultural land conversion is needed. Based on this information, land conversion that occurs intensely in certain areas can be detected earlier, so that the right policy can be formulated. This study attempt to assess the spatial variability of agricultural land conversion at urban fringe area of Yogyakarta and identify its affecting factors.

2. Study Area

The study area was located at urban fringe area of Yogyakarta. According to administrative divisions, the study area are a part of Sleman district and Bantul district. Study area within Sleman district consist of 3 sub-district namely Gamping, Mlati and Depok. Study area within Bantul district also consist of 3 sub-district namely Kasihan, Sewon and Banguntapan. The study area totally covers 28 villages. The location of the study area is shown in Figure 1.

Figure 1. Location of Study Area

3. Methodology

Spatial variability of agricultural land conversion, in this study was assessed trough map and spatial statistical analysis. The analysis was applied against different type of data, which obtained from various sources. Results of analyses were presented as maps to provide understanding about spatial distribution of agricultural land conversion and spatial clustering as well.

3.1. Data
The data used in this study were area and location of agricultural land conversion, number of population, population growth and distance of agricultural land to the city. Area and location of agriculture land conversion were derived from multi temporal land use maps. Those maps were obtained from 1: 25.000 Indonesian topographic map (RBI) and interpretation of satellite imageries. Statistical data published by Central Bureau of Statistics (BPS) were used to acquire number of population. The population data is then used to calculate population density as well as annual population growth. Distance of agricultural lands to the City of Yogyakarta were obtained through spatial analysis the so-called proximity analysis.

3.2. Spatial Variability Analysis of Agricultural Land Conversion
Variability of agricultural land conversions were analyzed based on the intensity of such conversion within each village. Intensity of agricultural land conversions was quantified using an index called Relative Conversion Index (RCI). The calculations principle of RCI referring to the Location Quotients LQ analysis techniques, commonly applied in the analysis to determine the level of industrial agglomeration. Formula or equation to calculate the RCI can be written as follows:

$$RCI = \frac{R_{Dk}}{R_{Tk}} \quad (1)$$

$$R_{Dk} = \frac{A_{Dk}}{A_D} \quad (2)$$

$$R_{Tk} = \frac{A_{Tk}}{A_T} \quad (3)$$

Relative Conversion Index (RCI), is the ratio between the proportion of conversion of agricultural land in each village (R_{Dk}), with the proportion of agricultural land conversion in the entire study area (R_{Tk}). The proportion of agricultural land conversion in each village (R_{Dk}) is the ratio between area of converted agricultural land (A_{Dk}) with total area of agricultural land in the village (AD). The proportion of agricultural land conversion in the entire study area (R_{Tk}) is the ratio between area of converted agricultural land with total area of agricultural land in the entire study area.
RCI values indicate the intensity of agricultural land conversion in a village, relative to other villages in the entire study area. If the RCI value is 1, it indicates that the intensity of the of agricultural land conversion in a village is equal to the intensity of the entire study area. The higher the value of RCI, the higher the intensity of agricultural land conversion.

3.3. Spatial Cluster Analysis of Agricultural Land Conversion
Spatial cluster analysis was performed to identify the spatial center of agricultural land conversion. The spatial statistic technique so-called hot spot analysis was used. The cornerstone of technique is Getis-Ord (Gi*) statistic. Areas of agricultural land conversions for each village were used as input of the analysis. Through this analysis, the existence of hot spots was identified based on area of agricultural land conversion in a village and its surrounding. Getis-Ord statistics are calculated based on the equation or the following formula:

$$G_i^* = \frac{\sum_{j=1}^{n} w_{i,j} x_j - \bar{X} \sum_{j=1}^{n} w_{i,j}}{S \sqrt{\frac{\left[n \sum_{j=1}^{n} w_{i,j}^2 - \left(\sum_{j=1}^{n} w_{i,j} \right)^2 \right]}{n-1}}} \quad (5)$$

Where Gi* (pronounced G-i-star) is Getis-Ord statistics, $\boldsymbol{x}_j$ denote area of agricultural land conversion in village ***j***, $\boldsymbol{w}_{ij}$ is spatial weight between village ***i*** and village ***j***, ***n*** is total number of villages in study area.

Hot spot analysis provide information about spatial cluster and spatial center of agricultural land conversion in study area. In spatial statistic, a village which has a vast land conversion not necessarily the center of land conversion. The spatial center of land conversion will be the village with a vast land conversion and is surrounded by villages with similar conditions.

3.4. Correlation Analysis

Factors affecting the agricultural land conversion were analyzed using regression techniques. Previous studies were used as reference to identify presumably factors related to the intensity of agricultural land conversion. The intensity of agricultural land conversion, serves as the dependent variable. The number and population growth as well as the distance to the city serves as the independent variable. Simple regression equation can be written as follows:

$$Y = \alpha + \beta X \quad (8)$$

Where Y denote the intensity of agricultural land conversion, α is a regression constant, $\beta 1$ denote regression coefficients of independent variable X.

4. Result and Discussion

4.1. Agricultural Land Conversion at Urban Fringe Area of Yogyakarta

Spatial analysis against maps of agricultural land for years 1993, 2000 and 2007, generate new map referred to as agricultural land conversion maps. There were two agricultural land conversion maps i.e. 1993-2000 map and 2000-2007 map, as shown in Figure 2. Those maps provide an overview about spatial distribution of such land conversions in the study area. Further analysis against those maps, by means of map overlay with administrative division, provide more detail information in terms of area (ha) and proportion (%) of agricultural land conversion for each village. Those information are shown in Table 1.

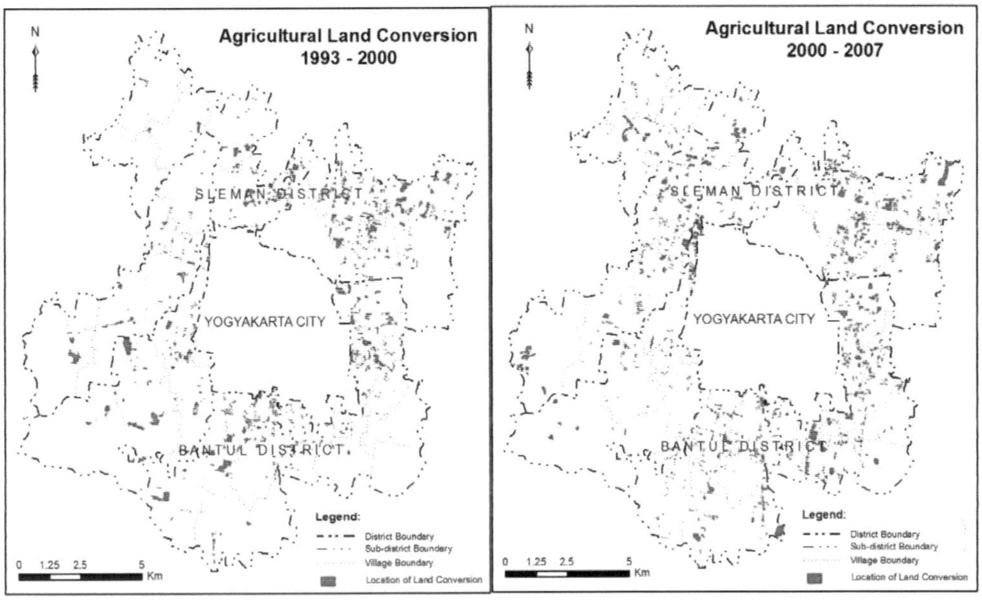

Figure 2. Spatial distribution of agricultural land conversions 1993-2000 and 2000-2007

Table 1.Area and proportion of agricultural land conversion for each village in the study area

District	Sub-District	Village	Agricultural Land Conversion			
			1993-2000		2000-2007	
			(Ha)	(%)	(Ha)	(%)
Bantul	Banguntapan	Banguntapan	52.43	12.70	55,07	15,49
		Baturetno	18.79	7.03	25,98	10,45
		Jambidan	0	0.00	5,5	2,25
		Potorono	0.52	0.19	11,78	4,38
		Singosaren	3.13	10.03	4,7	16,73
		Tamanan	10.25	4.12	8,26	3,47
		Wirokerten	3.14	1.03	28,87	9,65
	Kasihan	Bangunjiwo	14.96	3.48	6,4	1,54
		Ngestiharjo	14.18	8.81	21,37	15,15
		Tamantirto	16.47	5.81	6,22	2,42
		Tirtonirmolo	6.28	2.42	11,36	4,56
	Sewon	Bangunharjo	30.37	6.62	30,01	7,09
		Panggungharjo	21.58	6.96	28,78	10,09
		Pendowoharjo	10.4	2.67	6,92	1,86
		Timbulharjo	5.76	0.95	34,72	5,79
Sleman	Depok	Catur Tunggal	27.28	12.41	39,07	20,34
		CondongCatur	51.59	14.21	50,38	16,72
		Maguwoharjo	43.87	6.35	57,27	9,11
	Gamping	Ambarketawang	10.15	3.45	6,15	2,17
		Balecatur	13.17	2.97	19,38	4,55
		Banyuraden	8.24	3.91	19,09	9,49
		Nogotirto	7.1	3.88	14,56	8,38

	Trihanggo	4.61	1.24	36,88	10,04
Mlati	Sendangadi	11.89	3.75	25,04	8,27
	Sinduadi	25.55	9.61	18	7,61
	Sumberadi	2.48	0.78	10,49	3,33
	Tirtoadi	1.37	0.44	3,52	1,15
	Tlogoadi	6.39	2.15	10,5	3,61
Study Area		421.95	4.71	596.27	7.05

4.1.1. Relative Conversion Index (RCI). In this study, RCI was used as indicator to assess intensity of agricultural land conversion. The higher the value of RCI, the higher the intensity of agricultural land conversion. RCI values were calculated using several parameters, as already explained in the methodology. Area and proportion of agricultural land conversion for each village, as shown in Table 1, were used as a basis for calculation of RCI. The calculations show the value of RCI in the study area ranged from 0 to 3.02. Detail calculation of the RCI for each village in the study area, are shown in Table 2. It should be noted that not all of the parameters used in calculation are shown in the table.

Table 2. Relative conversion index (RCI) of agricultural land for each village in the study area

District	Sub-District	Village	RCI Parameters			RCI
			A_{Dk}	A_D	R_{Dk}	
Bantul	Banguntapan	Banguntapan	55.07	355.55	15.49	2.20
		Baturetno	25.98	248.66	10.45	1.48
		Jagalan	0.00	0.00	0.00	0
		Jambidan	5.50	244.19	2.25	0.32
		Potorono	11.78	269.14	4.38	0.62
		Singosaren	4.70	28.09	16.73	2.37
		Tamanan	8.26	238.30	3.47	0.49
		Wirokerten	28.87	299.03	9.65	1.37
	Kasihan	Bangunjiwo	6.40	414.73	1.54	0.22
		Ngestiharjo	21.37	141.02	15.15	2.15
		Tamantirto	6.22	257.00	2.42	0.34
		Tirtonirmolo	11.36	249.38	4.56	0.65
	Sewon	Bangunharjo	30.01	423.01	7.09	1.01
		Panggungharjo	28.78	285.16	10.09	1.43
		Pendowoharjo	6.92	372.20	1.86	0.26
		Timbulharjo	34.72	600.07	5.79	0.82
Sleman	Depok	Catur Tunggal	39.07	192.07	20.34	2.88
		CondongCatur	50.38	301.36	16.72	2.37
		Maguwoharjo	57.27	628.57	9.11	1.29
	Gamping	Ambarketawang	6.15	283.72	2.17	0.31
		Balecatur	19.38	426.09	4.55	0.64
		Banyuraden	19.09	201.18	9.49	1.35
		Nogotirto	14.56	173.85	8.38	1.19
		Trihanggo	36.88	367.23	10.04	1.42
	Mlati	Sendangadi	25.04	302.92	8.27	1.17

Sinduadi	18.00	236.42	7.61	1.08
Sumberadi	10.49	315.43	3.33	0.47
Tirtoadi	3.52	307.34	1.15	0.16
Tlogoadi	10.50	290.59	3.61	0.51

For the purpose of visualization, the RCI values then grouped or classified into three categories using equal interval method. This method was used because standard classification of RCI was not available yet. Table 3 shows the classification of intensity of agricultural land conversion based on the value of RCI. Furthermore, this classification was used to mapping or visualizes spatial distribution of the intensity. Spatial distribution of intensities of agricultural land conversion is shown in figure 3.

Table 3. Classification of the intensity of agricultural land conversion based on RCI

No	RCI	Intensity of Conversion
1	< 1,01	Low
2	1,01 – 2,02	Moderate
3	> 2,02	High

Figure 3 provide a better visual insight about spatial as well as temporal variability of agricultural land conversion in the study area. High intensity of agricultural land conversion, in the 1993-2000 periods, mostly situated at north eastern part of study area. In the 2000-2007 periods, areas of high intensity grow toward western part of study area.

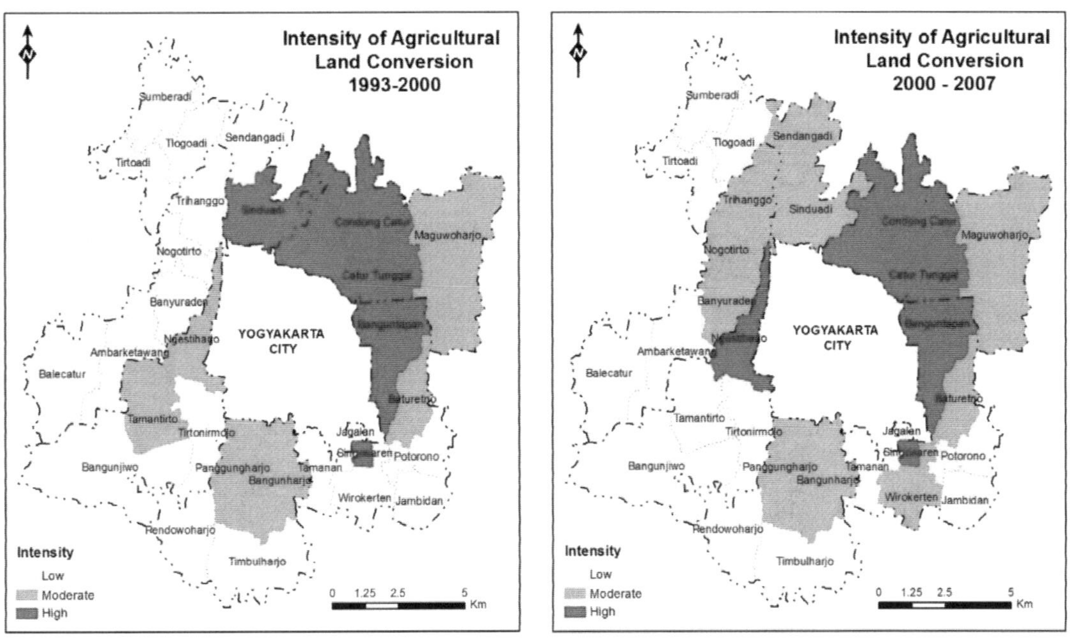

Figure 3. Spatial distribution of the intensity of agricultural land conversions based on RCI

4.1.2. Center of agricultural land conversion (hot spot). The term center of agricultural land conservation in this study was adopted from a spatial analysis method called hot spot analysis. This method is used to identify the presence of spatial clusters of agricultural land conversion. Analysis was performed using two different unit of analysis that are village and point location. Hot spot analysis based on Getis-OrdGi*, produce a value referred to as GiZscore (z-score) and GiPvalue (p-value). The GiZscore is the standard deviation which indicates the grouping intensity of the analysed values. In spatial pattern analysis, GiPvalue shows a probability of random spatial patterns. The smaller the value GiPvalue, the less likely random the spatial patterns of agricultural land conversion. Based on GiZscore and GiPvalue, village which significantly regarded as spatial centre of agricultural land conversion could be identified. Significance of analysis was determined using three level of confidence that are 99% (α = 0.01), 95% (α = 0.05) and 90% (α = 0.1).If a village has a GiPvalue less or equal to 0.001 then it could be significantly regarded, at 99% level of confidence, as a spatial center or hot spot agricultural land conversion. Table 4 shows the complete result of village-based hot spot analysis in the study area.

Village-based hot spot analysis revealed interesting pattern about agricultural land conversion in the study area. Hot spot of agricultural land conversion for 1993-2000 periods is slightly different from hot spot for 2000-2007 periods. It can be seen that within 1993-2000 periods, there were three villages which considered as significant hot spot at 95% level of confidence. Those villages were Caturtunggal, Condongcatur, Maguwoharjo and Banguntapan. The first three villages are located in Depok sub district and part of Sleman district, while the last village are part of Bantul District.. Using the same level of confidence, there were five villages which considered as significant hot spot within 2000-2007 periods. Those two additional villages were Caturtunggal and Maguwoharjo. It also interesting to note the existence of what so-called cold spot of the conversion within 2000-2007 periods. Those cold spot were located in Bangunjiwo village, part of Kasihan sub district and within Bantul district. Comprehensive information about spatial distribution of village-based hot spotscould be seen on maps shown in figure 4.

Table 4. Results of village-based hot spot analysis of agricultural land conversion

District	Sub-District	Village	Getis-OrdGi* Statistic			
			1993-2000		P2000-2007	
			GiZScore	*GiPValue*	*GiZScore*	*GiPValue*
Bantul	Banguntapan	Banguntapan	1.490	0.136	1.284	0.199
		Baturetno	-0.710	0.478	-0.329	0.743
		Jagalan	-0.339	0.734	-0.376	0.707
		Jambidan	-1.967	0.049	-1.607	0.108
		Potorono	-1.942	0.052	-1.343	0.179
		Singosaren	-1.432	0.152	-1.292	0.196
		Tamanan	-0.994	0.320	-1.033	0.302
		Wirokerten	-1.432	0.152	-1.292	0.196
	Kasihan	Bangunjiwo	-0.387	0.699	-1.625	0.104
		Ngestiharjo	-0.783	0.434	-1.287	0.198
		Tamantirto	-0.548	0.584	-1.474	0.141
		Tirtonirmolo	0.369	0.712	-0.898	0.369
	Sewon	Bangunharjo	-1.024	0.306	-0.465	0.642
		Panggungharjo	-0.084	0.933	-0.096	0.924
		Pendowoharjo	-0.465	0.642	-0.453	0.651
		Timbulharjo	0.366	0.714	0.616	0.538
Sleman	Depok	Catur Tunggal	4.319	0.000	3.620	0.000
		CondongCatur	2.544	0.011	1.757	0.079
		Maguwoharjo	2.115	0.034	2.547	0.011
	Gamping	Ambarketawang	-0.356	0.722	-0.947	0.344

	Balecatur	-0.128	0.899	-1.495	0.135
	Banyuraden	-0.835	0.403	-0.551	0.582
	Nogotirto	-0.826	0.409	-0.287	0.774
	Trihanggo	-1.093	0.274	-0.446	0.655
Mlati	Sendangadi	-0.737	0.461	-0.059	0.953
	Sinduadi	0.970	0.332	1.380	0.168
	Sumberadi	-1.333	0.182	-1.108	0.268
	Tirtoadi	-1.714	0.086	-0.831	0.406
	Tlogoadi	-1.102	0.271	-0.546	0.585

4.2. Factors related to spatial variability of agricultural land conversion

Agricultural land conversion in the study area are varies spatially and that why its referred as spatial variability. Previous researches have indicated that spatial variability of agricultural land conversion has relationship with several factors. Distance to urban area, population growth, and land fragmentation are among of these factors.

4.2.1. Distance of agricultural land to urban area. Study area could be divided into several zone base on certain distance to urban area. In this research, distance to urban area is defined as a distance to Yogyakarta city. Using buffer analysis with 500 m interval, the study areas were divided into 15 zone. Area of agricultural land conversion was calculated within each zone as shown in Table 5. Based on this calculation, relationship between agricultural land conversion and distance to urban area were analysed. The analysis shows, there are strong correlation between distance of agricultural land to urban area and the intensity of conversion, i.e. area of agricultural land conversion ($R = 0.915$, $R^2 = 0.837$)The closer the agricultural land to urban area, i.e. Yogyakarta city, the higher the intensity of conversion. Figure 5 shows graphic correlation between distance to urban area or city and the intensity of agricultural land conversion.

Tabel 5. Distance to Yogyakarta city and agricultural land conversion

Zone	Distance to City (m)	Agr. Land Conversion (Ha)	
		1993-200	2000-2007
1	500	69.48	104.08
2	1,000	70.49	77.41
3	1,500	70.85	73.99
4	2,000	45.05	82.94
5	2.500	47.24	65.75
6	3,000	66.38	46.75
7	3,500	31.09	45.23
8	4,000	27.37	62.43
9	4,500	28.36	35.79
10	5,000	44.28	27.5
11	5,500	10.08	9.97
12	6,000	8.64	24.83
13	6,500	7.47	29.42
14	7,000	1.33	10.43
15	7,500	0	0.56

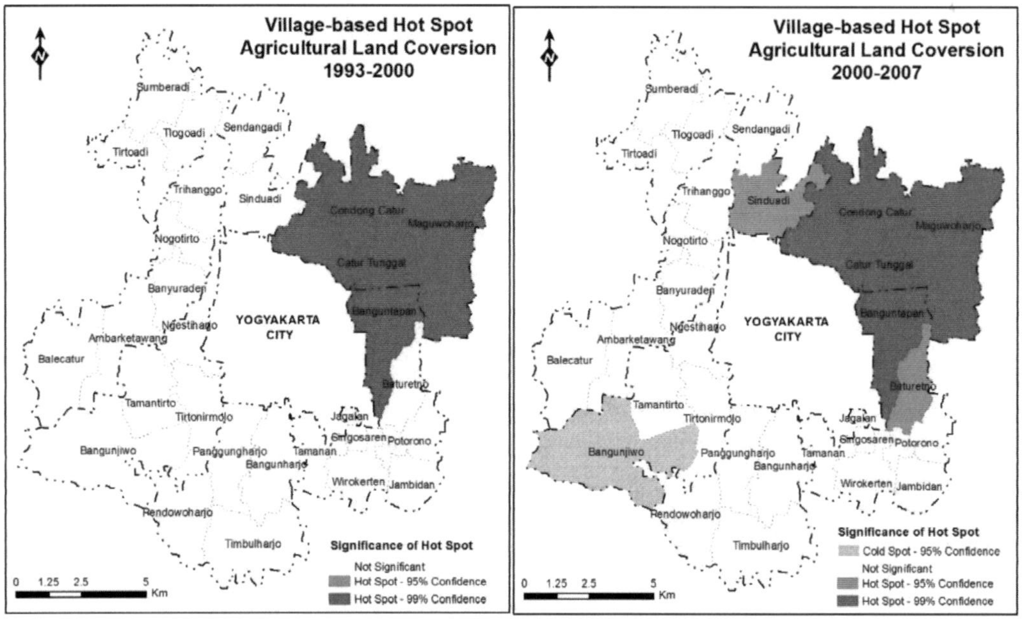

Figure 4. Spatial distribution of village-based hot spot of agricultural land conversions

4.2.2. Population. Demographic condition in the study area is considered dynamic in general. During 1993-2007 periods, number of population had increase significantly. Analysis against population data shows, the number of population increased by 47,039 during 1993-2000 periods while during 2000-2007 periods it's increased by 55.094. Detail information about population in the study area are shown in Table 6. Population growth in the study area considered has relation with the intensity of agricultural land conversion. Degree of correlation between population and intensity of agricultural land conversion is considered moderate ($R = 0.592$, $R^2 = 0.352$). The conversion of agricultural land tends to be more intense as population increase. Graphic correlation between population and agricultural land conversion is shown in figure 5.

Tabel 6. Number of population and population growth in the study area during 1993-2007 periods

District	Sub-District	Village	Population			Population Growth	
			Year				
			1993	2000	2007	1993-2000	2000 - 2007
Bantul	Banguntapan	Banguntapan	22,281	24,767	32,618	2,486	7,851
		Baturetno	8,571	9,874	11,863	1,303	1,989
		Jagalan	3,275	3,365	3,466	90	101
		Jambidan	7,040	7,508	7,938	468	430
		Potorono	7,964	8,872	9,977	908	1,105
		Singosaren	2,338	2,659	3,136	321	477
		Tamanan	7,330	7,876	8,525	546	649
		Wirokerten	8,059	9,016	10,713	957	1,697
	Kasihan	Bangunjiwo	17,223	18,916	20,585	1,693	1,669
		Ngestiharjo	20,789	24,138	26,265	3,349	2,127
		Tamantirto	13,224	14,264	15,421	1,040	1,157
		Tirtonirmolo	16,500	17,938	19,107	1,438	1,169

	Sewon	Bangunharjo	17,079	17,631	18,723	552	1,092
		Panggungharjo	20,413	22,672	25,505	2,259	2,833
		Pendowoharjo	15,807	17,096	18,378	1,289	1,282
		Timbulharjo	15,833	16,344	16,718	511	374
Sleman	Depok	Catur Tunggal	47,411	53,978	60,045	6,567	6,067
		CondongCatur	27,586	31,502	34,574	3,916	3,072
		Maguwoharjo	20,097	23,612	26,686	3,515	3,074
	Gamping	Ambarketawang	14,396	16,200	18,781	1,804	2,581
		Balecatur	12,148	13,614	16,256	1,466	2,642
		Banyuraden	9,721	10,854	12,692	1,133	1,838
		Nogotirto	11,361	12,924	14,650	1,563	1,726
		Trihanggo	11,471	12,196	13,096	725	900
	Mlati	Sendangadi	10,192	10,722	11,061	530	339
		Sinduadi	23,140	27,381	31,824	4,241	4,443
		Sumberadi	10,515	11,272	11,979	757	707
		Tirtoadi	7,523	7,960	8,387	437	427
		Tlogoadi	8,527	9,702	11,078	1,175	1,376

4.2.3. Fragmentation of agricultural land. Agricultural land considered fragmented if separated parcel owned by single ownership. In more technical manner, fragmentation exists if there are so many small polygon with the same characteristic exist, i.e. polygon of agricultural land. This technical definition was used in this study because data of parcel ownership was not available or difficult to be obtained. This study used patch density, which is known as a measure of landscape metrics, to measure agricultural land fragmentation.Patch density was calculated using village as a basis of calculation. Further analysis shows the relationship between patch density and the intensity of agricultural land conversion.The correlation considered strong enough ($R = 0.624$, $R^2 = 0.389$). Fragmented agricultural land is more susceptible to be converted to other uses. Graphic correlation between patch density, as a measure of land fragmentation, and agricultural land conversion is shown in figure 5.

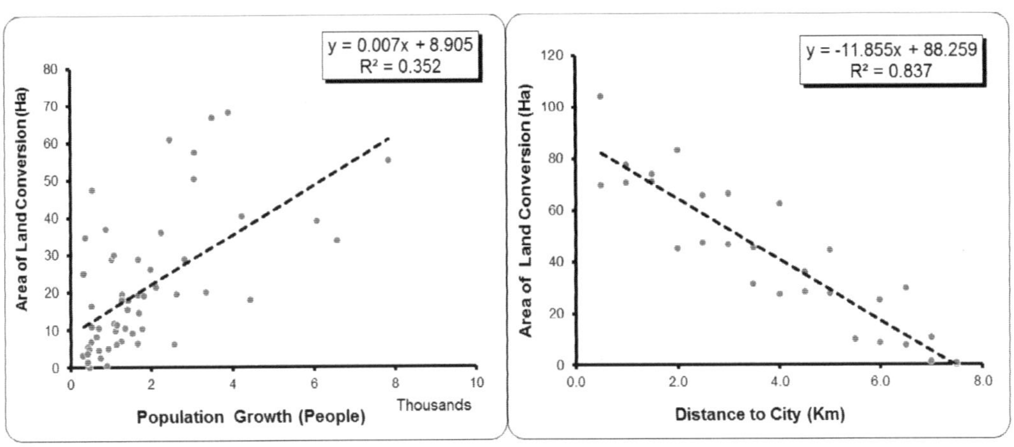

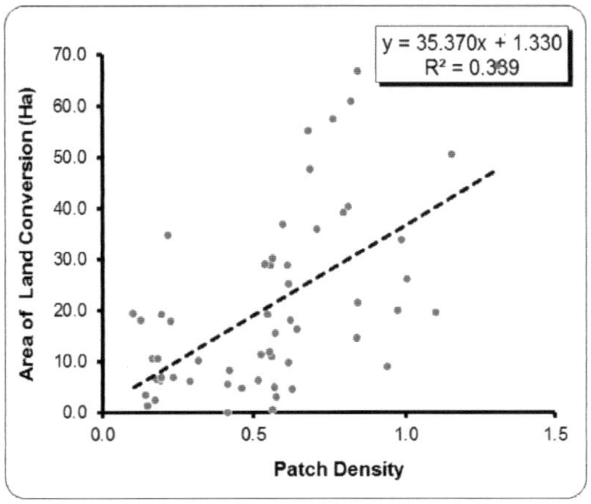

Figure 5. Graphic correlation between conversion of agricultural land and relating factors

5. Conclusion

Conversion of agricultural land cannot be avoided but it should be controlled. Spatial distribution and intensity of the conversion should be identified early, so that appropriate policy could be formulated. Map analysis and spatial statistic can be utilized to obtain information related to spatial variability of agricultural land conversion. This kind of spatial analysis provides a better visual insight about spatial as well as temporal variability. During temporal framework of this study, which is 1993 to 2007, conversion of agricultural land had taken place in various locations. Instead of using absolute location as well as absolute quantity, this study use what so called relative conversion index (RCI) and spatial cluster or hot spot, to provide quick spatial understanding related to distribution of agricultural land conversion. Significant spatial cluster of agricultural land conversion were located in north eastern part of the study area. Policy maker related to land use, has to pay more attention to those area particularly in relation to fragmentation of agricultural land. Fragmented agricultural land is susceptible to be converted to other uses.

Acknowledgements

This research was supported by Faculty of Geography Universitas Gadjah Mada trough research grant for educational staff (2015). It's also considered as part of research supported by Ministry of Research, Technology and Higher Education of the Republic of Indonesia, trough doctoral research grant (2016).

References

[1] Jiang, L., Deng, X., and Seto, K.C. 2013 The impact of urban expansion on agricultural land use intensity in China *Land Use Policy***35** pp 33-99

[2] Mazzocchi, C., Sali, G. and Corsi, S. 2013 Land use conversion in metropolitan areas and the permanence of agriculture: Sensitivity Index of Agricultural Land (SIAL), a tool for territorial analysis, *Land Use Policy***35** pp 155-162.

[3] Morrison, K.T., Nelson, T.A., and Ostry, A.S. 2011 Mapping spatial variation in food consumption, *Applied Geography* **31** pp 1262-126

[4] Mutoko, M.C., Hein, L., and Bartholomeus, H. 2014 Integrated analysis of land use changes and their impacts on agrarian livelihoods in the western highlands of Kenya*Agricultural System* **128**pp 1-12

[5] Susilo, B 2015 Analisis Peta dan Statistik Spasial untuk Kajian Variabilitas Keruangan

Konversi Lahan Pertanian di Daerah Pinggiran Kota Yogyakarta*Reseacrh Report (unpublished)* Faculty of Geography Universitas Gadjah Mada.

[6] Tai-Yang, Z., Xian-Jina, H., Xiu-Ying, Z., and Wang Ke 2011 Temporal and spatial variability of agricultural land loss in relation to policy and accessibility in a low hilly region of southeast China*Land Use Policy***28** pp 762-769

Image Mining in Remote Sensing for Coastal Wetlands Mapping: from Pixel Based to Object Based Approach

N M Farda[1*], P Danoedoro[1], Hartono[1], A Harjoko[2]

[1]Faculty of Geography, Universitas Gadjah Mada, Sekip Utara Bulaksumur, Yogyakarta, 55281, Indonesia
[2]Faculty of Mathematics and Natural Sciences, Universitas Gadjah Mada, Sekip Utara Bulaksumur, Yogyakarta, 55281, Indonesia

*email: farda@geo.ugm.ac.id

Abstract. The availably of remote sensing image data is numerous now, and with a large amount of data it makes "knowledge gap" in extraction of selected information, especially coastal wetlands. Coastal wetlands provide ecosystem services essential to people and the environment. The aim of this research is to extract coastal wetlands information from satellite data using pixel based and object based image mining approach. Landsat MSS, Landsat 5 TM, Landsat 7 ETM+, and Landsat 8 OLI images located in Segara Anakan lagoon are selected to represent data at various multi temporal images. The input for image mining are visible and near infrared bands, PCA band, invers PCA bands, mean shift segmentation bands, bare soil index, vegetation index, wetness index, elevation from SRTM and ASTER GDEM, and GLCM (Harralick) or variability texture. There is three methods were applied to extract coastal wetlands using image mining: pixel based - Decision Tree C4.5, pixel based - Back Propagation Neural Network, and object based - Mean Shift segmentation and Decision Tree C4.5. The results show that remote sensing image mining can be used to map coastal wetlands ecosystem. Decision Tree C4.5 can be mapped with highest accuracy (0.75 overall kappa). The availability of remote sensing image mining for mapping coastal wetlands is very important to provide better understanding about their spatiotemporal coastal wetlands dynamics distribution.

1. Introduction

During the last decade, satellite remote sensing imagery has become the source of the most significant recent image database in monitoring the Earth's surface. Image database have diversity in spatial, spectral, and also temporal resolutions. In addition, remote sensing image database is an archive of spatial information that is growing fast enough.Since Landsat-1 was launched in 1972 and operates, this development makes an experience in collecting, processing and analyzing the data around the world. Although such a limitation in the interpretation of information from the database with a large number still remains.Today almost all image processing techniques designed to run on a single recording image and only a few algorithms and techniques to handle multi-time image data [1]. Humans will also experience difficulties and even almost impossible to find information on the knowledge and the underlying pattern in the image when dealing with large data collections. This situation makes the "knowledge gap" in the process of extracting information of digital remote sensing satellite imagery. Knowledge gap arises because it is currently limited image mining techniques and information extraction in large image data sets [2].

Image mining [3], [4] now has a trend and development in framework, technique, level of information, and also applied in the real world. Framework developed into two categories, namely information-oriented [5] and function-oriented. Techniques also developed from object recognition [6], image retrieval and indexing [7], image classification and image clustering [2], association rule mining [8], until the neural network. In addition, level of information developed from the pixel level, object, semantic concept, until the pattern and knowledge [5]. Pixel level is the lowest layer in an image mining system. It is raw image information such as image pixels and a simple image features (color, texture, and shape). Object Level, the focus of the Object level is to identify domain- specific features such as objects and homogeneous regions in the images. Clustering algorithm, along with the domain knowledge can help the image segmentation into several regions/objects.

Field studies of wetlands have been developed since the 1971 Ramsar Convention, because of the importance of wetland ecosystems role in contributing to biodiversity, regulating climate, food sources, sources of water circulation, fishery resources, and medicines for local communities (http://www.ramsar.org accessed April 16, 2016). Wetlands in Indonesia are experiencing pressures toward destruction and deforestation drastically. Degradation rate of about 2 million hectares per year which include illegal logging, forest fires, and large-scale forest conversion such as monoculture oil palm plantations, Industrial Plantation Forest (HTI), mining, and aquaculture shrimp (www.walhi.or.id accessed April 17, 2016). The symptoms of climate change are also influenced by changes that occur in wetlands, with the release of carbon from wetland slash and burn, wetland drying, peat mining, and changes in land use that are not controlled in the wetlands[9].

Due to the importance of wetlands and the growing satellite image database, the aim of this research is to extract coastal wetlands information from satellite data using pixel based and object based image mining approach.

2. Image mining in remote sensing

Image mining is a synergy of data mining technology with image processing to assist analysis and understanding in the image domain [3]. Image mining is also "an interdisciplinary science that includes computer vision, image processing, image retrieval, data mining, machine learning, database and artificial intelligent" [6].

The focus of the image mining is the pattern extraction of image data collection with a large amount. The patterns are in a variety of image mining such as classification patterns, description patterns, correlation patterns, temporal patterns, and spatial patterns [3].

The process of image mining (Figure 1) starts from the image database, followed by pre-processing, transformation and feature extraction, mining, interpretation and evaluation, and discover of knowledge. Digital image processing for Remote Sensing in general has similarity stage with image mining (Figure 1) in pre-processing stage, transformation and extraction of information (features). After transformation and feature extraction, image-mining algorithms will work extracting important pattern to generate knowledge. A series of processes above are called Knowledge Discovery in Image Database[10].

In reality, earth observation data analysis is still performed in a very hard way at the end of continual cycles of trial and error [11]. To overwhelm this, several researches are being managed to introduce several innovative ideas. These are broadly known as knowledge-driven information mining that is based on human-centered concepts, which implements new features and functions allowing improved feature extraction, search on a semantic level, the availability of collected knowledge, interactive knowledge discovery, and new visual user interfaces [12].

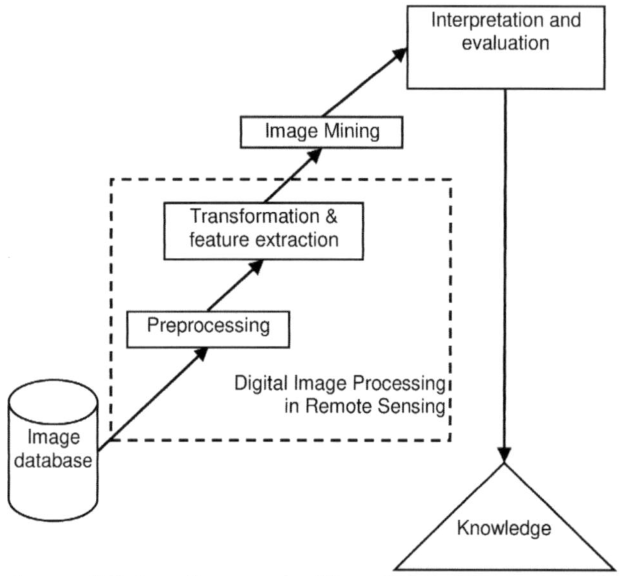

Figure 1. Image Mining Process in Knowledge Discovery related to Remote Sensing Digital Image Processing, modified from [3].

3. Coastal wetlands

Wetlands according to [13] are divided into three components of the ecosystem, namely hydrology, geomorphology, and climate. Hydrology has influence and can change the physiochemical environment, which in turn, along with hydrology, determining the biotic communities found in these wetlands. Overall the forcing functions of wetlands, ecosystem or landscape in this case, including climate, solar energy, temperature patterns, and precipitation. Climate coupled with the landscape geomorphology affect where and when water is long enough then this causes the wetlands that exist.

One of the objects on the earth is water, where two-thirds of the earth is water (sea water). Image mining on water objects will become complex when mixed with land, in this case is an object of wetlands. Wetlands areas under the Ramsar Convention on wetlands are brackish, peat lands and waters; natural or artificial; permanent or temporary; with stagnant or flowing water; fresh, brackish or salt, and including marine areas that depth is not more than six meters at a time low tide. This definition includes the plains of coral reefs and sea grass beds in coastal areas, mud plains, mangrove swamps, estuaries, rivers, freshwater marshes, swamp forests and lakes, and swamps and salty lakes [14]. Wetlands in general by BAKOSURTANAL can be classified as mangrove swamps, estuaries, sea grass, seaweed, coral reefs, lakes, rivers, rice fields and ponds (http://atlasnasional.bakosurtanal.go.id).

4. Materials and methods

The materials and methods described (Figure 2) in this part cover: (a) study area, (b) selection of materials and data, (c) design of field sampling and data collection, (d) transformation and feature (information) extraction from moderate spatial resolution imagery, (e) image mining, and (h) accuracy assessment.

Location of the study area is the Segara Anakan Lagoon (SAL), which is included in the administrative area of Kampung Laut district, Cilacap regency, Central Java province, Indonesia. Kampung Laut sub-district consists of 4 villages, which are Ujung Alang, Ujung Gagak, Panikel and Klaces.

A proportional random sampling pattern applied to collect the wetland data. Proportional based on relief and wetland heterogeneity, the tentative relief map is needed in order to stratify the area and unsupervised classification using k-mean clustering is performed in order to obtain preliminary spectral-related wetland map, from above proportion, location selected randomly.

This study used medium spatial resolution satellite imagery of Landsat MSS (25 April 1978), Landsat 5 TM (5 July 1991), Landsat 7 ETM (22 June 2001), and Landsat 8 OLI (1 May 2014) as primary images and DEM data from SRTM imagery and ASTER GDEM as additional image. Topographic maps (Peta Rupabumi Digital Indonesia) at scale 1:25.000 with 12.5 m contour interval produced by Bakosurtanal (now is Badan Informasi Geospasial) in 2000, covering research area in the Segara Anakan lagoon, Central Java. These included map sheets Kalipucang (1308-241) and Pengolahan (1308-242).

Open-source software for image processing and GIS process used Quantum GIS (with plugin: GRASS, Orfeo Toolbox, and SAGA). Weka - Adams (Algorithm Development and Mining) version 0.4.4, Weka (Java Programming for Machine Learning) version 3.6.0, and Idrisi TerrSet used for image mining process.

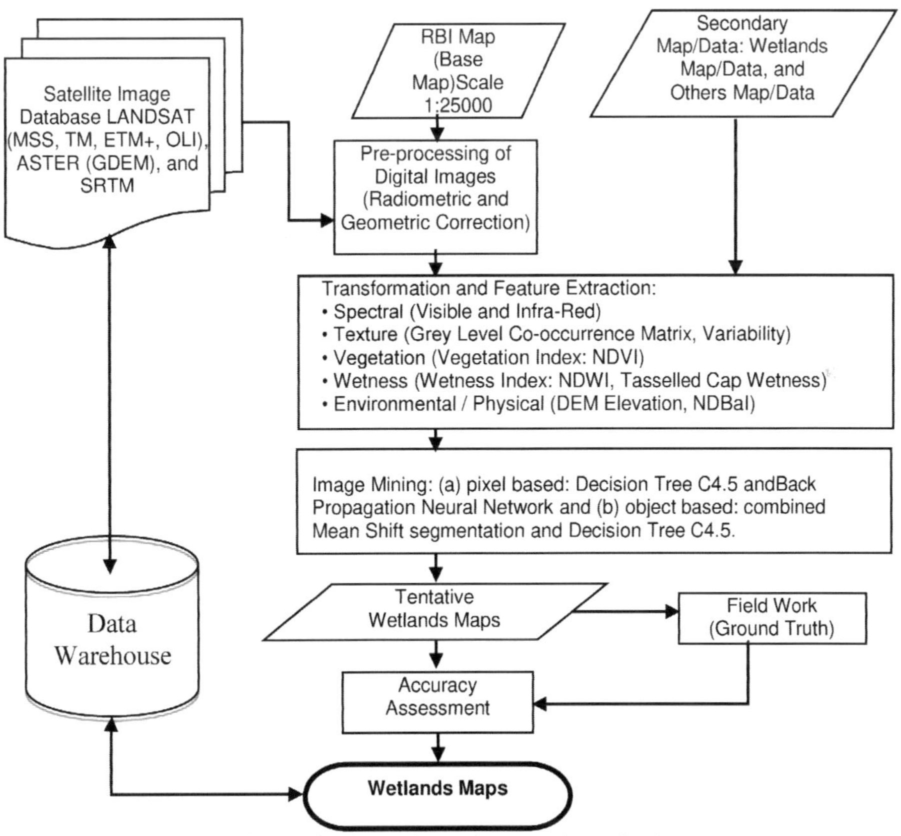

Figure 2. Flow chart of research methods.

Feature extraction and transformation required for image mining input is divided into four sections, namely: (1) The spectral factor, (2) textural factors, (3) factors of vegetation, and (4) physical factors. Spectral factors include spectral channels that exist on Landsat imagery of the MSS, TM, ETM, until the OLI. Texture factors using Haralick (GLCM) texture. Vegetation factors utilizing NDVI as vegetation index derived from the red and infrared channel. Physical factors using elevation values from ASTER image data, and also several indices such as wetness index, and bare soil index. Detailed explanation regarding these factors exists in the following section.

Amount of input bands varies according to the imagery used (Landsat MSS, TM, ETM, or OLI). Landsat MSS used 4 bands, Landsat TM and ETM used 6 bands, and Landsat OLI used 7 bands.

Variations in the input data are divided into spectral channel, vegetation index, wetness index, bare soil index, DEM (elevation), texture, and PCA (Table 1).

Table 1. Input data and the number of bands used in image mining.

No.	Input Data	Quantity	Band Name and Description)*
1.	Landsat 3 MSS, Landsat 5 TM, Landsat 7 ETM, and Landsat 8 OLI	4 to 7	Band 1: Coastal, Band 2: Blue, Band 3: Green, Band 4: Red, Band 5: Near Infrared (NIR), Band 6: Middle Infrared 1, and Band 7: Middle Infrared 2.
2.	Mean Shift image segmentation from Landsat 3 MSS, Landsat 5 TM, Landsat 7 ETM, and Landsat 8 OLI	4 to 7	Band 1: Coastal, Band 2: Blue, Band 3: Green, Band 4: Red, Band 5: Near Infrared (NIR), Band 6: Middle Infrared 1, and Band 7: Middle Infrared 2.
3.	*Inverse* PCA image from Landsat 3 MSS, Landsat 5 TM, Landsat 7 ETM, and Landsat 8 OLI	4 to 7	Band 1: Coastal, Band 2: Blue, Band 3: Green, Band 4: Red, Band 5: Near Infrared (NIR), Band 6: Middle Infrared 1, and Band 7: Middle Infrared 2.
4.	*Bare Soil Index*	1	Band 8: NDBaI
5.	Vegetation Index	1	Band 9: NDVI
6.	*Wetness Index*	1	Band 10: NDWI
7.	ASTER GDEM, and SRTM	1	Band 11: DEM (Elevasi)
8.	GLCM (Harralick) or *Variability* Textures	8 1	Band 12 to Band 19: HT Or Band 12: TexVR
9.	PCA image from Landsat 3 MSS, Landsat 5 TM, Landsat 7 ETM, and Landsat 8 OLI	3 to 5	Band 20 to Band 24 Or Band 13 to Band 17

Note:)* The number of bands used in the image depends on Landsat data (MSS, TM, ETM, or OLI).

Image mining process applied: (a) pixel based: Decision Tree C4.5 and Back Propagation Neural Network and (b) object based: combined Mean Shift segmentation and Decision Tree C4.5.Decision Tree C4.5 used in this study is a non-parametric univariate technique for classifying remotely sensed data. Using training site data, Decision Tree C4.5 in sequence splits the data to form homogenous subsets resulting in a hierarchical tree of decision rules. Three splitting algorithms were used: entropy (information gain), gini, and ratio. Split type is the separation algorithm in form of a decision tree, while pruning (trimming) decision tree is used to trim trees that have been formed. Auto pruning has been done to prune the leaves in the proportion of 1%, 10% and 30%.

Multi-Layer Perceptron (MLP) in this research attempts the classification of remotely sensed imagery through a Multi-Layer Perceptron neural system classifier utilizing the back propagation (BP) calculation. The calculation is based on information from sampling area. MLP also undertake a non-parametric regression analysis between input variables and one dependent variable with the output containing one output neuron, i.e., the predicted memberships[15].

Geographic Object Based Image Analysis (GEOBIA) with mean shift segmentation and decision tree image mining[16] applied to extract coastal wetlands ecosystem map.Image segmentation based on mean shift procedure [17]–[19] is an extension of discontinuity preserving smoothing algorithm. Each pixel is associated with a significant mode of joint density domain located in its neighbor, after the fashions nearby trimmed in generic feature space analysis.

Accuracy assessment of the results of image mining is done by creating confusion between the maps of the image matrix with the data mining field or a reference map that has been recognized truth. Accuracy assessment method used is based on [20]–[22].

5. Results and discussion

5.1. Pixel based: Decision Tree C4.5 and Back Propagation Neural Network (BPNN)

The decision tree C4.5 used input parameters: entropy, gini, ratio, and auto-pruning have a significant influence on the results. Pruning has an influence on the resulting object of detail, the more detailed the object the smaller auto-pruning proportion value, and vice versa. When auto-pruning leaves proportion the greater, then the map seen having a generalization, but it can also be classes that are small in size to be lost because of the class combined with a more dominant class (Figure 3).

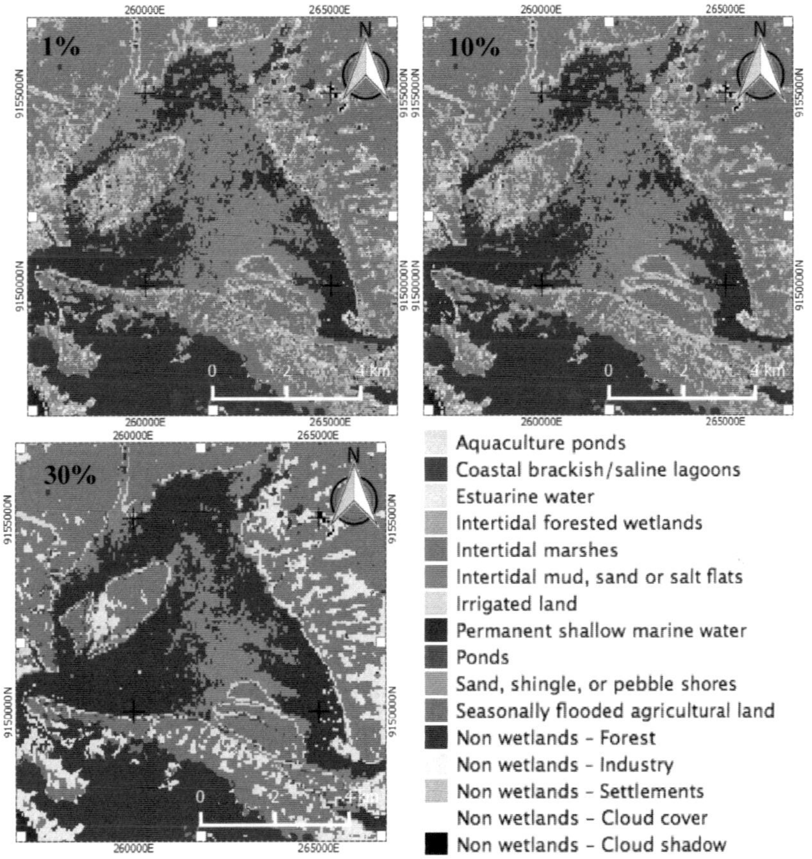

Figure 3. Image mining using Decision Tree C45 with pruning 1%, 10% and 30% of entropy applied to Landsat MSS 1978.The greater pruning the map seen having a generalization.

Entropy, ratio, and gini also have different effect on each application on Landsat imagery in 1978, 1991, 2001 and 2014. Entropy used in decision tree to calculate the homogeneity of a sample, if the sample is completely homogeneous the entropy is zero and if the sample is an equally divided it has entropy of one. Ratio is modification of the information gain (entropy) that reduces its bias towards multi-valued attributes. Gini index of a pure table (consist of single class) is zero because the probability is one. Similar to entropy, gini index also reaches maximum value when all classes in the table have equal probability.

The complexity of a decision tree is usuallymeasured by one of the following: (1) the number of nodes, (2) number of leaf, (3) tree depth and (4) a number of attributes that are used. Therefore, the example simulation on Figure 4 shows a fairly complex decision tree. There issome things are important to know the results of the decision tree: (1) pixels,(2) percentage of pixels, and (3) the purity index. Pixels are the number of pixels in the leaf.Value in percent is the percentage of pixels that are included in the total pixels in the class. Purity index on the decision tree is the percentage of correct pixels that really belongs to a certain class in the leaf of the total number of pixels in the leaf. Details of the leaf on the decision tree calculated from the training site data.

Figure 4 show example of the decision tree generated from C4.5 algorithm applied entropy and 30% pruning. There are three setting on pruning: 30%, 10%, and 1% each simulation (entropy, gini, and ratio) with decision tree C45 algorithm.The result from decision tree C45 entropy 30%: number of input attributes 17, number of node 14, number of leaf 15, tree depth 6, and bands applied (9): 1, 2, 5, 7, 13, 15, and 17. Decision tree C45 entropy 10%: number of input attributes 17, number of node 42, number of leaf 43, tree depth 9, and bandsapplied (11): 4, 16, 2, 1, 17, 15, 16, 3, 5, 13 and 7. Decision tree C45 entropy 1%: number of input attributes 17, number of node 59, number of leaf 60, tree depth 13, and bands applied (11): 4, 16, 2, 1, 17, 15, 16, 3, 5, 13 and 7.The higher pruning impact on decreasing the number of node, leaf, depth, and bands applied.

```
+—— Root
+—— Band 4 < 17.5
¦       +—— Band 16 < 9.46799993515015
¦       ¦       +—— Band 2 < 13.5
¦       ¦       ¦       +—— Band 1 < 21.5 Leaf Class 16 [Pixels 4 : 100.00%]   Purity :  75.00%
¦       ¦       ¦       +—— Band 1 > 21.5 Leaf Class 8 [Pixels 6 :  66.67%]   Purity :  66.67%
¦       ¦       +—— Band 2 > 13.5
¦       ¦               +—— Band 16 < -0.465999990701675 Leaf Class 3 [Pixels 9 :  70.00%]   Purity :  77.78%
¦       ¦               +—— Band 16 > -0.465999990701675 Leaf Class 2 [Pixels 16 :  65.00%]   Purity :  81.25%
¦       +—— Band 16 > 9.46799993515015
¦               +—— Band 5 < 0.69879999756813 Leaf Class 10 [Pixels 2 :  50.00%]
¦               +—— Band 5 > 0.69879999756813 Leaf Class 6 [Pixels 8 :  60.00%]   Purity :  75.00%
+—— Band 4 > 17.5
        +—— Band 2 < 19
        ¦       +—— Band 17 < 29.5
        ¦       ¦       +—— Band 15 < 46.7690010070801 Leaf Class 1 [Pixels 11 :  60.00%]   Purity :  54.55%
        ¦       ¦       +—— Band 15 > 46.7690010070801
        ¦       ¦               +—— Band 13 < 0.352040827274323 Leaf Class 9 [Pixels 5 :  75.00%]   Purity :  60.00%
        ¦       ¦               +—— Band 13 > 0.352040827274323
        ¦       ¦                       +—— Band 4 < 38.5 Leaf Class 5 [Pixels 12 :  53.33%]   Purity :  66.67%
        ¦       ¦                       +—— Band 4 > 38.5 Leaf Class 11 [Pixels 4 :  33.33%]   Purity :  50.00%
        ¦       +—— Band 17 > 29.5 Leaf Class 12 [Pixels 6 : 100.00%]   Purity :  83.33%
        +—— Band 2 > 19
                +—— Band 1 < 33
                ¦       +—— Band 7 < -1.2993540763855 Leaf Class 13 [Pixels 1 :  33.33%]
                ¦       +—— Band 7 > -1.2993540763855 Leaf Class 7 [Pixels 11 :  91.67%]
                +—— Band 1 > 33
                        +—— Band 1 < 48 Leaf Class 13 [Pixels 2 :  66.67%]
                        +—— Band 1 > 48 Leaf Class 15 [Pixels 4 : 100.00%]
```

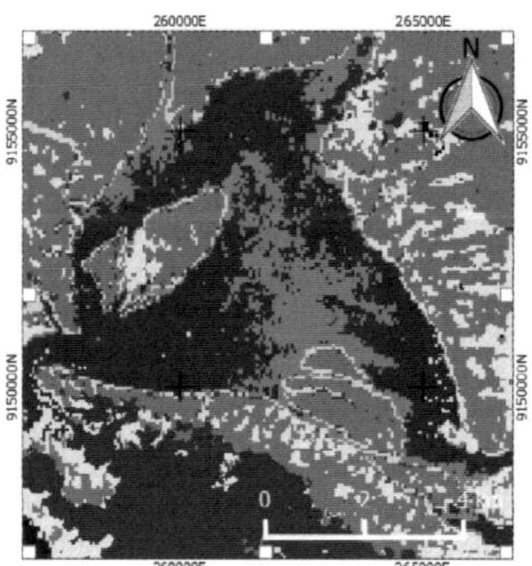

Value	Color	Label
1		Aquaculture ponds
2		Coastal brackish/saline lagoons
3		Estuarine water
4		Intertidal forested wetlands
5		Intertidal marshes
6		Intertidal mud, sand or salt flats
7		Irrigated land
8		Permanent shallow marine water
9		Ponds
10		Sand, shingle, or pebble shores
11		Seasonally flooded agricultural land
12		Non wetlands – Forest
13		Non wetlands – Industry
14		Non wetlands – Settlements
15		Non wetlands - Cloud cover
16		Non wetlands - Cloud shadow

Figure 4. Example of a decision tree is generated from Decision Tree C4.5 image mining algorithm on Landsat 3 MSS 1978 applied entropy and 30% pruning. Result map and legend from this decision tree. Above show leaf class in decision tree equal to value in legend map.

The Back Propagation Neural Network (BPNN) algorithm was run on Landsat satellite image in 1978, 1991, 2001 and 2014 (Figure 5). The setting parameter on BPNN: 17 – 21 inputs, 100000 iterations, 1 hidden layer with 16 nodes, and 16 outputs.The experimental results using algorithms BPNN shows the results of image classification in 1978, 1991, 2001 and 2014 were able to separate the object of vegetation quite well. The object of water body like the coast brackish water / saline lagoon, estuarine water, and the permanent shallow water marine also easy to classify but ponds class look over classified. Objects associated with open land such as intertidal mud, sand or salts flats, and sand, shingle, or pebble shores classified quite well, although the image looks over classified on the image of 1978. Non wetlands object like the class of industry and settlements are more prevalent misclassification in some locations especially on the image of 2001.

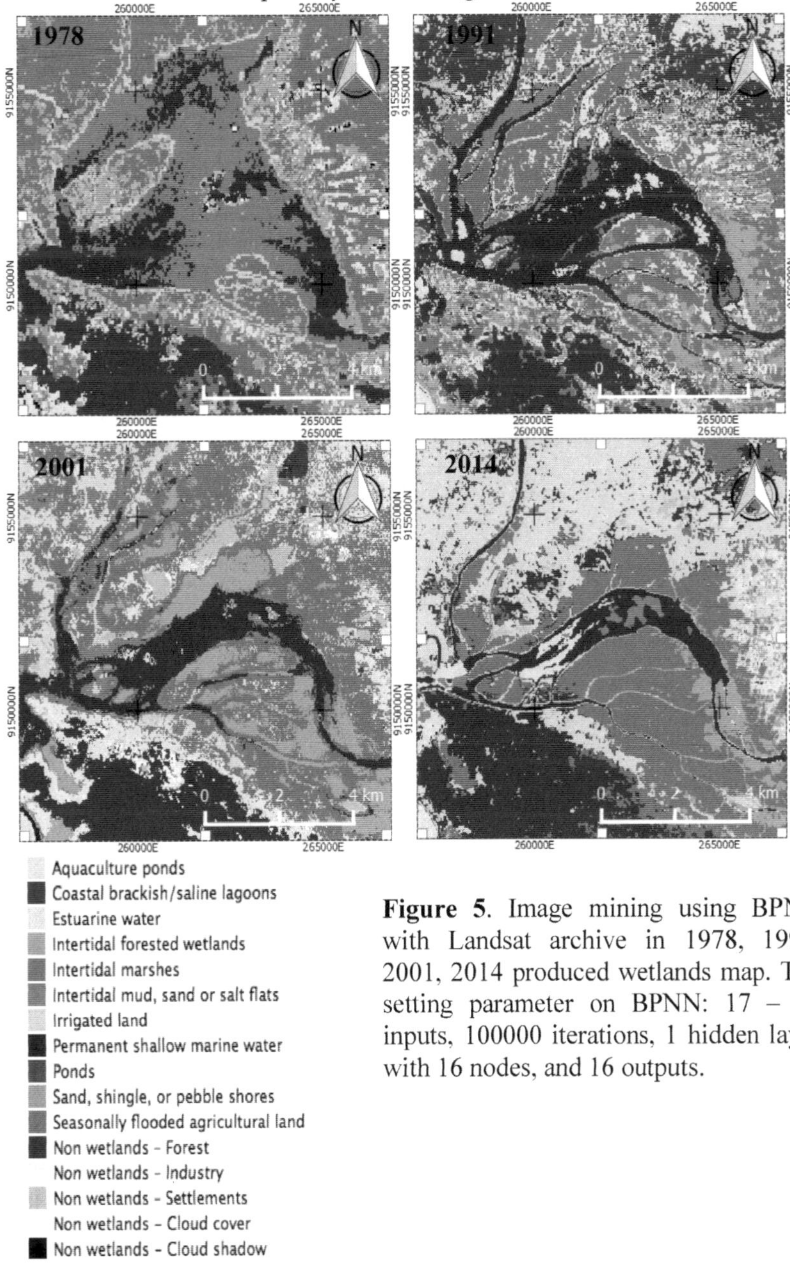

Figure 5. Image mining using BPNN with Landsat archive in 1978, 1991, 2001, 2014 produced wetlands map. The setting parameter on BPNN: 17 – 21 inputs, 100000 iterations, 1 hidden layer with 16 nodes, and 16 outputs.

5.2. Object based: combine Mean Shift segmentation and Decision Ttree C4.5 classification

The results of mining for example in the Landsat OLI image in 2014 using GEOBIA (GEOgraphic-Object-Based Image Analysis) approach, with the first stage is segmentation using Mean Shift clustering algorithm (Figure 6) and the second stage is classification using C4.5 Decision Tree algorithm (Figure 7). Application entropy, gini, and ratio on mean shift image have similarity in delineation result on wetlands class: coastal brackish/saline lagoons seasonally flooded agricultural land, and intertidal mud. Entropy makes the result map too dominate in classifying estuarine water (Figure 7). Gini create the map result quite natural in classifying each class. And ratio built too many irrigated land class.

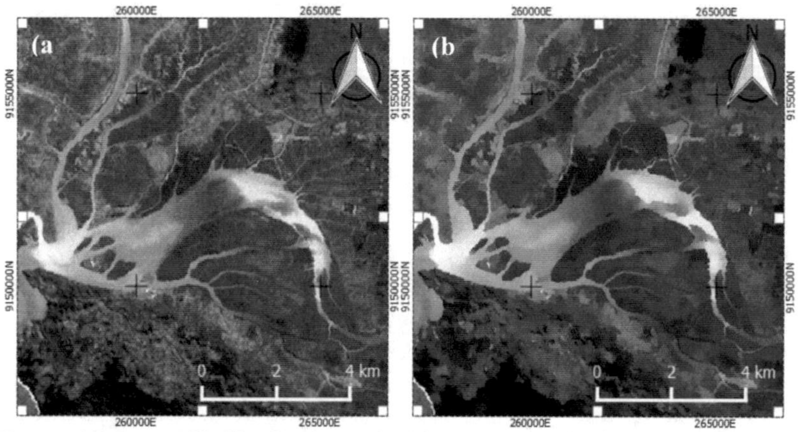

Figure 6. Mean shift filtering is applied in (a) composite Landsat 7 ETM 321 in 2001 to (b) segmenting the image into homogeneous objects.

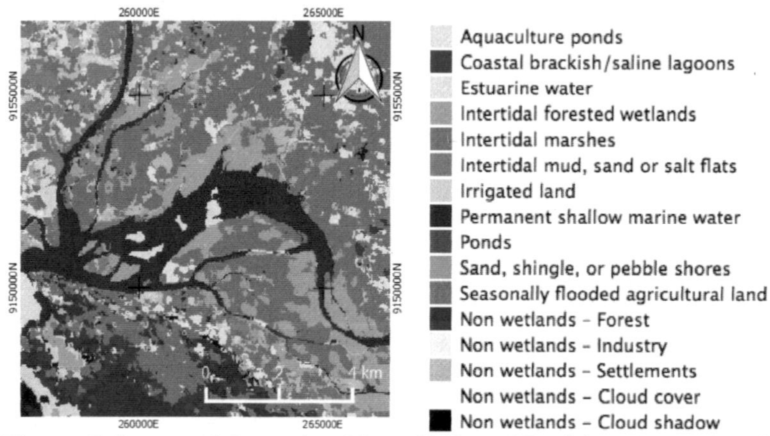

Figure 7. Image mining using Mean Shift and Decision Tree C45 with 1% pruning of entropy applied to Landsat 7 ETM 2001

5.3 Accuracy assessment

Simulation is broadly divided into a simulation using the input value DN (Digital Number) and Spectral Reflectance. Data for training samples are amounted to 140 samples and samples for accuracy test is also the same number (140 samples). Simulation of two large groups, divided into 6 groups with each group there are 76 simulations, so that the total simulation is 456. From the simulation results can be concluded the assessment accuracy on some of the following example (Table 2 and Figure 8).

The accuracy results of the auto-pruning with a proportion of 1% in general have the highest accuracy, followed by the proportion of 10%, and the last with a proportion of 30%. The greater the proportion of pruning makes the results to be generalized on wetland maps. This resulted in the generalization there are several classes of wetlands with small size will be lost and join with the dominant class (the class with greater extents).

The combined input data from DN (Digital Number), inverse PCA and PCA on C4.5 Decision Tree algorithm has the highest overall kappa (0.75). Data cleaning is very influential here, where the value of the inverse PCA and PCA is one way to reduce or eliminate the data that is not as significant as in the input image mining.

Geographic-Object-Based Image Analysis approach does not promise highest accuracy but can increase accuracy. Medium spatial resolution imagery on Landsat imagery (30 meters) has been the object generalizes wetlands, so at this stage of segmentation using Mean Shift clustering to group homogeneity made under segmented image. Objects that minor objects joined with major objects, this applies as majority filter. The decrease in the accuracy of the image mining method is also used due to training area measuring 1 pixel. Training area in future studies supposed to use more than one pixel or using polygons.

Reflectance value has no effect on improving accuracy, precisely DN higher degree of accuracy. DN value is derived from the image of the high-level correction is L1T, with this correction is already an adequate level for coastal wetlands mapping applications.

Table 2. Evaluation of accuracy assessment on the results of mining Landsat image with various algorithms, using 140 samples test accuracy on the data input from DN (Digital Number), inverse PCA and PCA.

No.	**Accuracy Assessment** (*Overall Kappa*)	Landsat MSS (1978)	Landsat TM (1991)	Landsat ETM+ (2001)	Landsat OLI (2014)
	Decision Tree C4.5				
1	Entropy, Pruning 1%	0.539551	0.585049	0.599222	0.750487
2	Entropy, Pruning 10%	0.52666	0.504686	0.625376	0.750487
3	Entropy, Pruning 30%	0.450118	0.496368	0.60115	0.695822
4	Gini, Pruning 1%	0.515382	0.592866	0.607694	0.641465
5	Gini, Pruning 10%	0.507895	0.601324	0.633678	0.656844
6	Gini, Pruning 30%	0.483741	0.542984	0.602862	0.640163
7	Ratio, Pruning 1%	0.500139	0.570195	0.593703	0.665778
8	Ratio, Pruning 10%	0.501973	0.545172	0.603724	0.689303
9	Ratio, Pruning 30%	0.52716	0.591398	0.607448	0.688109
	Mean Shift and Decision Tree C4.5 (GEOBIA)				
11	Entropy, Pruning 1%	0.648261	0.656179	0.692975	0.694631
12	Entropy, Pruning 10%	0.6323	0.648037	0.636477	0.679294
13	Entropy, Pruning 30%	0.557895	0.521286	0.54675	0.63974
14	Gini, Pruning 1%	0.608194	0.633769	0.66606	0.703819
15	Gini, Pruning 10%	0.56206	0.609971	0.600576	0.703291
16	Gini, P.30%	0.53491	0.627521	0.511029	0.624895
17	Ratio, P.1%	0.617806	0.586443	0.683075	0.704132
18	Ratio, P.10%	0.610028	0.586854	0.665934	0.680241
19	Ratio, P.30%	0.589905	0.566259	0.698906	0.686011
	Back Propagation Neural Network (BPNN)				
21	MLP, 100000 iteration	0.527054	0.497544	0.674715	0.64555

Note: the highlight of green color mean the highest accuracy in each column.

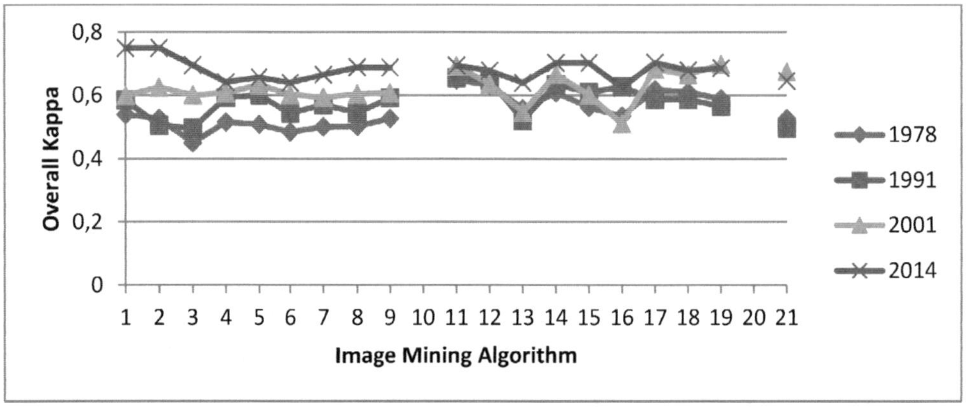

Figure 8. Overall Kappa at various image mining algorithms (Table 2) with the data input from DN (Digital Number), inverse PCA and PCA.

6. Conclusions

Image mining using remote sensing data can be used to map coastal wetlands ecosystem. The combined input data from DN (Digital Number), inverse PCA and PCA on C4.5 Decision Tree algorithm has the highest overall kappa (0.75). The accuracy results of the auto-pruning with a proportion of 1% in general have the highest accuracy, followed by the proportion of 10%, and the last with a proportion of 30%. Geographic-Object-Based Image Analysis approach does not promise increase accuracy in image mining to map coastal wetlands ecosystem. Reflectance value has no effect on improving accuracy, precisely DN (Digital Number) higher degree of accuracy.

This study has recently started image mining for mapping coastal wetlands using temporal Landsat imagery. Knowledge gap of research is still much to be filled, especially the utilization of remote sensing for monitoring coastal wetlands using multiresolution and multitemporal satellite image.

Acknowledgements

The authors would like to convey a gratitude to the Ministry of Research, Technology and Higher education of the Republic of Indonesia and the Board of Executive Universitas Gadjah Mada (UGM), CNRD (Center for Natural Resources and Development) and DAAD (German Academic Exchange Service) for funding and support this research. This paper is part from PhD thesis of the first author entitled "Image Mining in Remote Sensing for Coastal Wetlands Monitoring: an Information-driven framework perspective".

References

[1] M Datcu, H Daschiel, a Pelizzari, M Quartulli, a Galoppo, a Colapicchioni, M Pastori, K Seidel, P G Marchetti, and S D'Elia 2003Information mining in remote sensing image archives: system concepts*IEEE Trans. Geosci. Remote Sens.***41**(12) pp. 2923–2936

[2] M P dos Santos Silva, G Câmara, M I S Escada, and R C M de Souza 2008 Remote Sensing Image Mining: detecting agents of land-use change in tropical forest areas*International Journal of Remote Sensing***29** pp. 4803–4822

[3] W Hsu, M L M Lee, and J Zhang 2002Image Mining: Trends and Developments*J. Intell. Inf. Syst.***19** (1) pp. 7–23

[4] N Dey, W B A Karâa, S Chakraborty, S Banerjee, M A M Salem, and A T Azar 2015Image mining framework and techniques: a review*Int. J. Image Min.***1**(1) p. 45

[5] J Zhang, W Hsu, and M Lee 2001*An Information-driven Framework for Image Mining* pp. 232–242,.

[6] M C Burl, C Fowlkes, and J Roden 1999 Mining for Image Content *Syst. Cybern. Informatics/ ...*, pp. 1–8

[7] H Daschiel and M Datcu 2005Information mining in remote sensing image archives: System evaluation*IEEE Trans. Geosci. Remote Sens.* 43(1) pp. 188–199
[8] C Ordonez and E Omiecinski 1998*Image mining: A new approach for data mining*
[9] J S Bhatti, R Lal, M J Apps, and M A Price 2006*Climate change and managed ecosystems* (Boca Raton, FL: CRC Press, Taylor & Francis Group)
[10] H J Miller and J Han 2009*Geographic Data Mining and Knowledge Discovery*, Second. (Boca Raton, FL: Taylor & Francis Group, LLC)
[11] B Bhatta, *Research Methods in Remote Sensing*. Dordrecht: Springer Netherlands, 2013.
[12] M Datcu and K Seidel 2005 Human-centered Concepts for Exploration and Understanding of Earth Observation Images in *IEEE Transactions on Geoscience and Remote Sensing***43** pp. 601–609
[13] W J Mitsch and J. G. Gosselink 2007*Wetlands* (Wiley)
[14] J Davis 1997*Manfaat Lahan Basah*(Jakarta: Direktorat Jendral PHPA-AWB)
[15] P MAtkinson and A R L Tatnall 1997Introduction Neural Networks in Remote Sensing *http://dx.doi.org/10.1080/014311697218700***18**(4) pp. 699–709
[16] M AVieira, A R Formaggio, C D Rennó, C. Atzberger, D. A. Aguiar, and M. P. Mello 2012Object Based Image Analysis and Data Mining Applied to a Remotely Sensed Landsat Time-series to Map Sugarcane Over Large Areas *Remote Sens. Environ.***123**, pp. 553–562
[17] D Comaniciu and P Meer 2002Mean Shift: A Robust Approach Toward Feature Space Analysis*IEEE Trans. Pattern Anal. Mach. Intell.***24**, pp. 603–619
[18] N M Farda 2008*Klasifikasi Berorientasi Obyek Berdasarkan Segmentasi untuk Analisis Citra Penginderaan Jauh Resolusi Spasial Tinggi*(Universitas Gadjah Mada)
[19] P Wicaksono and N. M. Farda 2015Aplikasi Algoritma Klasifikasi Mean Shift untuk Pemetaan Habitat Bentik Studi Kasus Kepulauan Karimunjawain *Prosiding PIT MAPIN XX 2015* pp. 664–672.
[20] J R. Jensen 2004*Introductory Digital Image Processing: A Remote Sensing Perspective. Third edition*, 3rd ed. (Prentice Hall)
[21] T M. Lillesand, R W Kiefer, and J W Chipman 2004*Remote Sensing and Image Interpretation*, vol. 3rd..
[22] R G Congalton and K Green 2009Assessing the Accuracy of Remotely Sensed Data: Principles and Practices*Photogramm. Rec.***2**(130) p. 183

Application of OpenStreetMap (OSM) to Support the Mapping Village in Indonesia

Nurin Swasti Kanthi[1] and Taufik Hery Purwanto[2]

[1]Student of Remote Sensing and GIS Program, Universitas Gadjah Mada, Yogyakarta, Indonesia
[2]Heads of Remote Sensing and GIS Program, Universitas Gadjah Mada, Yogyakarta, Indonesia

nurin.sk95@gmail.com, taufik@ugm.ac.id

Abstract. Geospatial Information is a important thing in this era, because the need for location information is needed to know the condition of a region. In 2015 the Indonesian government release detailed mapping in village level and their Parent maps Indonesian state regulatory standards set forth in Rule form Norm Standards, Procedures and Criteria for Mapping Village (NSPK).Over time Web and Mobile GIS was developed with a wide range of applications. The merger between detailed mapping and Web GIS is still rarely performed and not used optimally. OpenStreetMap (OSM) is a WebGIS which can be utilized as Mobile GIS providing sufficient information to the representative levels of the building and can be used for mapping the village.Mapping Village using OSM was conducted using remote sensing approach and Geographical Information Systems (GIS), which's to interpret remote sensing imagery from OSM. The study was conducted to analyzed how far the role of OSM to support the mapping of the village, it's done by entering the house number data, administrative boundaries, public facilities and land use into OSM with reference data and data image Village Plan. The results of the mapping portion villages in OSM as a reference map-making village and analyzed in accordance with NSPK for detailed mapping Rukun Warga (RW) is part of the village mapping. The use of OSM greatly assists the process of mapping the details of the region with data sources in the form of images and can be accessed for Open Source. But still need their care and updating the data source to maintain the validity of the data.

Keywords: Open Street Map (OSM), NSPK, Mapping Village

1. Introduction

Indonesia is a big country with many province and many island. Because of that, in 2015 the Indonesian government has launched unified geospatial information for the entire province of Indonesia with making Indonesian Parent Map [1]. It aims to integrate geospatial information Indonesia, so the location information in Indonesia are no longer different perception or confusing.

Geospatial information displayed to date the most detailed to the limit of the Village Administration. Geospatial Information Agency began planning the administrative boundary mapping to RW (RukunWarga) and RT (RukunTetangga). Detailed mapping of the area aims to present accurate boundary information. This is done in order to avoid conflicts over land boundaries. However, more detailed mapping such as building limit has not been done by the government, where as the mapping building is very important for the manufacture of a detailed plan of an area.

Over time, the mapping is not only done by manual survey, but there have been new digital technologies, namely with Web GIS and Mobile GIS. It is influenced by the developing knowledge about Information and technology. Both in terms of the Internet network that is easily accessible and increasing smartphone users in the world as well as in Indonesia. The increasing use of these technologies can be a bonus e-Commerce and e-Technology. But it would be a loss if it is not utilized properly.

Web GIS is one of geography technology which use the Web to display map information, so it can be accessed anytime, anywhere and by anyone. OpenStreetMap (OSM) is a Web GIS product. OpenStreetMap with base Websites can be applied to the smartphone so that it can also be classified in the category MobileGIS. Use of Mobile GIS mapping process is very helpful because it can be taken to any area. Open source systems make anyone can edit, download it and upload the real map, certainly the results shown will be based on the consent Admin.

The merger between detailed mapping and Web GIS is still rarely performed and are fully utilized. In fact, it can really help search spatial information of a region or place in a rapid, precise and accurate. Even now public knowledge of Web GIS and Mobile GIS is a free (Open Source) is still not too many, the most widely known today is Google Maps. In fact, in addition to Google Maps there is a wide range of Web and Mobile GIS such as one featuring the OSM building data in detail.

1.1. Aim of the research

This research was conducted with the aim of:

1. Knowing the concept of using OpenStreetMap mapping.
2. Apply OpenStreetMap for mapping village based on NSPK.

2. Methodology

Village Mapping need a 2 general methods, that is by remote sensing and GIS approach. OpenStreetMap have the source for remote sensing approach like image satellite from bing map, and have many progam to do GIS analyze. For mapping the village we choose Kwarasan RW 05, Nogotirto, Gamping, Sleman. The choice of location is based on several reasons. The first reason is Nogotirto located on the border village of Sleman and Yogyakarta are expected to conflict areas in the region. The second reason is that the village had their plans RW 05 which can be used as a reference. Plan options include local street names and numbers eraser existing homes in the area. The second reason is a detailed mapping requires good local knowledge, so that the local elections are based also by researchers in the area residence Kwarasan RW 05, Nogotirto, Gamping, Sleman.

The implementation stage is a stage core research in the study. Stages of implementation include:

2.1. Input a Village Map to OSM

The process input and editing OSM data must go through the account creation first. Making do with @nurinsk name. OpenStreetMap provide flexibility for the owner of the account to edit and input data. The results of each editing OSM user need not wait for the approval of the central admin. Data will be instantly displayed in the OSM website. Input data to OpenStreetMap can generally be done in two ways : directly through the window *OpenStreetMap.org* (in-browser editor) and using the application JOSM (Java OSM). The difference lies in the use of the input method application or not.

Input data RW 05 Kwarasan done with directly through the window *OpenStreetMap.org*, choose "Edit with ID", that because the map data plans that exist in the form of .pdf and no form of data shp, and Edit with ID is very simple and easy to understand in its application. Input data were digitized point line and area suitable appearance OSM imagery available. Graphical data input point, line and area accompanied with a description of the attribute is a general facility name and house number. In addition to editing data input data is also performed on existing data, but requires justification form or description (ID). The process after the data input is editing the attributes, which include giving the name or number on the object, depending on the type of object.

2.2. Mapping using OSM Administrative Boundary

Administrative boundary mapping is an important element in the village because it shows the boundaries of the zone administration smallest RT. Limit Administrative Mapping involves the role of related people such as village heads or a native who has local knowledge of the area of study. Administrative boundary can be either natural objects such as rivers or artificial objects such as roads and boundaries of the house. Object identification administrative boundaries performed using the image of the OSM.

After input and edit all object and mapping the administrative boundary in OSM save the results by choose the save icon on OSM, so the results of digitization stored on OSM and may appear on the main page of the website.

2.3. Survey / Field Test

The survey was conducted to validate the existing data in the OSM and the condition of the object in the field. The survey was conducted on a wide range of objects that exist, the equipment used when the field survey covering Smartphone Sony Experia U is already integrated with GPS so it can use the geo-tagging to determine the coordinates of the location.

2.4. Completion Phase

The completion stage or the final stage of research includes the presentation of the data export results data from the website OpenStreetMap. Completion phase of the study as follows:

2.4.1. Map Village on OSM's proven conflict. OpenStreetMap data validation process performed by the OSM validator. Validation is done to avoid the perception of conflict due to differences in study areas. This is because their freedom to enter and change the data, then it is necessary validator to investigate different data in the same area and choose a more precise data. OpenStreetMap also has a variety of topology requirements that must be met when performing digitization. The topology requirement will cause the error and warning if it is not met. Terms include overlaying digitized, naming, polygon boundaries, etc.

2.4.2. An analysis of the data on OSM with NSPK. Results of entering data into OSM to have various objects types that can be displayed in a map that came out of the website. NSPK contains various details of the content and requirements that must be included on a map of the village, and then there is need for adjustment of writing, visualization and the benefits of OSM maps with existing NSPK.

3. Result and Discussion

3.1. Input Map Village in Browser OpenStreetMap

Results can be viewed in a browser input OpenStreetMap (*http://OpenStreetMap.org*)

Figure 1. Input Results in Website OSM [2]

Results digitized and input the data in OSM generates village map can be used by anyone at any time. In addition there is the OSM map that can be edited in accordance with the data and knowledge of the user. Results digitized main distinguishing Mobile OSM with other GIS is on the building, the number of buildings digitized in the study area village Kwarasan RW 05 reached 124 houses and one barn and industry. Map results have an accuracy of a shift of 1.01 meters. OpenStreetMap also provide data attributes that can be directly processed after export to .shp. Kwarasan RW 05 village mapping including roads, points, land use and building area. Attribute data from OSM on landuse show objects such as land use graveyard (Cemetary), paddy (Farmland) etc. While in the building area include home (house) and industry.

3.2. Errors and Warnings solve from OSM

Justification errors and warnings made by reference tutorial spatial data handling intermediate and advanced editing techniques of Humanitarian OpenStreetMap Team Indonesia. Data that have been through the process of justification is valid and can be used according to the rules of the OSM.

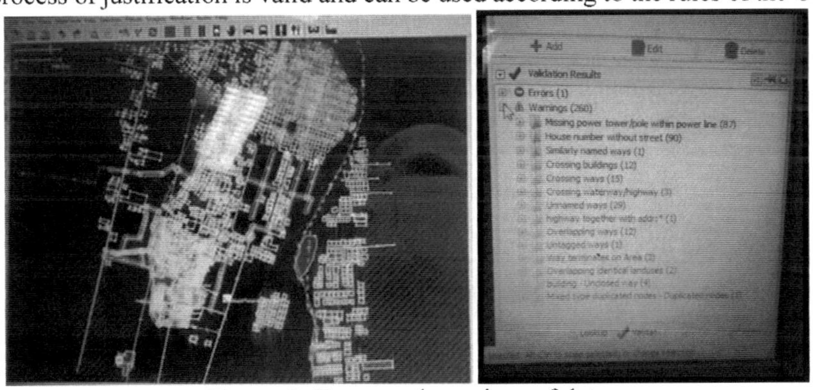

Figure 2. Errors and warnings of data

Inspection is done checking for conflicts and validation of geometry and attributes. Map of RW 05 Kwarasan avoid conflict because no editing of the same area in the same time. Preliminary examination of the geometry and attributes shows there is one error and 260 warnings. Errors occur due to incorrect administration of attribute types. Warning that there is caused by incomplete data and confusion data. Warning majority (90 warnings) occurs in the naming of house number without attribution road. In addition to the warning there is also caused by overlapping land use, roads and buildings that cuts another building, etc.

3.3. Role of OSM to Support Mapping Village According NSPK

RukunWarga mapping is one part of village mapping. Mapping of reference of the village in Indonesia spiked on NSPK standards (Norms, Standards, Procedures and Criteria for Mapping Village). The use OSM is one application that can be used to facilitate the mapping of the village. Mapping can be performed in more detail than the OSM mapping villages, reaching mapping the buildings house. Mapping is done only until RW coverage, but this research can be developed later into wider area coverage.

Analysis of image interpretation is done on a high-resolution image OSM who have true color composite. OpenStreetMap image interpretation results in the hamlet Kwarasan RW 05, Nogotirto, Gamping, Sleman, Yogyakarta includes objects points, lines and areas. The object point consists of a mosque, a line object includes Roads, Rivers and administrative boundaries, the object area, including houses, cemeteries, courts, gardens and fields.

The maps have provisions that cover the content of the information the village level. According to the detailed NSPK content must be present on the map village include toponymy, administrative boundaries, infrastructure, network hydrographic and land use. Below is the ability to display content OSM mapping the village:

3.3.1. Toponym. OpenStreetMap provide completeness toponymy to provide information in the form of street names, village names, etc. OpenStreetMap can display the names.

Figure 3. Toponym [2]

3.3.2. Limit Region Administration. Administration limit shown on mapping the village of dusun Kwarasan include RW 05 and RT/RW limits. Limits valid administration needed to show administratively appropriate area. Limit the administration is in the village Kwarasan RW 05 includes roads, rivers and borders the house. Input data using OSM administration limit sign in the form of lines and the selected type is a special line and named "Limits and Boundaries RT RW". Type of line used to map the administrative boundaries of the interpreter because the significantly OSM types are not available in administrative boundary line only Special Point.

Figure 4. Administrative Boundaries [2]

3.3.3. Infrastructure (Lines, Points). Infrastructure includes roads, public facilities / social and rural assets. On the Map RW 05 Infrastructure includes roads and public facilities. General facilities in RW 05 include mosque, substations and warehouses. Symbols that can be raised on OSM just mosque, substations and warehouses are not on OSM symbol.

Figure 5. Infrastructure [2]

3.3.4. Hydrographic Network. Hydrographical network an irrigation canal and river network include rivers, irrigation, and drainage. Hydrographical network in RW 05 is the river.

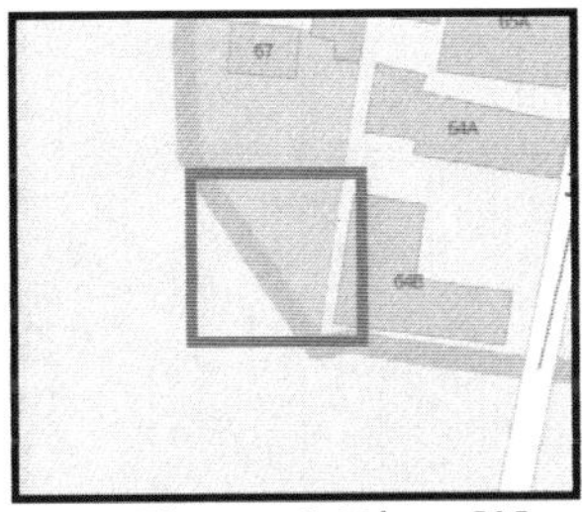

Figure 6. River [2]

3.3.5. Land Use. Land use is any area or place used for the purpose and benefit of man. There are several types of land use were recorded which is building houses, fields, gardens, cemeteries and fields.

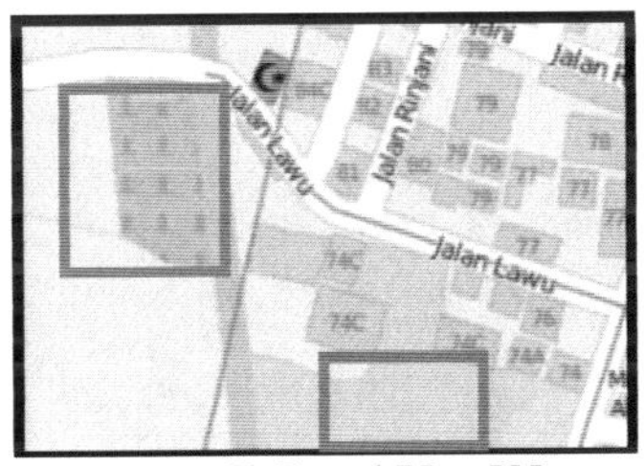

Figure 7. Land Use [2]

3.4. Concept Mapping on OSM

Mapping study of regional differences would affect existing land use types. For instance mapping in urban areas is certainly packed with buildings and industry, and in remote areas in Indonesia is dominated by forest. This makes the OSM can be used for various types of mapping, but the imagery available in the different regions have different renewal. Remote areas tend not to have many significant changes in contrast to cities that rampant development.

Mapping Village with OSM relied on imagery from bing. Most image quality when used OSM mapping is quite good, but the resolution is lower than that of Google Earth imagery. When the results of OSM maps digitized image overlaid with bing it was right and appropriate, but when overlap to Google Earth imagery there is a slight shift, but the image is more noticeable.

Input data to the OSM Open Source, so pul in the export process, OSM provides a variety of ways to download the data to the OSM. Disclosure of data on the causes OSM mapping can be done very easily and can perform mapping an area by adjusting the local knowledge interpreter. The digitized map data to OSM is currently exported to shp shapefile that there are eight types of waterways, roads, points, railroad, places, nature, land use and field. The exported map from OSM can be process to detailed map based on NSPK or many more thematic map.

Figure 8. Detailed Map Building Kwarasan RW 05, Nogotirto, Gamping, Sleman.

4. Conclusion

OpenStreetMap is one GIS application integrated with GPS, Web and Mobile, is open source in terms of input, editing and export data, so the user can input and process data map detail easy to level the building. OpenStreetMap can be a good alternative for mapping the village, because OSM provides a data source in the form of images, the level of detail of the building and export data in various formats such as shapefiles, and the result map on OSM already based on NSPK village mapping role.

References

[1] Badan Informasi Geospasial2016*NSPK Pemetaan Desa*(Yogyakarta: BIG) pp 1-53
[2] OSM, H 2016*Training OpenStreetMap*(Yogyakarta: Tim OSM) pp 1-10

Combined aerial and terrestrial images for complete 3D documentation of Singosari Temple based on Structure from Motion algorithm

Husnul Hidayat and A B Cahyono

Department of Geomatics Engineering, Institut Teknologi Sepuluh November, Sukolilo, Surabaya, 60111, Indonesia

hidayat@geodesy.its.ac.id

Abstract. Singosaritemple is one of cultural heritage building in East Java, Indonesia which was built in 1300s and restorated in 1934-1937. Because of its history and importance, complete documentation of this temple is required. Nowadays with the advent of low cost UAVs combining aerial photography with terrestrial photogrammetry gives more complete data for 3D documentation. This research aims to make complete 3D model of this landmark from aerial and terrestrial photographs with Structure from Motion algorithm. To establish correct scale, position, and orientation, the final 3D model was georeferenced with Ground Control Points in UTM 49S coordinate system. The result shows that all façades, floor, and upper structures can be modeled completely in 3D. In terms of 3D coordinate accuracy, the Root Mean Square Errors (RMSEs) are $RMSE_X$=0,041 m; $RMSE_Y$=0,031 m; $RMSE_Z$=0,049 m which represent 0.071 m displacement in 3D space. In addition the mean difference of lenght measurements of the object is 0,057 m. With this accuracy, this method can be used to map the site up to 1:237 scale. Although the accuracy level is still in centimeters, the combined aerial and terrestrial photographs with Structure from Motion algorithm can provide complete and visually interesting 3D model.

1. Introduction

As civilization evolves, it leaves many things as its evidences. One type of these evidences is cultural heritage building. These buildings are very important in the context of archaeology and architecture because they show the footprints on the development of art and construction technology. But many of these buildings are threatened. The heirs of these buildings may dissolved over the time, leaving these buildings without any proper maintenance and preservation. On the other hands, these buildings are also threatened for example by fire, earthquake, war, or erosion. Therefore, documenting these buildings becomes a fundamental task for restoration and reconstruction in the future [1]. One of cultural heritage building in Indonesia is Singosari Temple. This temple is located at Singosari District, Malang Regency, East Java. The archaeologist predicts that this temple was built in 1300s. In 1930s, the Dutch colonial government restored this temple [2].

Documenting cultural heritage building can be done by using terrestrial laser scanner. This device can record million point cloud data of the building quickly. In early stage of development, laser technology has surpassed traditional close range photogrammetry. But in past years by development of digital photogrammetry and computer vision, the extraction of 3D point data from images can be done automatically. Nowadays the generation of detailed 3D models from images is a fact, eroding the superiority of laser scanning [3]. The Structure from Motion (SfM) algorithm which based on multiview stereovision enables the extraction of 3D point cloud cloud from unordered image sequences from low cost camera. This novel method is low cost, effective, and can be an alternative to traditional surveying or photogrammetric method [4].

Nowadays, existing terrestrial photogrammetry technique utilizes this algorithm and greatly improves the result and productivity. The building façades can be modeled perfectly while the roof and upper structure are difficult to be modeled only by using terrestrial images [5]. By the advent of low cost UAV system, nowadays aerial photography can be done easily. Combining this technology with existing terrestrial photogrammetry will give more complete coverage data of building structure, especially the roof and upper structures. Therefore a complete model of cultural heritage site and its surrounding environment can be modeled completely without any missing parts. Several projects have shown that employing UAVs for cultural heritage documentation gives satisfactory result [6,7, 8]. UAV can obtain fast overview of the site, being a standard tool for excavation, or even to do 3D modeling which make it a universal tool for archaeologists [9]. This research aims to make a 3D model of Singosari Temple by combining aerial and terrestrial photographs as shown in figure 1. The site was selected because currently there is no digital 3D archive of this landmark. The incomplete top structure becomes a good case study on how aerial images play important roles ini cultural heritage modeling.

Figure 1. Singosari Temple in terrestrial and obliqueaerial photo.

Figure 2. Singosari Temple in oblique aerial photo.

2. Methodology

2.1. Research location

The Singosari Temple is located at Candi Renggo Village, Malang Regency, 9 kilometers northward from Malang City (figure 2). The site is located in the middle of settlement, but still accessible by many vehicles since it is located only about 500 meters westward from national road. This temple is 14,10 high with the footprint measured about 8×8 meters [10]. Unlike other temples, this temple has very little reliefs. The top structure is also incompleted. These facts suggest that the construction process was not finished. Today, the temple is still used as a tourism object and sometimes the cultural events are held there.

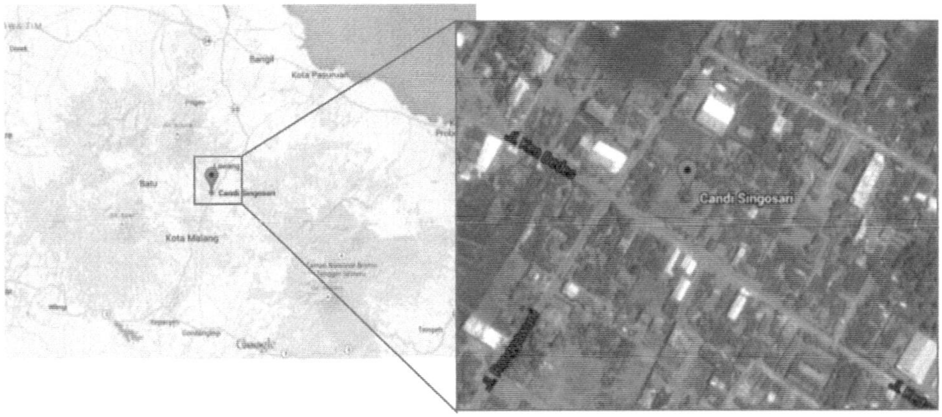

Figure3. Research location (source : http://maps.google.com).

2.2. Terrestrial image acquisition

Terrestrial image acquisition was done by covering the object from all directions. This is done to ensure that all façades of the object are well-captured as shown in figure 3. A 12-MP resolution consumer grade camera Canon Powershot SX260 HS with 4-mm focal length was used for this task. There are 39 terrestrial images in this project that were taken from all directions. The images were taken from roughly 15-meter distance. Since the images cover the object in 360-degrees, therefore the intersection angle between image's optical axis is roughly 9 degrees. The short baseline ensures large overlap between images which the SfM algorithm requires [4].

Figure 4. Example of terrestrial images of Singosari Temple.

2.3. Aerial image acquisition

The aerial image acquisition was done mainly to cover upper structures and building floor by employing the same camera used in terrestrial image acquisition. The camera was mounted on a lightweight quadcopter DJI Phantom 2. In order to ensure each façades covered completely, the vehicle moved in circular trajectory around the temple while the camera takes the images every 5 seconds. The camera was tilted about 45 degrees to the ground to take oblique imagery (figure 4). To cover the building completely from the air, the vertical aerial imagery was also taken (figure 5).

Figure 5. Example of oblique aerial images of Singosari Temple.

Figure 6. Example of vertical aerial images of Singosari Temple.

2.4. Ground control survey and measurement

The SfM algorithm is very efficient in shape reconstruction. The level of detail of 3D point cloud depends on image quality. It can generate comparable results with Terrestrial Laser Scanning does [11]. But the main drawback of SfM algorithm is it only recovers the shape of an object. The SfM algorithm cannot recover absolute scale, position, orientation of an object unless there are some control measurements. In order to achieve properly georeferenced model, the Ground Control Points (GCPs) were established by terrestrial surveying method. The reference frame consists of 4 benchmarks whose coordinates were measured using total station instrument. As the control, two of these benchmarks were surveyed using differential Global Positioning System. Once the reference frame have been established, the GCPs can be surveyed. There were 5 GCPs located at some parts of the temple. In order to check the accuracy of georeferencing, some Independent Control Points (ICPs) were also measured. Additional length measurements of some line segments of the temple were also conducted to check the geometric accuracy of the model.

2.5. Image matching and alignment

This is the first step in SfM algorithm pipeline which becomes a key for the next step. Three dimensional restitution from 2D images requires information about camera position and orientation in

3D space. The camera position and orientation in 3D space can be obtained from relative orientation between images. The relative orientation can be done by selecting corresponding tie points between the images. This process can be done automatically with image matching technique. As the correspondance between images has been established, then camera position during image acquisition can be estimated.

In this research, aerial and terrestrial image alignment were done separately for hardware efficiency. Beside that, between terrestrial and aerial images there are wide baselines. This causes the aerial and terrestrial image alignment cannot be done simultaneously with SfM unless there are some tie points that link both types of image blocks.

Terrestrial image alignment gives image position and orientation in arbitrary coordinate system. Beside that, this process also delivers sparse point cloud which reveals wall structures of the building. Since floor and top structures have lack visibility in terrestrial images, these structures are less revealed in sparse point cloud. To overcome this problem the aerial image alignment was done. This alignment forms photogrammetric block of aerial images. This block reveals floor and top structures although the wall structures were partially revealed. Combining these blocks could reveal whole structure of the temple which is represented in sparse point cloud (figure 6). Some tie points were manually marked on both image blocks, which further these block were joined together based on coordinate transformation using these tie points.

(a) (b)

(c)

Figure 7. Terrestrial (a), aerial (b), and combined image block (c).

2.6. Dense reconstruction

After image alignment and sparse reconstruction, the next step is dense reconstruction. In this process the object geometry are revealed with higher detail. The dense reconstruction mainly relies on space intersection and digital image matching technique. The exterior orientation parameters of each image were used to estimate 3D position of point cloud. These parameters also play important role in digital image matching. Epipolar geometry of a point between many images can be generated from exterior orientation parameter. Having epipolar line between the images, the digital image matching process can be done only by searching along the epipolar line instead searching in entire image. Therefore the matching process can be done faster [12].

Combined aerial and and terrestrial image block gives complete 3D point cloud of the temple. The point cloud of walls, floors, and top structures shows clearly the entire shape of the temple with intricate details (figure 7). But however the 3D point cloud only represents discrete points of continuous surfaces. Therefore in order to make continuous surfaces of the temple the meshing process must be carried out.

Figure 8. 3D point cloud of the temple with high detail.

2.7. Meshing and texturing

Meshing process was done to turn discrete point cloud data into continuous 3D surface of the building. Before beginning the process, some points which were considered as outliers were removed. These points are located far from predicted model surface which was caused by inaccurate dense matching process. The surfaces were constructed in the form of Triangulated Irregular Network (TIN).

After the surface have been constructed, the next step is texturing. In this process the texture from original images is projected to model surfaces. If there are some particular surfaces imaged at more than one image, then the projected texture is mixed from these images.

Figure 9. 3D mesh and textured model.

2.8. Georeferencing

Since the SfM algorithm only reveals 3D shape of an object with arbitrary coordinate system, it is obvious that georeferencing process is required. This process can be done by using Ground Control Points whose coordinate have been surveyed before. To check geometric accuracy of reconstructed 3D model, some Independent Check Point were also surveyed and marked (figure 9).

The Ground Control Points consist of 5 points in UTM Zone 49S with WGS 1984 ellipsoid. The elevation uses ellipsoidal height as reference. The Independent Check Points were placed at various parts of the temple. The ICP placement is only based on natural target that identifiable on the images. This was done since target placement on the upper parts of the temple were not allowed. The coordinate of Ground Control Points and Independent Check Points were acquired by terrestrial surveying.

Figure 10. Ground Control Points position (red dots) and Independent Check Point (yellow dots)distribution.

3. Results and discussion

3.1. A 3D model of Singosari Temple

From SfM pipeline applied in this research, a complete 3D model of Singosari Temple was reconstructed. This model not only shows the building, but also its surrounding environment. All façades were reconstructed with good detail except the rear façade. This was caused by backlighting which causes this façade was not seen clearly which affect the accuracy of 3D point clouds. Therefore the resulting façade has rough surface compared to other façades.

In terms of model completeness, the combination of aerial and terrestrial images gives satisfactory results. Terrestrial imaging could not provide complete 3D reconstruction because the lack of visibility in floor and roof structure. With the aid from aerial images, the floor and top structure can be modeled very good forming complete 3D model with its surrounding environment.

(a) (b)

(c) (d)

Figure 11. 3D model of Singosari Temple.
(a) North view.(b) South view. (c) Top view. (d) East view.

This 3D model also shows good details of top structure. From the model it has been shown that the upper structure were not formed completely as shown in figure 11.

Figure 12. Incomplete top structure of the temple.

3.2. Accuracy analysis
Accuracy analysis was done by comparing 3D coordinates of Independent Check Points (ICPs) and length of line segments with reference data obtained from terrestrial surveying. There are 12 ICPs which used for this accuracy test. The results show that the mean absolute errors of 3D coordinates are dX=0.034 m; dY=0.026 m; anddZ=0.042 m, with *Root Mean Square Error* (RMSE) values $RMSE_X$=0,041 m; $RMSE_Y$=0,031 m; and $RMSE_Z$=0,049 m. These RMSE values correspond to 0.071 m displacement in 3D space. The planimetric accuracy of this project tends to be better than its vertical accuracy. This is caused that most of control points lie on horizontal plane.

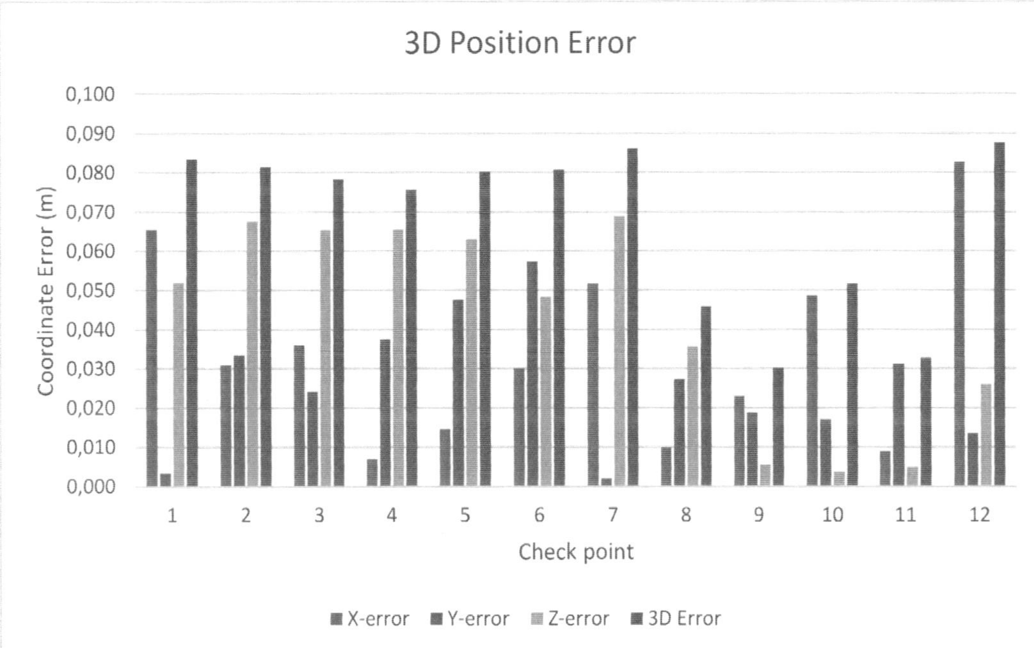

Figure 13. Error in 3D position.

To check the geometric accuracy of 3D model, length measurements were also conducted. There are 43 line segments used for accuracy test. These lines were measured on the site using distometer. For the line segments that lie on the upper part of the temple their lengths were determined by well known intersection method in terrestrial surveying. The result shows that the mean absolute error of line measurements is 0.057 m with RMSE=0.065 m. This result is roughly equal to the RMSE of 3D position.

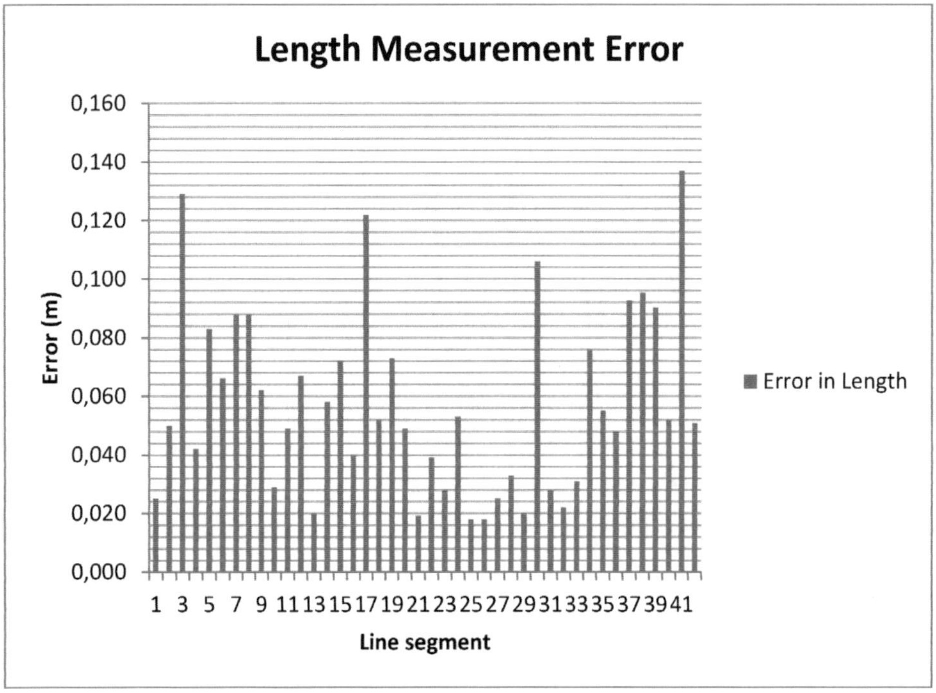

Figure 14. Error in length measurement.

According to Technical Specification of Topographic Mapping for 1:25000 scale by Bakosurtanal (now Geospatial Information Agency: BIG) [13], the cartometric accuracy for mapping is 0,3 mm. Based on largest achieved RMSE, i.e. 0,071 m, the 3D modeling by combining UAV aerial and terrestrial images with SfM processing pipeline can be used for mapping the building up to 1:237 scale.

4. Conclusion

This research has found that combination of terrestrial and aerial UAV images with Structure from Motion processing method can be used to make a 3D model of Singosari Temple completely. All façades, structures, floors, and its surrounding environment can be modeled without any missing part contrary to the same method that relies only to terrestrial images. From the accuracy analysis it has been shown that the Root Mean Square Errors (RMSEs) are $RMSE_X$=0,041 m; $RMSE_Y$=0,031 m; $RMSE_Z$=0,049 m which represent 0.071 m displacement in 3D space.With this accuracy, this method can be used to map the site up to 1:237 scale. The use of natural target in this research can reach centimeter-grade accuracy. Finally it can be stated that UAV images are suitable for complete 3D documentation of cultural heritage site.

Acknowledgement

Authors would like thank to Lembaga Penelitian dan Pengabdian Masyarakat (LPPM) Institut Teknologi Sepuluh November for supporting the research funding, and also for Balai Pelestarian Cagar Budaya Jawa Timur, Mojokerto who has permitted field data collection at research area.

References

[1] Alshawabkeh Integration of laser scanning and photogrammetry for heritage documentation [dissertation on the internet]. [Stuttgart (Germany)]: Universität Stuttgart; 2006 [cited 2016 Sep 10]. Available from: http://www.ifp.uni-stuttgart.de/publications/dissertationen/phd_thesis_alshawabkeh.pdf

[2] Perpustakaan Nasional Republik Indonesia. Candi Singasari [Internet]. Jakarta: Perpustakaan Nasional Republik Indonesia; [cited 2016 Sep 10]. Available from: http://candi.perpusnas.go.id/temples/deskripsi-jawa_timur-candi_singasari

[3] Skarlatos D and Kiparissi S 2012 Comparison of laser scanning, photogrammetry, and SfM-MVS pipeline applied in structures and artificial surfaces *ISPRS Annals of the Photogrammetry, Remote Sensing and Spatial Information Sciences* **I3**

[4] Westoby M J, Brasington J, Glasser N F, Hambrey M J and Reynolds J M 2012 Structure from Motion photogrammetry: a low cost, effective tool for geoscience applications *Geomorphology* **179** 300-14

[5] Hidayat H and Cahyono A B 2015 3D Reconstruction of Singosari Temple Using Terrestrial Images and Structure from Motion Algorithm*Proc. of Int. Conf. of Indonesian Society for Remote Sensing (ICOIRS)* (Surabaya: Department of Geomatics Engineering)

[6] Püschel H, Sauerbier M, Eisenbeiss H 2008 A 3D model of Castle Landenberg (CH) from combined photogrammetric processing of terrestrial UAV-based images*The International Archives of the Photogrammetry, Remote Sensing and Spatial Information Sciences* B6b **37**

[7] Mozas-Calvache A T, Perez-Garcia J L, Cardenal-Escarcena F J, Mata-Castro E and Delgado-Garcia J 2012 Method for photogrammetric surveying of archaeological sites with light aerial platforms *Journal of Archaeological Science* Issue 2 **39** 521-30

[8] Lo Brutto M, Garraffa A, Meli P 2014 UAV platforms for cultural heritage: first results *ISPRS Annals of the Photogrammetry, Remote Sensing and Spatial Information Sciences* II **5**

[9] Seitz C and Altenbach H 2011 Project Archeye-the quadcopter as the archaeologist's eye *The International Archives of the Photogrammetry, Remote Sensing and Spatial Information Sciences* **38-1** C22

[10] Balai Pelestarian Cagar Budaya Jawa Timur. Candi Singosari [Internet]. Mojokerto: Balai Pelestarian Cagar Budaya Jawa Timur; [cited 2015 Mar 15]. Available from: http://kebudayaan.kemdikbud.go.id/bpcbtrowulan/2015/01/23/candi-singosari-2

[11] Mancini F, Dubbini M, Gattelli M, Stecchi F, Fabbri S and Gabbianelli G 2013 Using Unmanned Aerial Vehicles (UAV) for high resolution reconstruction of topography: the Structure from Motion approach on coastal environments *Remote Sensing* **5** 6880-98

[12] Wolf P R and Dewitt B A 2004 *Elements of Photogrammetry: with Application in GIS* (Columbus: McGraw Hill)

[13] Bakosurtanal. SNI 6502.2:2010 - Spesifikasi Penyajian Peta Rupa Bumi – Bagian 2: Skala 1: 25000 [Internet]. 2010 [cited 2016 Sep 10]. Available from http://www.bakosurtanal.go.id/assets/download/sni/SNI/18.%20SNI%206502.22010%20Spesifikasi%20penyajian%20peta%20rupa%20bumi%2025.000.pdf

The Low-Cost UAV-Based Remote Sensing System Capabilities for Large Scale Cadaster Mapping

Catur Aries Rokhmana and Susilo Utomo

Department of Geodetic Engineering, Universitas Gadjah Mada, Jl. Grafika No.2, Yogyakarta 55281, Indonesia

caris@ugm.ac.id

Abstract. There is at least 190 million hectare of Indonesia's land part area that should be mapped in large scale cadaster maps. The completion of cadaster maps up to scale 1/2500 are still an open problems. The very high resolution images with spatial resolution less than 10cm can be a good choice to be able to see a parcel boundary. The use of Unmanned Aerial Vehicle-based remote sensing system can produce aerial photograph with spatial resolution less than 10 cm. For keeping UAV-based system low-cost, the system architecture consists of (1) aerial platform from hobby aeromodelling; (2) consumers grade camera sensor; (3) a low-cost GPS logger for ground control survey; (4) an open source structure from motion processing; and (5) an open source GIS software. This system was tested for producing a cadaster base map in a paddy field area at Trimulyo village at Bantul Region. The high resolution image with Ground Sampling Distance up to 7 cm can easily to see a parcel boundary and some of boundary markers. Some of the parcel areas were used for geometric evaluation. The bundle adjustment with 9 Ground Control Points (GCP) shows the error less than 12 cm in all coordinates component. Meanwhile, the percentage of area differences from some parcel samples show less than 5% of differences. In future, the potential use of a GPS kinematic assisted photo flight should be explored in accordance to reduce the need for GCP survey.

Keywords: UAV-Based remote sensing; large scale cadaster maps, parcel area calculation

1. Introduction

Indonesia region is one of the biggest archipelago country which has land part and ocean part areas. This will be a challenge for the completion of cadastral mapping. There is at least 190 million hectare of Indonesia's land part area should be mapped in large scale cadaster maps. Until 2006 only less than 10 percent of the area has been plotted into maps with a scale of 1:1000 and 1:2500 [1]. The lack of national large scale base map is one of the conditions that have to be resolved for a good land registration. The acceleration of cadastral mapping is an imperative need to be resolved. So, the very high resolution imagery with spatial resolution less than 10cm from remote sensing technique can be a good choice to be able to see a parcel boundary situation. The use of Unmanned Aerial Vehicle-based remote sensing system (UAV-based RS system) can produce aerial photograph with spatial resolution less than 10 cm. in this paper, the UAV-based RS system define as an UAV platform that can carrying a point and shoot digital camera for capturing an aerial photograph in visible band.

The UAV-Based system has not been optimally usedfor the land parcel mapping, since it less accurate than terrestrial mapping survey. This because in general, the UAV-Based system is obtained by non-metric camera which has some geometric limitation. However, the use of UAV has some advantages, such as: (1) cost-effective, (2) the accuracy can be improved by camera calibration, (3) fast in mapping production, and (4) The UAV platform has less weight that make it more portable (see [2], [3]).

This paper shows one experience of using the UAV-Based system for cadaster mapping in the rice fields' area of Trimulyo Village, Jetis, Bantul, and D.I. Yogyakarta. The accuracy of orthophoto product in this study has been tested with the existing land registration map from the National Land Agency and the GPS stop and go survey measurements methods also. This study focused on the land parcel area measurement accuracy from the UAV-Based system. More than 27 sample parcels selected among the sites of ± 20 ha. The parcel objectsare an area open area of rice fields which a flat topography with little vegetation has covered. Furthermore, The UAV-basedsystem architecture and how it works is also described in this paper.

2. Indonesia Cadaster Survey Needs

According to the regulation of the State Minister of Agrarian Affair/Head of National Land Agency number 3, 1997 [4], the cadaster boundary data acquisition procedure consist of some activities such as registration basic mapping, delineation cadaster boundary, cadaster survey and mapping and registration mapping. The following paragraph shows the need for cadaster survey activities, which are:

1. Cadaster mapping used TM-3 maps projection coordinates system
2. The registration basic mapping use map scale 1/1.000 for residential area; 1/2.500 for agriculture area; and 1/10.000 for Plantation area.
3. The parcel boundary corner should be marking physically with pegs/pipe (monument). The size of the pipe can be varies between 5cm – 20cm for parcel less than 10Ha.
4. The photogrammetric survey product such as photo maps; orthophoto; and blow-up photo can be used for registration basic mapping, especiallyin open area.

So, to be able to use the UAV-Based system for cadaster survey, it should has capabilities to produce the aerial photo that meet some perquisite, such as

1. The aerial photo should be able to see the boundary and produce a large scale maps up to 1/1.000. So, the aerial photo should has Ground Sampling Distance (GSD) 5cm-10cm to be able to see the boundary marking.
2. The operation in the field of the system should be as simply as possible and cost effective due to the limited infrastructure supporting in Indonesia.
3. Regarding to the large volume of parcel number, the system should fast in production up to 350Ha per day.
4. The UAV-based system should be able to work and collaborate with other existing systems. One of the existing system such as CORS-BPN (Continuous Operation Reference System) can be used for supporting a Ground Control survey during filed data acquisition.
5. The system should work safely.

3. THE UAV-BasedRemote Sensing System Architecture

The use of Unmanned Aerial Vehicle-based remote sensing system can produce aerial photograph with spatial resolution less than 10 cm. The system consists of instrumentation and personnel who run this system. Personnel team consists of (1) a pilot who operates an aerial platform (model airplane), (2) Assistant navigator that helps pilot, and (3) data processing operator is responsible for survey control points, and processing of digital photogrammetry. For keeping UAV-based system low-cost, the system architecture consists of (1) aerial platform from hobby aeromodelling; (2) consumers grade

IOP Conf. Series: Earth and Environmental Science **47** (2016) 012005 doi:10.1088/1755-1315/47/1/012005

camera sensor; (3) a low-cost GPS logger for ground control survey; (4) an open source structure from motion processing; and (5) an open source GIS software.In the last ten years a number of studies using unmanned aerial vehicle for mapping from the air to produce large-scale map. This system is intended to fill the need for low cost and implementing for small areas (see [5], [6]. [7]). Basically, there is no fundamental difference between conventional and UAV-based aerial mapping procedure, except in instrumentation and automation in data processing. The UAV platform do not require high operational costs. In addition, the weight of the aircraft less than 3kg can easily stored and carried. The use of electric UAVPlatform is also safer and environmentally friendly. The UAV platform can fly below the cloud, then atmospheric disturbances can be minimized. However, the lightweight UAV platform is not quite able to maintain stable flight against the wind with a speed of over 40 km per hour. This making it difficult to be able to follow the flight plan. Limitations of aerial photos to identify objects that covered byhigh vegetation or other objects are also of particular concern.

Table 1 shows the characteristics of the instrumentation system utilizing aeromodelling aircraft to carry a digital camera and avionics systems with a total payload weight <750 grams. Avionics system consists of GPS 10Hz, and Autopilot (Ardupilot: ATMEGA 8bit). The camera remote shutter release is controlled by script program to be able to run interval shutter.

Table 1. The characteristic of UAV for mapping system

Aerial Platform Type	 a. RC high wing and Portable Backpack, b. Hand launch takeoff and Net landing c. Electric DC brushless lipo battery d. Ground speed 50– 66 km/h e. Flight time 20min effective
Avionic System Weight < 350gr	a. R/C 7 Ch with 1W UHF 433MHz for range < 10 km b. Autopilot (open source Ardupilot) c. GPS receiver 10Hz, d. RF modem 1W 900Mhz range < 10km
Imaging sensor Weight < 350gr	a. Pocket Digital Camera > 12MPix field of view > 60° b. Anti-vibration mounting system (foam-based or silica gel)
Ground Station (Portable)	a. Laptop Win 32Bit (XP or Win7) b. Ground Station Software (Ardupilot Mission Planner) c. Antenna receiver with Omni-directional path 12dB d. Database maps for moving map
Data Processing and Visualization	a. Structure from motion software for automatic bundle adjustment; DSM generation; and orthophoto b. MeshLab (open source) for point cloud visualization

In practice, fly parallel with the winds direction can be easier than trying to fly in a cross wind direction. Photograph with areas overlapping greater than 80% should be taken for several advantages such as (1) reduce the coverage lost data due to turbulence, (2) a high correlation between adjacent photos can be advantages for automation process. The data processing can produce two types of products, namely (1) orthophoto image mosaic, and (2) digital elevation model (DEM). The orthophoto and DEM products are processed by digital photogrammetric procedures. This procedure requires at least four 3D control points (X, Y, Z) in a bundle adjustment algorithm with self-calibration.

4. Results and Discussion

The single photo flight mission was running to coveralmost 80Ha paddy field area at Trimulyo Village, Jetis, Bantul.The UAV platform carrying a point shoot camera Nikon COOLPIX AW-120, with the flying high 250m above the ground level and produce up to 68 number of aerial photo with GSD 7.77 cm. The picture resolution is sharp enough to be able to see the boundary marking (see Figure 1). The Pre-mark color should striking colors of the surrounding area, the shape and size of the pre-mark must be considered. Pre marks were clearly visible on the photo will help reduce the error of defining the position of GCP and the determination of the boundaries of paddy fields were measured. The boundary markers use 20 cm x 20 cm and 12 cm x 12 cm rectangular shape paper.

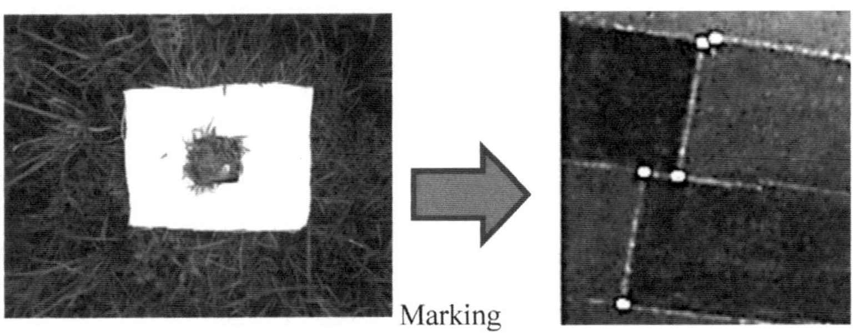

Figure 1. The Boundary marking image resolution

The post processing aerial photo processing by structure from motion algorithm and bundle adjustment with self-calibrationhas been done with 9 GCP and 6 check points. The Bundle adjustment reported the error precision 12 cm in all componentor less than 2 pixel. While the Check point error is less than 2pixel also. This error can fulfill for the 1/2.500 map scale accuracy. Furthermore, the evaluation of area calculation has been done for comparing the area calculation differences between existing parcel maps, GPS Survey, and orthophoto image. There are 27 parcel of rice paddies as sample. All of the 27 paddy fields have been registered or certified. While the GPS stop and go positioning survey have a centimeter to decimeter accuracy. Based on the post-processing of data using software GNSS Solutions, the boundary point consisting of 60 points with a fixed solution that has an accuracy of 1.00 cm - 12.9 cm and 41 points the rest of the solution float with a precision of 6.90 cm –68.3cm. Coordinate data cutoff point used must have accuracy less than 0.25 m in accordance with the tolerance allowed by the shift point PMNA / Head of BPN No. 3, 1997. Thus, only 88 meet the vantage point of tolerance. Of the 88 point limit was formed 18 field paddy field.

Table 2 shows the comparison of area calculation between the three methods. The comparison between the existing maps and orthophoto shows that there are some significant differences exist in the parcel boundary with the NIB 05 230, 03 070 and 03 065 with a different area of 70.696 m^2, 38.427 m^2 and 18.954 m^2 respectively. Some of the causes significant area difference, among others; (1) the one side of the corner boundary grow large trees that pre-mark is not visible, (2) some of the boundary monument has been move, and (3) the physical parcel of paddies rice in the field is not clear or not match with the existing boundary. Regarding to the statistical t-test, there areonly two parcels with significant area difference as mention earlier.

Table 2. Area calculation comparison

No.	No. Parcel (NIB)	Area Calculation (m^2)		
		Existing Maps	***Orthophoto***	**GPS Stop and Go**
1	02925	507,842	516,888	516,208
2	02931	437,971	447,046	447,957
3	02937	1589,657	1583,033	-
4	02960	511,641	526,937	534,790
5	02961	516,481	503,349	497,028
6	02963	242,357	250,653	250,367
7	02964	163,739	167,485	164,955
8	03019	532,207	526,082	505,861
9	03039	252,703	258,019	256,068
10	03065	251,839	232,885	221,206
11	03070	82,793	121,220	127,554
12	03078	237,867	243,687	254,580
13	03079	270,373	252,362	-
14	03104	207,603	210,302	212,625
15	03120	238,934	243,171	237,230
16	03135	300,584	306,760	-
17	03138	601,881	612,687	-
18	03147	254,790	238,923	-
19	04278	333,865	337,780	339,143
20	04606	391,810	397,871	394,067
21	04607	771,396	773,168	776,393
22	05048	706,375	722,272	-
23	05049	553,173	544,079	-
24	05053	1060,595	1058,324	-
25	05060	864,094	862,801	877,663
26	05061	380,097	389,304	394,808
27	05230	410,150	339,454	-

Furthermore, the area calculation comparison between the GPS Stop and Go survey and orthophoto shows that only two parcel has significant area differences. The Biggest area differences is 20.220 m^2. The biggest parcel error at parcel number (NIB) 03 019, this caused by there is no boundary pre-marking exist. So, that the main error due to the miss-alignment of boundary point picking (digitizing).

The comparisonof percentage area difference between existing maps and the digitization orthophoto and GPS measurement shows a good result. Regarding to the regulation, the area differences should less than 5%. Based on the evaluation of different percentage of the average area from orthophoto maps and theexisting mapsis 4.512%. Meanwhile, the percentage difference between the average area of orthophoto and GPS measurements is 1.804%.

5. Conclusion

This paper shows a successful of using UAV-based remote sensing system for producing the orthophoto as a basic cadaster maps. The orthophoto product accuracy is meet the 1/2.500 map scale accuracy. Some notes should be emphasized, regarding to the reason of biggest area differences, among others: (1) pre-mark is not visible or covered by vegetation, (2) some of the boundary monument has been move or missing, and (3) the physical parcel boundary of paddies rice in the field

IOP Conf. Series: Earth and Environmental Science **47** (2016) 012005 doi:10.1088/1755-1315/47/1/012005

is not clear or not match with the existing maps boundary.The bundle adjustment shows the error less than 12 cm in all coordinates component. Meanwhile, the percentage of area differences from some parcel samples show less than 5% of differences. In future, the potential use of a GPS kinematic assisted photo flight should be explore in accordance to reduce the need for GCP survey.

References

[1] Sumarto, I, W.R.Idrus, Virgo Eresta Jaya, R. Eko, D. K. Gindow, B., Adhi, E. Putranty, E. Pintadi, P. Hadi, Y. Aziz, A. Giyanto, Rahardjo, F., H. Feryandi, I. Herawati, Firman A.S. and S. Yusra 2008 Cadastral Base Mapping Activity in Indonesia, FIG Working Week 2008, Stockholm, Sweden 14-19 June 2008

[2] Rokhmana, C. A. 2014 Percepatan Pemetaan Kadaster Memanfaatkan Teknologi Wahana Udara Tanpa Awak *Jurnal Bhumi* No 39 April 2014

[3] Rokhmana, C. A. 2015 The Potential of UAV – Based Remote Sensing for Supporting Precision Agriculture in Indonesia The 1st International Symposium on LAPAN-IPB Satellite for Food Security and Environmental Monitoring *Procedia Environmental Sciences* **24** pp 245–253.

[4] Regulation of The State Minister of Agrarian Affairs/Head of National Land Agency on Provision of Land Registration, Pub. L. No.3 1997.

[5] Rokhmana, C.A. 2009 The Potential Applications of Ballon Photogrammetry for Cadaster Mapping (Bali: *Proceeding SEASC*)

[6] Clark, A.F., J. C. Woods and O. Oechsle 2010 A Low-Cost Airbone Platform for Ecological Monitoring *International Archives of Photogrammetry, Remote Sensing and Spatial Information Sciences* **XXXVIII** Part 5 Commission V Symposium Newcastle upon Tyne UK

[7] Niethammer, U., S. Rothmund , M. R. James , J. Travelletti , M. Joswig 2010 UAV-Based Remote Sensingof Landslides *International Archives of Photogrammetry, Remote Sensing and Spatial Information Sciences* **XXXVIII** Part 5 Commission V Symposium Newcastle upon Tyne UK

Aerial Photographs of Landslide on Clapar, Madukara District of Banjarnegara as a Spatial Geodatabase

Anggi Permata, Anwar Juniansyah, Eka Nurcahyati, Mousafi Dimas Afrizal, Muhammad Adnan Shafry Untoro, Na'ima Arifatha, Raden Ramadhani Yudha Adiwijaya, Barandi Sapta Widartono, and Taufik Hery Purwanto

Department of Geographic Information Science, Faculty of Geography, Universitas Gadjah Mada, Yogyakarta, Indonesia

muhammad.adnan.s@mail.ugm.ac.id

Abstract. Landslide is a natural disaster that commonly happen in Banjarnegara Regency. Spatial geodatabase is required to help understanding the landslide, planning for post-disaster recovery, even helping for prepare the upcoming disaster that may happen. The purpose of this reserach is for providing an orthophoto mosaic as high-resolution imagery and a Digital Surface Model (DSM), and compare it with Digital Model Model (DEM) generated from terrestrial surveying. The DSM and DEM were compared to see which method was the best to provide a precise spatial geodatabase in natural disaster spot, especially at Clapar's Landslide.

Keywords: Aerial Photography, Digital Surface Model, Landslide, Spatial Geodatabase

1. Introduction

Landslide is defined as mass movement of material, such as rock, earth, or debris. This movement is primary affected by gravitation force. A rapid mass movement can cause human injury and loss in infrastructure, life, and assets. This event is described as disaster which frequently happens in area with hilly topography.

Landslide occurs in area with significant steepness of slope (around >15°). A slope with developedsoils, sufficient intensity of rainfall, and type or arrangement of rocks with dip angle will increase the possibilities of landslide event [6].

Banjarnegara is one of the region which has frequent events of landslide. Since January 2015 until July 2017, 17 cases of landslide had been happening and causing more than 1200 people forcedly evacuating [2]. Topography with moderately steep condition and relatively high-intensity of rainfall become main factors for landslide in Banjarnegara.

In March 25th 2016, a landslide attacked the small part of this region called Clapar Village, destroying houses, infrastructures, and plantations. This event causes a necessity of data and information of the landslide, yet due to the debris the access to the location is difficult to reach. Therefore, a technology is needed to facilitate the access to the location and data collecting effort.

Remote sensing has many methods and technologies that can be used to collect, map and extract information of landslide. Aerial photography is one of the propitious methods to gain landslide imagery with high-spatial resolution formatted in small scale area. UAVs (Unmanned Air Vehicles)

have become popular platform to facilitate this method by using sensor that has greater results thansatellite imagery. UAVs also provide low altitude flight, effectiveness of time in gaining images, low cost budget than using satellite imagery, and less cloud disruption [5]. Recent advances show that the result of using UAVs also can be used to build Digital Model in surface or elevation basis based on overlapping imageries [8]. This method has been used in many researches related in various hazards mapping, such as validation of flood susceptibility map around Kudus region by[7]. Hence, using this method is suitable for accessing, extracting and collecting data and information about Clapar's landslide.

This paper presents the use of aerial photography imageries provided by UAV to collect data of landslide area. The data processed into Digital Surface Model (DSM) which provides elevation information of the materials accumulated by the debris. The result is compared with DEM provided by terrestrial surveying in same study area. The comparison is done by looking at 3D models built from both models, and implementing map algebra to identify height differences between two digital models.

2. Study Site

Clapar village is located in an area with a high-level intensity of landslide hazard. This level can be seen from how the landslide is occurred frequently in this region, Banjarnegara. Data provided by Badan Nasional Penanggulangan Bencana (BNPB) shows that there were 25 landslide locations had been recorded, including Clapar Village, in range from 11-12 December 2014. The largest landslide in Clapar Village itself was occurred in March 25, 2016, swallowing almost half of the village's region. The hazard happened due the spring that should be flowing down form the river but because the lack of space, the water seeped onto the ground, making the soil very soft, and causing the ground to flow as gravitation pull it down. The physical characteristics of the site itself also had quite-impacting factors to the landslide, such as low-stability type of soil and steep slope, making the motion to move faster.

Topology area with steep slope makes the condition of soil moving faster. Rainfall reaching in Clapar is calculaed for about 301 to 400 millimeters per month. Clapar has a specific rock formationcalled Rambatan Formation. Clapar landslides Crown in the village is located at an elevation of 837 meters above sea level with a slope of 23%. While the tip of the tongue landslide Clapar located at an elevation of 705 meters above sea level with a slope of 12%. The crown part of thelandslidewas known to be used in plantation of *Salacca zalacca* or Salak. Clapar possible landslides village lied in fault zones or fault of the north-south direction.

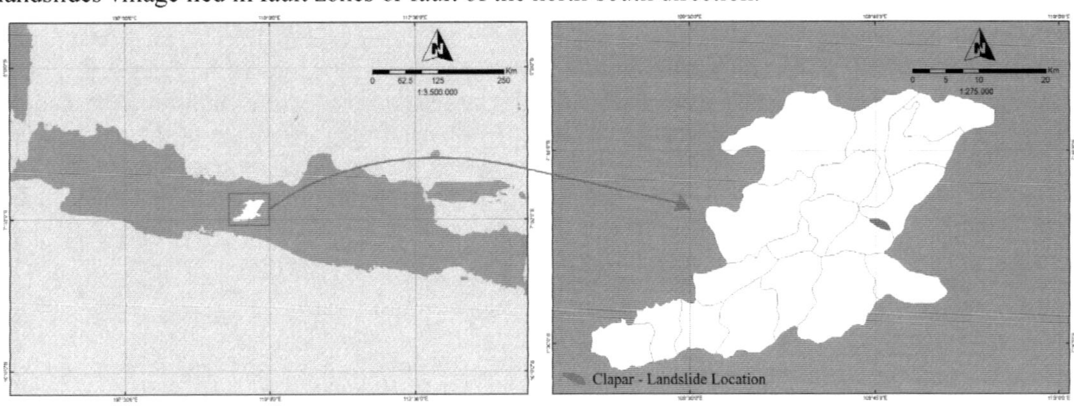

Figure 1. Location of landlides on Clapar, Madukara District of Banjarnegara

3. Data and Methods

3.1. Data

3.1.1. UAV. Aerial photographs used to build the orthophoto mosaic and Digital Surface Model were taken with DJI Phantom 3 Professional. DJI Phantom 3 Professional is a drone that well-known for its easy-to-fly so everyone just can fly it and take any aerial photos or videos. DJI 3 Professional is equipped with 1/2.3" CMOS camera with 12.4 megapixel effective pixels that paired with 3-axis gimbal [3]. This features make DJI Phantom series is widely-used in aerial photography and videography.

The aerial photographs were taken in one days, but need two days to measure all of the GCPs. The landslide was photographed at 25th May 2016 around 01.00 P.M.. The GCPs measured at 25th-26^{h} May 2016 around 09.00 P.M. until 03.00 P.M..

The photos were taken manually by the pilot. Pilot estimate how big the endlap and sidelap oh each photos. The landslide were taken twice with different direction to overcome problems that may occur because of the pilot mistook the photo and lead to a photo without proper endlap and sidelap causing its cant be arranged into a mosaic. The first flight direction is horizontally to landslide, or north-east to south-west. The second flight direction is vertically to landslide, or north-west to south-east.

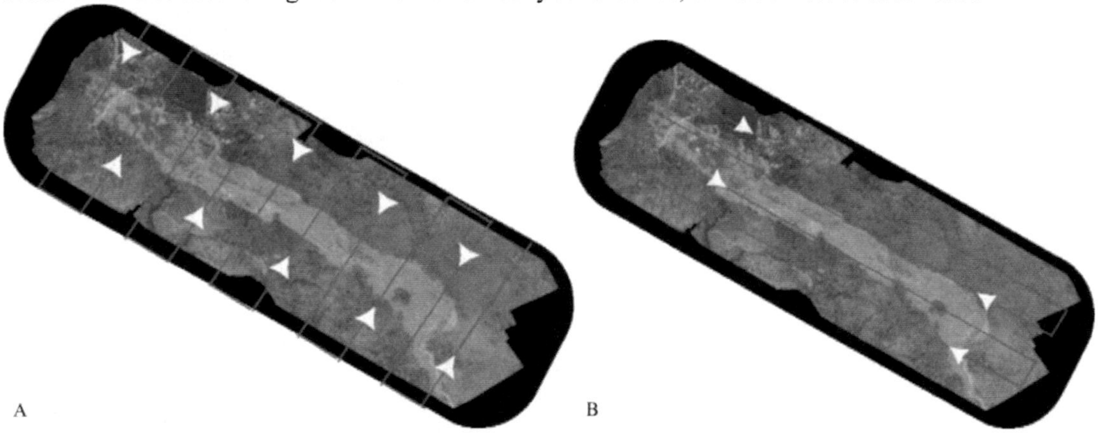

Figure 2. Flight paths of the drone when took the photos above landslide horizontally (A) and vertically (B)

3.1.2. Ground Control Points. Ground Control Points or GCPs are well-defined natural features or pre-marked artificial features that appear on the photographs and whose location in reference system are known [1]. GCPs that used were cross-shaped artificial features made from cardboard and painted orange. This crosses were ±1 meter long. The GCPs distributed in the main area of the landslide so they represent the different terrain heights and can be seen easily from above because there is no other object in the landslide that will cover it. The coordinates of the GCPs were determined using Leica Viva GS08 Global Navigation Satellite System (GNSS) Receiver with Real Time Kinematic (RTK)-Network Transport of RTCM via Internet Protocol (NTRIP) methods

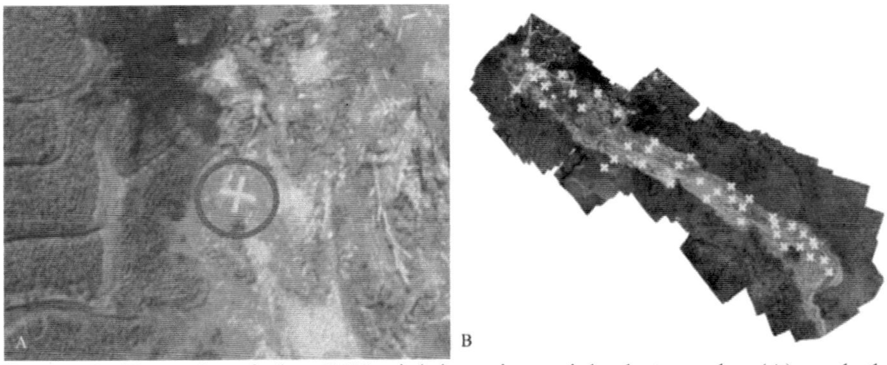

Figure 3. Example of the GCP sightings in aerial photographs (A) and the distribution of the GCPs at the area of landslide (B).

3.2. Methods

3.2.1. Aerial Photography. The making of Elevation Differences Map from aerial photographs and terrestrial survey consists of three steps: 1) generate a orthophoto mosaic and DSM from aerial photographs and GCPs coordinate that measured by GNSS using Agisoft Photo Scan Professional; 2) generate a DEM from elevation points by terrestrial survey (Total Station and Theodolite) using Kriging interpolation in ArcMap 10.3; and 3) generate Elevation Differences Map using Raster Calculator in ArcMap 10.3.

Agisoft Photo Scan Professional was used to generate orthophoto and DSM from aerial photograph and GCPs coordinate that measured by GNSS. Agisoft Photo Scan Professional used beacause it is widely known for its easy to use for generating orthophoto and DSM in many previous studies[4].

Kriging interpolation method has applications in spatial prediction and automatic contouring that measure distance and orientation between data (Shekhar and Xiong, 2008). Elevation points from terrestrial survey interpolated using automatic contouring feature in Kriging.

Elevation Differences Map is a map that represents the differences of height from DSM that generated from aerial photographs and DEM that generated from elevation points by terrestrial survey. Elevation Differences Map is actually a raster map that every pixel value generated from an equation DSM pixel value minus DEM pixel value. This operation can be done using Raster Calculator in ArcMap.

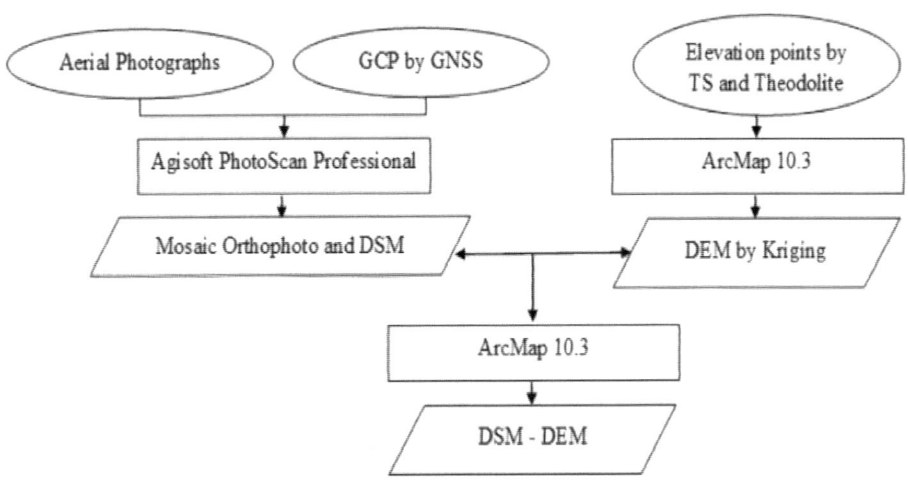

Figure 4. Flow chart of the project.

IOP Conf. Series: Earth and Environmental Science **47** (2016) 012006 doi:10.1088/1755-1315/47/1/012006

3.2.2. Terrestrial surveying. Landslide study area has topography or the appearance of the high and low clumps of landslide. Measurements of terrestrial survey were intended to collect altitude data. These measurements were conducted over three days from May 24-26,2016. The measurementsprocess took approximately 6 hours from 09 pm to 15 pm o'clock. The study area is divided into three parts to be measured effectively. Each section has a relationship and reference the same with each other. Theodolites were the main tool and is located in on of the baseline, followed by Total Station as secondary tool. Baseline is made from the west which is at the higher slopes and eastward toward the lower slope height.

Measurements using Theodolith were done manually, while the Total Station had more practical measurement methods than theodolite. Total Station was assisted by a prism with a laser system that was able to determine the angle, distance and point coordinates automatically. All information collected by the total station stored in an external computer where the data could be manipulated and added to the CAD program. Using data collected and stored by surveying tools, the DEM were processed with kriging interpolation methods.

4. Results

4.1. Orthophoto Mosaic

Orthophoto mosaic is generated from aerial photographs which has spatial resolution for about 0.054 meters. Orthophoto mosaic itself shows the wide of area damaged by landslide disaster. The landslide area extends for about 1096 meters length and 132 meter width, with orientation from north west to south east. Most of the area damaged by landslide was identified as *Salacca zalacca*, also known as Salak. This landslide also destroyed few local houses and a main local road.

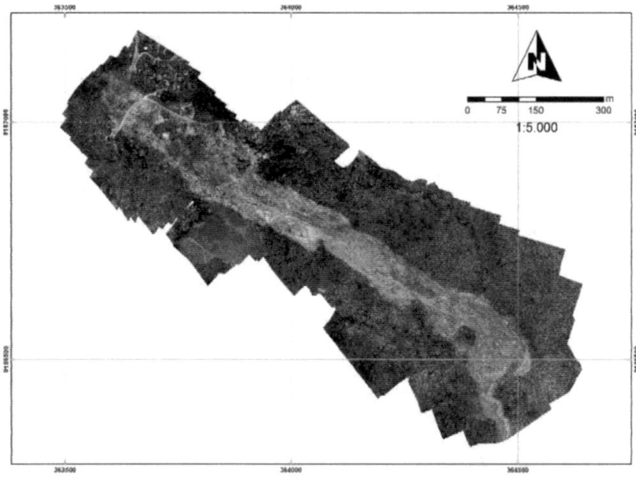

Figure 5. Mosaic orthophoto of the landslide.

4.2. DEM generated from terrestrial surveying

DEM (Digital Elevation Model) generated from interpolation between points from terrestrial measurements represents shape of landslide's aftermath. The amount of points used to generate DEM is 349 points. The points itself were not distributed equally in landslide area. To compare DSM provided from aerial photography with DEM provided from terrestrial surveying, we only choose 221points. The result from the comparison shows differences of height variation between 670 meters to 628 meters, covering area which extends 52400.66 m^2.

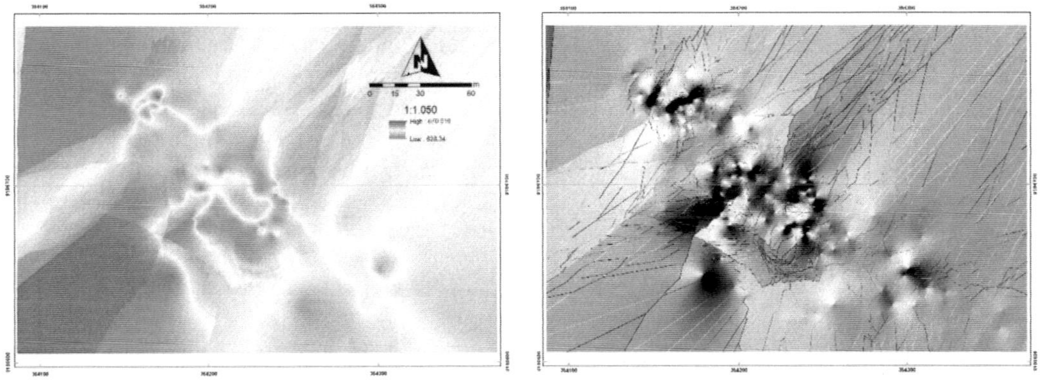

Figure 6. Digital Elevation Model (DEM) generated from terrestrial (right) and its hill shade model.

4.3 *DSM generated from aerial photographs*
DSM generated from aerial photograph imageries has spatial resolution for 0.1 meters. To compare it with DEM accumulated by terrestrial surveying and aerial photography, DSM was intersected in identical area with DEM generated by terrestrial surveying.

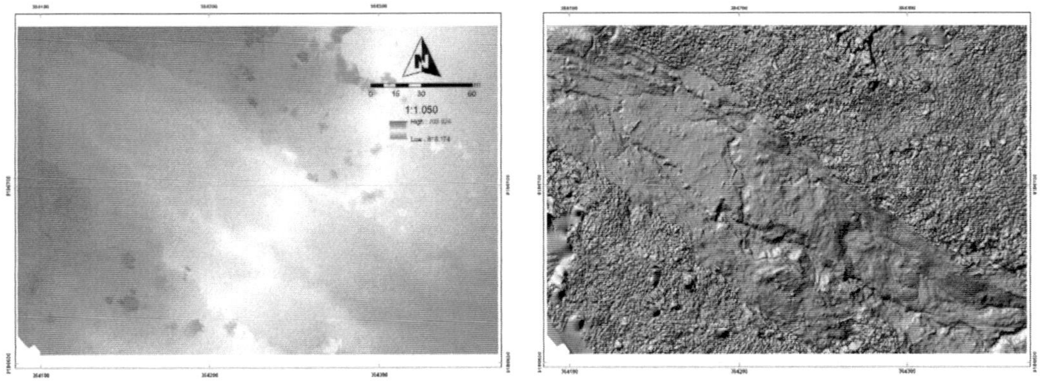

Figure 7. Digital Surface Model (DSM) generated from aerial photographs (right) and its hill shade model.

4.4 Elevation Differences Map
Elevation Differences Map generates from the differences of pixel value from DSM generates from aerial photographs and DEM generates from terrestrial surveying. Pixel value from DSM and DEM represent elevation points of the landslide therefore Elevation Differences Map represent the differences from these two methods of surveying.

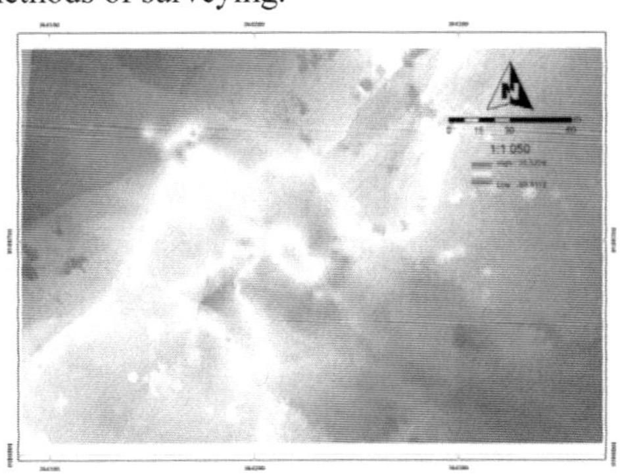

Figure 8. Elevation Differences Map from DSM and DEM

5. Discussion

DEM generated from aerial photography results differences from terrestrial surveying. Both model are intersected in the middle of study area, which is shown by figure 9. The first DEM shows higher elevation in north-west landslide area yet lower elevation in south east area. Both models have maximum difference for about 68min the northwest. The differences between both DEM are narrowing in the middle, and back enlarging in south-east part. DEM generated from aerial surveying has more similarities with the real form of landslide than the DEM generated from terrestrial surveying.

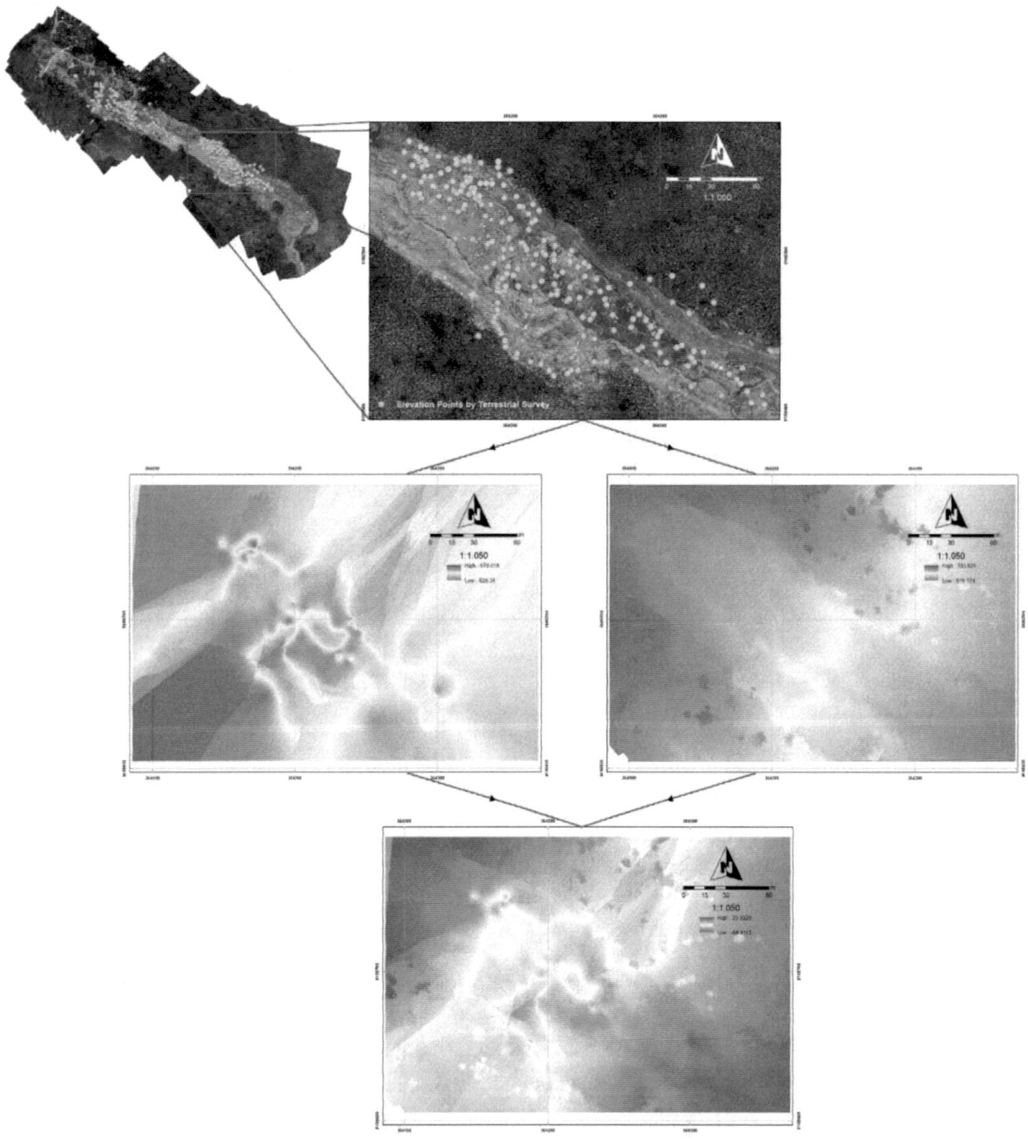

Figure 9. Map Differences Elevation making process

The big differences between DEM based on aerial photography and DEM based on terrestrial surveying can be caused by the amount of measured points that were used in terrestrial surveying hadn't been distributed equally around the study area. There were many points that weren't accessible because of study area's condition which was dangerous and not stable to be stepped on. The southwest part of the area was formed by high slope with unstable soils, which was caused by the presence of small river. The south-east part was also inaccessible because the terrain's condition was very dangerous to be stepped on. Those factors caused difficulties for terrestrial surveying to measure points that were planned in those two area. Most of the part of landslide area weren't stable to be observed, with soils' condition that weren't quite solid to hold the instrument for surveying. The different situation occurred in aerial photography process. Most of the part of landslide area was accessible and can be collected as aerial imageries data. The differences of height based on DEM can be also caused by the accuracy in measuring coordinate of surveyed points.

DEM generated by aerial photographs was more efficient in process that terrestrial surveying. It had taken three days to collect data from all parts of landslide area by terrestrial surveying with the effective time of 6 hours per day, while it only needed two days by aerial photography to collect all data and process it. The result generated from aerial photographs also shwesown more representing form of landslide area than terrestrial surveying.

6. Conclusion

Digital Surface Model that generated from aerial photographs and GCPs by GNSS give better represntation of landslide's terrain than the terrestrial surveying's DEM. DSM can represent the variation of the terrain surface with more impressive visualisation because aerial photographs can easily cover all of the area of landslide so every inch of terrain's data can be extracted become elevation points. Actual elevation and location was generated from GCPs coordinate by hight-precision GNSS. DSM and orthophoto mosaic are more suitable as spatial geodatabase than DEM from terrestrial surveying because its more presenting the real terrain in the disaster spot so it can be more representative for the reader of the geodatabase.

References

[1] Aber J S, Marzolff I, and Ries J B 2010 *Small-Format Aerial Photography: Principles, Techniques, and Geoscience Applications* (Amsterdam:Elsevier) p 124

[2] Badan Nasional Penanggulangan Bencana 2016 *Data dan Informasi Bencana Indonesia* - http://dibi.bnpb.go.id/data-bencana/lihat-data/

[3] DJI Phantom 3 Professional Specifications http://www.dji.com/phantom-3-pro/info

[4] Ishiguro S, Hiroya Y, and Hiroyuki O 2016 Evaluation of DSMs generated from multi-temporal aerial photographs using emerging structure from motion-multi-view stereo technology *Geomorphology* **268** 67

[5] Shofiyanti R. 2011.*Teknologi Pesawat Tanpa Awak Untuk Pemetaan dan Pemantauan Tanaman dan Lahan Pertanian*. (Informatika Pertanian) p 20-62

[6] Siegel F R 2016 *Mitigation of Dangers from Natural Anthropogenic Hazard: Prediction, Prevention, and Preparedness* (Switzerland: Springer) p 37

[7] Suryanta J 2014 *Penggunaan Unmanned Aerial Vehicle Untuk Validasi Peta Rawan Banjir Di Kabupaten Kudus dan Pati* (Jakarta:Pusat Pemetaan dan Integrasi Tematik, Badan Informasi Geospasial)

[8] Westoby M J, Brasington J, Glasser N F, Hambrey M J, Reynolds J M 2012 'Structure-from-Motion' photogrammetry: A low-cost, effective tool for geoscience applications *Geomorphology* **179** 301

Orthorectification of Sentinel-1 SAR (Synthetic Aperture Radar) Data in Some Parts Of South-eastern Sulawesi Using Sentinel-1 Toolbox

Athar Abdurahman Bayanudin[1] and Retnadi Heru Jatmiko[2,3]

[1]Department of Remote Sensing and GIS, Vocational School, Universitas Gadjah Mada, Yogyakarta, Indonesia.
[2]Cartography and Remote Sensing, Faculty of Geography, Universitas Gadjah Mada,Sekip Utara, 55281 Yogyakarta, Indonesia

retnadi_geougm@yahoo.com

Abstract. Sentinel-1 satellite uses active microwave sensor that has a better ability than the optical system in generate satellite image without clouds cover and can operate in day or night in all weather conditions. However, the side looking viewing geometry of SAR system and the object height variation or topography on the surface of the earth causing various distortions of geometry in the SAR (Synthetic Aperture Radar) image. It needed to be corrected through orthorectifiation process before the further implementation. This research aims to Orthorectifying the Sentinel-1 of SAR data in some regions of Southern Sulawesiusing Sentinel-1 Toolbox software and assessing the geometrical accuracy of orthorectification's results. Two orthorectify methods have been used, namely Range-Doppler Terrain Correction and SAR-Simulation Terrain Correction. Each SAR data orthorectify results (resampling to 30 m resolution) compared qualitatively and quantitatively. Qualitative by visual comparison between the results orthorectify. Quantitative by the calculation of the accuracy results orthorectify both methods on the GLS (Global Land Survey) 2000 Landsat image using 99 test points coordinate then calculated the RMSEr (Root Mean Square Error). The results showed that, there is no significant differences in visual geometry between the orthorectify results. Orthorectified SAR image in some parts of South-eastern Sulawesi from both method showed the same RMSEr value is 25.27 m which means the horizontal accuracy is less than one pixel.

Keywords: SAR Sentinel-1, Sentinel-1 Toolbox, Orthorectification, RMSEr; Southeast Sulawesi

1.Introduction

Indonesia as one of the widest countries, needs information sources which are fast, accurate, and also can cover the entire area for monitoring purposes. Remote sensing data is one of the data sources that is able to meet these basic requirements for extractingthe physical phenomena information on the Earth's surface rather than to do a field survey.

European Commission (EC) in collaboration with the European Space Agency (ESA) triggers a program of earth observation, called Copernicus which aims to provide information that is accurate, up to date and accessible for enhancing the environmental management, understanding, and reducing the climate change's impacts, and also guaranting the civil security [1]. Sentinel 1-A satellite is the first of seven Missions of remote sensing satellite imagery which was launched on April 3, 2014, and it has an active sensor in the Band-C. SAR imagery data can be downloaded for free throughout the data portal of Sentinel-1 on https://scihub.esa.int/.

Sentinel-1 SAR data availability can be optimally used due to the active microwave sensor which has a better ability than the optical system in generate imagery without clouds cover, and also it can be operated both day and night in all weather conditions [2]. However, the side looking viewing geometry of SAR system and other conditions of topography and object height, can cause geometric distortions such as foreshortening, layover, and shadow [3]. Orthorectification is needed to reduce those geometric distortions and especially to generate the information in accurate position.

Moreover, ESA also develop an open source software called Sentinel-1 Toolbox (S1Tbx) for processing SAR data including the orthorectification process especially Sentinel-1 data. Orthorectification known as a process to conversion of the two-dimensional image from a sensor-coordinate to a map-coordinate [4]. Radar image geometry correction is quite difficult to do. The availability of this software can be the solution to generate corrected SAR data especially for Sentinel-1 SAR data. There are two orthorectification methods in this software, namely Range Doppler Terrain Correction and SAR-Simulation Terrain Correction. The results of both methods are need to be evaluated for digital image processing.

Researchers wanted to do the orthorectification process on one of the Sentinel-1 data which covers some parts of South-eastern Sulawesi using S1 Tbx. This area to be area of study because of its position that close to the equator and optical data available on the scope of the region disturbed by cloud cover. The writers expect that this study can be used as a reference on Sentinel-1 application.This research aims to orthorectification of Sentinel-1 SAR data in some parts of South-eastern Sulawesi using Sentinel-1 Toolbox software and assess geometrical accuracy of the orthorectify results.

2. Data and Methods

2.1. Study Area

South-eastern Sulawesi were selected as the study area. The study area is administratively located on Southeast Sulawesi Province, Indonesia. Figure 1 shows the data coverage Sentinel-1 SAR in the research area on google earth.

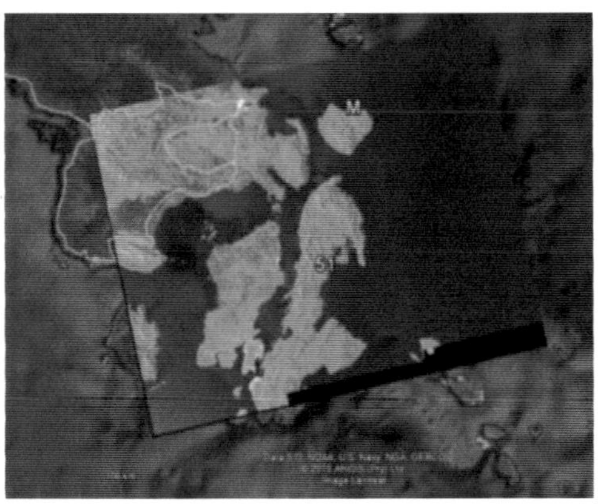

Figure 1. Display of sentinel-1 map overlay data on google earth.

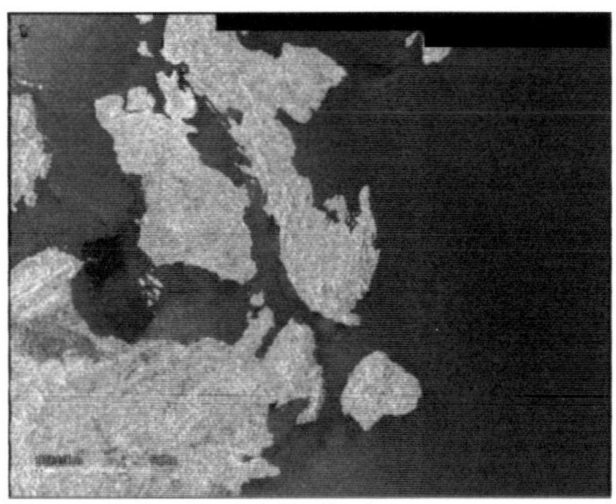

Figure 2. Sentinel-1 SAR data some parts of South-eastern Sulawesi before the orthorectification.

2.2. Data

Figure 2 shows the data that used in this research. The chosen data of South-eastern Sulawesi was orbit/track 3086/39 which obtained on November 1st 2014, Ascending.The SAR data used in this research was Sentinel-1 data with Band-C (5250-5570 Mhz), Level 1 GRDH (Ground Range Detected

High Resolution), Interferrometric Wide-Swath Mode (IW), VV polarization, pixel size of 10 meter, azimuth x range resolution = 20 x 22 m.

When a DEM (Digital Elevation Model) of the observed geographical area is available, a procedure called orthorectification can be applied on allowing the correction of some geometric distortions [5]. Ideally, DEM data has a same resolution toward SAR data, so it would produce a better orthorectified imagery. Because of limited DEM data in high resolution, DEM SRTM 3Sec V.4 on Sentinel-1 Tbx was chosen as an input for processing the elevation data. GLS-2000 of Landsat imagery was used as data reference on determining the Ground Control Points (GCPs). Path/ row of scenes that used in this study were P112/R64, P113/63, P112/R63, P113/R64.

2.3 Methods

Figure 3shows the orthorectification steps of Sentinel-1 SAR data using Sentinel-1 Toolbox and its accuracy assessment.The data were imported into BEAM-DIMAP format then separately processed using orthorectification operator on S1Tbx. Imported data would be processed using Terrain Correction operator for orthorectification process, there are Range Dopler Terrain Correction (RDTC) and SAR-Simulation Terrain Correction (SARSimTC) on Sentinel-1 Toolbox version 1.0.3 (released date: November 3rd, 2014). Process running on computer with Linux operating system and 16Gb RAM(Random Access Memmory). The output is made in the format of BEAM-Dimap.

*2.3.1. Range Doppler Terrain Correction.*The Range Doppler Terrain Correction (RDTC) operator implements the Range Doppler orthorectification method (Small and Schubert, 2008) for geocoding SARimagery from single 2D of raster radar geometry. It uses available orbit state vector information in the metadata, the radar timing annotations, the slant to ground range conversion parameters together with the reference DEM data to derive the precise geolocation information [6]. Detail algorithm of this method can be found in Small and Schubert[7]. First step is input/output parameter, using the imported SAR data as an input in orthorectification and choose the directory output for save the orthorectified image. Input and output data saved in one project file. The second step is defining the parameters and band amplitude VV which will be processed using SRTM 3 Sec as inputs to the DEM data, it will be automatic downloaded, both of DEM and Image Resampling use bilinear interpolation method. Define pixel spacing (m) of 30 m, corrected image resampled from 10 m to 30 mthus this size will fit both to the orthorectified SAR data size in the GLS-2000 and the map projection that is used, UTM (Universal Transfer Mercator) and also the datum, WGS (World Geodetic System) 1984. The areas without elevation will be masked based on DEM data. After defining the parameters, then the process will be running automatically.

*2.3.2. The SAR-Simulation Terrain Correction.*The SAR-Simulation Terrain Correction (SARSimTC) operator generates orthorectified image using rigorous SAR Simulation. There are some major steps in this operation[6].

- SAR simulation: Generate simulated SAR image using DEM, the geocoding and orbit state vectors from the original SAR image, and mathematical modeling of SAR imaging geometry. The simulated SAR image will have the same dimension and resolution as the original image;
- Co-registration: The simulated SAR image (master) and the original SAR image (slave) are co-registered and a WARP function is produced. The WARP function maps each pixel in the simulated SAR image to its corresponding position in the original SAR image;
- Terrain correction: Traverse DEM grid that covers the imaging area. For each cell in the DEM grid, compute its corresponding pixel position in the simulated SAR image using SAR model. Then its corresponding pixel position in the original SAR image can be found with the help of the WARP function. Finally the pixel value for the orthorectified image can be obtained from the original SAR image using interpolation.

Some parameters on SAR-Simulation Terrain Correction Operator are defined like on Range Doppler Terrain Correction Operator because we would compare the orthorectification result from both of method. SAR simulation steps are using Amplitude as source band, SRTM 3 Sec as DEM data, and bi-linear interpolation method for DEM resampling.GCP selection is done by set the parameters as default and define the RMS Threshold 1.0. Next step on SARSim-Terrain Correction we define RMS Threshold 1.0 with 1 WARP polynominal Order, bilinear interpolation method for image resampling, 30 m pixel spacing and projected to UTM Datum WGS1984.

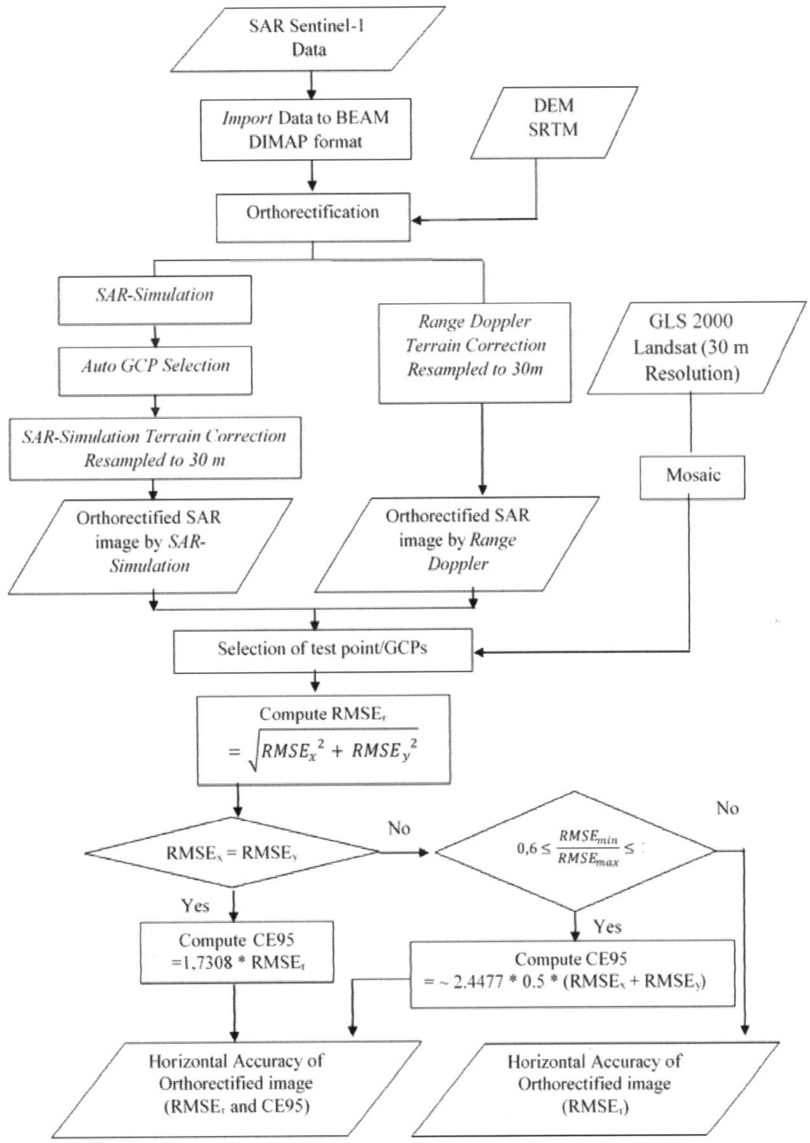

Figure 3. Flow chart of the orthorectification and accuracy assesment.

2.3.3. Accuracy Assessment. Each orthorectified data (resampled 30 m) will be compared qualitatively and quantitatively. The qualitative analysis is conducted by making an RG composite of the orthorectified imagery. The orthorectified data (Range Doppler method) is displayed in red channels, meanwhile the SAR-simulation method is displayed in the green channel. When there was a difference of significant registration occured, the distinct edge of the features boundaries are being observed.

Qualitative analysis was conducted by calculating the accuracy both of orthorectified data on the mosaic of GLS-2000 using GCPs.The GLS-2000 of Landsat imagery will be used as basic imagery on coordinates acquisition of GCPs in ENVI software. The GCPs will be located separately across the plain area at the linear or any objetcts intersections such as crossroads, irrigation canal and bridge. On the mountainous area at the ridge which can easy identified. The number of GCPs on mountainous area is more than on plain area. The total number of points refers on Lopez [7] for the calculation of the horizontal accuracy using a circular error (CE) 95% is requires about 95 to 100 sample check points. So, for the montainous area, the samples were 57 points and for the plain area were 42 points. The same location of GCPs is used on orthorectified image from both method. We calculate the residual error between the GCPs to generate the residual vector that represent shift direction between GLS 2000 landsat and orthorectified image. It is represented by vector line using XY to Line facilities on ArcGIS 10.1 Software.

Accuracy assesment is conducted by calculating RMSE (Root Mean Square Error) such as error value on X and Y coordinate (RMSEx and RMSEy), accumulation of roots RMSEx squares and RMSEy Squares (RMSEr), and CE95 (Circular Error 95%) value, based onhorizontal accuracy of NSSDA (*National Standard for Spatial Data Accuracy*) [9].Calculation performed on both of orthorectified image (Range Doppler and SAR Simulation method), on the plain area coordinate GCPs, mountainous area, and all coordinate GCPs.Error value on X axis ($RMSE_x$) and Y axis ($RMSE_y$) are calculated using these equations below:

$$RMSE_x = \left(\sum \frac{(x_{data\,i} - x_{ref\,i})^2}{n}\right)^{1/2} \quad (1)$$

$$RMSE_y = \left(\sum \frac{(y_{data\,i} - y_{ref\,i})^2}{n}\right)^{1/2} \quad (2)$$

Where: $x_{data\,i}$, $y_{data\,i}$ = coordinates of the ith check point in the dataset
$x_{ref\,i}$, $y_{ref\,i}$ = coordinates of the ith check point in the independent source of higher accuracy
n = the number of tested check points

The RMSEx and RMSEy values are used as inputs in the RMSEr as showed on the equation (3) below.

$$RMSE_r = \left(RMSE_x{}^2 + RMSE_y{}^2\right)^{1/2} \quad (3)$$

We compute circular error at the 95% confidence levelto obtain horizontal accuracy of the orthorectified image from Range Doppler and SAR Simulation method. There were 2 cases for calculating the accuracy,

Case1: when $RMSE_x$ = $RMSE_y$. RMSEy. The accuracy of the value should be calculated using equation (4):

$$Accuracy_r = 1.7308 * RMSEr \quad (4)$$

Case 2: Approximating circular standard error when $RMSE_x \neq RMSE_y$. If $RMSE_{min}/RMSE_{max}$ is between 0.6 and 1.0 (where $RMSE_{min}$ is the smaller value between $RMSE_x$ and $RMSE_y$ and $RMSE_{max}$ is the larger value), circular standard error (at 39.35% confidence) may be approximated as 0.5*(RMSEx + $RMSE_y$). If error is normallydistributed and independent in each the x- and y-component and error, the accuracy value accordingto NSSDA may be approximated according to the following formula (5):

$$Accuracy_r = \sim 2.4477 * 0.5 * (RMSE_x + RMSE_y) \quad (5)$$

3. Results and Discussion

3.1. Visual Comparison of Orthorectification Results.

SAR imagery (Figure 2) is not projected on the map coordinates of each pixel but still in the original coordinate position of data (rows/columns) in the field of ground range. As the recording data in ascending way, the objects are visible on the upper side is the South side.

In the Figure 4 it showed the orthorectified imagery in South-eastern Sulawesi. Generally, the imageery is in the actual position based on the orthorectification process of RDTC and SAR-SimTC methods. Each of pixels that have been corrected and projected will visually appear in the actual position. Visually it appears that mainland of Southeast Sulawesi as well as other islands beneath it has been stretched to the actual position which is to the southeast. Pixels that were previously in the position coordinates of the original data has been located on the actual position. We can observed that the top side of the SAR image before orthorectification has been located on the south side after corrected. The appearance of marine waters are initially dark grey to bright grey. It turned into grey after the orthorectification process. This masking process happened automatically when the pixels have no radiometric information (value = 0).

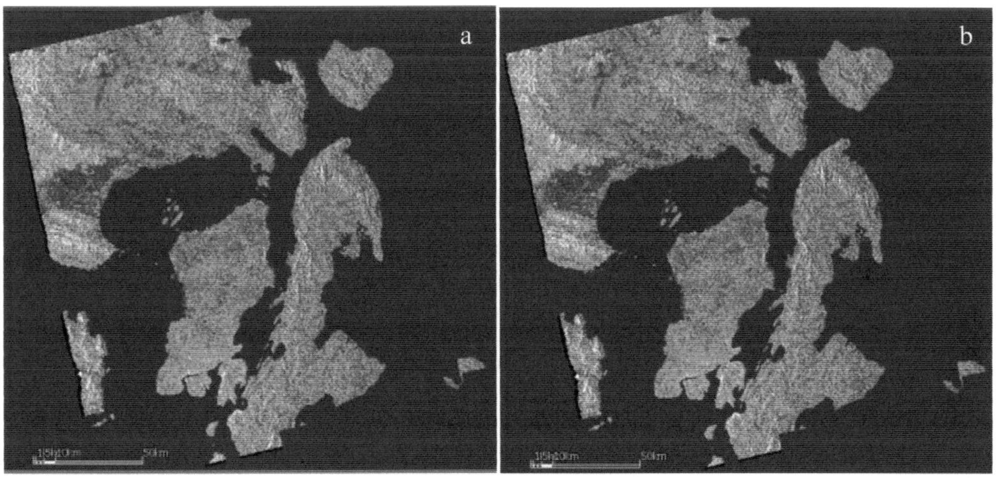

Figure 4. Orthorectified SAR ImageOrthorectified SAR imagery in some parts of South-eastern Sulawesiby Range Doppler Method (a); by SAR-Simulation Method (b).

Visual land observation can be conducted after the orthorectification process is done as showed in Figures 4.There were some changes in the hue color that turned into darker color, it was because of the resampling pocess and interpolation analysis (bilinear interpolation).Speckle noise has been slightly reduced, but based on figure 4 generally there are no significant visual difference between orthorectified image from Range Doppler and SAR Simulationas well as the differences in the appearance of the geometry of the image including image rotation.

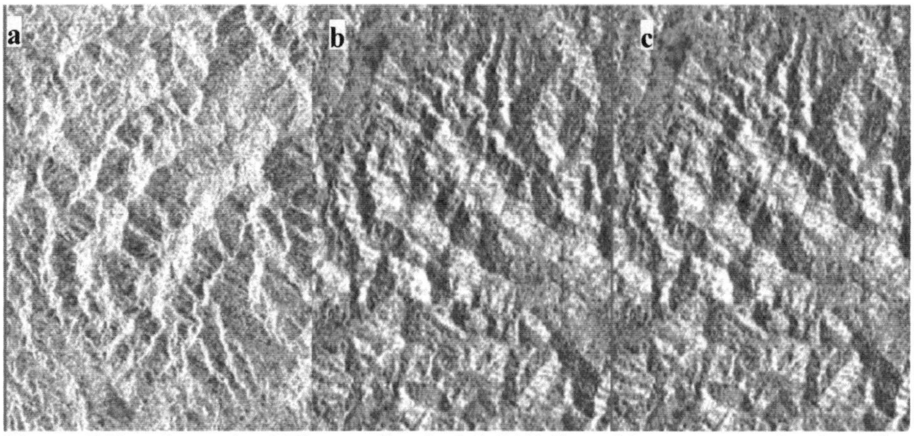

Figure 5. Foreshortening Effect on Sentinel-1 SAR Data (a) compare to the data after Range Doppler Orthorectification (b) and after SAR Simulation Orthorectification.

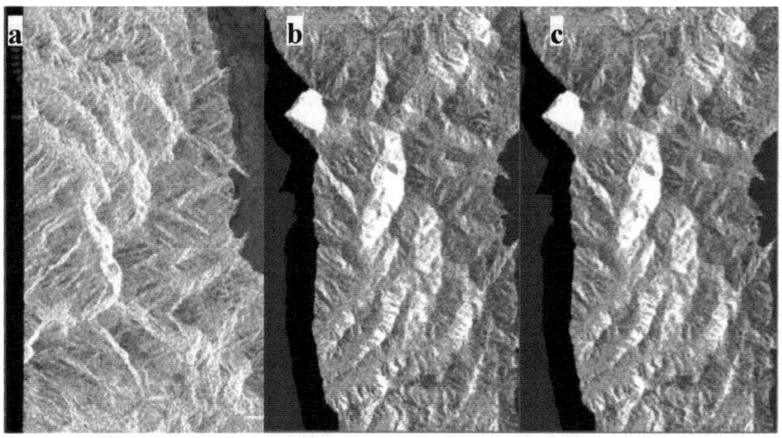

Figure 6. Layover Effect on Sentinel-1 SAR Data (a) compare to the data after Range Doppler Orthorectification (b) and after SAR Simulation Orthorectification.

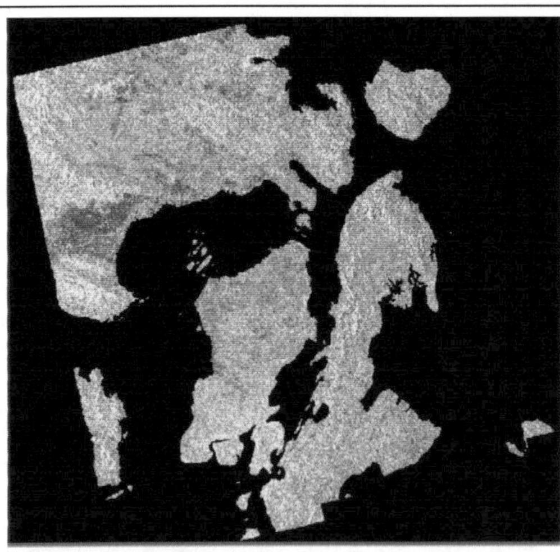

Figure 7. Red and Green Composite Image from orthorectified Southeast Sulawesi SAR data (R: Range Doppler, G: SAR-Simulation)

Based on Figure 5 and 6 we can summarize that both of methods are visually effective in reducing the geometrical distortions on the forshortening and layover effects. Before the orthorectification process, the hilly topogaphy which has a foreshortening effect is shorten than the actual size. After the correction process it has been returned to the actual size. The bright pixels were represent the streched slopes. The mountainous topography with layover effect looked like close to the nadir point, but after the orthorectification process, the hill tops are back to the actual position..

Figure 7show the orthorectified image from both of methods that had been Red and Green composited. Orthorectified SAR image by RDTC method on the red channel and orthorectified SAR image by SAR-SimTC method on the green channel. The yellow color shows the pixels match (has same pixel position and its values are almostequal), meanwhile the green and red colours were indicated the shift between the images. Based on figure 7 we can observe that there are no significant color changes or the red or the green color for the entire coverage of the image, which should be identified easily on the edge of the masking results between sea and land.

Figure 8both on plain terrains (*a* and *b*) and mountainous (*c* and *d*) showed visually that there were no differences after orthorectification process was done both from RDTC and SAR-SimTC methods. In detail the shift between the imagegery is not significant. Based on Figure 8 There were no distinctions of edge effects which showed by Red and Green colours, meanwhile, in the border and adjacent objects, there were differences.Figure8(*a*) showed that there was no shift between crossroads and rivers, as well as runway (picture 8(*b*)), and also ridge on mountainous and hilly area(picture 8(*c*)). Red and Green pixels which appeared in the imagery was Radiometric aspect of orthorectification results.

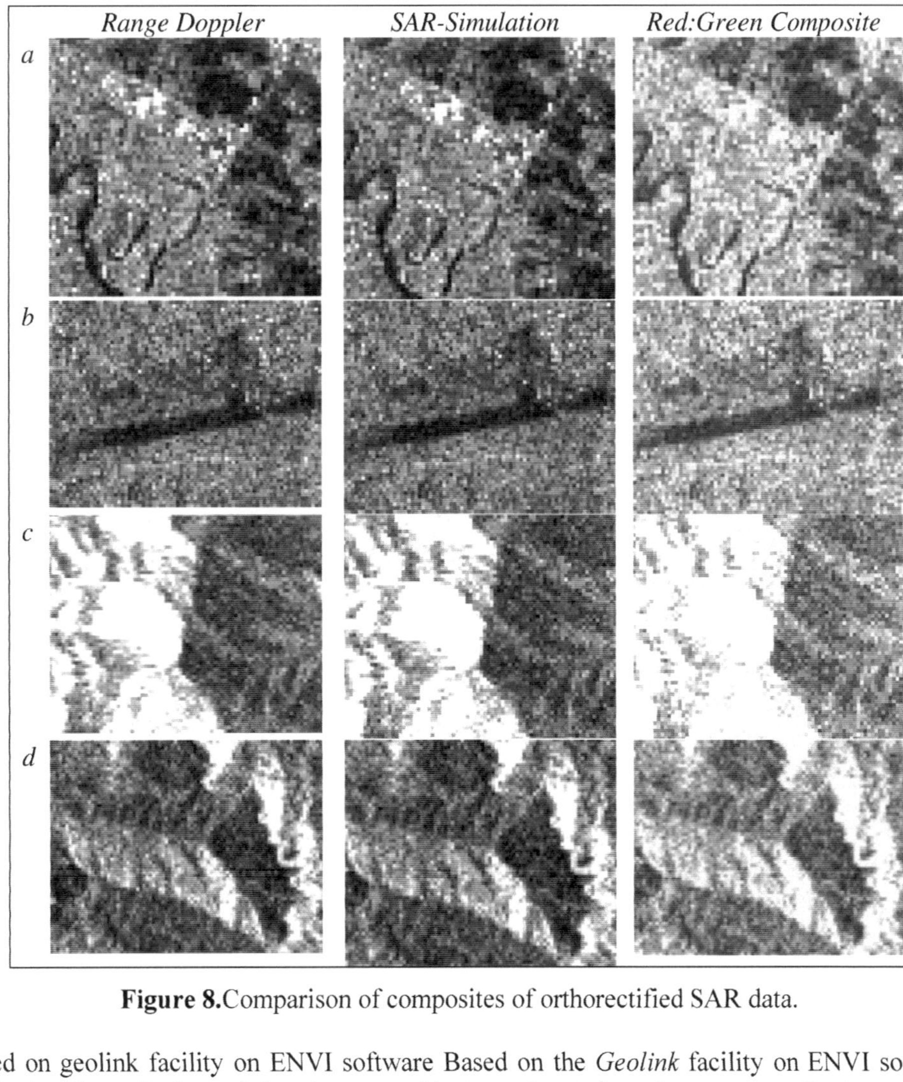

Figure 8. Comparison of composites of orthorectified SAR data.

Based on geolink facility on ENVI software Based on the *Geolink* facility on ENVI software as shown in the Figure 9, showed that the same objects are located on the same coordinates eventhough using different methods. Picture *e* was indicated object as vegetation coverage and Picture *f*was indicated as the ridge of mountainous area.

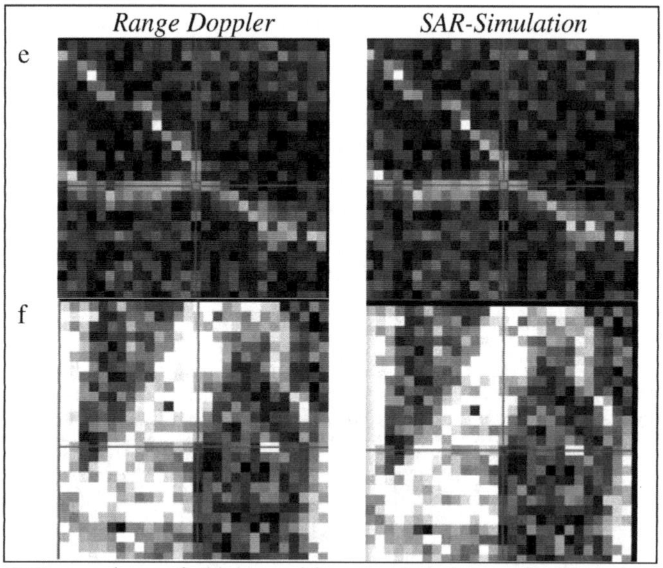

Figure 9. Comparison of object positions on the plain and mountainous area.

3.2. Comparison Results Between Orthorectified SAR data and GLS-2000 of Landsat Imagery.
Residual vector show the shift between orthorectified results of both methods,direction of the arrow is a residual vector between the coordinates of a reference to the coordinates on the same object on orthoretified SAR image, which there were 99 test points/GCPs.

The length and direction of the arrows Length and direction of the arrows are indicated the length and direction shifts that of the arthorectified data in the actual positions.The distances of arrow points from the reference into the GCPs are showed in theFigure 10 and 11.This distance is not the actual distance but it has been magnified 400 times.

Based on Figure 10the shifts which has happened on the orthorectified data showed the random pattern as slightly tilted to the eastern on plain area and the hilly area on the top of image. south side of the South-eastern island (Kabaena Island). Besides, the shifts were also tend to the south-eastern part with the pixel was less than one pixel on the mountainous area (Kabaena Island) (the left bottom corner of the image).Moreover, the shift direction of plain terrain in the mainland of Muna District was lead to the northeast. The magnitude of shift direction which has more than one pixels was located on the east side of the image (Mainland of Baubau District) and especially on the hilly or mountainous areas.

Based on the residual vectors, we can analyse the shift of the orthorectified data. The shift of the plain terrain was smaller than the mountainous one. It could be indicated by the short green arrows which distributed around the plain terrain, meanwhile the longer green arrows were distributed around the undulating, hilly, and mountainous areas. On the areas which have big and steep slopes, the amplitudes are indicated by the red arrow and it has an shift more than 30 meters.

There were different pattern of arrows that lead into northwest direction especially in the mountainous areas. This is caused by the possibility of morphological changes, for example erosion on hilly and mountainous areas, since the difference recording years of based image (GLS-2000)and SAR data. Based on figure 10 and 11there were no significant differences of two results both in geometrical orthorectification and Range Doppler and SAR-SimTC methods, either on the magnitude of shift or the direction

IOP Conf. Series: Earth and Environmental Science **47** (2016) 012007 doi:10.1088/1755-1315/47/1/012007

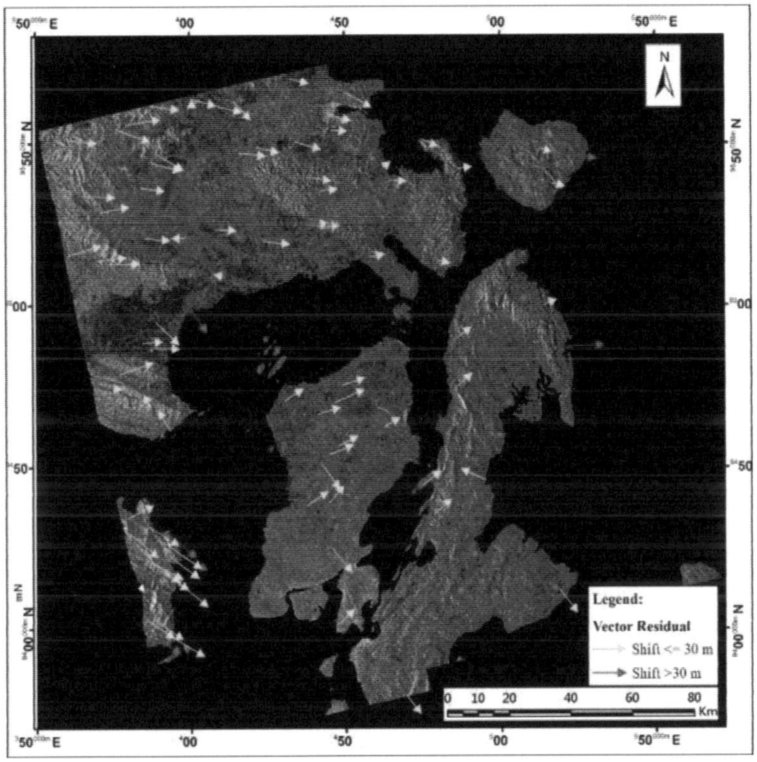

Figure 10. Residual Vector Scene: 3086/39; Orthorectified Sentinel-1 SAR by Range Doppler Method in Some Parts of South-eastern Sulawesi using multiple factor of 400 times.

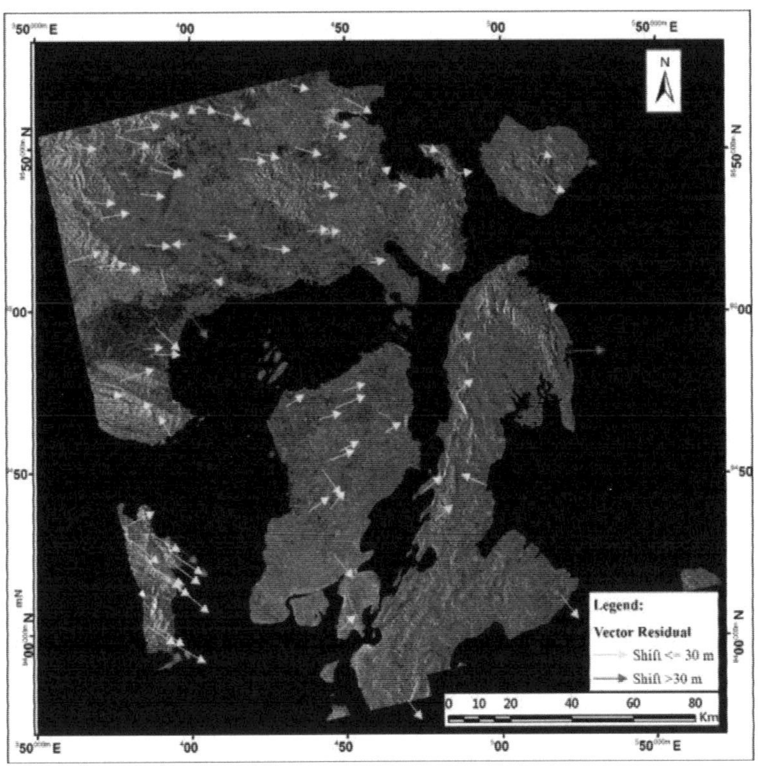

Figure 11. Residual Vector Scene: 3086/39; Orthorectified Sentinel-1 SAR by SAR-Simulation Method in Some Parts of South-eastern Sulawesi using multiple factor of 400 times.

Table 1 was indicate the quantitative analysis of Orthorectified SAR Sentinel-1 in some parts of South-eastern Sulawesi. The RMSE value was determine the horizontal accuracy. The value of RMSE and CE95 will be used as inputs on calculating the horizontal accuracy based on the NSSDA using 99 GCPs, but in this study, only RMSEr value was counted. Since the value of the ratio RMSEx and RMSEy not circular (RMSEx ≠ RMSEy) and the ratio was 0.46 (RMSEmin/ RMSEmax) are not on a value between 0.6 and 1, which is a requirement to calculations of CE95.

Table 1. Accuracy calculation results of Orthorectified SAR Sentinel-1 data based on RMSEr

Method	Topography	Number of GCPs	$RMSE_x$	$RMSE_y$	$RMSE_r$
Range Doppler Terrain Correction	Plain	42	17.77	10.58	20.68
	Mountainous	57	26.06	10.71	28.18
	All Areas	99	22.91	10.68	25.27
SAR-Simulation Terrain Correction	Plain	42	17.77	10.58	20.68
	Mountainous	57	26.06	10.71	28.18
	All Areas	99	22.91	10.68	2527

The results of RMSE calculation both on the orthorectification and Range Doppler and SAR-SimTC methods showed the similar values (see Table 1). In the plain area, the error value was smaller than in the mountainous area. Both of shifts on the RMSEx and RMSEy are 17.77 m and 10.58 m, meanwhile, in the mountainous area are 26.06 m and 10.71 m. The value of RMSEr on the plain terrain with 42 GCPs is 20.68 m, which is smaller than the RMSEr value on the mountainous area with 57 GCPs as 28.18 m.

Based on Table 1, shows that the value of RMSEx always bigger than RMSEy. This is indicates that the shift results of orthorectified image are biggeronthe range direction than the azimuth direction [10]. RMSEr values based on Table 1 shows the same value (25.27 m) on both of the RDTC and SAR-SimTCBoth RDTC and SAR-SimTC methods have the same value based on the 99 GCPs as 25.57 m. the same value is generated based on the GCPs taken on the same objects. This showed that there was no significant geometrical difference between two results of orthorectification. Each pixel represents the same object which has the same coordinates.

The shift value was caused by the orthorectification process. DEM data has a low spatial resolution which is less than 90 meters and the resampling process is at 30 meters, thus the process of resampling is conducted from 10 m to 30 m in the GCPs determination. Because of the difference of recording time of GLS-2000 of Landsat imagery and Sentinel-1 SAR imagery, it is possible to find not matched existing object because the change of land cover and terrain in observation area.

It was interesting to see there is no significant differences in visual geometry between the orthorectified imagery on Range Doppler and SAR-Simulation method. There should be generate different geometry of orthorectified image, regarding the difference in principle between orthorectified Range Doppler method and SAR-Simulation.But those differences still can be found in the case of Radiometric value.

3.3. Radiometric Comparison of Orthorectification Results.

Figure 12showed the enlarged orthorectified imagery and composites on the plain area (*p* and *q*), undulating terrain (*r*), and the hilly-mountainous terrain (*s* and *t*). Each picture shows the green and red pixels which is spread up in the different intensities. Picture p and q are displayed as linear objects, thus the appearance of red and green colours in the surrounding pixels are indicated the heterogenous objects. The pixels which are formed of the red and green pixels in the composite have different intensity values of grey that can be considered as noise of SAR imagery, meanwhile the yellow colours are formed from the pixels which have the similar or equal value of intensities.

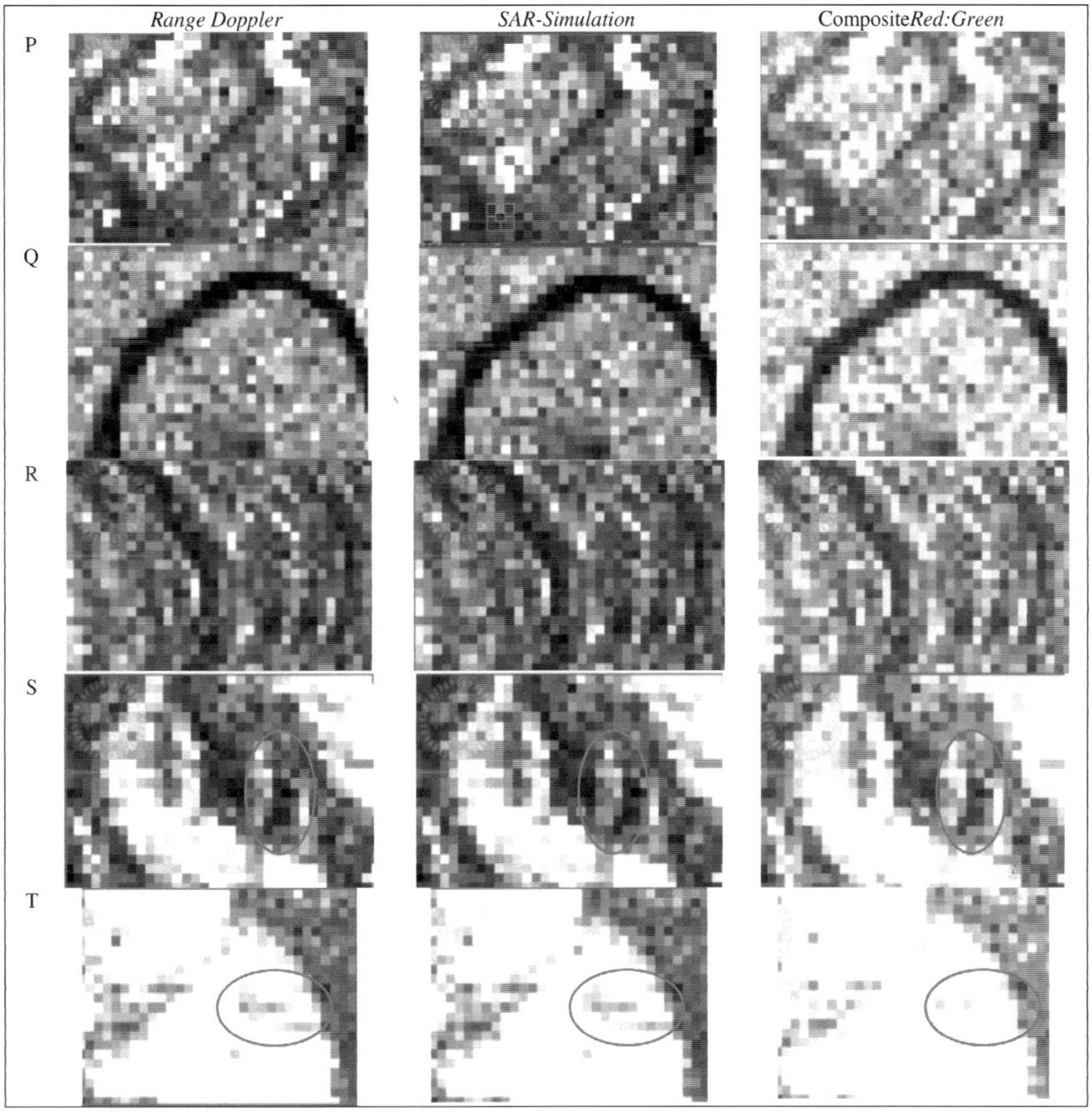

Figure 12. Detail comparison on Composited Orthorectified SAR Image

The green and red composite images are caused by the pixels that having different intensities in the undulating areasshown on the object (r). Same as the picture of *s* and *t*, the mountainous area showed in the red circled, which has a different intensity or brightness values but still in the same geometrical patterns. The picture *t* showed that most of the pixels in the yellow colour are indicated the same value. Moreover, there are some green and red pixels which appeared in the areas, this means that the pixel has different value.

These results showed that both methods can do generate the orthorectified SAR data in different pixel value intensity or backscatter value but not significant.But in the geometrical terms, it still has the same object form which is not much different. So, it may be a necessity to do radiometric calibration process while doing the orthorectification process using Sentinel-1 Tbx.

Those accuracy of orthorectified Sentinel-1 SAR data in some parts of South-eastern Sulawesi from each method may be due to issues during orthorectification process such as DEM data input have low spatial resolution (90 m) than Sentinel-1 SAR data, the resampled Sentinel-1 SAR data to a lower pixel spacing. The quality of the DEM is important during orthorectification, particularly in relation to the location of features, stretching effects and generation of local incidence angle information, but also for radiometric calibration [11]. Moreover, the geometry distortion of GLS-2000 landsat data and human error factor such as confusion in selecting GCPs which used to accuracy assesment should be considered.

4. Conclusion

The orthorectification process of SAR data in some parts of South-eastern Sulawesi has been done by using Range Doppler and SAR-SimTC methods on Sentinel-1 Tbx. There were 99 GCPs taken using GLS-2000 of Landsat Imagery which is used as the referenced imagery. The results of horizontal accuracy showed there is no significant differences in the geometrical visual of orthorectified image from both methods of Range Doppler and SAR-SimTC. The RMSEr value is 25.27 m which means that the horizontal accuracy is less than one pixel. There is still a difference that can be found in the radiometric aspect between the orthorectified SAR data from each method.

5. Acknowledgment

The authors are very grateful to PUTEKDATA-LAPAN (Indonesian National Institute of Aeronautics and Space) for provided opportunities and various facilities to process the SAR Sentinel-1 data. Finally, we would like to thank Dr. Ari Katmoko Sambodo, M.Eng. for the suggestion and discussion in the appointment of the research topic and the data processing for this paper.

References

[1] European Space Agency 2015*Copernicus Overview* [Online] Available: http://www.esa.int/Our_Activities/Observing_the_Earth/Copernicus/Overview3 Accessed: February 17, 2015

[2] CCRS (*Canada Centre for Remote Sensing*) 2014 *Fundamental of Remote Sensing* [online] Available:http://www.nrcan.gc.ca/sites/www.nrcan.gc.ca/files/earthsciences/pdf/resource/tutor/fundam/pdf/fundamentals_e.pdf Accessed: January 23, 2015

[3] Sutanto 1987*Penginderaan Jauh:***2** (Yogyakarta: Gadjah Mada University Press) p 238

[4] Masanobu S 2010 Ortho-Rectification and Slope Correction of SAR Data Using DEM and Its Accuracy Evaluation *IEEE Journal of Selected Topics In Applied Earth Observations and Remote Sensing* **3**(4) pp 647-71

[5] Cossu R, F Brito, O Colin, L Fusco, P Goncalves, MLavalle, and M Paces 2007 *Global Automatic Orthorectification of ASAR Products In Esrin G-Pod.* Proc. 'Envisat Symposium' (Switzerland: ESA SP-636)

[6] S1Tbx 1.0.3 Software, help document 2014

[7] David S and Adrian S 2008 *Guide To ASAR Geocoding, Ref. RSL-ASAR-GC-AD, Issue: 1.01* (Swiss: Remote Snsing Laboratory/University of Zurich) p36

[8] Dianovita and Riyan M 2014 *Kajian Ketelitian Geometri Citra Landsat 8 Level 1T. Proc. Seminar Nasional Penginderaan Jauh* (Jakarta: Lapan) pp 254-260

[9] Federal Geodetic Control Committee 1998 *Geospatial Positioning Accuracy Standards Part 3: National Standard for Spatial Data Accuracy, FGDC-STD-007.3-1998.* (Washington, D.C.: Federal Geographic Data Committee) pp 10-11

[10] Erich M, Urs F and Daniel N 1993 *SAR Geocoding: Data and Systems*ed G Schreier (Karlsruhe: Wichmann) chapter 7 pp 173-85

[11] Anthea L M, Ian T, Anthony K M, Mark W, Kim L, Peter C, Eric L, Zheng-Shu Zhou, and A H 2012 *Radar Processing Methodologies For The Generation of Wall-to-Wall Mosaics*vol1,edA.L. Mitchell et al (Australia: International Forest Carbon Initiative Research Alliance (IFCI RA))

Identifying Sugarcane Plantation using LANDSAT-8 Images with Support Vector Machines

Sidik Mulyono[1], Nadirah[1]

[1]Agency for the Assessment and Application of Technology (BPPT), Republic of Indonesia

Email: sidik.mulyono@bppt.go.id, nadirah@bppt.go.id

Abstract. The use of remote sensing has been highly beneficial in the identification and also mapping and monitoring of plantations. The identification of plantations includes the physiology, disease, environmental conditions, and also the production and time of harvesting. It can be done by doing satellite imagery classification. However, to reach the final result of identification, it could be carried out by getting the solid ground truth information. This paper will discuss about detection of sugarcane plantation in Magetan district of East Java province area by using LANDSAT-8 image with specific approach of phenology profile using EVI (Enhanced Vegetation Index) value from satellite data, as an alternative vegetation index to address some of the limitation of the NDVI (Normalized Difference Vegetation Index). Method of classification used for detecting sugarcane plantation is Support Vector machines (SVM), which is a promising machine learning methodology. It has the ability to generalize well even with limited training samples and complex data. A number of samples of phenology profile for training purpose using SVMs are obtained from the area that identified as sugarcane plantation during field campaign in 2015. The same manner is also done for the objects instead of sugarcane plantation with relatively the same number of samples. The result of the research shows that Remote Sensing is able to detect the sugarcane plantation cross the district with good accuracy.

1. Introduction

Sugarcane (*Saccharum officinarum*) is a special kind of grass that allows multiple cuts throughout several years and is considered as a semi-perennial. It grows quickly with an abundant reproduction and allows for economic exploitation of many parts of the plant. It has a 4-phases during its growing, i.e. germination phase, tillering phase, stem elongation phase, maturity and ripening phase. From planting until the first cut it is called cane plant, whose cycle lasts between 12 to 18 months, depending on the season and region of planting. After the first cut, the regrowth of sugarcane is replaced by the ratoon cane, with a normal cycle of 12 months. This planting cycletypically happens 4 to 5 times before it completely ploughed out and replanted by a new plant crop.

In Indonesia, sugar which is one of the basic needs for the community and industry, is still continuing to be a problem due to lack of domestic production, whilst the demand of sugar tends to be increased continuously. A large proportion of sugar produced by the sugarcane is imported from overseas countries. The condition of existing sugarcane mills that mostly located on Java and Sumatera island is too old, low productivity, and highly dependent on sugarcane farmers that only have very limited area for sugarcane crop on Java and Sumatera island. The rapid increment of sugar demand compared to the production improvement is relatively not balanced, makes Indonesia be a big importer country for raw crystal sugar as well as refined sugar. The development of sugar industry must be conducted integrally which starts from plantation, processing, marketing, and distribution supported by stakeholders, supporting agencies such as human resources development, finance and banking, and transport.

Since sugar production capacity in Indonesia is still not sufficient to fulfil the national demand on sugar, therefore the Ministry of Agriculture had been revised the target of production capacity in 2014 from 5.7 million tons to 3.1 tons (agroindonesia.co.id, 2012). The target decline was triggered by the difficulty to acquire new land for expansion of sugarcane field. To achieve the production of 5.7 million tons, it needs a new sugarcane field about 350,000 hectares. But in reality, extending program for expansion of sugarcane field does not run smoothly, and until now the required land has not been able to be acquired S. Sinunggila, 2015). However, this paper will discuss about detection of sugarcane plantation in Magetan district of East Java province area by using LANDSAT-8 images with specific approach of phenology profile using EVI (Enhanced Vegetation Index) value from satellite data, as an alternative vegetation index to address some of the limitation of the NDVI (Normalized Difference Vegetation Index).

Recent studies have used remote sensing images with both passive sensor (reflectance based measurement) and active sensor (backscatter based measurement) to map sugarcane fields through a various classification methods applied to remote sensing image time series, since spatial information can be provided in detail by satellite imagery. The amount of detail contained in the imagery is dependent on the attributes of the satellite. The intensity of the reflected electromagnetic radiation from agricultural vegetation such as sugarcane can be measured and recorded by satellite remote sensing instruments as a digital image. EO-1 Hyperion Hyperspectral imagery has been used to develop prediction model for detecting variety and plant cycle of sugarcane crop with support vector machines and random forests [1]. Understanding the backscattering behavior of sugarcane over its planting cycle and its relationships with sugarcane growth-related parameters is prerequisite for developing effective methods to map sugarcane plantation area and monitor sugarcane growth using ENVISAT Advanced SAR data. The alternating polarization precision (APP) data was proposed for mapping sugarcane growth area and retrieve LAI of sugarcane crop derived from the polarization ratio of ENVISAT ASAR HV/HH data [2]. In addition, Antunes et al. (2011) proposed the potential of data mining techniques to analyze MODIS/Terra remote sensing data is proposed for classifying sugarcane crop using the decision tree induction technique with occupying some rules [3]. The constellation data of both SPOT-4 and SPOT-5 are applied to predict a productivity of sugarcane crop using an empirical relationship with a growing season-integrated Normalized Difference Vegetation Index NDVI, which is compared with the Kumar-Monteith efficiency model and a forced-coupling method [4]. Meanwhile, NDVI derived from LANDSAT-8 has great ability for detecting crop type, crop conditions (harvested or growing) and mapping sugarcane crop field for medium sized farms, and normalized difference water index (NDWI) derived from LANDSAT-8 also has great potential to distinguish for green and burnt harvest by segregating dry and humid surfaces [5]. This paper addresses on how to identify sugarcane plantation through several time series of EVI derived from LANDSAT-8 images that represents phenology profile of 1 planting cycle for sugarcane crop. Support vector machines (SVM) is adopted to distinguish phenology profile for non-sugarcane field and sugarcane plantation area.

Figure 1. Sugarcane crop

2. Study area

The study area is located in Magetan district of East Java Province (Figure 1), which centered at 7° 38' 30" S and 111° 20' 30" E and altitude between 660 - 1,660 meters above sea level. Geographically, Magetan district is suitable for the development of agricultural products, therefore a superior product of this district is agricultural products, both food crops and other horticulture. Plantation crops in this district is mostly dominated by sugarcane, which are used to supply materials for the sugar crystal to sugar mills in Magetan district. Sugarcane plantation area in this district is obtained from GPS field survey during fiscal year of 2015 managed by PTPN-XI, which is shown in figure 2 (a) indicated by red shading. This area is then used as a reference to evaluate the accuracy of the model for identifying sugarcane plantation in all over Magetan district.

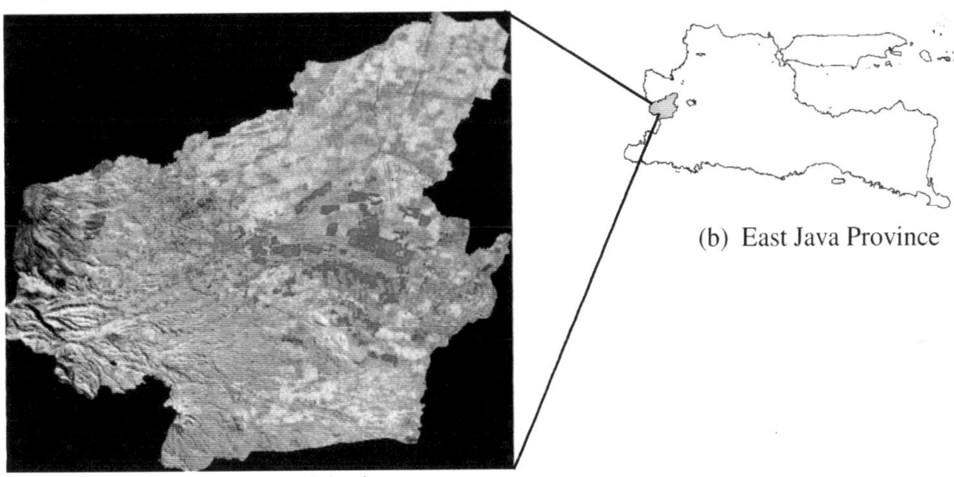

(b) East Java Province

(a) Magetan District

Figure 2. Study area

3. Data and Methodology

3.1 Data samples

LANDSAT-8 satellite launched on February 11, 2013, belongs to U.S government is dedicated for earth observation. This eighth satellite in the LANDSAT program is built by a collaboration between National Aeronautics and Space Administration (NASA) and the United States Geological Survey (USGS), which provides moderate-resolution (15 m -100 m, depending on spectral frequency) measurements of the Earth's terrestrial and polar regions in the visible, near-infrared, short wave infrared, and thermal infrared. The images used in this study are obtained by downloading them from website http://glovis.usgs.gov/ for free.

Table 3. LANDSAT-8 images used in this study

No. image	Date of image	No. image	Date of image	No. image	Date of image
1	2015-01-14	9	2015-05-22	17	2015-09-15
2	2015-01-30	10	2015-06-07	18	2015-09-27
3	2015-02-15	11	2015-06-23	19	2015-10-13
4	2015-03-03	12	2015-07-09	20	2015-10-29
5	2015-03-19	13	2015-07-25	21	2015-11-14
6	2015-04-04	14	2015-08-10	22	2015-11-30
7	2015-04-20	15	2015-08-26		
8	2015-05-06	16	2015-09-11		

A number of LANDSAT-8 images used in this study are 22 series acquired from January 14 until November 30, 2015 with having 7-bands spectral respectively. After doing radiometric, geometric, and atmospheric correction, the calculation of enhanced vegetation index (EVI) is applied to all images to perform EVI images using IDL code program in ENVI. Finally, the entire images of EVI is merged into a cubic image of EVI by using layer stacking technique in ENVI, whereas one pixel will present phenology profile during 2015 in all over Magetan district.

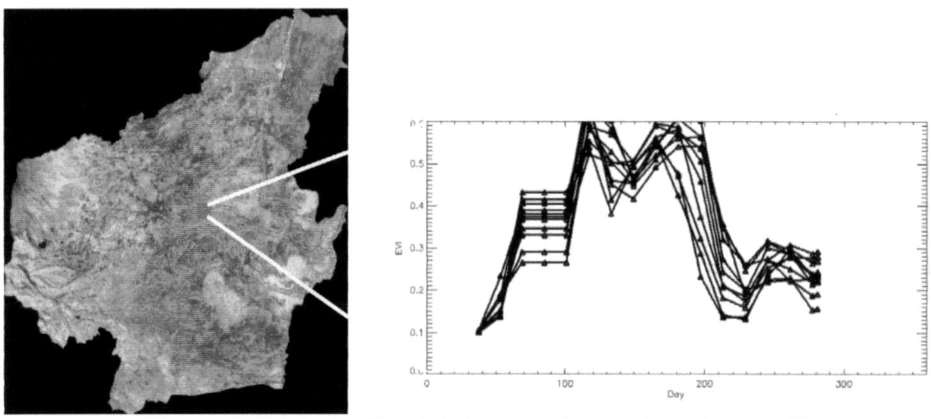

Figure 3. Original EVI cubic image and crop phenology profiles

Since LANDSAT-8 image is sensitively affected by cloud, then the profile looks like having irregular fluctuations, as shown in figure 3, which is it will lead difficulties in advance analysis. In order to avoid something that are not desirable in the analysis, it should be done by smoothing technique called Savitzky-Golay digital filter method with 2 order of polynomial and 5 points of subsets. This method is defined for the purpose of smoothing the data, that is, to increase the signal-to-noise ratio greatly distorting the curves in general.

In this study, instead of conducting the field survey, we selected training sample areas from smoothed EVI cubic image for 6 classes of the object surrounding Magetan district area. One class for sugarcane crop area is derived from the sugarcane plantation area managed by PTPN XI, whereas the rest of them (5 classes) are for the objects represent instead of sugarcane crop that are derived arbitrary from the images, as shown in figure 5. From this figure, we can understand that through phenology profile derived from EVI, each object can be determined as a unique profile.

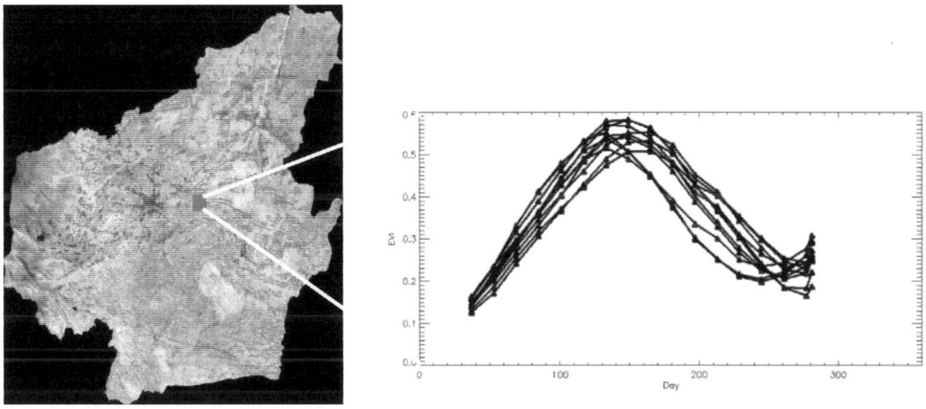

Figure 4. Smoothed EVI cubic image and crop phenology profiles

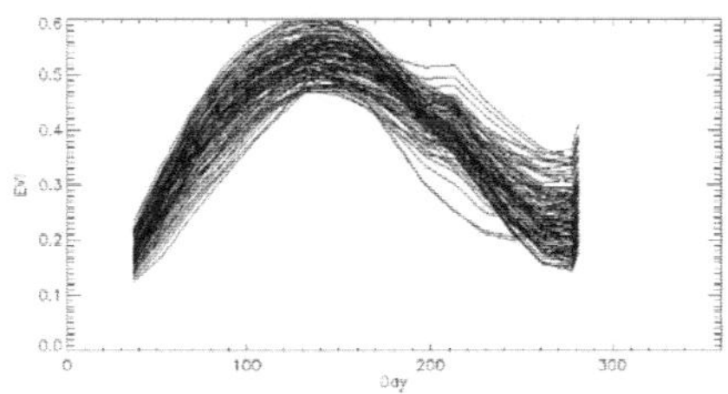

(a) Sugarcane: 2,506 pixels

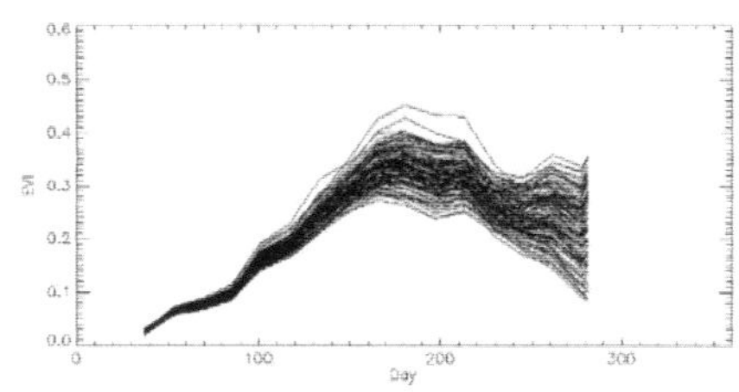

(b) Non Sugarcane 01: 1,567 pixels

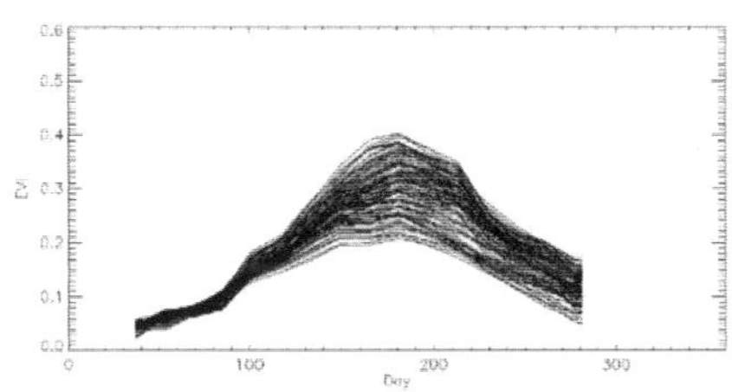

(c) Non Sugarcane 02: 1,640 pixels

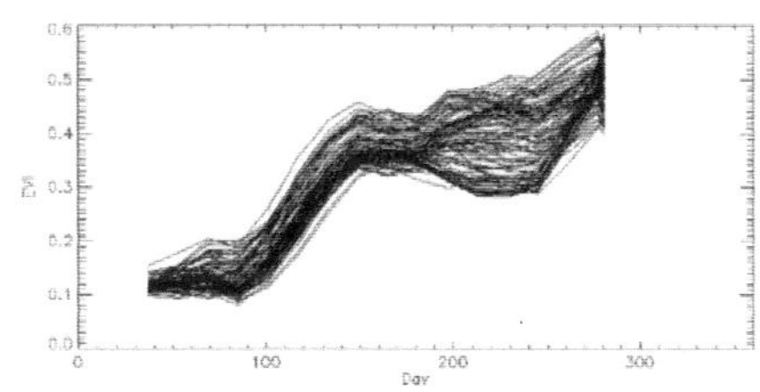

(d) Non Sugarcane 03: 1,618 pixels

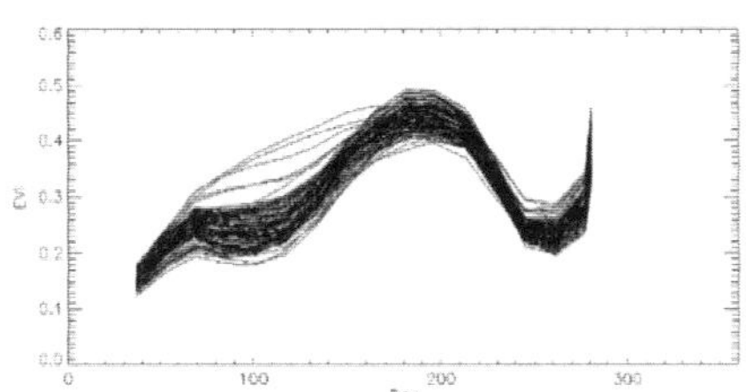

(e) Non Sugarcane 04: 2,559 pixels

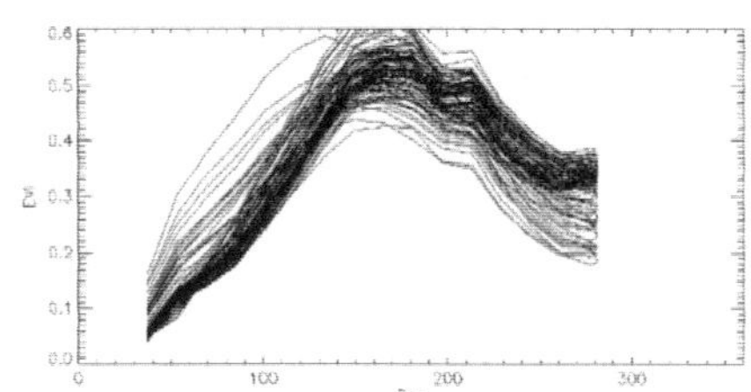

(f) Non Sugarcane 05: 2,437 pixels

Figure 5. Training sample data through phenology profile

3.2. Support Vector Machines

The support vector machine (SVM) is a popular technique and useful for data classification. Basically, SVM classifies binary data by determining the separating hyper-plane or decision surface, which maximizes the margin between the two classes in the training data [6]. Kernel function provides SVMs with powerful ability of efficiently determining the nonlinear decision surfaces after transforming the input space into kernel space. In addition, SVM is able to transform a non-convex problem into a convex problem with a single global minima and thanks to its quadratic optimization scheme. Given a training set of instance-label pairs (x_i, y_i), where $i = 1, \ldots, l$, $\mathrm{x}_i \in R^n$, and $y \in \{1, -1\}^l$, the SVM requires the solution of the following optimization problem:

$$\begin{aligned} \min_{\mathrm{w},b,\xi} \quad & \frac{1}{2} w^T w + C \sum_{i=l}^{l} \xi_i \\ \text{Subject to} \quad & y_i(\mathrm{w}^T \phi(\mathrm{x}_i) + b) \geq 1 - \xi_i \\ & \xi_i \geq 0 \end{aligned} \quad (1)$$

Here training vectors x_i are mapped into a higher dimensional space by the functionϕ. SVM finds a linear separating hyper-plane with the maximal margin this higher dimensional space, whereas $C > 0$ is the penalty parameter of the error term.

The development of SVMs classification for sugarcane field detection is done by using a library for support vector machines (LIBSVM) based on sequence minimal optimization (SMO) [7] installed in ENVI software. The kernel trick used in this paper is radial basis function (RBF).

$$K(\mathbf{x}_i, \mathbf{x}_j) = \exp\left(-\gamma \|\mathbf{x}_i - \mathbf{x}_j\|^2\right), \gamma > 0 \quad (2)$$

Compared to other kernel tricks (linear, polynomial, and sigmoid), the γ parameter of RBF kernel provides greater flexibility in controlling the desired classification accuracy. Such accuracy control mechanism is highly important when we want to combine the SVM as weak classifiers in boosting learning to overcome the over-fitting problem.

3.3. Multi class strategy for SVM

One major disadvantage of SVM, however, is that SVM classification is binary. This problem can be resolved in a simple and robust procedure by training several SVMs simultaneously in any multiclass strategy. In this study, we use pairwise classification strategy that performed in pyramid structure to classify 6 classes of object, which will consist of at least 15 classifiers. This strategy is already involved in ENVI software platform, and we set value for γ and C as the default, i.e. 0.045 and 1.0 respectively.

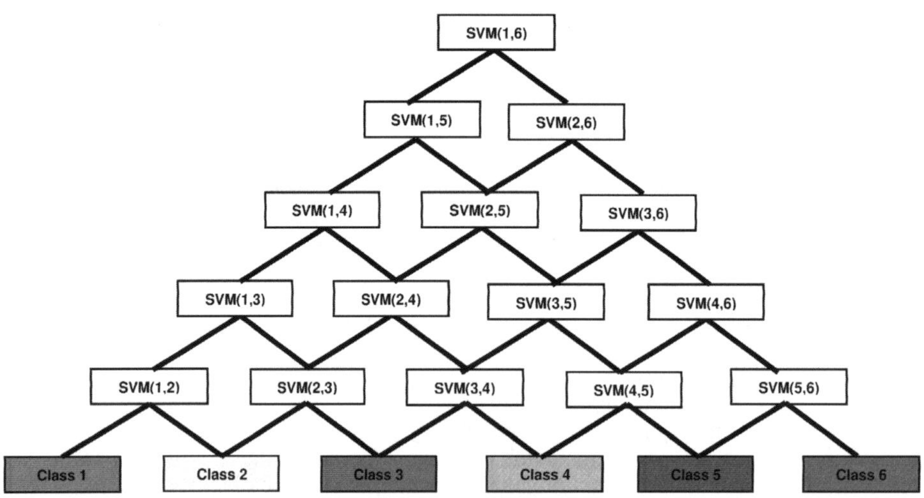

Figure 6. Pairwise multiclass strategy

4. Result and discussion

We compared the result obtained from SVM with conventional classification method in remote sensing that is called spectral angle mapper (SAM) that provided in ENVI software as well. In case of the pairwise multiclass strategy of SVM, the entire training sample data are employed as an input for learning. Meanwhile, in case of SAM, each training sample data are used as reference vectors with threshold set at 0.2 radian against the target in general. Both classification results are evaluated against the sugarcane plantation area managed by PTPN-XI as a reference area. Total accuracy can be determined by calculating a total number of sugarcane's pixel predicted by both classification methods allocated in reference area, divided by entire pixel of reference area. Total accuracy for SVM reached about 90.78%, whereas SAM reached only at 78.92% (figure 7).

By using the same manner, we then can identify the sugarcane plantation in entire area of Magetan district as shown by red shading in figure 4.2. The total sugarcane plantation can be detected by SVM is reached about 20,410 ha (37.40%), whereas in case of SAM is reached about 11,143 ha (20.95%). Both methods give the promising results on identification of sugarcane plantation in Magetan district that can provide important information regarding to the potential of sugarcane material for supplying to sugarcane mill, beside the existing area that have been managing.

(a) SVM (b) SAM

Figure 7. Classification result in reference area of PTPN-XI

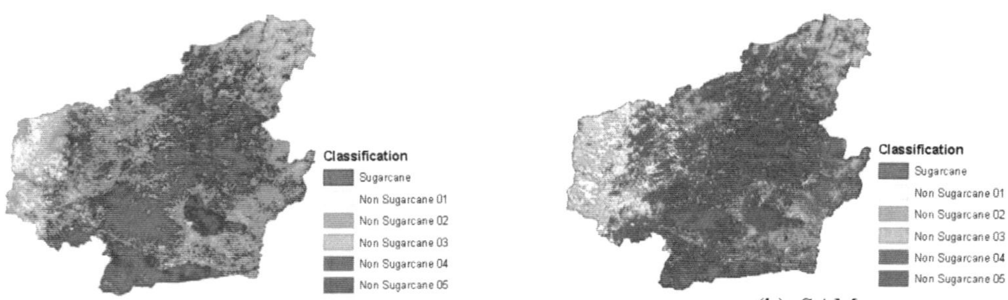

(a) SVM (b) SAM

Figure 8. Sugarcane plantation in Magetan district

5. Conclusion

In this paper we proposed the novelty method for identifying sugarcane plantation area through crop phenology profile of EVI time series derived from LANDSAT-8 images.In term of classification, the data set of phenology profile derived from vegetation index have much better approach instead of we use only spectral data set, because they represent the whole of characteristic of an object in a certain time range.

In addition, this method is also suitable applied for classifying other crops/plantations or objects that have phase change, such as palm oil plantation, acacia plantation, etc. Using this classification model, we can identify another sugarcane plantation outside area managed by PTPN-XI.By detecting more sugarcane plantation, it will give an information on sugarcane crop stock in order to increase the material supply to the sugarcane mill to achieve national need on sugar.

6. Acknowledgement

The authors would like to thank the people from the PTPN-XI for their efforts in collecting the *in situ* data for sugarcane plantation used as reference area in this study, and colleagues from Center of Technology for Natural Resources Development, Indonesia Agency for the Assessment and Application of Technology for theirhelp for receiving and processing LANDSAT-8 images.

Reference

[1]. YL Everingham, KH Lowe, DA Donald, DH Coomans, and J Markley 2007Advanced Satellit Imagery to Classify Sugarcane Crop Characteristic *Agron. Sustain. Dev.***27** 111-117

[2]. Hui Lin, Jinsong Chen, Zhiyuan Pei, Songling Zhang, and Xianzhi Hu 2009Monitoring Sugarcane Growth Using ENVISAT ASAR Data *IEEE Transactions on Geoscience and Remote Sensing* **47**(8) August 2009

[3]. JFG Antunes, LHA Rodrigues, and S R de M Oliveira 2011 Data Mining for Sugarcane Crop Classification Using MODIS Data *EFITA/WCCA* '11

[4]. Julien Morel, Pierre Todoroff, AgnèsBégué, Aurore Bury, Jean-François Martiné 4and Michel Petit 2014Toward a Satellite-Based System of Sugarcane Yield Estimation and Forecasting in Smallholder Farming Conditions: A Case Study on Reunion Island *Remote Sens.***6** 6620-6635; doi:10.3390/rs6076620

[5]. Betty Mulianga, AgnèsBégué, Pascal Clouvel, and Pierre Todoroff 2015Mapping Cropping Practices of a Sugarcane-Based Cropping System in Kenya Using Remote Sensing *Remote Sens.***7** 14428-14444; doi:10.3390/rs71114428

[6]. Rick Archibald and George Fann October 2007Feature Selection and Classification of Hyperspectral Images with Support Vector Machines *IEEE Geoscience and Remote Sensing Letter* **4**(4) October 2007

[7]. Chih-Wei Hsu, Chih-Chung Chang, and Chih-Jen Lin 2010 *A Practical Guide to Support Vector Classification*, Software available at http://www.csie.ntu.edu.tw/~cjlin, last updated April 15, 2010

GEOBIA for Land Use Mapping Using Worldview-2 Image in Bengkak Village Coastal, Banyuwangi Regency, East Java

Fitzastri Alrassi[1], Emil Salim[1], Anastasia Nina[1], Luthfi Alwi[1], Projo Danoedoro[2], Muhammad Kamal[2]

[1] Post Graduate Student of Remote Sensing Program, Faculty of Geography, Universitas Gadjah Mada, Indonesia

[2] Lecturer of The Faculty of Geography, Universitas Gadjah Mada, Indonesia

fitzas@gmail.com

Abstract. The east coast of Banyuwangi regency has a diverse variety of land use such as ponds, mangroves, agricultural fields and settlements. WorldView-2 is a multispectral image with high spatial resolution that can display detailed information of land use. Geographic Object Based Image Analysis (GEOBIA) classification technique uses object segments as the smallest unit of analysis. The segmentation and classification process is not only based on spectral value of the image but also considering other elements of the image interpretation. This gives GEOBIA an opportunities and challenges in the mapping and monitoring of land use. This research aims to assess the GEOBIA classification method for generating the classification of land use in coastal areas of Banyuwangi. The result of this study is land use classification map produced by GEOBIA classification. We verified the accuracy of the resulted land use map by comparing the map with result from visual interpretation of the image that have been validated through field surveys. Variation of land use in most of the east coast of Banyuwangi regency is dominated by mangrove, agricultural fields, mixed farms, settlements and ponds.

Keywords: GEOBIA, Land use, coastal, WorldView-2

1. Introduction

Since the last fifteen years, the use of object-based approach for classifying remotely-sensed images are getting more attention from researchers. Unlike the pixel-based approach which heavily relies on the spectral value of the individual pixels, the object-based approach uses segments as the smallest mapping unit and incorporates some visual interpretation keys in the classification. This includes colour, shape, pattern, size, association, etc. [2]. Thus, object-based classification are able to define the object classes based on spectral and spatial aspects of the image at once [2]

Object-based approach for semi-automated LULC mapping have been a major focus area of remote sensing and image processing research since the past decade [11]. The object-based classification works in the same way as a pixel-based classification. However, the difference is that it does not classify each pixel individually, but classifying the image based on the segment or group of pixels that compose an object. The training areas for the classification of the objects are derived from the existing database in order to avoid a time-consuming manual acquisition [10]. GEOBIA provides an opportunity to improve the extraction of land cover and use information from remote sensing image data.

Land is a delineable area of the earth's terrestrial surface, encompassing all attributes of the biosphere immediately above or below this surface including those of the near-surface climate the soil and terrain forms, the surface hydrology (including shallow lakes, rivers, marshes, and swamps), the near-surface sedimentary layers and associated groundwater reserve, the plant and animal populations, the human settlement pattern and physical results of past and present human activity (terracing, water storage or drainage structures, roads, buildings, etc. [4]. Land use is characterized by the arrangements, activities and inputs by people to produce, change or maintain a certain land cover [3] while land cover refers more to "the vegetational and artificial constructions covering the land surface [6]. Construction is entirely visible directly from remote sensing imagery.

This research aims are to apply the GEOBIA classification method for generating the classification of land use in coastal areas, and assess the accuracy of the map result. The result of this research is land use classification maps using GEOBIA.

2. Methodology and Materials

2.1 Study Area

The study area in this research is Bengkak Village, in the coastal area of Wongsorejo District, Banyuwangi Regency, East Java with sample area extent of about 289,796.20 m^2 (Figure 1). In this area we found high variation of land uses including sea water, fishpond, built-up area, river, mangrove area, open field, etc. This high land uses variation in this study area gives us an ideal location to test the land use classification using GEOBIA approach.

Figure 1. Localization of area study (Source: Adapted from WorldView-2 image dated July 2015. (DIGITALGLOBE, 2016)

2.2 Materials

This study is conducted using WorldView 2 image data. This data launched October 2009 with 11 bit per pixel and repeat time 1.1 days. WorldView 2 provide 8 multispectral bands (4 new band and 4 standard bands) and 1 panchromatic. The multispectral bands are coastal (400 - 450nm), red (630 -690 nm), blue (450 - 510 nm), red edge (705 - 745 nm), green (510 - 580 nm), near-IR1 (770 - 895 nm), yellow (585 - 625 nm), near-IR2 (860 - 1040 nm). Every bands has a different specific applied one to each other. Multispectral bands has spatial resolution 1.85 m GSD at nadir, 2.07 m GSD at 20° off-nadir.

In other hand, panchromatic band (450 - 800 nm) has 0.46 m GSD at nadir, 0.52 m GSD at 20° off-nadir spatial resolution. WorldView 2 with system orbit sun- synchronous has an average revisit time of 1.1 days. This image data of WorldView 2 is impressed on July 15th 2015 with 0% cloud cover (Handbook WorldView 2)

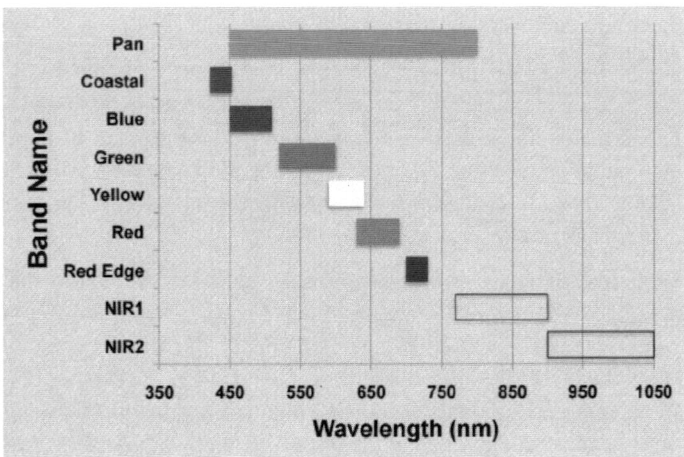

Figure 2. WorldView 2 Spectral Bands (Source: DigitalGlobe Proprietor)

2.3 Methodology

The method used in this study include:

2.3.1 Pre-processing. This process is started with geometric correction or generally known as georeferenced. Geometrically corrected imagery can be used to extract accurate distance, polygon area, and direction (bearing) information [5].

2.3.2 Processing. In this step we use two step classification:

2.3.1.1 Visual classification

In this classification we interpret the remote sensing image using a visual method image remote sensing with the visual method. The result of this process is map reference to validation.

2.3.2.2 GEOBIA classification

In this step, we begin with segmentation image and then classification. This second step produces a tentative map of land use. Image segmentation algorithm is a crucial step in image processing and analysis. The goal of segmentation is to separate an image into some regions of feature and to pick up the interesting objects [7]. The use GEOBIA (Geographic Object Based Image Analysis) in the process of remote sensing image classification complements previous techniques. GEOBIA have a better ability for not only consider the pixel values in the classification process, but also other elements of interpretation, especially for images with high resolution [1]. One software that is able to perform GEOBIA process is eCognition issued by Trimble. GEOBIA starts with the segmentation process using multiresolution segmentation method with a parameter of value scale, compactness, smoothness, shape and color [2] Multi resolution segmentation is an optimization procedure which, for a given number of image objects, minimizes the average heterogeneity and maximizes their respective homogeneity [9. Furthermore, the classification is done through two phases: the first is to select a sample object which is then classified using the nearest neighbor method. The second stage is incorporate elements of other interpretations that are specific to each object such as the size and associations.

After we have map reference and tentative map of land use, we can start to accuracy assessment. We can overlay both of two data to know the differences. Percent error accuracy assessment means value of overshoot and undershoot between them. To accuracy assessment we use sampling area with purposive random sampling.

3. Result and Discussion

3.1 Segmentation Process

Segmentation algorithms are used to subdivide entire images at a pixel level, or specific image objects from other domains into smaller image objects. In the segmentation process for determining the scale is quite an important aspect, because of this will affect the next process in the classification. The determination of scale will affect the segmentation on the image). (In this research the segmentation procces of worldview-2 images by inserting 10 in the settings on scale value with a weight on the bands of costal, blue, green, and yellow are 1 and Near infra-red 1, Near infra-red 2, Red, and Red Edge are 3. Meanwhile, the shape value is 0.1 and the compactness is 0.5. The result of this process is obtained from the segmentation of each segments that have a proximity of homogeneity such as figure 4.

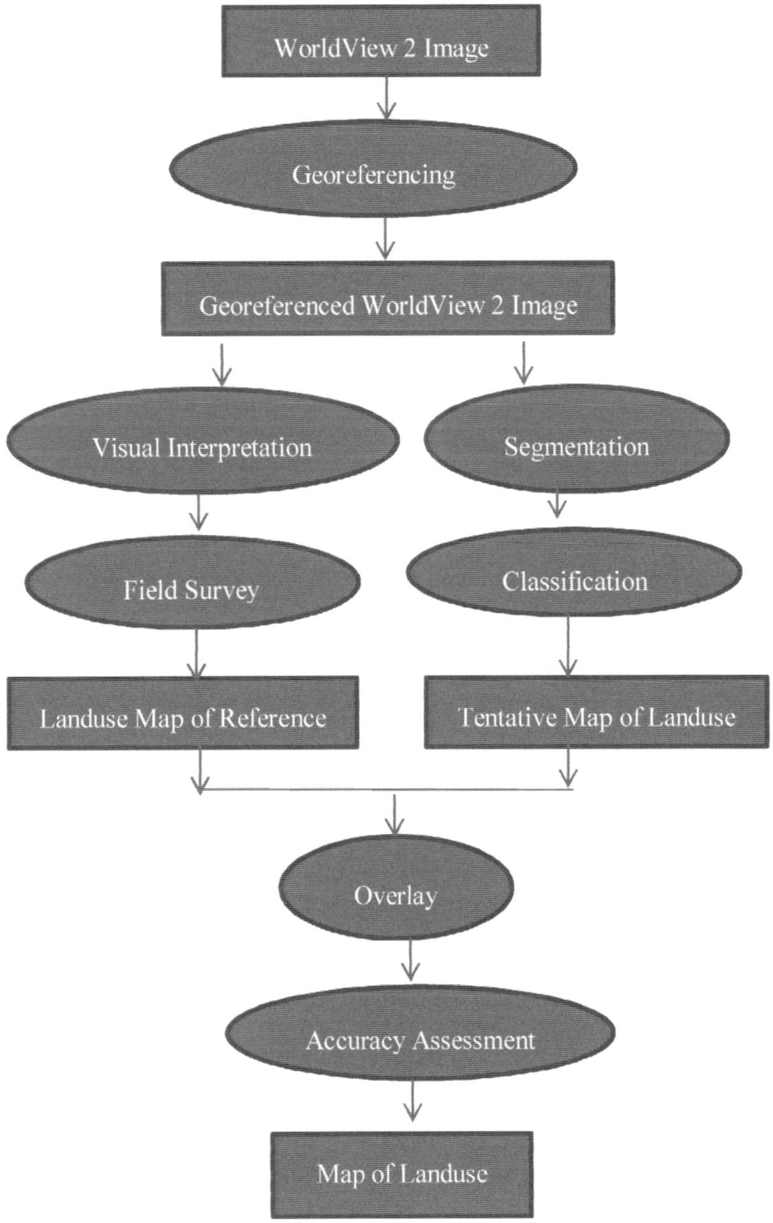

Figure 3. Flow Chart

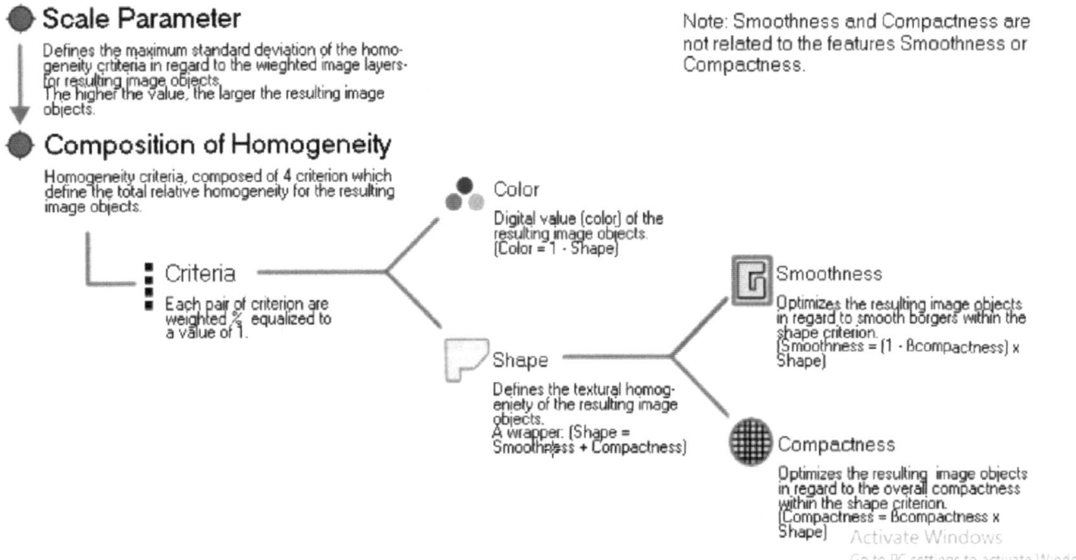

Figure 4. Flow diagram of Multiresolution segmentation (Source: reference book of ecognition developer [9]

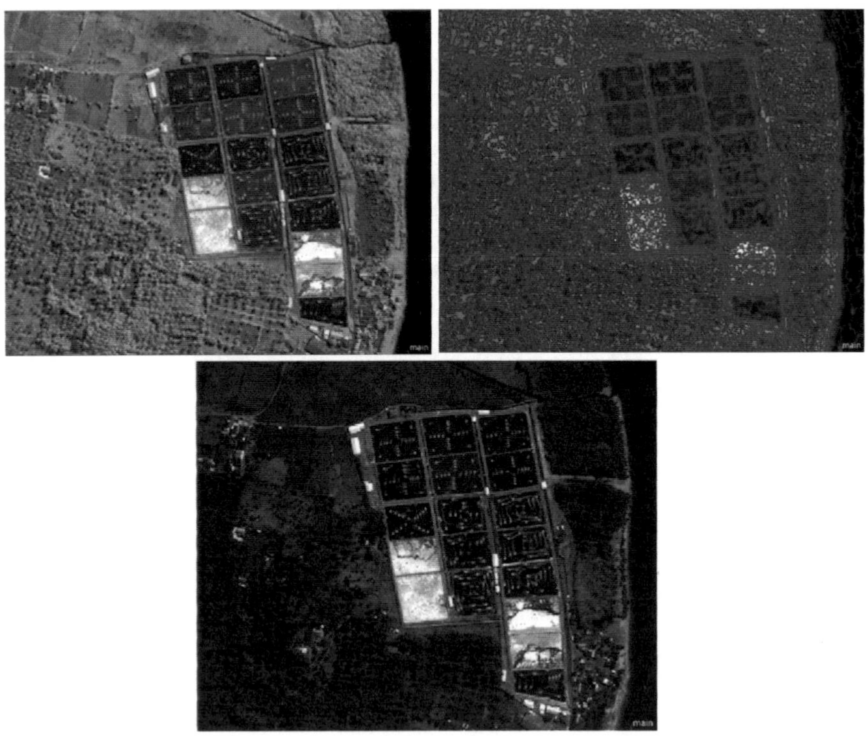

Figure 5. (a) WV-2 image composite 762; (b) images segmentation scale 10; (c) WV-2 image RGB

The use of value as a segmentation base is to consider spatial resolution that affects the value scale. Value of shape and color that is inversely proportional that has shape value higher than colour value makes colour value become lower. While the value of compactness is inversely proportional to the value of subtlety. The compactness value higer than value of its subtlety become lower [2].

3.2 Classification Process

The classification process using the nearest neighbor algorithm (NN) taking into account the value of the channel spectral Near infra-red 1 (NIR1), Near infra-red 2 (NIR2), Red and using The normalized difference vegetation index (NDVI) values, brightness and maximal difference. NDVI is a simple graphical indicator that can be used to analyze remote sensing measurements, typically but not necessarily from a space platform, and assess whether the target being observed contains live green vegetation or not.The selection of a sample of the object segmentation produces some variations including : Sea, settlements, swamps, mixed crops, river, shrubs, ponds, moor, open land and mangrove. Samples were selected based on class training and on the terms of a visual interpretation of the value of the object in the image histogram. This technique considers the aspect of sample classes with some of the features extracted into information that can be able to apply it to the image, resulting in a classification of land use in the image.

A simple NN classification was performed using all potential training sites as input data. The second classification procedure uses the estimated class thresholds to set up an object-based classification rule. Determination of classes using the estimated threshold class is used to obtain the best classification of several objects which are mixed with other objects and errors in the determination of class by NN classification.

Grading errors occurred in the NN when entering class of mixed crops, because only considering the value of any spectral objects that should be in the classification of mangrove as mangrove itself is generally identified as mixed crops class. To get the correct grade of classification threshold is using estimate distance by sea class, which means that the class mangrove is in a certain distance to the sea class. So class mixed crops near the sea is then converted into mangrove class using estimation and classification threshold distance by sea class.

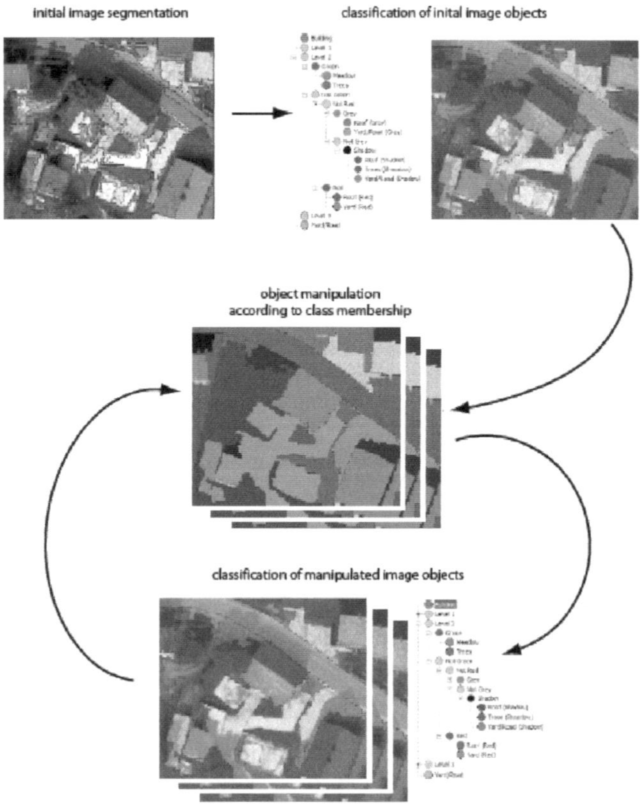

Figure 6. Iterative process of (knowledge-based) image segmentation and classification

Obtaining the classification results based on the sample is carried out in detail using grade separation element which is a measure of other interpretations and associations. For example the sea has great segmentation size because it has a high homogeneity in order that the sea will be easier to identify. Similarly, mangrove object has an association that is very close to the sea so, although having a spectral reflectance curve is almost the same as other vegetation objects, elements of this interpretation will be able to separate the mangrove class of its own. This process is performed to separate the elements of interpretation which can be entered each object class.

3.3 Accuracy Assessment

Accuracy assessment of GEOBIA is another hot issue [10]. Some studies use point data to assess GEOBIA results, while others use polygons (i.e., objects) as reference data [10]. Unlike the pixel-based image analysis methods, accuracy assessment for OBIA method involves some aspects of geometry that consists of shape, location, and symmetry between extracted objects and reference objects [8]. In this research we use polygons to be a reference data from visual interpretation. We asses the accuracy of every object by using stratified random sampling.

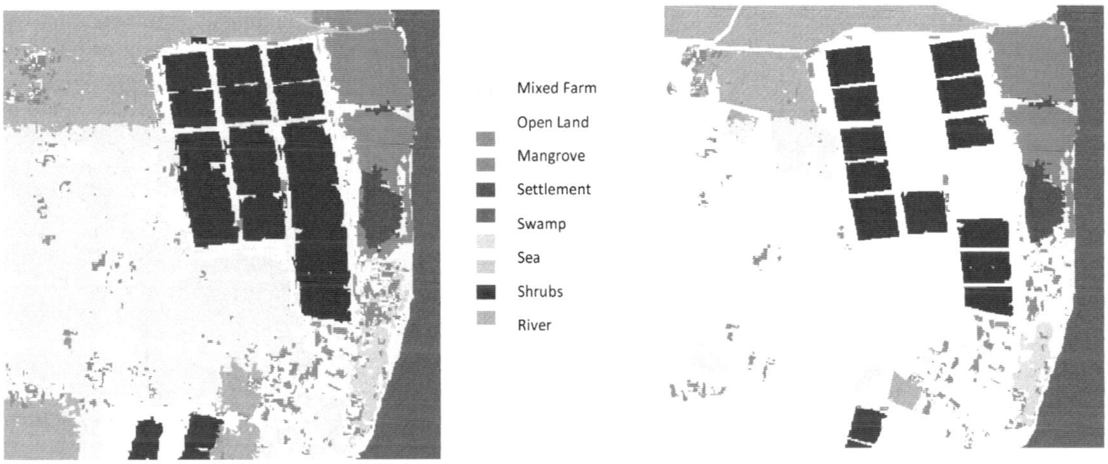

Figure 7. (a) GEOBIA classification, (b) Visual Classification (Reference Map)

In this research we have 9 classes of landuse in coastal area in Bengkak Village. They are open land, sea, mangrove, settlement, river, fishpond, Accuracy assessment is done for every class of landuse (Table 1). We have a new information about true or false classification in m^2 and in % for every class and overall landuse.

Table 1. Accuracy Assessment

Landuse	True (m^2)	False (m^2)	True (%)	False (%)
Open land	11157.56	10303.62	51.99	48.01
Sea	47668.43	69.67	99.85	0.15
Mangrove	26530.55	1476.67	94.73	5.27
Settlement	9391.80	16665.01	36.04	63.96
River	398.98	37.11	91.49	8.51
Swamp	6519.32	484.50	93.08	6.92
Shrubs	5521.31	1283.40	81.14	18.86
Fishpond	79946.56	1371.79	98.31	1.69
Crop land	49028.05	6837.39	87.76	12.24
Mixed Farm	11968.27	3136.20	79.24	20.76
Surface Area	**248130.83**	**41665.37**	85.62%	14.38%
OVERALL	**289796.20**			

From Table 1 above, we know that overall accuracy landuse of GEOBIA classification is 85.62%. Although this accuracy value is quite high, a closer observation of the accuracy of each object shows that the gaps are very significant. For all samples areas, it was found that the highest accuracy was obtained for the sea with 99.85%. Mangrove, swamp, river and fishpond are all have higher than 90%. On the other hand, the lowest accuracy was obtained for open land and settlement.

Although spectral element is not the only basis for classification, but it is still considered in the classification process GEOBIA. It affects the accuracy values between settlement and open land is relatively low. As it is known, the roof of the settlements in Indonesia, including in Banyuwangi is constructed with tiles made from raw material red soil. Wavelengths in the WorldView-2 imagery only to NIR-2 so as NDBI spectral transformation can not be applied (Figure 8).

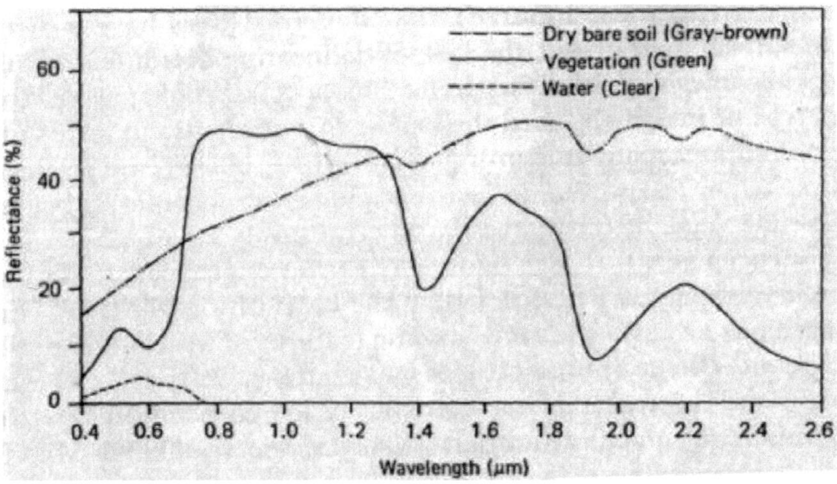

Figure 8. Spectral Curve (Source: Lillesand & Kiefer, 1984)

4. Conclusions

GEOBIA techniques combined with high-resolution images such as WorldView 2 can be used for land use classification including in the coastal area. This classification technique does have several advantages when compared with the classification of visual and multispectral classification. Some of these advantages include the ability of automation as well as elements outside the spectral interpretation that can be taken into account in this classification. In this study, image interpretation based solely on the elements of color, size, and spectral association alone but to some particular object requires another interpretation elements such as shape, texture, and pattern of the site in object recognition. The accuracy of the use of the element will greatly affect the accuracy of classification.

References

[1] Blaschke, T., & Strobl, 2001 What' s wrong with pixels? Some recent developments interfacing remote sensing and GIS. *Interfacing Remote Sensing and GIS*, (October), 1–7.

[2] Danoedoro, Projo 2012 *Pengantar Penginderaan Jauh Digital* (Penerbit Andi : Yogyakarta)

[3] Di Gregorio, A. and Jansen, L.J.M 1998 A New Concept For A Land Cover Classification System *The Land* **2**(1): 55-65.

[4] FAO 1997 Negotiating a Sustainable Future for Land. Structural and Institutional Guidelines for Land Resources Management in the 21st Century (FAO/UNEP, Rome)

[5] JR 2005 *Introductory Digital Image Processing* (*3th ed*) (Pearson Prentice Hall, New Jersey)

[6] Lindgreen, D.T 1984 *Landuse Plannning and Remote Sensing* (Martinus Nijhoff Publisher, Dolrecht)

[7] Rao, Shweta Lawanya & Dolley Shukla 2014 *International Journal of Advanced Research in Computer Science and Software Engineering* Volume **4** Issue 4, April 2014 : 49 – 56.

[8] Schopfer, E.; Lang, S., Object fate analysis—A virtual overlay method for the categorisation of object transition and object-based accuracy assessment. In Proceedings of 1st International Conference on Object-based Image Analysis (OBIA 2006), Salzburg, Austria, 4–5 July 2006.

[9] Trimble 2014 *Trimble eCognition Developer User Guide*, 1–266.

[10] Weng, Qihao 2014 *Scale Issues in Remote Sensing* (Wiley : Canada)

[11] Yang, Xiaojun 2011 Urban Remote Sensing : Monitoring, Synthesis, and Modelling in the Urban Environment (Wiley : UK)

Comparison of Chlorophyll-a Measurement Using Multi Spatial Imagery and Numerical Model in Bali Strait

Rizki Hanintyo[1], Eko susilo[1]

[1]Balai Penelitian dan Observasi Laut - KKP

Jalan Baru Perancak, Jembrana, Negara, Bali, Indonesia

rizki.hanintyo@yahoo.com

Abstract. One of abundance of nutrient in coastal area is chlorophyll-a, where chlorophyll-a is a driving factor to derive fishing ground location. There's several method has been used to derive chlorophyll-a concentration from insitu data, remote sensing satellite data, and ocean dynamics models, but we are not decide those data for certain area in coastal area and open ocean. In this paper, we will use Aqua-Terra MODIS, Landsat -8 and INDESO Satellite observation dataset to detect chlorophyll-a and also using dynamic modelling produced by INDESO and MyOcean/Marine Copernicus. Using a lot variation of data to understand the differences of spatial resolution and accuracy with observed insitu data. After calculating data comparison with insitu data in bali strait on 2014, we can conclude that Landsat-8 data has the highest r value r=0.41. For model data, comparison between insitu data and MyOcean data has r=0.3. in the other hand, insitu data and INDESO data model has r=0.26. In coastal areas (case 2 water), Landsat-8 data performed to give detail information of chlorophyll-a. Furthermore for open ocean areas (case 1 water), we can use numerical model from MyOcean/Marine Copernicus because it has lower resolution and able to forecast. For those data, it can help non artisanal fisheries to predict chlorophyll-a area in their fishing ground areas

1. Introduction

Chlorophyll is one of oceanographic parameters that able to define a fishing ground location [19]. In the other hand, Bahrenfield (2006) mention that chlorophyll-a is supporting factor of phytoplankton number and productivity which is correspond with fisheries. Kunarso, et al [13] observe oceanographic parameters in southern bali to understand the highly connection between fish abundance and chlorophyll-a. Peak season of thunnus catchment when there's higher chlorophyll-a concentration. Therefore, we need precise and accurate method to determine the distribution of chlorophyll-a ocncentration in the ocean either on the coastal area or in the open ocean area.

We would like to understand how the best method are suitable for coastal and open ocean between satellite and numerical model data, so it could reduce cost of insitu measurement data retrieval. According to [16], from few past decade, an earth observation programme using satellite technology has been prioritized in research activities to investigate the quality of remote sensing data. Remote sensing technology has been used to detect chlorophyll-a, [2] using MODIS Satellite, [15] using MODIS, SEAWIFS and MERIS to explain reflectance uncertainty of chlorophyll-a value, [11] with ADEOS/OCTS satellite and [6] using Landsat-8 LDCM Satellite. Those satellite dataset has various spatial resolution. MODIS satellite with 1 km spatial resolution and Landsat-8 data with 30 m spatial resolution. Based on these studies, algorithm and Landsat-8 satellite effectively identify chlorophyll-a distribution for coastal areas.

The weakness of remote sensing satellite to monitor chlorophyll-a daily is cloud [8]. Aqua-Terra MODIS Satellite and LANDSAT-8 are remote sensing satellite with passive sensor. Passive sensor needs energy source from the sun to observe object on earth. The sensor can not capture earth object with cloud.

Numerical models approach, which is combining satellite data and insitu data, is one of new approach to create oceanographic parameters data spatially without cloud. One of modelling data provider of chlorophyll-a in global areas has been launched by MyOcean/Marine Copernius (http://marine.copernicus.eu/). Indonesia also has National Oceanographic Data Center located in Institute for Marine Research and Observation in INDESO Project (www.indeso.web.id). NODC INDESO Project produce oceanographic models of physical and biogeochemical which consist of chlorophyll-a models. Both models created by INDESO and MyOcean has 9 km grid resolution with 1 day temporal resolution forecasted for 10 days ahead. Using large scale modelling approach and forecasted, the model is expected to explain variation of chlorophyll-a in case-1 water and able to predict to help fisherman catch the fish. In the other hand, for case-2 water, the spatial resolution of models is not enough to describe dynamics of physical and biogeochemical parameters in case-2 water.

Both method to observe chlorophyll-a use several approach to get the data. In Kahru et al. [9] has been evaluate 5 satellite data to retrieve chlorophyll-a using standard algorithm from both satellite. In the other hand, the difference between multi spatial and multi sensor has been explained by [1] using airborne sensor hyperpectral CASI with 1 meter spatial resolution, Worldview-2 with 1,8 meter, Sentinel-2 with 20 meter, Landsat-8 with 30 meter and MODIS with 250 meter spatial resolution [17]. has been created simulation to understand the difference between temporal and spatial resolution from MODIS and 3 dimensional physical and biogeochemical oceanographic parameters developed by Hamburg University called ECOHAM 3 Model.

We would like to compare remote sensing satellite data and numerical model of cholorphyll-a parameter in case 2 water / coastal and case - 1 water / open ocean. Bali strait choosed as our research location because there's abundance of fisheries potential. Using this research to observe cholorphyll-a for fisheries, it has higher implication to artisanal fisheries and non artisanal fisheries in bali strait. Furthermore, it could be the direction for fishing ground location in coastal and open ocean.

2. Method

We are using 3 differences satellite remote sensing, Landsat-8, Aqua-Terra MODIS and INDESO Satellite Observation / INDESO Satobs. INDESO Satobs is re-analysis data from NPP Suomi Satellite VIIRS Sensor. Table 1 explain spatial, temporal and radiometrik from each satellite. And also we're using 2 different models came from INDESO Models using optimum interpolation and using open boundary condition, and also Marine Copernius Models using optimum interpolation - regional kriging method.

Table 1. Satellite Remote Sensing Characteristic Comparison

	Landsat 8	**Aqua-Terra MODIS**	**INDESO Satobs / NPP-Suomi VIIRS**
Spatial	Multi : 30 m Pan : 15 m	Multi : 250 m, 500 m, 1000 m	Imagery band : 375 m Moderate band : 750 m
Temporal	16 day	1 day	1 day
Radiometric	16 bit	12 bit	12 it

2.1 *Landsat 8*

Landsat 8 data used in this paper is path row 117/066 downloaded from http://earthexplorer.usgs.gov using Bulk Download Application. Date of acquisition and imagery scene used in this paper mention in table 2.

Table 2. date and scene ID of Landsat 8

Path Row	Nama File	Tanggal	Waktu Perekaman
Path 117 Row 066	LC81170662014157LGN00	06.06.2014	10.29.38 WITA
	LC81170662014173LGN00	22.06.2014	10.29.41 WITA
	LC81170662014237LGN00	25.08.2014	10.30.03 WITA
	LC81170662014253LGN00	10.09.2014	10.30.07 WITA
	LC81170662014269LGN00	26.09.2014	10.30.04 WITA
	LC81170662014285LGN00	12.10.2014	10.30.11 WITA
	LC81170662014301LGN00	28.10.2014	10.30.09 WITA

7 scene imagery of Landsat -8 used to observe chlorophyll-a in bali strait has identic date of acquisition arround 10.30 WITA and water sample acquisition arround 07.00 - 11.00 WITA. The true colour of Landsat - 8 imagery shown in figure 1 below. Bali strait lies from 2,3 km to 53 km.

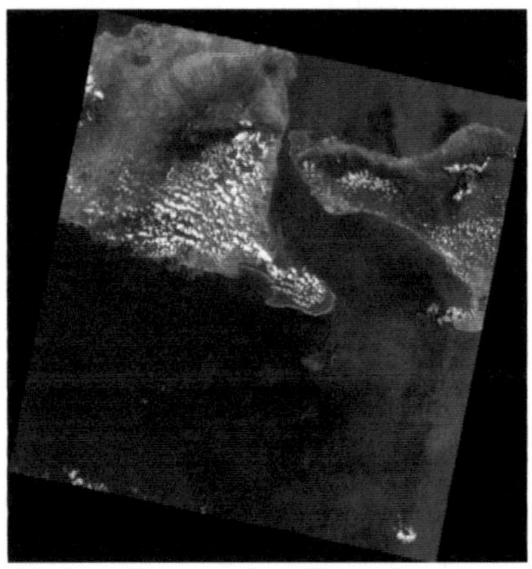

Figure 1. True Colour Composite of Landsat -8 path row 117/066 date acquisition september 26th 2014

Extraction of chlorophyll-a from Landsat-8 using Motoaki (1998) algorithm. This algorithm cames from ADEOS-OCTS satellite and modified in Hanintyo (2015) which has same wavelength with Landsat-8. Motoaki algorithm that has been modified shows below

$$Chl = 0.2818\left(\frac{L_3}{L_2}\right)^{3.497}$$

Where :

L2 = Reflectance value of Band 2 Landsat 8

L3 = Reflectance value of Band 3 Landsat 8

2.2 Aqua-Terra MODIS

The Aqua-Terra MODIS used in this paper came from Level 2 data of MODIS with Chlorophyll-a parameters. Those Level 2 data refers to ATDB MOD 22 of MODIS. The data downloaded from www.oceancolor.gsfc.nasa.gov with same acquisition date of Landsat-8 data. In figure 2 shows us the level 2 data of Aqua-Terra MODIS for Chlorophyll-a. The left image is an image before reprojection and the right image is an image after reprojection

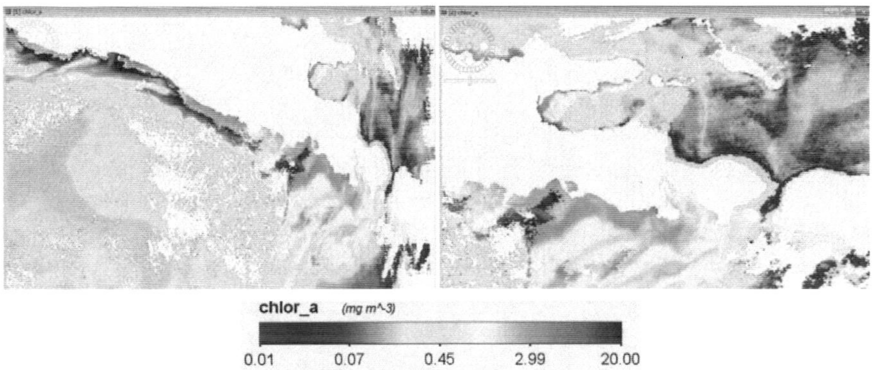

Figure 2. MODIS Level 2 chlorophyll september 26th 2014 before reprojection (left) and after reprojection (right)

2.3 *INDESO Satellite Observation*

The dataset of INDESO Satellite Observation / Satobs came from multiple satellite imagery to Chlorophyll-a parameters derive from VIIRS sensor on Suomi NPP Satellite [7]. There's 3 process has been done to derive INDESO Satobs data, which is level - 2 processing, level 2 editing and level 3 processing. This dataset has 0.02^0 spatial resolution or equal to 2,2 km. The dataset downloaded from INDESO webportal in http://www.indeso.web.id. In Figure 3 show us INDESO Satobs dataset for the whole data. The coverage of INDESO Satobs lies on Indonesia area and cover some part of south east asia, australia and papua new guinea. Some part of ocean areas, there is a blank area without data that indicate that areas are cloud. This is happening because this data was a reanalysis of satellite remote sensing imagery.

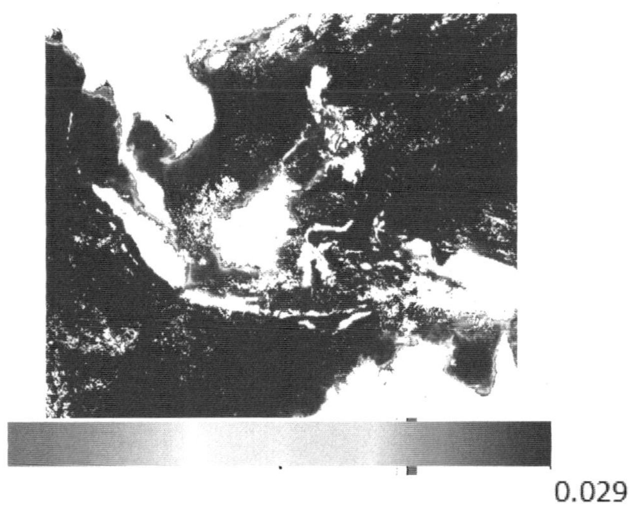

Figure 3. INDESO Satobs dataset for chlorophyll-a on februari 1st 2015

2.4 INDESO Optimum Interpolation Models

INDESO has been started operational since september 2014. Every thursday at 00.00 Central Indonesia Time (WITA), INDESO start creating numerical model for physical and biogeochemical oceanographic parameter and forecasted for 10 days ahead. SST_CHLA_NPP_cloudless will be used in this paper as INDESO Models data of Chlorophyll-a. This dataset will be renewed every thursday for 10 days ahead.

In the framework of INDESO Project, coupled model of physical and biogeochemical oceanographic parameter developed from 90-144 E and 20 S - 25 N which cover all Indonesia Areas [5]. Physical model (SST Parameter) based on OPA Model (NEMO2.3) will be called as INDO12 Model [18] In the other hand, Biogeochemical model simulated by PISCES Model version 3.2 [5]. Models simulated by Open Boundary Condition (OBC) to understand the condition of models. For biogeochemical models, OBC derived from a lot of climatology database such as nitrate, phosphate, dissolved silicate, oxygen, dissolved inorganic carbon and alkalinity [5]. For Physical oceanogprahic parameters, derived from the daily output of global ocean forecasting system on $1/4^0$ produce by Mercator Ocean [18]. INDESO Model for chlorophyll-a show in figure 5. The data range lies from 0.04 ,g/m3 until 7.05 mg/m3. For this research, we will match up the dataset with the same date as Landsat-8 dataset passing.

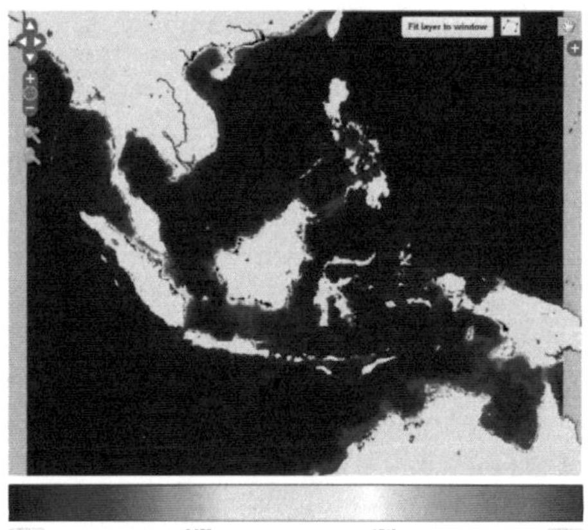

Figure 4. INDESO Chlorophyll-a models from INDESO (date : februari 1st 2015)

2.5 Marine Copernicus models

Marine Copernicus is one of project under European Comission which observe oceanographic parameters using remote sensing technology or using numerical model. In this research, we will use optimum interpolation - numerical model from Marine Copernicus downloaded from http://www. Marine.copernius.eu_ with dataset called Global_ocean_color_chlorophyll_Optimal_Interpolation. We are using same timeframe as Landsat-8 passing date. To download Marine Copernius Model, we are using Motu Python Script because there will be a large dataset and the possibility to subset the global dataset using Motu Python Script. Figure 5 shows the models of Marine Copernius - chlorophyll-a paremeters. This dataset has been cropped from global model with 25 N - 20 S and 90-160 E.

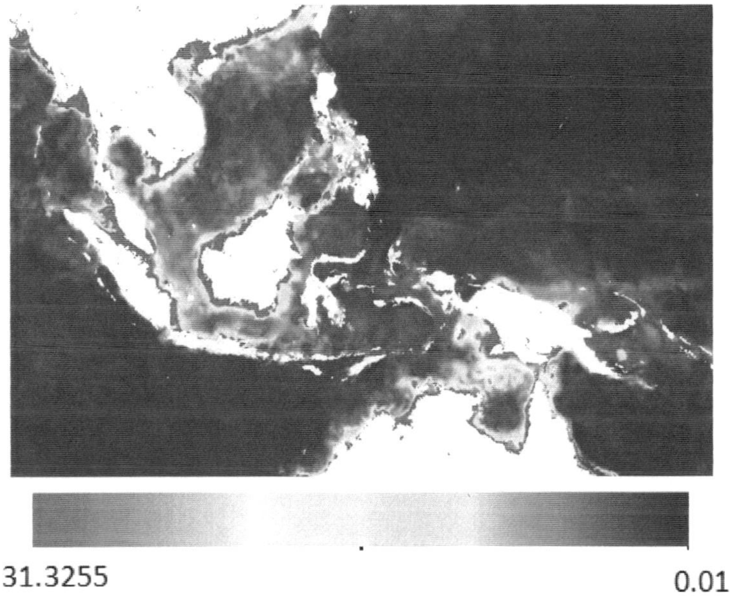

Figure 5. Marine Copernicus Models of Chlorophyll-a parameters

2.7 Field Observation Data

We has been held a field survey in bali strait started from june 1st 2014 until November 3rd 2014 to collect physical and biological oceanographic parameters. One of biological parameter is chlorophyll-a. For each day of survey, we has been collected 2 different location. For this paper, we're using 7 days or 14 field dataset. For each field dataset, we will match up the coordinate with remote sensing imagery and models to read each chlorophyll-a value in the pixels.

The water sample that has been collected each day will be analyzed in water quality laboratory in IMRO (LKP BPOL). LKP BPOL has been accredited by KAN no. LP-454-DN for ISO.IEV 17025-2005. Chlorophyll-a filtering using GFF filter paper and using UV-VIS Spektrometer Scotech SPUV-24. Figure 6 show us the location of field survey for the whole date. Spatial location of field survey based on sardinella lemuru fisherman in bali strait. When fisherman catch sardinella lemuru fish, we are collecting the water sample for each site 5 liters. The sample will be delivered to LKP-BPOL to be analyzed using Standard Method.

Figure 6. Location of field survey in bali strait from june until november 2014

3. Result

3.1 Landsat-8

The algorithm that has been used to detect chlorophyll-a on Landsat-8 satellite data, adopted from chlorophyll-a alogorithm from Motoaki Kishino [11]. Those algorithm used for ADEOS-OCTS and has been modified as mention on Rizki, 2015. On Rizki [6] explained that there's a huge differences between ADEOS-OCTS and Landsat-8 data on green wavelength. ADEOS-OCTS has 2 band of green wavelength, in the other hand Landsat-8 just has 1 band of green wavelength. if we take a look deeper on the wavelength, ADEOS-OCTS lies between 0.511-0.575 μm, therefore on Landsat-8 lies between 0.525-0.6 μm.

Figure 7 is the result of chlorophyll-a algorithm of Landsat-8 on september 26th 2014 using modified algorithm from Rizki [6]. The chlorophyll-a has range from 0.0152 $\mu g\ L^{-1}$ until 0.2818 $\mu g\ L^{-1}$ with mean value 0.09622 $\mu g\ L^{-1}$.

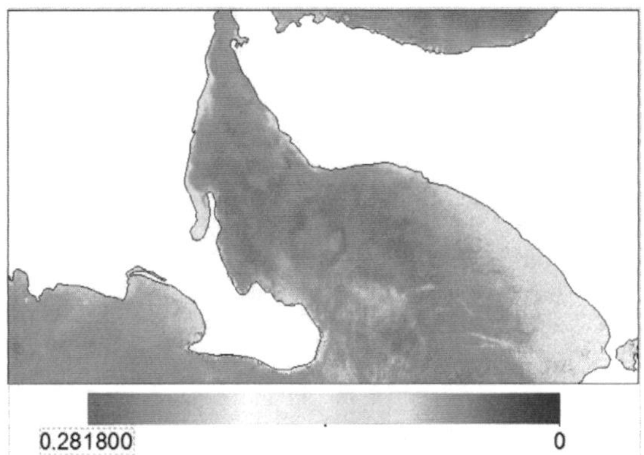

Figure 7. Chlorophyll - a data from Landsat 8 acquisition date September 26th 2014

In figure 8, we're able to see the negative correlation trends between satellite data from Landsat 8 and insitu data with r = 0.41. If we take look deeper on spatial distribution between insitu data position, we're able to see that on one pixels of Landsat data represent from one insitu data location.

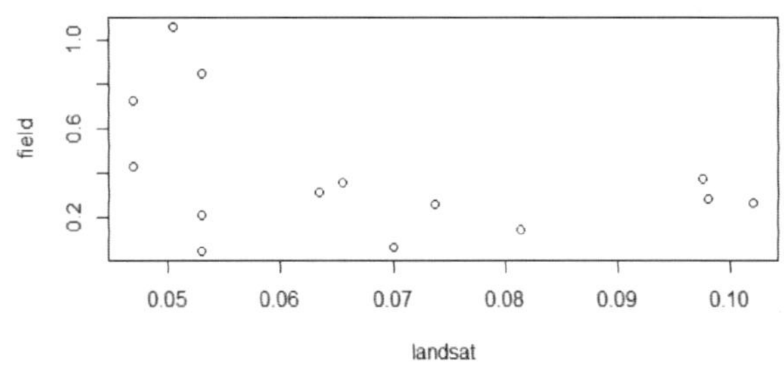

Figure 8. scatter plot of Landsat 8 data and insitu data

3.2 Aqua-Terra MODIS

The Aqua Terra MODIS satellite has a narrow wavelength (15 nm) to observe ocean color properties. Compared to Landsat 8 data, it has a wide wavelength (100nm) to derive chlorophyll-a. Chlorophyll-a algorithm for Aqua-Terra MODIS has been documented on Algorithm Theoretical Base Document / ATDB 19 [10]. Aqua Terra MODIS on figure 9 has lower spatial resolution (1km) compared with Landsat 8 data (30m). The coastline of bali unable to defined very well because of coarse spatial resolution. Un-mixing pixels will be occured on some coastal object and the result of chlorophyll-a parameter will be vary. On some location, we are able to see some cloud formation with white colour. On those area, we can not identify chlorophyll-a and other parameters.

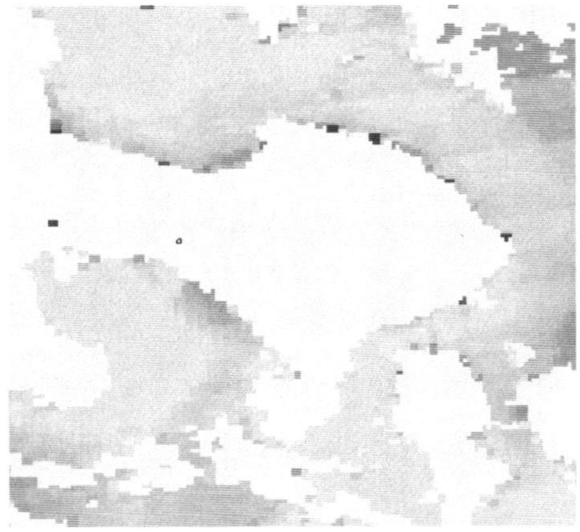

Figure 9. Level 2 chlorophyll-a imagery of Aqua Terra MODIS (acq date September 26th 2014)

Scatter plot on figure 10 shows us separability of modis and field data. There is 2 outlier with chlorophhyll-a 2 μg L^{-1} concentration. Aqua Terra MODIS has low correlation with r = -0.02

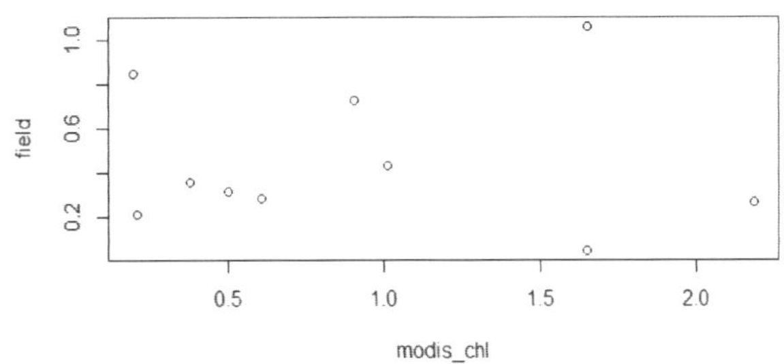

Figure 10. scatter plot of Aqua Terra MODIS

3.3 INDESO Satobs

The INDESO satobs data has 2 product on it, Sea Surface Temperature data and Chlorophyll-a data. The sea surface temperature data collected from 4 satellite data, which is

- Aqua MODIS Satellite
- Terra MODIS Satellite
- AVHRR Sensor on METOP-A Satellite
- AVHRR Sensor on METOP-B Satellite

In the other hand, for Chlorophyll-a data collected from VIIRS Sensor on NPP-Suomi satellite. The VIIRS data of INDESO collected from OceanColor Website on near real time with 5 hours delay from acquisition. The VIIRS sensor has 750m spatial resolution (Vandemuelen, 2015). There is 3 step to finalize level 1 VIIRS data to be INDESO Satobs data, which is :

- Level 2 processing, is a deriving phytoplankton data using Polymer Algorithm (Steinmetz, et al, 2011). The benefit of using that algorithm is able to compute phytoplankton on a thin cloud, dust and some area with sunglint. The level 2 processing of VIIRS on figure 11.

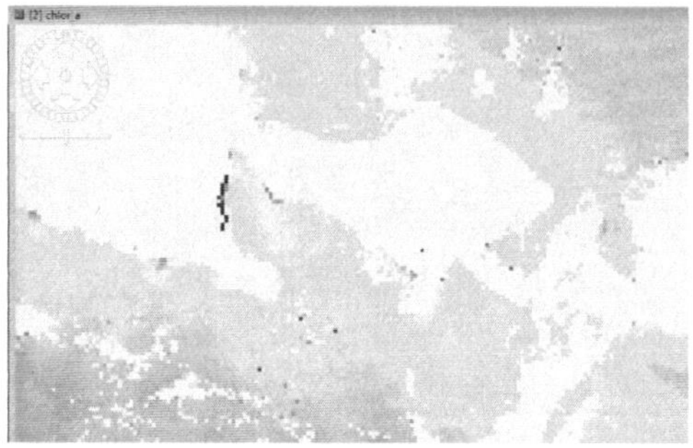

Figure 11. VIIRS level 2 processing over Bali Strait Area

- Level 2 editing, is a process to correct low quality pixel, de-striping process using gaussian filtering and change grid size output with 0.01^0 spatial resolution.
- Level 3 processing, is a final data processing. Using imagery from 0.01^0 resolution and converted to 0.02^0 resolution / 2,2 km to get the INDESO Satobs data. Figure 12 is a result of INDESO Satobs data, which is a re-analysis data from 750m to 2,2 km. If we compare between figure 12 and figure 11, spatial resolution converting process reduce some details on coastal object.

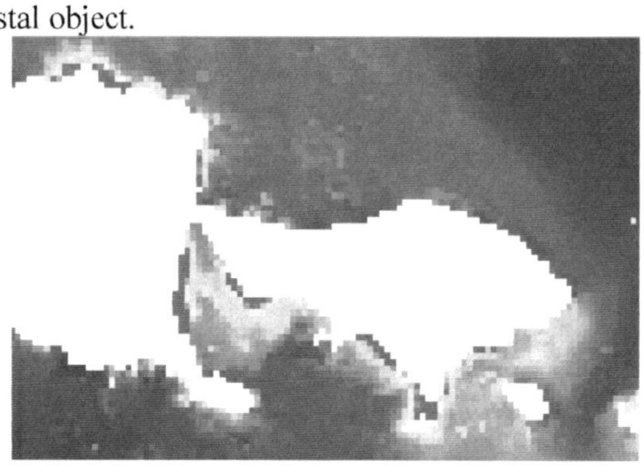

Figure 12. Result of Level 3 Processing of INDESO Satobs data.

3 dimensional scatter plot on figure 13 shows data of INDESO Satobs, Landsat data and field data. The effect of spatial resolution occur between INDESO Satobs data with Landsat data, which is INDESO Satobs data seems overvalue compared with Landsat and field data. Using up scaling of VIIRS data on INDESO Satobs data, the pixel value that has been filtered using gaussian filter. The gaussian filter is a low pass filter that will make a pixel smoother [3]. On 2 dimensional scatter plot (figure 13), there is a negative correlation between INDESO Satobs data and field data with r value r=-0.25.

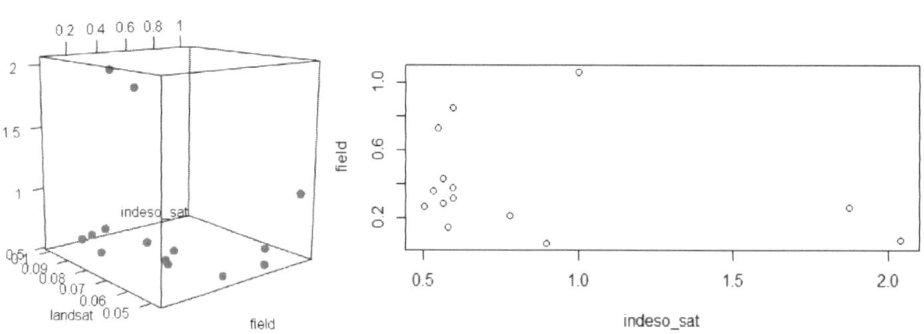

Figure 13. 3 dimensioan chart and scatter plot of INDESO Satobs data

3.4 INDESO Optimum Interpolation

INDESO Optimun Interpolation / INDESO OI downloaded directly from INDESO web portal on www.indeso.web.id. the downloaded data is a gridded data with a netCDF format. "Multidimension Tools" on ArcGIS has been used to extract this data. Extraction of INDESO OI data on specific location of field data using "extract multi value to point" tools on ArcGIS.

On figure 14, we can not see a cloud formation on INDESO OI data. That is a benefit using numerical model method. Therefore, using numerical model will resulted a coarse spatial resolution and a lower information of coastal object. For example, we can not extract information near Benoa port of Bali, because the spatial resolution is coarse.

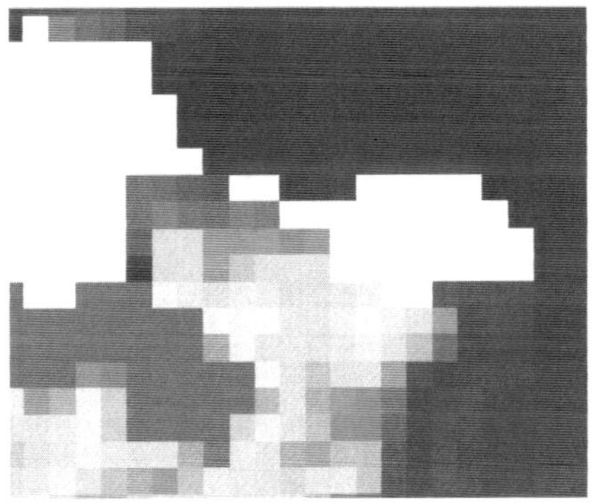

Figure 14, Raster view of INDESO OI data over bali strait on September 26th 2014.

3 dimensional scatter plot on figure 15 resulted a difference between field observation with INDESO OI data and Marine copernius / MyOcean Models. On 3d scatterplot, there a differences of maximum and minimum value from ach data. INDESO OI data tend to be overestimate compared with field data, even MyOcean data is more overestimate. On 2d scatterplot, there is no linear regression and has negative correlation with r = -0.26.

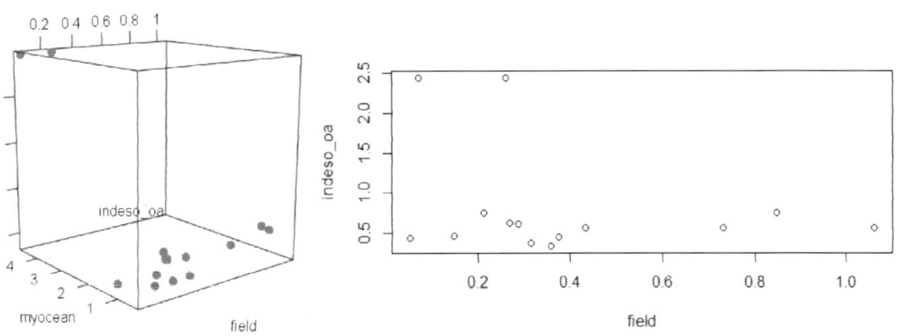

Figure 15. 3dimensional and plot data of INDESO OI

3.5 Marine Copernicus Model

Marine Copernicus / MyOcean Models developed using Optimum Interpolation Method. The models has 2 input from level 2 chlorophyll-a VIIRS sensor of NPP suomi Satellite and Level 2 chlorophyll-a from MODIS. OI Models developed using "Regional Kriging Anisthropic Covariance Models". Compared to INDESO OI data, Marine copernius has higher spatial resolution of input data. Using a higher spatial resolution input data, this model has a better spatial resolution (compared with INDESO OI) with 4,4 km. Figure 16 below is a result of Marine Copernicus Model. Details of coastal object has been better defined.

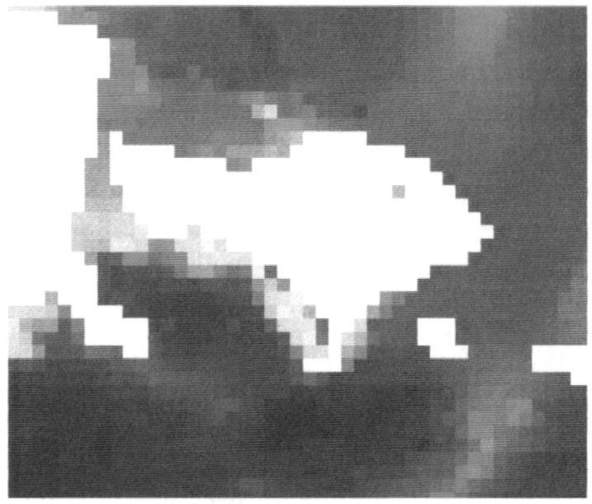

Figure 16. Raster view of Marine Copernicus model over bali strait on Februari 1st 2015.

On figure 17, scatter plot between Marine Copernicus data and field data show us that there's a huge different range between those data. As mentioned on INDESO OI, Marine Copernicus data tend to be overestimated where Marine Copernicus data has a data range from 1 - 4 µg L^{-1} . In the other hand, the field data range from 0 - 1 µg L^{-1} and also has low r correlation with r = - 0.30.

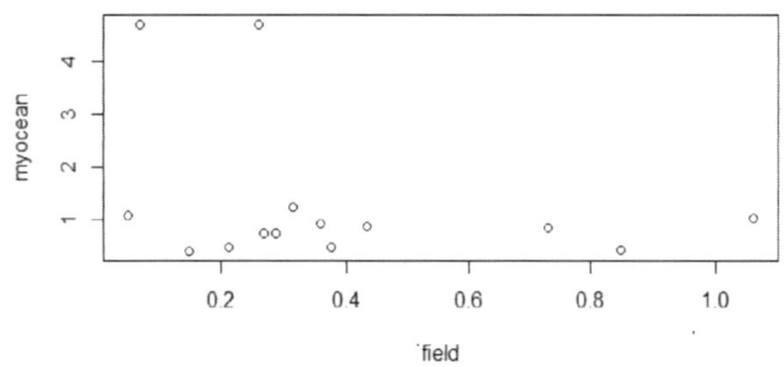

Figure 17. scatter plot between Marine Copernicus data and field data.

After extracting pixel value on field survey, these chart below on figure 18 show r value of each data on one chart. Figure 18 is a plot with r value for remote sensing and field data. We are using R script and a code from [20] to create this chart. Based on chart, Landsat data has a higher correlation with field data. Spatial resolution as a main factor of it. 30 meter spatial resolution able to describe chlorophyll-a value over coastal areas. The second factor is a depth of bit data where Landsat 8 has 16 bit data will has a pixel range from 0 - 65535. Pixel value of Landsat was so detail, so does chlorophyll - a value. Landsat 8 data able to help artisanal fisheries to understand where is the highest chlorophyll-a concentration.

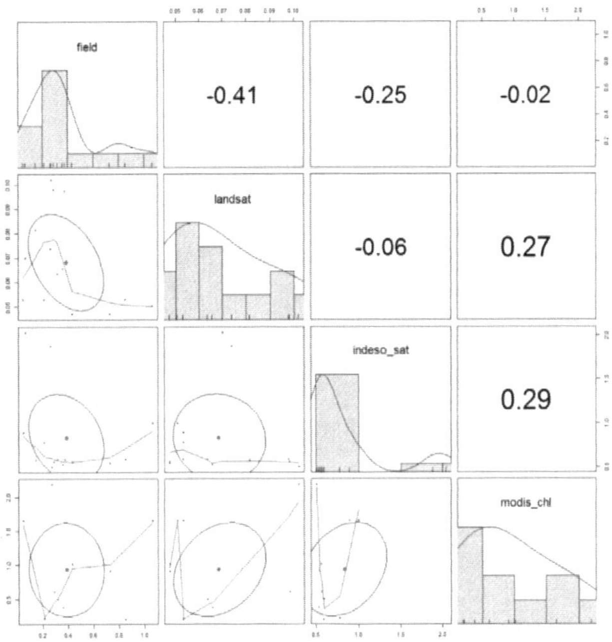

Figure 18. scatter plot and r value of remote sensing data.

For next figure, on figure 19, is chart for numerical model data. The chart show us that Marine Copernicus Model has higher correlation value thatn INDESO OI data. As stated above, spatial resolution is a main factor where Marine Copernicus has 4,4 km resolution and INDESO OI has 9 km resolution. Numerical model has a forecasting benefit and also a cloud free data. As marine copernicus has a good correlation value, it will be a benefit for bigger fisherman (more than 30 GT ships) to understand chlorophyll-a concentration on case - 1 water with a good prediction.

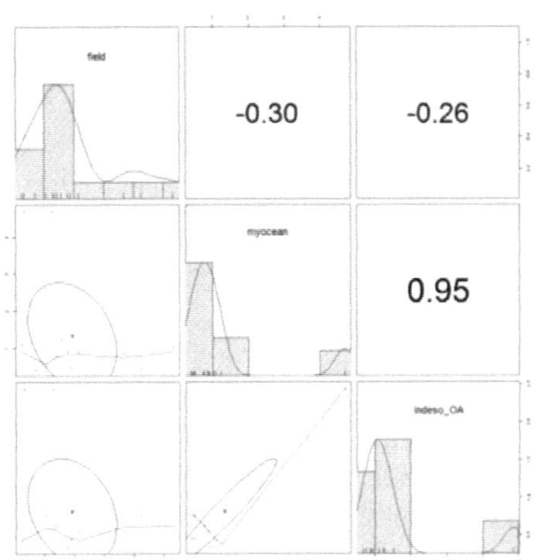

Figure 19. scatter plot and r value of numerical model data

4. Discussion

Based on measurement above, Landsat 8 data has the highest correlation value than the others. On Landsat data, one field data represent one pixels. But if we plot field data to model or remote sensing imagery with lower spatial resolution, there will be one pixel on more than 2 field data, as we can see on figure 20.

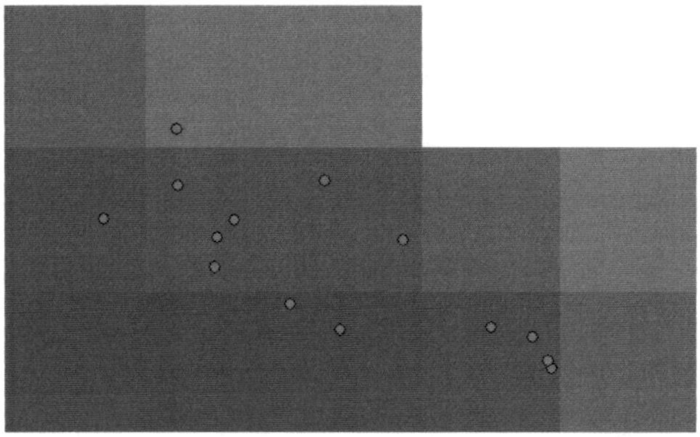

Figure 20. field data on INDESO OI data models.

Differences of spatial resolution on remote sensing data is one of challanges to observe earth object. Landsat 8 data will be fitted to used by artisanal fisheries on case - 2 water because of higher spatial resolution but lack of temporal resolution (16 days). For prediction capability, Marine copernicus data will be a good data for fisheries on case - 1 water because it has a good spatial resolution then INDESO OI data. We need a field observation data with a different range between field data to observe pixel per pixel. As mentioned on [18] and [5], insitu data for numerical model has a sparse distribution, so they can able to describe pixel per pixel observation.

5. Conclusion

Landsat 8 data has a tendency to approach with the result of field observation. Landsat 8 data good to be used for monitoring chlorophyll -a on coastal areas with a good spatial resolution (30 meter). In the other hand, Landsat 8 data has laxity of temporal resolution (16 days). Marine Copernicus data model good to be used for monitoring chlorophyll-a concentration on case -1 water because it has lower spatial resolution and prediction capability.

We need more analysis of numerical model data. The point sof representative field observation data of pixel values of numerical model is strongly recommended. It also necessary to develop algorithms for Landsat 8 Chlorophyll-a. The algorithm given that we used have a undervalue the estimated field data.

References

[1] Beck, Richard, Shengan Zhan, Hongxing Liu, Susanna Tong, Bo Yang, Min Xu, Zhaoxia Ye, et al. 2016. "Comparison of Satellite Reflectance Algorithms for Estimating Chlorophyll-a in a Temperate Reservoir Using Coincident Hyperspectral Aircraft Imagery and Dense Coincident Surface Observations *Remote Sensing of Environment* **178** Elsevier Inc.: 15–30. doi:10.1016/j.rse.2016.03.002.

[2] Cannizzaro, Jennifer Patch., Kendall L Carder 2006 Estimating chlorophyll a concentration from remote sensing reflectance in optically shallow waters *J Remote Sensing of Environment* (101): 13 -24

[3] Danoedoro, Projo 2012 *Pengantar Penginderaan Jauh Digital* (Andi Publishing : Yogyakarta)

[4] Department of the Interior- US Geological Survey 2015 *Landsat 8 (L8) data user hand book.* (USGS. Sioux Fall South Dakota)

[5] Gutknecht, E, dkk 2015 Evaluation of an operational ocean model configuration at 1/12^0 spatial resolution for the Indonesian seas (NEMO2.3/INDO12) - part1:Biogeochemistry. Geoscientific Model Development vol 8.

[6] Hanintyo, Rizki, dkk. 2015 Variabilitas dan validasi algoritma klorofil-a daerah selat bali menggunakan data OLI Landsat-8 *Semnaskan UGM 2015* : Yogyakarta.

[7] INDESO 2015 *Product User Manual - SST and CHL-a Composite Satellite Observation* (INDESO-CLS : France)

[8] Joshi, Neha, Matthias Baumann, Andrea Ehammer, Rasmus Fensholt, Kenneth Grogan, Patrick Hostert, Martin Rudbeck Jepsen, et al. 2016 A Review of the Application of Optical and Radar Remote Sensing Data Fusion to Land Use Mapping and Monitoring *Remote Sensing* **8** (1): 1–23. doi:10.3390/rs8010070

[9] Kahru, Mati, Raphael M. Kudela, Clarissa R. Anderson, Marlenne Manzano-Sarabia, and B. Greg Mitchell 2014 Evaluation of Satellite Retrievals of Ocean Chlorophyll-a in the California Current *Remote Sensing* **6** (9): 8524–40. doi:10.3390/rs6098524.

[10] Kendall, L, dkk. 2003 *MODIS Ocean Science Team Algorithm Theoretical Basis Document : ATBD 19 Case 2 Chlorophyll a Version 7* (University of South Florida - College of Marine Science)

[11] Kishino, Motoaki., Takashi Ishimaru, Ken Furuya, Tomohiko Oishi, Kiyoshi Kawasaki 1998 In-Water Algorithms for ADEOS/OCTS *Journal of Oceanography* Vol **54** pp 431 to 436.

[12] Kishino, Motoaki,. Akihiko Tanaka, Joji Ishizaka 2005 Retrieval of chlorophyll a, suspended solids, and colored dissolved organic matter in tokyo bay using ASTER data *J Remote Sensing of Environment* **99**: 66 – 74.

[13] Kunarso, Supangat, A., & Wiweka 2008 Keunggulan Aplikasi Peramalan Fishing ground Tuna di Lokasi Upwelling dengan Bantuan Citra Satelit Harian, 13 (September), 127–132.

[14] Kunarso, dkk. 2011 Variabilitas suhu dan klorofil di daerah upwelling pada variasi kejadian ENSO dan IOD di perairan selatan jawa sampai timor *J Ilmu Kelautan* **16** (3) : 171-180

[15] Mélin, F, G Sclep, T Jackson, and S Sathyendranath 2016 Remote Sensing of Environment Uncertainty Estimates of Remote Sensing Re Fl Ectance Derived from Comparison of Ocean Color Satellite Data Sets *Remote Sensing of Environment* **177** The Authors: 107–24. doi:10.1016/j.rse.2016.02.014.

[16] Sá, C., D. D'Alimonte, A. C. Brito, T. Kajiyama, C. R. Mendes, J. Vitorino, P. B. Oliveira, J. C B da Silva, and V. Brotas 2015 Validation of Standard and Alternative Satellite Ocean-Color Chlorophyll Products off Western Iberia *Remote Sensing of Environment* **168**: 403–19. doi:10.1016/j.rse.2015.07.018.

[17] Tiedje, Bente, Andreas Moll, and Lars Kaleschke 2010 Comparison of Temporal and Spatial Structures of Chlorophyll Derived from MODIS Satellite Data and ECOHAM3 Model Data in the North Sea *Journal of Sea Research* **64** (3). Elsevier B.V.: 250–59. doi:10.1016/j.seares.2010.03.003.

[18] Tranchant, B, G Reffray, E Greiner, D Nugroho, Ramonville Saint-agne, and Ramonville Saint-agne 2016 Evaluation of an Operational Ocean Model Configuration at 1 / 12 ° Spatial Resolution for the Indonesian Seas . Part I : Ocean Physics *Geoscientific Model Development* **9**: 1037–64. doi:10.5194/gmdd-8-6669-2015.

[19[Zainuddin, Mukti 2011 Skipjack Tuna in Relation To Sea Surface Temperature and Chlorophyll-a Concentration of Bone Bay Using Remotely Sensed Satellite Data *Jurnal Ilmu Dan*

Teknologi Kelautan Tropis **3** (1): 82–90.

[20] Zuur, Alain F, Elena N Ieno, Neil J Walker, Anatoly A Saveliev, and Graham M Smith 2009 Mixed Effects Models and Extensions in Ecology with R *Journal of Chemical Information and Modeling* Vol. **53**. Springer. doi:10.1017/CBO9781107415324.004.

Coral reef destruction of Small island in 44 years and destructive fishing in Spermonde Archipelago, Indonesia

Nurjannah Nurdin[1,3*], Teruhisa Komatsu[2], Chair Rani[1], Supriadi[1], Sitti Fakhriyyah[1], Agus[3]

[1]Marine Science Department, Hasanuddin University, Jl. Perintis Kemerdekaan 10, Makassar, 90245, South Sulawesi, Indonesia.
[2]Atmosphere and Ocean Research Institute (AORI), The University of Tokyo, 1-15-1, Kashiwanoha, Kashiwa, 277-8564, Japan.
[3]Center for Regional Development and Spatial Information, Jl. Perintis Kemerdekaan 10, Makassar, 90245. Indonesia.

*email: nurj_din@yahoo.com

Abstract: Coral reefs are among the most diverse and threatened ecosystems on the planet. The most commonly stated for developing coral reef remote sensing techniques is to asses and or to monitor the status of these ecosystems. The study site was selected one of small island in inner zone Spermonde archipelago, Indonesia. We used Landsat MSS, Landsat TM, Landsat ETM, and Landsat OLI data to examine changes in the coral reefs of inner zone island in the Spermonde Archipelago from 1972 to 2016. The image processing are gap fills, atmospheric correction, geometric corrections, image composites, water column corrections, unsupervised classifications, and reclassification. Some of component change detection procedure was applied to define change. The results showed significant changes in 44 years. Disturbed coral reefs are typically characterized by loss of coral cover by increase in the abundance of dead corals and rubble. Local factors such as destructive fishing is direct destruction of inner zone island. While the impact of local threats may be reduced through management action, global threats to coral reefs are likely to increase in severity in the coming years.

Keywords: Coral reef, destruction, small island, Spermonde archipelago, Landsat

1. Introduction

1.1 Background

Coral reefs ecosystem is very sensitive to environmental impacts related to human activity on coastal and marine areas. Human impacts, direct and indirect have been recognized as a higher threat than the natural disturbances. The major causes of coral reef degradation in Indonesia are blast fishing, sewage, industrial pollution and cyanide fishing [13]. One indication of a good ecosystem health is coral reef ecosystem resilience rate, explicitly the ability of an ecosystem to recover from a damaged condition to be better [12]. Resilience rate of coral reef ecosystem is one important indication to the manager in well and right managing coral reef from disturbances and estimate the length of the process of rstoration and rehabilitation of coral reefs ecosystem [4], especially for coral reef ecosystem, one of the factor that influence the resilience rate is the availability of hard substrate in a bottom waters as a

IOP Conf. Series: Earth and Environmental Science **47** (2016) 012011 doi:10.1088/1755-1315/47/1/012011

settlement for coral larvae. Spermonde islands located in the Makassar Strait, South-West side of the peninsula of Sulawesi Island. It is known as island communities Sangkarang islands consisting of ± 121 islands. Spermonde islands divided into four zones e.g. outer zone, middle outer zone, middle inner zone and inner zone. Ballang Lompo is one of small islands in inner zone, Spermonde Archipelago which has high potential ecosystem especially coral reef distribution. Therefore, spatial dynamic mapping of coral reef are needed to develop geodatabase and create good planning for coastal area. There are studies using medium spatial resolution data, e.g. Landsat [1, 3].This results is important for planners and scientists because it can be used to simplify the management planning and change detection in Ballang Lompo island.

1.2 Objectives

The objective of this research was to spatial analysis of coral reefs transformation for 44 years and anthropogenic impact.

2. Methodology

2.1 Study Area

Ballang Lompo is one of the islands in the Spermonde Archipelago, PangkepDistrict, Indonesia. Ballang Lompo has a land area of 8 km2.

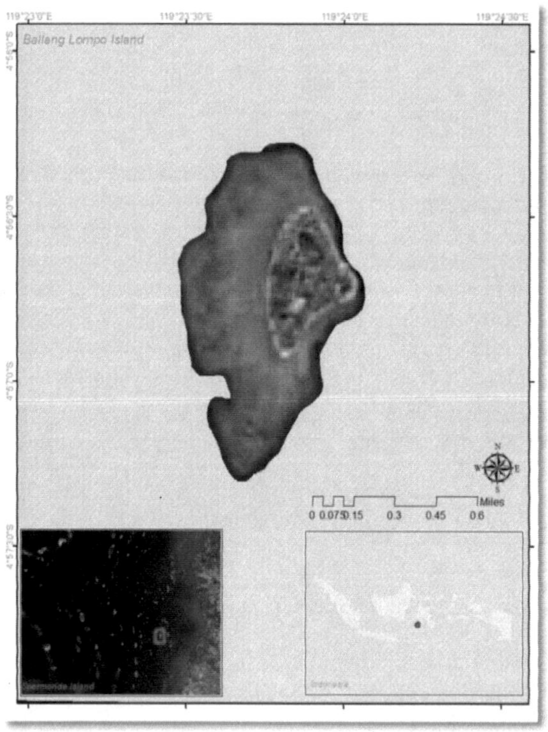

Figure 1. Study area in Ballang Lompo Island, one of the small islands in inner zone Spermonde Archipelago, Indonesia

2.2 Data

Identification of bottom types in Ballang Lompo Island using Landsat image selected Landsat MSS, TM, ETM, and OLI-TIRS to the years 1972, 1984, 1990, 1996, 2002 and 2016. The Landsat MSS data have four spectral bands, with a spatial resolution of 60 m. Landsat TM data have seven spectral bands, with a spatial resolution of 30 m for bands 1-5 and 7. The Landsat ETM data consisted of eight spectral bands with a spatial resolution of 30 m for bands 1-7. The Landsat 8 OLI-TIRS data have nine spectral bands with a spatial resolution of 30 m for bands 1-7 and 9. The following table shows the typesof data (Table 1).

Table 1. Types of data

No	Satelite	Sensor	Resolution (m)	Acquisition	Path/Row
1	Landsat_1	MSS	60	1972-10-28	122/063
2	Landsat_2	MSS	60	1984-10-26	122/063
3	Landsat_4	TM	30	1990-12-16	114/063
4	Landsat_5	TM	30	1996-04-28	114/063
5	Landsat_7	ETM	15	2002-06-24	114/063
6	Landsat_8	OLI-TIRS	15	2016-02-04	114/063

Classification bottom types consisting of five classes(live coral, seagrass, mix bottom, rubble and sand) were examined by Landsat MSSwhich have a 60 meter. There are six classes (dead coral, live coral, seagrass, mix bottom, rubble and sand) for Landsat TM, ETM, and OLI which have a 30 meter and 15 meter spatial resolution.

2.3 Image Processing

The image processing was composed of five main steps including: (1)geometric correction, (2) fusion, (3) preliminary analysis, and (4) image classification. Landsat satellite imagery t=1972, t=1984, t=1990, t=1996, t=2002 and t=2016 were being corrected in radiometric aspect and geometric aspect by rectify/resampling pixel position in image, and geo-referencing the image into the selected projection system. Optical band of imageries had 30 m and 60 of spatial resolution. Image fusion between multi-spectral and panchromatic Landsat 7 ETM, and Landsat 8 OLI-TIRS images was done to achieve higher spatial resolution (15m).After satellite images were geometrically corrected, preliminary analysis methods could be applied for image enhancement, composite RGB, and cropping. Image classification is a technique in processing a pixel of the image that has spectral reflectance appearance, which identified to separated objects that contained in the image satellite, and grouped then into certain characteristics in accordance with the real condition. Formerly, thiswas done by a masking technique to eliminate some of the area that was not needed. This research used unsupervised classification of ISOCLASS algorithm applying schema of habitat condition.

2.4 Ground Truth

Ground observation was important in order verify image interpretation result. Data from ground truth will increase the total image interpretation/classification accuracy [5]. Ground truth also occurred to collect field data such as area observation and interviewing local people about historical trend of previous land usage. Ground truth wasconducted to determine the actual habitat on the ground. The result of ground truth was compared with the result of classification image.

2.5 Accuracy Assessment

Accuracy assessment was being conducted by Kappa coefficient (κ) for accuracy assessment which relies on image training area. Accuracy test was used to make any calculation matrix for each error (confusion matrix) on any kind of coral reefs resulting from the analysis satellite imagery. The following is a table of confusion matrix form (Table 2).

Image validation was counted based on the above table such as *Overall Accuracy* (OA), *Producer's Accuracy* (PA), and *User's Accuracy* (UA). OA is a percentage of sample units that were classified accurately. PA and UA are ways of representing individual category accuracies instead of just the OA. PA is a percentage of probability average of a sample unit that refers to distribution of each class that had been classified in the field, while UA is a percentage of sample unit that actually represented the classes in the field. The confusion matrix in Table 2 helped make mathematical notation easy to understand.

Table 2. Confusion matrix

Classified to class	Reference (sample point)			Row Total (N_{i+})	User's Accuracy
	1	2	K		
1	N_{11}	N_{12}	N_{1K}	N_{1+}	N_{11}/N_{1+}
2	N_{21}	N_{22}	N_{2K}	N_{2+}	N_{22}/N_{2+}
K	N_{K1}	N_{K2}	N_{KK}	N_{K+}	N_{KK}/N_{K+}
Column Total (N_{+j})	N_{+1}	N_{+2}	N_{+K}	N	
Producer's Accuracy	N_{KK}/N_{+1}	N_{22}/N_{+2}	N_{KK}/N_{+K}		

2.6 Post Classification

A post-classification change matrix function was applied between 1972 to 1984, 1990 to 1996 and 2002 to 2016.

3. Results And Discussions

3.1 Image Interpretation of Coral Reef

Result of satellite image interpretation of coral reefs indicate six classes which were dead coral, live coral, seagrass, mix bottom, rubble and sand. Geometric correction process was done to obtain good accuracy image position of each class in coral reef and the real world. In this study, geometric correction process was performed using 7 point GCP. The result of geometric correction shows that the average RMSE obtained from Landsat images was 0.02. Our results show that for this shallow water, coral reef change can be seen by comparing the coral reef maps at different time series. The result of coral reef classification maps were checked with accuracy assessment method in order to obtain the level of map accuracy. As mentioned previously, standard map accuracy was more than 70%. It can be said that these map has been classified by representing good coral reef and ground area, which were then transformed into digital paper maps. confusion matrix of classification on image 2015 in Ballang Lompois 75.69%. Generally, the difference between the classification results and reality in the field is caused by water clarity and spatial resolution of satellite-derived data.

Time-series analysis of MSS, TM, ETM and OLI-TIRS data has revealed a net loss of 54.26 ha of live coral in the region from 1972 to 2016. The results of changes in the coral reef areas during the six acquisition years are represented in Table 3. During the periods of 1972 to 1984, the area of live coral decreased at 8.49 ha. Periods of 1990 to 1996 area of live coral decreased at 8.28 ha and followed in during periods of 2002 to 2016, the area of live coral decreased at 10.40 ha. The area of dead coral and rubble had increased. The resulst of coral reefs dynamics showed that live coral decrease in each year and probably this condition will continue in future year.

Table 3. Areal estimates of coral reefs in 1972, 1984, 1990, 1996, 2002,and 2016

Classes	Areal of BallangLompo Island (Ha)					
	1972	**1984**	**1990**	**1996**	**2002**	**2016**
Dead Coral	0,00	0,00	0,63	1,62	5,67	6,93
Live Coral	67,16	58,67	43,29	35,01	23,29	12,89
Seagrass	10,70	8,86	7,29	6,30	5,56	4,75
Mix Bottom	4,43	9,96	22,68	23,67	25,48	28,65
Rubble	5,17	8,12	9,54	15,57	21,77	28,45
Sand	16,97	18,82	18,45	19,71	19,64	19,73

Based on observations in the field, the coral reef habitat destruction in Ballang Lompo island largely from human activity.It proofed by many fragments of rock (rubble), andthe activities by explosives fishing such as bombs and chemicalsin shallow waters Ballang Lompo [2].

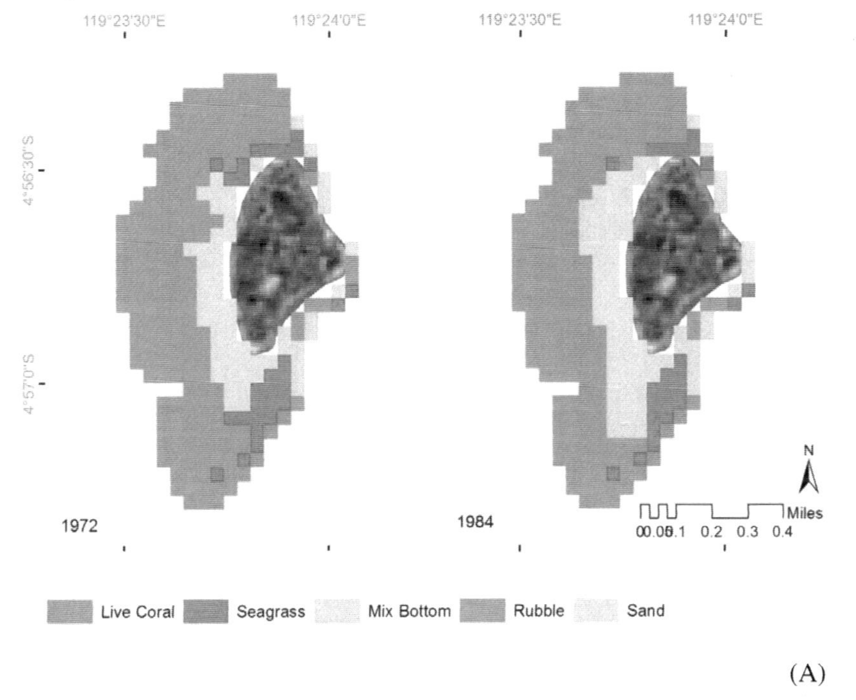

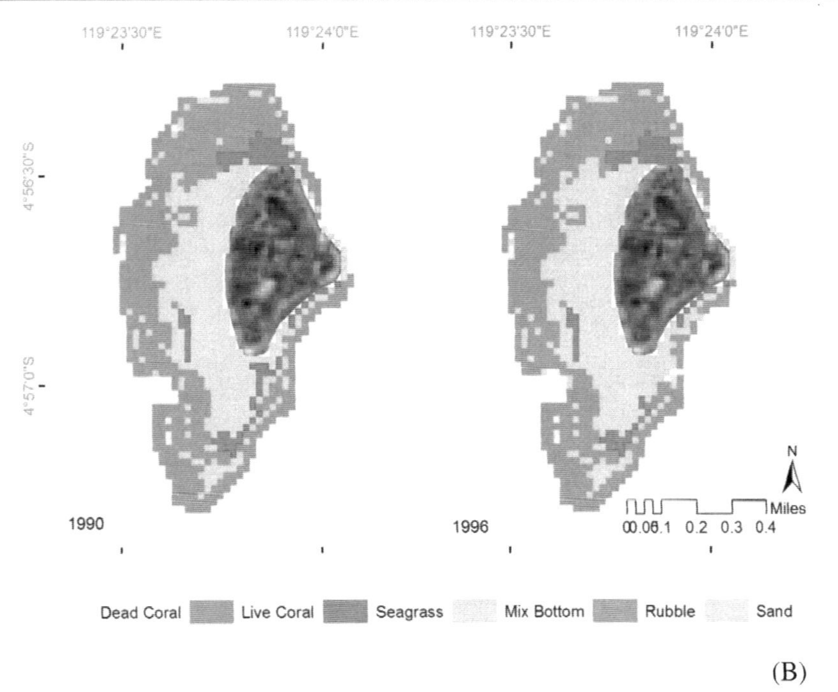

IOP Conf. Series: Earth and Environmental Science **47** (2016) 012011 doi:10.1088/1755-1315/47/1/012011

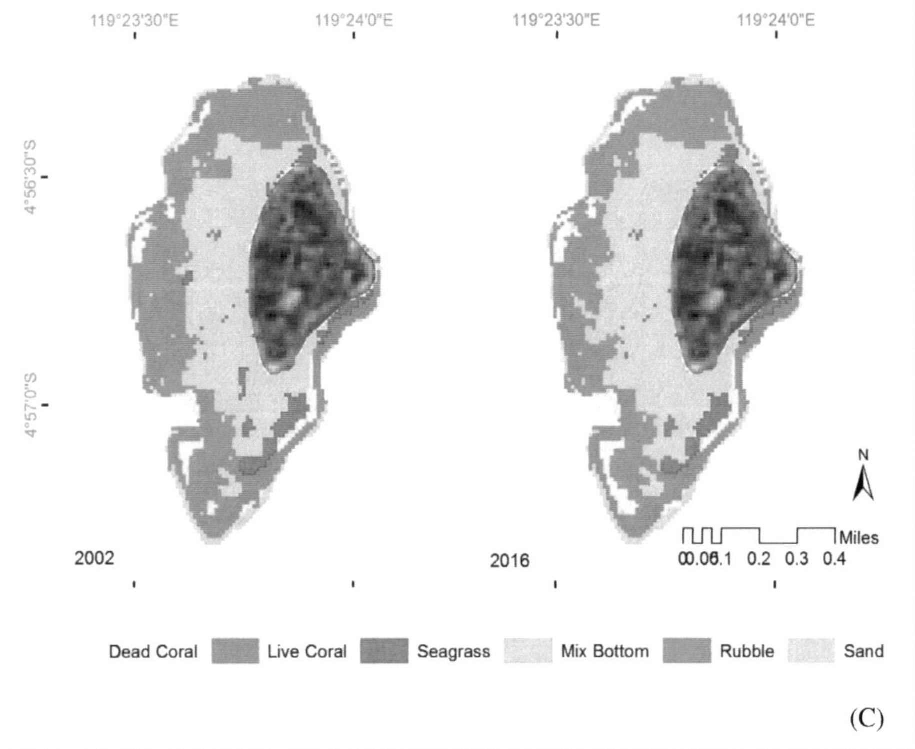

Figure 2. Coral reef classification maps: (A) 1972 and 1984 (Spatial resolution 60 meter), (B) 1990 and 1996 (spatial resolution 30 meter), (C) 2002 and 2016 (spatial resolution 15 meter).

3.2 Post Classification

The change matrix (Table 4) shows that 1972 to 1984 by 5.53 ha of live coral changed between to mix bottom, and 2.95 ha had changed to rubble. Deforestation rate during 2002 to 2016 was highest for live coral to rubble in Ballang Lompo Island by 7.06 ha. It could be observed that the changes from 1990 to 1996 show live coral and seagrass area decreased approximately 8.26 ha and 0.99 ha, while dead coral increased 0.99 ha, rubble increased 6.03 ha, and mix bottom increased 5.53 ha. Although the extent of shallow water may change from year to year due to varying human activity and oceanography parameters.

Table 4. Areal changes ofbottom types in 1972 to 1984, 1990 to 1996 and 2002 to 2016

Years	Areal change of coral (ha)					
	Live Coral	**Dead Coral**	**Rubble**	**Mix Bottom**	**Seagrass**	**Sand**
1972 to 1984		(C)				
Live Coral	58,67	0,00	2,95	5,53	0,00	0,00
Dead Coral	0	0,00	0	0	0	0
Rubble	0	0,00	5,17	0	0	0,00
Mix Bottom	0	0	0	4,43	0,00	0,00
Seagrass	0	0,00	0	0,00	8,86	1,84
Sand	0	0	0	0		16,97
1990 to 1996						
Live Coral	35,01	0,99	6,03	1,26	0	0

Years	Areal change of coral (ha)					
	Live Coral	**Dead Coral**	**Rubble**	**Mix Bottom**	**Seagrass**	**Sand**
Dead Coral	0	0,63	0	0	0	0
Rubble	0	0	9,54	0	0	0
Mix Bottom	0	0	0	22,41	0	0,27
Seagrass	0	0	0	0	6,30	0,99
Sand	0	0	0	0		18,45
2002 to 2016						
Live Coral	12,89	1,26	7,06	2,07	0	0
Dead Coral	0	5,67	0	0	0	0
Rubble	0	0	21,39	0,38	0	0
Mix Bottom	0	0	0	25,37	0	0,11
Seagrass	0	0	0	0,16	4,75	0,65
Sand	0	0	0	0,67		18,97

3.3 Coral Reefs Destruction

Coral reef damage are classified into three parts, specifically: (1) coral damage by biologial causes such as the existence of competition, predation, explosion of phytoplankton population, (2) coral damage by mechanism causes such as the existence of strong current, sedimentation, volcanism activities, change in temperature and salinity and sunlight penetration, (3) coral damage to human activities such as oil pollution, chemicals, taking the stony coral for industrial purposes and building, bombing, a collection of biota and etc [13]. Destructive fishing practices not only remove the resource itself (fish and invertebrate stocks) but also destroy the habitat [7,8, 9, 10, 11]. The detonated bomb's shock waves not only kill fish and other organisms within the 1-5 meter blast radius, but also pulverize the skeletons themsels [6]. There are a lot of damage of coral reefs as a result of people's behaviour in coastal areas of small islands Spermonde Archipelago; include Ballang Lompo island in inner zone. They use bombs and chemicals toxic to catch reef fish. In 2016 showed that live coral is decrease, as well as live coral while the dead coral, rubble, mix bottom and sand. Local people using corals and sand from coastal areas to be used as a material for building their homes. But the greatest threat to coral reefs on the island Ballang Lompo is damage caused by the activity of people who come from other islands around the Ballang Lompo island. Currently, the Society of Ballang Lompo island strive to prevent any activity of coral destruction that came from fisherman of another island in Spermonde Archipelago.Blast fishing is considered to be one of the most destructive anthropogenic threats to coral reefsbecause of its pernicious effects. Despite national and local government policy to control these illegal methods, destructive fishing is still widely practised by local fishermen. The most reason why these practices are difficult to eradicate is weak government control and monitoring. This a serious problemand while many blast fishermen are aware of the destruction their practices entail, they argue that they have no real alternatives [7, 8, 9, 10].

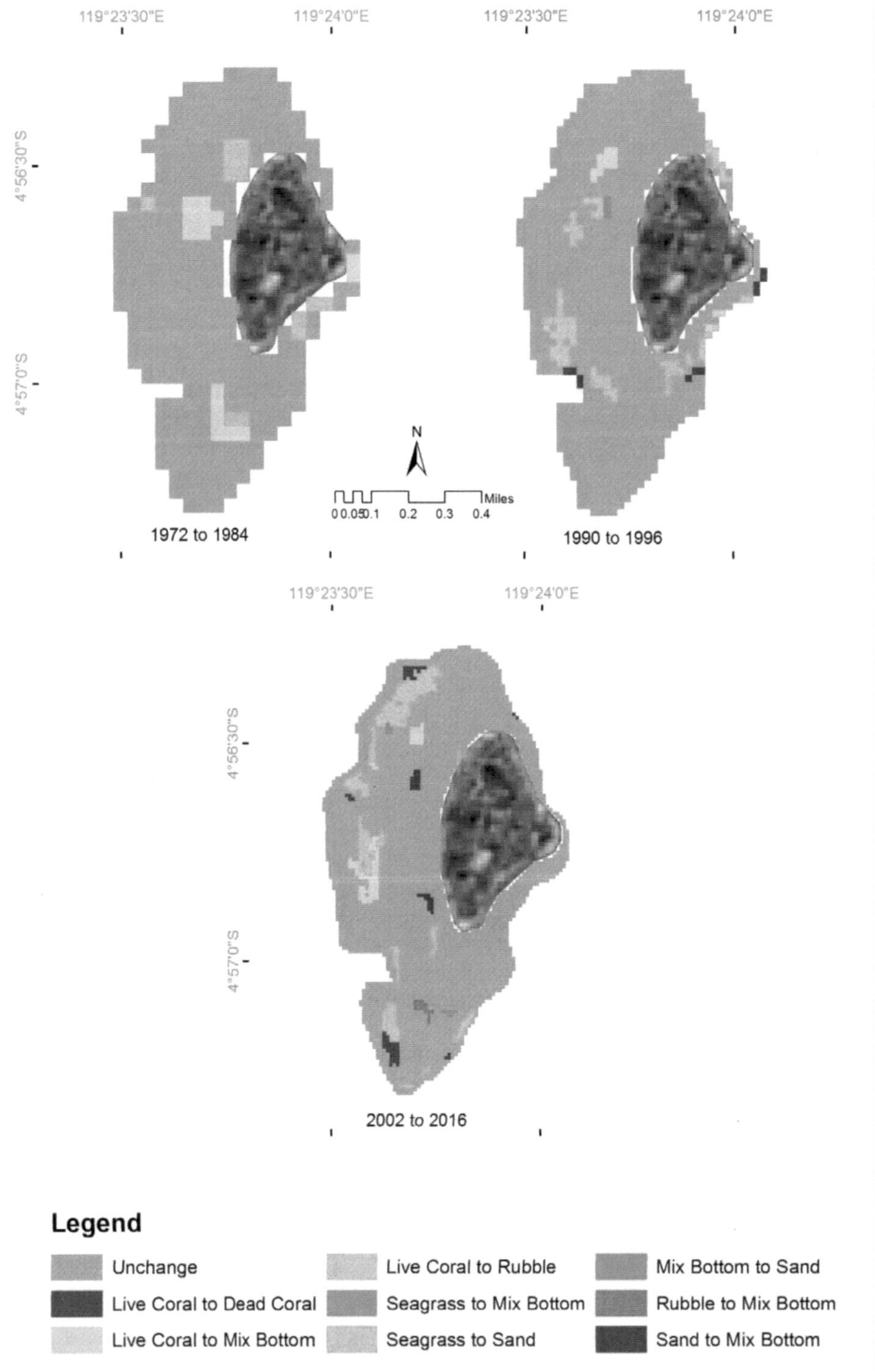

Figure 3.Post classification of coral reef of Ballang Lompo Island: from 1972 to 1984, 1990 to 1996 and 2002 to 2016

4. Conclusion

The results indicate that decreases in live coral areas resulted in increased areas of dead coral with algae and rubble during the forty fouryears live coral decreasing resulted in period from 1972 to 2016. Live coral from 1972 to 2016 was decreased, while dead coral and rubble were increased. Field observationson the shallow water of Ballang Lompo island revealed that coral reef habitat destruction was largely caused by human factors. This could be concluded because a lot of loose fragments of rubble that typically result from destructive activities such as explosive fishing (bombs) were found.

Acknowledgments

We would like to thank toResearch Center of Hasanuddin University, Indonesia, for funding sponsor. We gratefully acknowledge Regional Development and Spatial Information, Hasanuddin University for the field survey team support and some of images supported from the Ocean Remote Sensing Project for Coastal Habitat Mapping (WESTPAC-ORSP).

References

[1] B B Barnes, C Hu, K L Holekamp, S Blonski, B A Spiering, D Palandro, and B Lapointe 2014Use of Landsat data to track historical water quality changes in Florida Keys marine environments*Remote Sensing of Environment***140** 485–496

[2] Coral Reef Rehabilitation and Management Project [COREMAP]II 2010*Final Report: data base of Coral Reef Status in South Sulawesi*(Makassar (ID) Coremap II dan Dinas Kelautan dan Perikanan Provinsi Sulawesi Selatan)

[3] D Torres-Pulliza J R Wilson, ADarmawan, S J Campbell, and S Andréfouë 2013Ecoregional scale seagrass mapping: A tool to support resilient MPA network design in the Coral Triangle*Ocean & Coastal Management***80** 55e64.

[4] Grimsditch GD and Salm RV 2006 Coral Reef Resilience and Resistance to Bleaching(IUCN, Gland, Switzerland) 52 p.

[5] Liu JG and Mason PJ 2009*Essential Image Processing and GIS for Remote Sensing*(United Kingdom (GB): Wiley-Blackwell Publication)

[6] McManus JW, Reyes Jr RB, and Nanola Jr CL 1997 Effects of some destructive fishing methods on coral cover and potential rates of recovery*Environmental Management***21**:69-78.

[7] Nurdin N, Komatsu T, Agus, Akbar ASM, Djalil AR, and Amri K 2015 Multisensor and multitemporal data from Landsat images to detect damage to coral reefs, small island in the Spermonde Archipelago, Indonesia *Ocean Science Journal***50** (2) pp.317.

[8] Nurdin N, Amri K, Djalil AR, Jaya I, Aris A, and Akbar ASM 2014 Geospatial dynamic of seagrass in outer zone, Spermonde Archipelago, Indonesia using Landsat data from 1972-2013*Proceeding of SPIE-The International Society for Optical Engineering***9261**.

[9] Sawayama S, Nurdin N, Akbar ASM, Sakamoto SX, and Komatsu T 2015 Introduction of geospatial perspective to the ecology of fish-habitat relationships in Indonesia coral reffs: A remote sensing approach*Ocean Science Journal***50**(2) 22 june 2015pp 343-352

[10] Nurdin N, Komatsu T, Sawayama S, Amri K, Djalil A R, Patittingi F, Jaya I, Akbar ASM, and Aris A 2015 Coral reefs spatial transformation versus anthropogenic on small islands, outer zone spermonde archipelago *Proceedings*of *ACRS 2015-36th Asian Conference on Remote Sensing: Fostering Resilient Growth in Asia 2015*

[11] Pauly D, Silvestre G, and Smith IR 1989 On development, fisheries and dynamite: a brief review of tropical fisheries management*Natural Resources Modeling***3** 307-329

[12] Salm RV 2002 Building Survivability into Marine Protected Area Networks *The Nature Conservancy*

[13] Suharsono 1990Ecological and Physiological Implication of Coral Bleaching at Pari Island. Thousand Island Indonesia A Desertation(The University of New Castle Upon Tyne. Department of Ecological) 279 p.

IOP Conf. Series: Earth and Environmental Science **47** (2016) 012012 doi:10.1088/1755-1315/47/1/012012

Overview of National Thematic Data Integration (An Experience on One Map Mangrove Sulawesi)

A W Rudiastuti[1*], D M Yuwono[1], Niendyawati[1], G H Pramono[1], and B D Rahmanto[2]

[1]Geospatial Information Agency – Indonesia
[2]Ministry of Environment and Forestry Affair – Indonesia

*email: anindarudiastuti@gmail.com

Abstract. Playing role as coastal shield with enormous economic value and ecological functions, mangrove forest management is always challenging to be studied. As either the largest archipelagic countryor the largest mangrove forest habitat around the globe, Indonesia needs a national mangrove forest baseline data and its updating for coastal management. Many stakeholders and institutions, including Geospatial Information Agency (BIG), had conducted mangrove mapping and updating. However, in order to achieve one mangrove national data, coordination and synergy among stakeholders and institutions such as: the Ministry of Environment and Forestry as mangrove custodian, Indonesian National Institute of Aeronautics and Space, Ministry of Marine and Fisheries, and BIG aligned with the National Mangrove Working Group is needed. A fundamental step for national mangrove forest management is the establishment of National One Map Mangrove Program by means of coordination, synchronization, and integration of mangrove geospatial data from various stakeholders. This paper will discuss the technical process of data integration and field survey in order to produce One Map Mangrove Sulawesi with the same geo-reference, database, and also standard and specification. The result of One Map Mangrove Sulawesi Program comprises of information about mangrove current status, existing area, and its distribution in Sulawesi.Beside the geospatial data from Ministry of Environment and Forestry and other institutions, the primary data used to map mangrove forest in Sulawesi is SPOT 6 and SPOT 7(year 2014 – 2015) imageries yielded map scale of 1: 25,000. On screen digitation using NIR, Red and Green bands and Normalized Difference Vegetation Index (NDVI)image transformation are applied for the initial canopy density classification. Field survey was doneto obtain field data forvegetation analysis, image classification andre-interpretation. In 2015, the process of producing One Map Mangrove Sulawesi has been carried out. It was preceded by compiling geospatial data from BIG and other stakeholders. The integration process is based on national technical guidelines for mangrove mapping.Ground check also done by the One Map Mangrove team to ensure theaccuracy and novelty of data. The result shows that SulawesiTenggara has the largest mangrove forest area, it is 62,446.4 ha, with a crown density dominance in rare class. This paper illustrating the process of One Map Mangrove Sulawesi, where data from the local stakeholders are accommodated properly.

1. Backgrounds

Found almost at any type of coastal line and providing a variety of ecological and economic functions, mangrove stands is an interesting subject to be studied. Mangrove also play role on supporting the economy of local communities living in coastal area. Mangrove contribute to environment protection as well as local economic value such as (1) coastal protection from disasters such as tsunami, wind and wave damage[1,2]; (2) a place to live for a variety of aquatic fauna, especially because of its function as spawning area, nursery, living habitat[3,4] and nutrient sources for organisms, and (3) tourism. Provision of seafood [5] as well as timber are those of economic value for the coastal communities[6]. Mangrove-based tourism destination began a lot to be encouraged, such as mangrove tourist areas in North Jakarta or Bali[7]. Mangrove's existence is very important and several of its aspects has been studied and reviewed, yet these have received much less publicity[8].

Polidoro et al. [9] revealed that the highest diversity of mangrove existed in the Indonesia - Malaysia - Philippines areas, which is about 50- 60% of all types of mangrove species in the world are found in that region. However, this region also suffered high level of mangrove degradation since the 1980s - 2000s, particularly in Indonesia, with an estimated 25-30% [10,11]. In line with the FAO[12] which reveals that the rate of degradation of mangrove forests Asian region in the period 2005-2010 was 0.56%. Mangrove utilization as a source of timber and land cover change becoming the major cause of degradation[13]. Conversion of mangrove area into shrimp ponds, coastal tourism site, and also the settlements are the most[14]. It has struck the northern coast of Java in the 1990s where land conversion of mangrove often happens for shrimp farming and salt ponds. Other examples of world concern is the declining quality of the lagoon ecosystem Segara Anakan in Cilacap[15].

On 2009, Geospatial Information Agency (BIG) estimated the area of mangrove in Indonesia is to 3.2 million hectares [10]. The tendency to decline every year, where for 25 years (1980-2005) Indonesia has lost 30.1% or 1.3 million hectares of mangrove[11]. Those have made the importance of mangrove monitoring, both in terms of distribution, size, diversity and the current status of mangrove forest itself. In the era of geospatial information, the need for monitoring mangrove can be answered with a mangrove mapping technology that combines remote sensing and field surveys[16]. Remotely sensed data is one of the best source that can show the condition of mangrove forest, with many advantages such as synoptic coverage, provide information for inaccessible area, historical and near real time data [17–19].

Prior to the enactment of Act No. 4 of 2011 of Geospatial Information, various mangrove stakeholders in Indonesia issued the information of mangrove forest extent, causing on data discrepancies. This is describing on the difficulties on calculate the mangrove forests area accurately and could be agreed by all stakeholders. Referring to the national regulations on one spatial data, the entire mangrove stakeholders involved in KKMN initiate the process of uniting the mangrove thematic geospatial information. Both of central and local governments were also involved. Ministries or government agencies involved to perform these activities are Geospatial Information Agency (BIG), Ministry of Environment and Forestry, Ministry of Maritime Affairs and Fisheries. Also private stakeholders are involved, such as Wetlands International Indonesia and the Center for International Forestry Research (CIFOR).

The main purpose of this study is to describe the process of mangrove mapping and integration resulted in One Map Mangrove Sulawesi. Mapping method used in this paper is following the national regulation: Head of Indonesia Geospatial Information Agency Regulation No. 3 2014. This study used mid-resolution imagery and Normalized Difference Vegetation Index (NDVI) for mangrove canopy density interpretation. NDVI and simple image ratio between NIR and Red wavelengths can also be used to extract vegetation parameter and condition of the mangrove ecosystem[20]. There are several studies related to the delineation of mangrove area by using different imagery in Sulawesi[21–24]. Sulawesi is selected as one of islands that enriched by mangrove forest which is needed to be defined either per region or entirely. Furthermore, some mangrove areas in Sulawesi are belong to protected area, and many of mangroves areas has been threatened by human[13], so that it needs to be monitored. In addition, One Map Mangrove Sulawesi is a significant way to support land management, spatial planning, rehabilitation, and monitoring of coastal area especially on mangrove forests existence.

2. Methodology

This study focused on the two processes of mangrove data integration. The main process of this study is mangrove mapping using remote sensing technology assisted with ground truth or mangrove validation. Second, data integration comprised of geospatial data compilation and synchronization. Geospatial mangrove data were collected from stakeholders, mainly from Ministry of Environment and Forestry and Geospatial Information Agency. Image interpretation method occupied in this study were varied depend on the image datasets condition. In general, there are two major image interpretation methods: visual and digital interpretation. Brief and clear workflow of the mangrove mapping in Sulawesi was depicted in **Figure 1**.

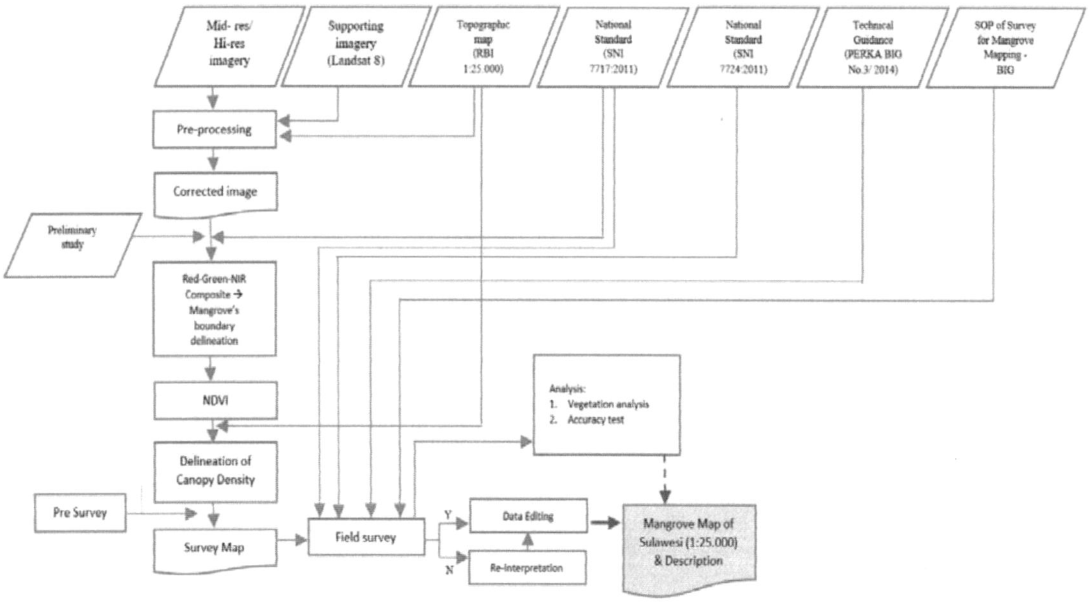

Figure 1. Flowchart of Mangrove Mapping

2.1. Study Area

Crossing the equatorial line, Sulawesi extends between the Wallace and Weber line that made it has several endemic flora and fauna. Administratively, Sulawesi divided into 6 provinces: Sulawesi Utara, Sulawesi Tengah, Gorontalo, Sulawesi Barat, Sulawesi Selatan and Sulawesi Tenggara (Figure 2). BIG has completed mangrove mapping process from 2014 to 2015.

Figure 2 shows the study area of mangrove mapping 2014 – 2015, scale of 1:25.000. Tan-yellow box revealed index where mapping has been done. It was based on the small-scale mangrove mapping for whole Indonesia in 2009[10]. The first phase of mid-scale mapping in 2014, it was completed northern part of Sulawesi (Sulawesi Selatan, Gorontalo, and Sulawesi Tengah province), while the rest were done in 2015.

2.2Materials and Methods

Base map used for image geo-rectification was Indonesia topographic map scale of 1:25.000. Satellite images used in this study are (Table 1):

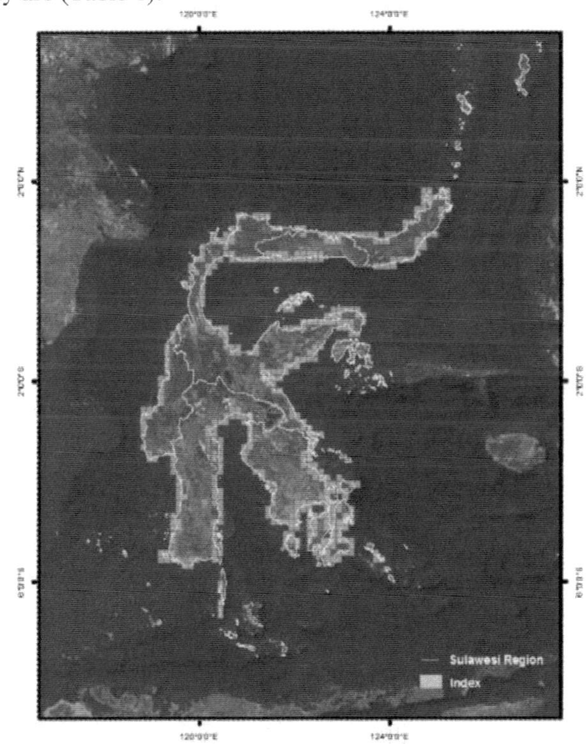

Figure 2. Study area

Table 1. Imagery for Mangrove Mapping

Year	Images	Acquisition date	Spatial Resolution	Analysis
2014	SPOT 5/6	October –	10 m; 5 m	Visual interpretation
	RGB Composite	December 2013	6 m; 1.5 m	
	ALOS	2011 - 2013	10 m	NDVI
	RGB Composite			
	Landsat 8	June 2014	30 m; 15 m	NDVI
2015	SPOT 5	November 2013	6 m; 1.5 m	Visual interpretation
	SPOT 6/7	Early 2015		& NDVI
	ALOS	2011 – 2013	10 m	Visual interpretation
	RGB Composite			
	Landsat 8	August 2015	30 m; 15 m	NDVI

2.3Image interpretation

Various data were used to complete the task (Table 1). Satellite imageries used based respectively on priority are: ALOS, SPOT 5/6/7, and Landsat 8. They are considered based on acquisition date, spatial resolution, cloud coverage, condition, and the availability of each image. This step was done to maintain the output scale of 1:25.000. However, from all the data, there were some drawbacks found, so that each type of image is used to cover up the shortcomings from another image.

Combination of Green – Red – NIR bands were used for on-screen digitation to determine the polygon of mangrove areas, carried out by referring to the result of preliminary study. In addition, delineation of mangrove extent and canopy density classes takes the advantages of hi-res imagery such as: SPOT 6/7. Using enhanced true and false color composite, canopy density classes can be easily differentiated. On the other hand, in order to classify canopy density using medium resolution imageries such as: Landsat and ALOS, digital interpretation using NDVI transformation (Eq. 1) were implemented[25].

Classification of canopy density referred to the national standard for Mangrove Surveying and Mapping (SNI 7717: 2011) scale of 1:25.000. The canopy density classes would be: high density, dense, medium, rare, and very rare. The result from image interpretation process becomes the main information to organize field survey, combined with the previous information from local people collected in pre-survey.

NDVI image transformation:

$$NDVI = \frac{Band\ (NIR) - Band\ (R)}{Band\ (NIR) + Band\ (R)} \quad (1)$$

Where:

- NIR : Near Infrared
- R : (visible) Red
- Range of NDVI : -1 to 1

2.4Field survey

The field survey was conducted for data retrieval in the sample points/ polygons that have been determined. To-do-list in field survey consists of first; validation form of existing mangrove. It is needed to verify the results of image interpretation especially in not-so-good image quality due to cloud existence or other disturbance. Second, inventory of mangrove stand parameters to determine: canopy density, density of trees, value index and mangrove species (dominancy). Third, fish eye lens were used to get 180° view of canopy. Lastly, secondary data collection to complete the information of mangrove areas.

Each sampling plot size of 10 x 10 m^2 were carried out along the coastal line (Figure 1), distributed up to 2 PS in each index box, resulted in 615 PS for total 438 mapping index at Sulawesi (Table 2). The number of PS was determined by considering the mangrove extent and canopy density classification, occupying stratified random sampling method. For canopy density and species dominancy also acquired in TS. It also used to test the accuracy mapping result.

Table 2. Mapping index and number of samples

Year	Σ Index	Areas/ Province	Number of PS	Number of TS
2014	209 NLP	Sulawesi Selatan, Gorontalo, Sulawesi Tengah	315	210
2015	229 NLP	Sulawesi Barat, Sulawesi Selatan, Sulawesi Tenggara	300	200

Two sampling types were held in every index, Plot Sample (PS) and Point Sample (TS). Inside the Plot Sample box, data inventory for trees (10 x 10 m), saplings (5 x 5 m), and seedlings (1 x 1m) were conducted while land use/ land cover was observed in TS sampling type. Illustration of sampling types design was presented on Figure 3 below:

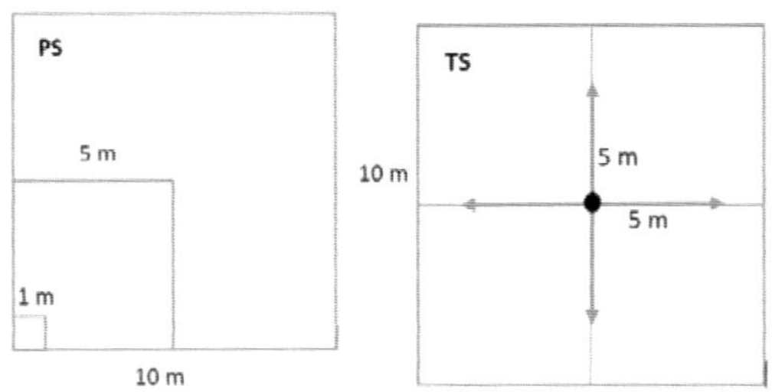

Figure 3. Plot Sample (left) and Point Sample (right)

2.5 Accuracy assessment

Image interpretation process resulted in two kind of maps. Firstly, land cover map which shows land cover class around mangrove forest area. This map explained that there are always other land cover existed surrounding mangrove forest area. In addition, interpreting mangrove will not always result in high classification accuracy in mangrove class, there are miss interpretations caused by the old image acquisition dates which is not relevant with the actual land cover condition. Second, canopy density map that shows classification of mangrove canopy density. After land cover map near mangrove forest area had been yielded, mangrove canopy density map was produced.Accuracy assessment of image interpretation was done using matrix accuracy developed by Short in 1982. Based on it, the overall accuracy can be calculated using a simple formula as follows:

$$Overall\ Accuracy = \left(\left(\textstyle\sum_{i=1}^{r} X_{ii}\right) / N \right) x\ 100\% \quad \textbf{(2)}$$

Where:

$$X_{ii} = diagonal\ matrix$$

$$N \ = sample$$

2.6 One Map Preparation

It started with many mangrove mapping activities organized by institutions, one map is aimed to integrate mangrove maps so that there is a single dataset that can be used by all institutions in mangrove resource management[26].First step on one map preparation started with data compilation from various institutions in a National Working Group of Mangrove (KKMN). Data that use to produce one map mangrove are form Geospatial Information Agency (BIG), Ministry of Environment and Forestry Affairs, Ministry of Marine and Fisheries Affairs, and other institutions such as Wetlands.

Determining standard and specification becoming the next phase to achieve One Map. Referred to standard Act 4/2011 about Geospatial Information, where thematic geospatial information must refer to base geospatial information, base geospatial information are vertical and horizontal geodetic control, bathymetric maps and topographic maps. Standard DGN-95/WGS- 84 also implemented. Standard DGN-95 is single reference in organize geospatial information in Indonesia, this standard used ellipsoid World Geodetic System 1984 (WGS-84).

Continued with integration spatial and attribute data, where these must be synchronized in order to join.Data integration stage is the stage of aggregation of data multiple data sources to produce a similar information of completeness, accuracy, scale and format data. The data which has integrated later verified. After joining the data, maps compiled become temporary one map for verification and ground check. Areas that reveal doubts, uncertainties and possibility of land changing into one of the groundcheck target, still local institutions were involved. Steps of one map mangrove depicted in Figure4.

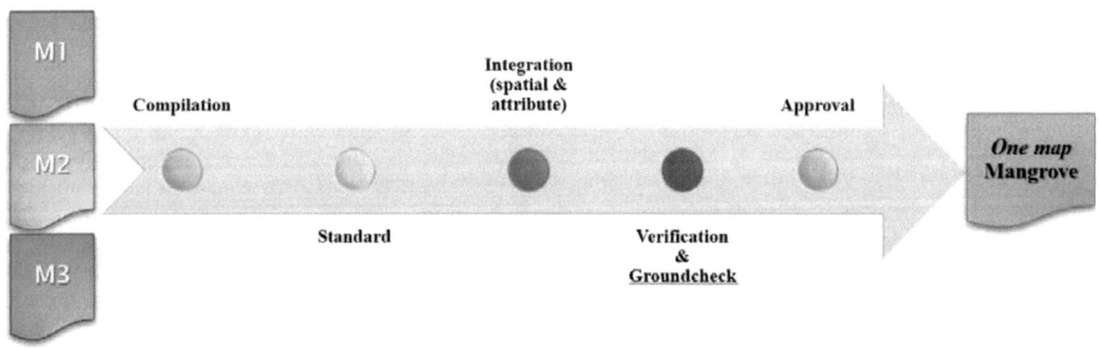

Figure 4. Steps on One Map Preparation

3. Result and Discussion

3.1 Mangrove map of Sulawesi

Unlike mangrove forest ecosystem in Java Island, mangrove forest ecosystem in Sulawesi is featured by narrow habitat area. In addition, geomorphologic condition of river and coastal area in Sulawesi support mangrove grow even denser than in Java Island. However, the important thing influencing mangrove forest condition in Sulawesi is population and development growth which is not as faster as Java. However, indication of mangrove degradation and deforestation in Sulawesi caused by population and local development growth also caught many researchers' attentions. One thing characterizes mangrove Sulawesi is there are many of mangrove forest grows well in small islands.

Figure 9 (a) showed the mapping results throughout the years 2014 - 2015, mangrove's extents sequentially to the six provinces are10,182.38 ha in Sulawesi Utara, 26,235.67 ha in Sulawesi Tengah, 7,918.58 ha in Gorontalo, 3,717.72 ha in Sulawesi Barat, 10,403.70ha in Sulawesi Selatan, and which has the largest area that is 62,426.42 ha in Sulawesi Tenggara (Table 4). Digital image interpretation using NDVI for obtaining themangrove canopy density classification were carried out, mostly in 2014 using Landsat 8, and for 2015 using SPOT 6/7 and Landsat 8. Both results were crosschecked with the real condition at field. The rare canopy density class was dominantly appeared in Sulawesi. Sequentially it was dominated in Sulawesi Tenggara (25,990.77 ha), Sulawesi Utara (6,065.39 ha), and Sulawesi Selatan (3,063.30 ha). Medium canopy density class (4,661.16 ha) dominated in Gorontalo and Sulawesi Tengah (10,416.87 ha), while dense canopy density dominated in Sulawesi Barat (1,786.06 ha). In general, *Rhizophora* and *Bruguiera* dominated almost throughout mangrove areas of Sulawesi.Overall accuracy both for land cover near mangrove forest area and mangrove canopy density are shown in Table 3. Overall accuracy for both results are above 85%.

Table 3. Overall accuracy (%) for land cover and canopy density interpretation

	Land cover	Canopy Density
Sulawesi Utara	88-95,83%	86-88,07%
Sulawesi Tengah		
Gorontalo		
Sulawesi Barat	98,81%	86,93%
Sulawesi Selatan		
Sulawesi Tenggara		

The overall accuracy both for coverage area and to NDVI interpretation are above 85%. Several reasons based on the steps on mapping process could probably be the reason.

1. Variations of imagery. Interpretation of land cover in 2014-2015 used the medium-to-high resolution imagery, so the appearance is quite clear to distinguish mangrove and non-mangrove. The average image used for the delineation of the boundaries (or class mangrove NDVI) has fairly good resolution as SPOT 6/7, SPOT 5 or ALOS, and the lowest was Landsat 8 (as complementary), so that the delineation of the boundary can be corrected through visual interpretation from the image.
2. The imagery used was relatively new. As seen in Table 2 that averagely the data were obtained in less than one year from the time of mapping process. In particular Landsat 8 acquisition date was the closest.
3. The plot sample (PS) position on field survey. The accuracy of canopy density could be affected by sampling. As has been done, the determination of the sample taken at the middle of the polygons interpreted as mangrove where the NDVI values tend to be homogeneous.
4. Delineation process shall refer to a lot of the imagery available. The process is carried back and forth between the main image and the supporting image. For the NDVI, the result was crosschecked with the green band.

3.2 One Map Mangrove Sulawesi

One Map is needed due to problems found in mangrove maps from institutions. Remotely sensed data as the common data for mangrove mapping have many resolutions. From mid- resolution imagery such as Landsat 8, which is adequate to give information of mangrove area and also canopy density[27], to high resolution imagery from SPOT 6 or SPOT 7[28,29]. Therefore, map with various scale were produced depending on the imagery. Difference georeferenced system also evokes data discrepancies. Last but not least, area coverage in mapping of mangrove. Figure 5 displays the problems found at the compilation stage of One Map Mangrove Sulawesi.In the first step, Mangrove map from BIG were compiled with other mangrove map from Ministry of Environment and Forestry Affairs, Ministry of Marine and Fisheries Affairs, and BPDAS as representation of local institution. Compiled data were crosscheck with the information from local knowledge. Not only the spatial aspects but also attribute data must be synchronized in order to enable join process. As it seen at Figure 5(b), where attribute tables from different institutions different fields.

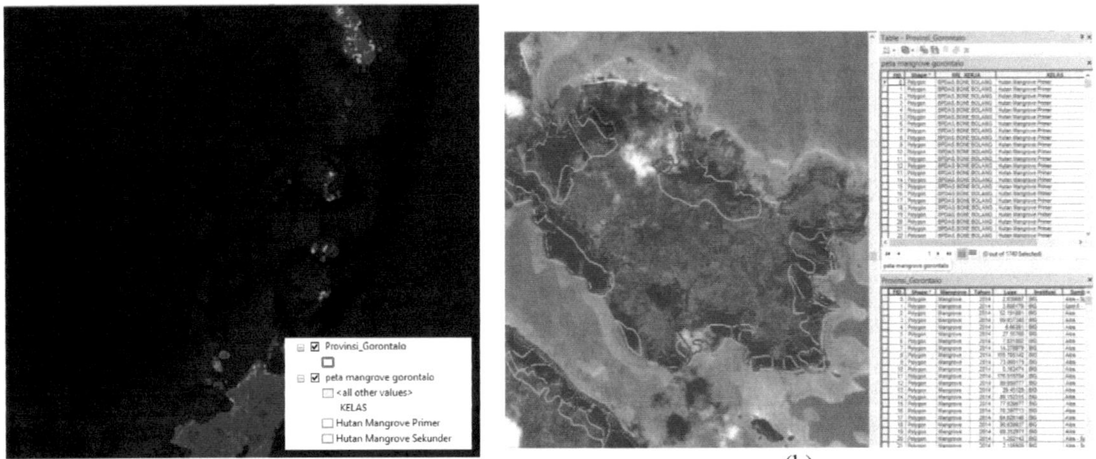

(a) (b)

Figure 5. Data discrepancies due to differences of (a) mapping area coverage and (b) imagery

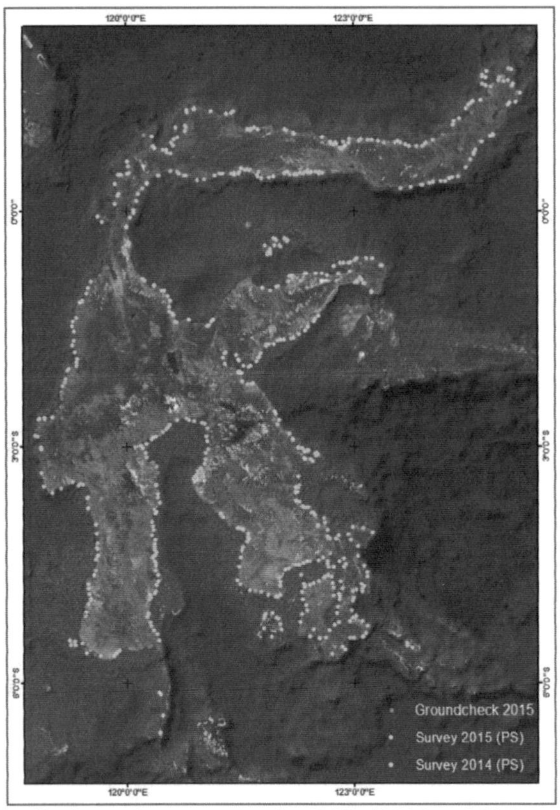

Figure 6. Distribution of survey plot by BIG (2014 – 2015) and ground check position by National Working Group of Mangrove (2015). Ground check for One Map Mangrove Sulawesi was done at October - November 2015, while field survey conducted by BIG was on October – November 2015 and September 2014

Areas with uncertainty of mangrove availability become a priority as ground check location. Further, areas where found only on one source of map need to validate, so that it was considered as ground check location. In spite of the time of mangrove mapping by BIG were organized in 2014 – 2015, survey did not cover all area and imagery that used as primary data for mapping were coming from various years between 2011- 2015. Therefore, some of them need to be validated. On One Map Mangrove Sulawesi, ground check location carried out in different position with survey plot conducted by BIG in 2014 - 2015, but several locations was nearby (Figure 6). All processes in composing One Map Mangrove Sulawesi were depicted in Figure 7.

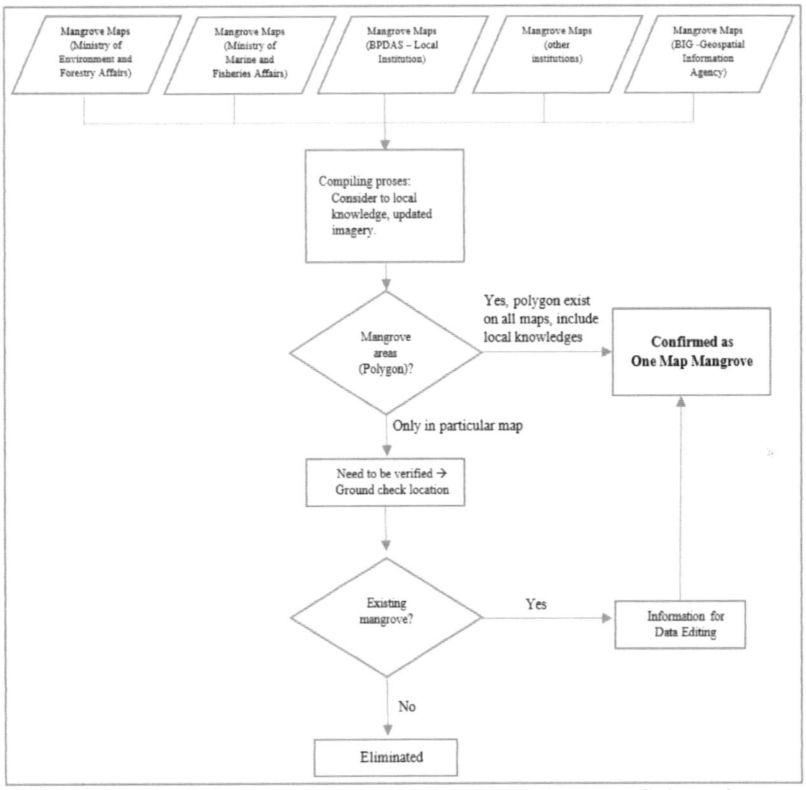

Figure. 7. Scheme on composing One Map Mangrove Sulawesi

Ground check were done not only to verify the data that have been collected, but also to check the delineation area as mangrove from the satellite image, since land conversion still becoming the main issues happened at mangrove areas. Groundcheck results diversify the condition of mangroves, particularly related community activities. In Buloa, Tallo– Sulawesi Selatan, the mangrove area has been slowly switching its function into garbage dump. Different from the area of Takalar - Jeneponto and Ujung Loe - Bulukumba, where mangrove exist as border from shoreline and river. At these locations, people have been aware of the importance of mangroves for coastal protection from abrasion. The presence of mangroves in addition to natural growth, also planted by civil society and through the rehabilitation program lead by Ministry of Environment and Forestry named Kebun Bibit Rakyat in 2014 (Figure 8).Information collected from ground check were used as guidance in data editing, as well as newest imagery available for such area.

IOP Conf. Series: Earth and Environmental Science **47** (2016) 012012 doi:10.1088/1755-1315/47/1/012012

Figure 8. Ground check in 2015, located in Sulawesi Selatan

BIG and other institutions have different area coverage on mapping, therefore One Map accommodate new information from other institution especially from local knowledge. Result in Figure9 showed the differences area coverage of mangrove mapping. Mangrove areas from small islands such as Banggai and Sangihe – Talaud Islands were added into One Map Mangrove. In mainland, information of mangrove areas around Donggala – Palu region were also added. Addition of mangrove areas on hectares from each province as result of one map process revealed in Table 4. It showed that mangrove map from BIG is accommodated as part of One Map Mangrove Sulawesi, and involvement of many stakeholders completed the information.

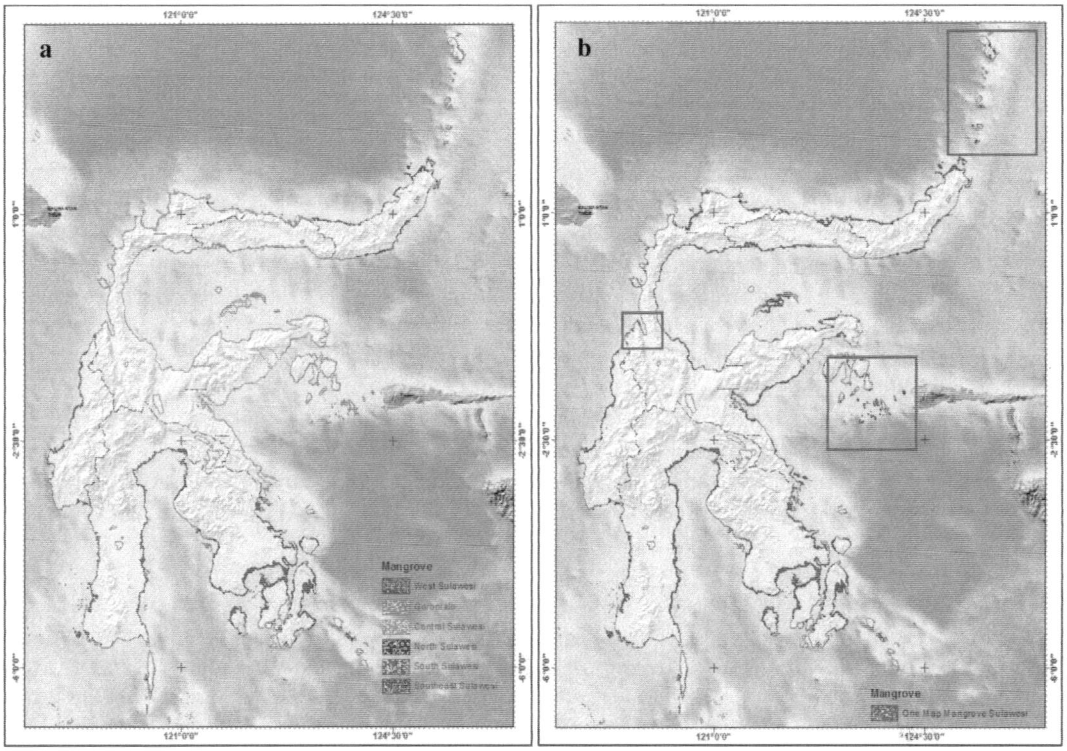

Figure 9. (a) Existing mangrove in Sulawesi (2014 – 2015); (b) Added mangrove areas (red boxes) as the result of One Map Mangrove Sulawesi

Table 4. Increasing of mangrove areas in Sulawesi as result of One Map Mangrove Sulawesi

	BIG (ha)	One Map (ha)
Sulawesi Utara	10,182.38	11,884.75
Sulawesi Tengah	26,235.67	33,538.95
Gorontalo	7,918.58	8,006.31
Sulawesi Barat	3,717.72	3,720.00
Sulawesi Selatan	10,403.70	10,412.00
Sulawesi Tenggara	62,426.42	62,446.83

4. Conclusions

In 2014 and 2015, BIG has completed mangrove mapping project at scale of 1:25.000. Field survey was carried out not only in the coastal of Sulawesi mainland, but also several small islands surround it. Using combination of ALOS – SPOT 5/6/7 – Landsat 8 imagery, mangrove extent was obtained for all parts of Sulawesi. Mangrove area in Sulawesi Barat, Gorontalo, Sulawesi Utara, Sulawesi Selatan, Sulawesi Tengah, and Sulawesi Tenggara yielded sequentially from the smallest to the largest between 3,717.72 and 62,426.42 ha. *Bruguiera sp.* and *Rhizophora sp.* becoming dominant species in Sulawesi. Rarecanopy density of mangrove dominated in Sulawesi Utara (59.57%), Sulawesi Tenggara (41.63%), and Sulawesi Selatan (29.44%). SulawesiBarat was dominated with dense canopy density area of mangrove (48.04%), while medium canopy densitydominated in Sulawesi Tengah (39.71%) and Gorontalo (58.86%).

The result of mangrove mapping 2014 - 2015 became the primary data on composing One Map Mangrove in 2015. In the process of composing, local and central institutions giving contribution on spatial data of mangrove from previous activity. Two steps of ground check were implemented for this study for validation. First, field survey was held for validation based on information from remote sensing data and interpreter, and the second was based on local knowledge. However, without prejudicing the advantages from the first field survey, ground check for One Map Mangrove are more reliable for yielding the updated information by involving local knowledge.

Mangrove mapping in Sulawesi which performed by BIG still relatively new, but through the process of One Map, by merging data and groundcheck, mapping results can be further refined, it can be seen from the addition of mangrove area in all provinces of Sulawesi around 2.28 – 7,303.28 ha. The largest one is for Central Sulawesi.Based on the survey results and groundcheck by KKMN team, found the tendency of mangrove land conversion into the pond, despite there also found the initiation of mangrove planting activities through local government and community.

One Map Mangrove resulted from collaborative works between governmental institutions (both central and local) and non-government organization. One map is an action to avoid ambiguity among users about the data. Technically to avoid redundancy and also increasing the contribution among stakeholders in composing one spatial information.

Acknowledgement
We thank to these following people and organization for giving any support in composing Mangrove Map of Sulawesi and also giving contribution on composing One Map Mangrove Sulawesi: Hardanto, Yayu SS, Siti Mumtazia, Aswin, Suseno Wangsit Wijaya, Rahmat Nugroho P., Ministry of Environment and Forestry Affairs, Ministry of Marine and Fisheries Affairs, LAPAN, Marine Mapping and Integration Division - BIG, Wetlands Intenational Indonesia, BPDAS Sampara, BPDAS Jeneberang – Walanae, BPDAS Saddang, BPDAS Lariang Mamasa, BPDAS Palu Poso, BPDAS Bone Bolango, BPDAS Tondano, and all members of National Working Group of Mangrove (KKMN).

References

[1] Danielsen F, Sørensen MK, Olwig MF, Selvam V, Parish F, Burgess ND, et al. 2005 The Asian Tsunami : A Protective Role for Coastal Vegetation *Science***310** 643. Available from: www.sciencemag.org/cgi/content/full/310/5748/643/ DC1

[2] Yanagisawa H, Koshimura S, Miyag T, Imamura F 2010 Tsunami damage reduction performance of a mangrove forest in Banda Aceh, Indonesia inferred from field data and a numerical model *J Geophys. Res. Ocean* **115**(6) 1–11.

[3] Kuenzer C, Bluemel A, Gebhardt S, Quoc TV, Dech S 2011 Remote sensing of mangrove ecosystems: A review *Remote Sens.***3**(5) 878–928.

[4] Nagelkerken I, Blaber SJM, Bouillon S, Green P, Haywood M, Kirton LG, et al. 2008 The habitat function of mangroves for terrestrial and marine fauna : A review *Aquatic Botany* **89**:155–85.

[5] Rönnbäck P 1999 The ecological basis for economic value of seafood production supported by mangrove ecosystems. *Ecol Econ***29**(2)235–52.

[6] Walters BB 2005 Ecological effects of small-scale cutting of Philippine mangrove forests. *For Ecol Manage***206**(1–3)331–48.

[7] Fitriana YR 2005 Diversity and abundance of macrozoobenthos in mangrove rehabilitation forest in Great Garden Forest Ngurah Rai Bali *Biodiversitas***7**(1)67–72.Available from: http://biodiversitas.mipa.uns.ac.id/D/D0701/D070117.pdf

[8] Valiela I, Bowen JL, York JK 2001 Mangrove Forests : One of the World ' s Threatened Major Tropical Environments *Bioscience* **51**(10) 807-15. Available from:http://bioscience.oxfordjournals.org/

[9] Polidoro BA, Carpenter KE, Collins L, Duke NC, Ellison AM, Ellison JC, et al 2010 The loss of species: Mangrove extinction risk and geographic areas of global concern. *PLoS One*5(4) e10095
[10] Hartini S, Saputro GB, Yulianto M, Suprajaka 2010 Assessing the used of remotely sensed data for mapping mangroves Indonesia*10th WSEAS/IASME Int. Conf. on Electric Power Systems, High Voltages, Electric Machines* p 210–5.
[11] FAO 2007 The world's mangroves 1980-2005. FAO *For Pap*;**153** 89.
[12] Giri C, Ochieng E, Tieszen LL, Zhu Z, Singh A, Loveland T, et al 2011 Status and distribution of mangrove forests of the world using earth observation satellite data. *Glob. Ecol. Biogeogr.* **20**(1)154–9.
[13] Nurkin B 1994. Degradation of mangrove forests in South Sulawesi, Indonesia. *Hydrobiologia***285**(1–3)271–6.
[14] Thu PM, Populus J 2007. Status and changes of mangrove forest in Mekong Delta: Case study in Tra Vinh, Vietnam. *Estuar. Coast. Shelf Sci.***71**(1–2)98–109.
[15] Jennerjahn TC, Yuwono E 2009. Segara Anakan, Java, Indonesia, a mangrove-fringed coastal lagoon affected by human activities. *Reg. Environ. Chang.* **9**(4)231–3.
[16] Satyanarayana B, Idris IF, Mohamad KA, Husain ML, Shazili NAM, Dahdouh-Guebas F 2010. Mangrove species distribution and abundance in relation to local environmental settings: A case-study at Tumpat, Kelantan Delta, east coast of peninsular Malaysia*Bot Mar***53**(1)79–88.
[17] Ibharim NA, Mustapha MA, Lihan T, Mazlan AG 2015 Mapping mangrove changes in the Matang Mangrove Forest using multi temporal satellite imageries. *Ocean Coast Manag.* **114**64–76.
[18] Long JB, Giri C 2011 Mapping the Philippines' mangrove forests using Landsat imagery. *Sensors***11**(3)2972–81.
[19] Agris RAALM, Arreto RAB 2010. Mapping and assessment of protection of mangrove habitats in Brazil *PANAMJAS* **5**(Ong 1995)546–56.
[20] Sari SP, Rosalina D. Mapping and monitoring of mangrove density changes on tin mining area. 2016;33:436–42.
[21] AS Akbar M, Saleh B, Nurdin N, Sofian I 2015Cellular automata-markov model of vegetation cover changes on a small island spermonde archipelago , *Indonesia 36th Asian Conf. on Remote Sensing 2015 (ACRS 2015): Fostering Resilient Growth in Asia* p 451- 64
[22] Laremba S 2014 Sebaran dan kerapatan mangrove di teluk Kota Kendari, Sulawesi Tenggara[Undergarduate theses of Hasanuddin University, Makasar]
[23] Jamili, Setiadi D, Qayim I, Guhardja E 2009 Struktur dan komposisi mangrove di Pulau Kaledupa Taman Nasional Wakatobi , Sulawesi Tenggara *Ilmu Kelautan* **14**(4)197–206.
[24] Arfan A, Toriman ME, Maru R, Nyompa S, Uca 2015Reflectance Characteristic of Mangrove Species using Spectroradiometer HR-1024 in Suppa Coast , Pinrang , South Sulawesi , Indonesia*Asian J. of App. Sci.***3(5)**642–8.
[25] Pastor-Guzman J, Atkinson PM, Dash J, Rioja-Nieto R 2015 Spatiotemporal variation in mangrove chlorophyll concentration using Landsat 8 *Remote Sens***7**(11)14530–58.
[26] Pramono GH, Wijaya W S, Nurwadjedi 2014. The one map initiative for mangrove in java island, Indonesia [Unpublished paper].
[27] Purwanto AD, Asriningrum W, Winarso G, Parwati E 2014. Analisis Sebaran dan Kerapatan Mangrove Menggunakan Citra Landsat 8 di Segara Anakan, Cilacap*Semin. Nas. Penginderaan Jauh* p 232–41.
[28] Jhonnerie, R 2015 Klasifikasi Mangrove Berbasis Objek dan Piksel Menggunakan Citra Satelit Multispektral di Sungai Kembung, Provinsi Riau [Theses of Institut Pertanian Bogor] 86p.
[29] Vetrita Y, Suwarsono SP 2014Pemanfaatan penginderaan jauh untuk pemantauan lingkungan

mangrove *Bunga Rampai Pemanfaatan Penginderaan Jauh untuk Pemantauan , Deteksi , dan Kajian Lingkungan.* edB Hasyim *et al* (Bogor: Crestpent·Press) page 77-86.

Remote sensing and GIS-based site suitability analysis for tourism development in Gili Indah, East Lombok

Debrina Agnes[1], Akbar Nandatama[1], Bagus Andi Isdyantoko[1], Fajri Aditya Nugraha[1], Giusti Ghivarry[1*], Perwira Putra Aghni[1], Renaldi ChandraWijaya[1], Prima Widayani[2]

[1]Undergraduate Students, Geographic Information Science, Faculty of Geography, Universitas Gadjah Mada

[2]Lecturer, Geographic Information Science, Faculty of Geography, Universitas Gadjah Mada

*email: giustighivarry@gmail.com

Abstract: Gili Indah area, located in Jerowaru, East Lombok Regency is a region that classified as farm area in spatial layout planning map of West Nusa Tenggara province. Gili Indah area has a potential as a new tourism attraction within its gilis (local term for 'small island'). Assessment should be done to prevent ecological disturbance and infringement towards spatial layout planning map caused by incorrect landuse. Land suitability assessment will be done using remote sensing approach whilst satellite imagery being used to get information about ocean ecology and land physical spatial distribution that will be the parameter of tourism land suitability, such as water clarity, ocean current, type of beaches' substrate, and beach typology. Field observation then will evaluate the accuracy of data extraction also as a material to do reinterpretation. The actual physical condition will be pictured after the spatial model built with GIS by tiered qualitative analysis approach. The result of assessment and mapping of tourism land suitability is that parts of Gili Indah Area (GiliMaringkik, Greater GiliBembeq, and Small GiliBembeq) are suitable for archipelago tourism while the others is not.

1. Introduction

Tourism is form of exploitation natural resource which prioritize on tour interset services for human satisfaction [4]. Ecotourism acording to Ceballos-Lascurain [1] is a trip to natural place that area ralatively undisturbed or contaminated wit objective of studying, admiring, and enjoying the scenery, vegetations, and wildlifes, as well as other form of cultural aspect from surounding society, both past and present. This definition continues to grow, from original purpose which was initialy only for studying, admiring, and enjoying the scenery becomes conserve the environment and to preserve the life and welfare of the local population [2]. Definition of ecotourism still evolving and create a concept for which travels to natural area aiming to build tourism industry (Eplerwood on [2]).

East Lombok regency is one of regencies located on West Nusa Tenggara Province and has local potential that coulb be developed. One is potential of the sea, but the utilization of natural resource is more focus on natural resource product than economic value in tourism aspect. Though West Lombok has quite appealing potential on tourism aspect like beach and dyke or small island. Based on available spaial map, majority of land in in the area of the dyke in East Lombok district is used as farmland.

2. Toursim in geographic study

2.1. Remote Sensing

Interpretation and object identification in remote sensing is done manuallu or digitally. Manual interpretation involves visual interpretation process that uses human eye to interpret. Interpretation by this method is performed by image with image or photo format, depending on the sensor used in data collection. Remote sensing image can also be displayed in digital format with structure of pixels with each of pixel representing digital number (DN). When the image presented in digital format, analysis process including interpretation can also be done with computer device. The analysis proses is digitally insert insert the object identification and information extraction of information automatically. However, the accuracy of digital interpretation can not be said can replace visual interpretation. Digital interpretation still serves as a supplement and assistance purpose for visual interpretation (Campbell, 1987 dalam CAMFER, 2013; NRCan, 2015).

2.2. Geographic Information System

Geographic information system is an outgrowth of the use of computer, where the computer originally used for map digitization. There are several definition of geographic information system, but Tmlinson reiterated in the International Geographical Union Commission on Geographical Data Processing and Sensing that GIS "is not a fieldthat stand alone but rather a similarity between information processing and various field using spatial analysis techniques." (Tomlinson, 1972 dalam Cowen, 1988) another definition of GIS, among others: (1) "information system with database of objects, aktivites or phenomenom that can be spatially observed, where it can be describedby a point, line, or polygon," (Dueker, 1979 on [3]), (2) "the comprehensive collection of tools for capturing, storing, calling, transformation, and visualization of spatial data from the real world for a particular use," (Borrough, 1979 on [3]).

2.3. Tourism

Tourism is considered as a geographical phenomenon because of the pattern that looks like the impact of tourism activites. Butler developed a model of regional development of tourism based on the impact of tourism activites taht occur in the area. Butler made models consists of seven phases, phase (1) exploration, when a small number of tourist visiting the area; (2) involvement, when the locals start providing facilites for tourist; (3) development, when the country of destination began to develop and promote the tourism area; (4) consolidation, as tensions between the country of destination and tourist also the number of the tourist who visit the area are not increased significantly; (5) stagnant, when the facilites in the tourist area is aging and damaged; (6) rejuvenation, during an investment from outside and the unkeep of the facility; and (7) decline of the current phase of rejuvenation failed, and people losing jobs related to tourism

3. Methodology

The process of assessing the level of suitability of land for tourism activities at Gili Indah Area divided into three stages; pre-processing stage, field observation, and post-analisys. At pre-processing stage value parameter water depth, type of beach and land use in order to obtain extensive information Gili Indah land use area and the beach area as well as shallow sea. The process of interpretation is done with the aid of imagery and topographical map of Indonesia. From this stage land use map, sea and beach area map, can be obtanied.

Field observation si done with the purpose of reviewing the location of the beach and shallow sea in the area of Gili Indah and to obtain information that can not be obtained from the process of interpretation. In this stage the data clarity of the water, flow velocity and depth of water measured. These data can be used to complete the assessment process to acquire land suitability value more accurately. Taking photos Gili Indah overview of the region also is done, for consideration and documentation.

The last stage is done asfter all field observation is done. At this stage all data measurement parameters processed as well as repeating the process of interpretation to fit with the results of field observations which is needed to processed a land use suitability map through assessment of each parameter and by utilizing geographic information systems or GIS.

4. Result

Figure 1. Land Tourism suitability map of Gili Belek

Figure 2. Land Tourism Suitability of Gili Maringkik

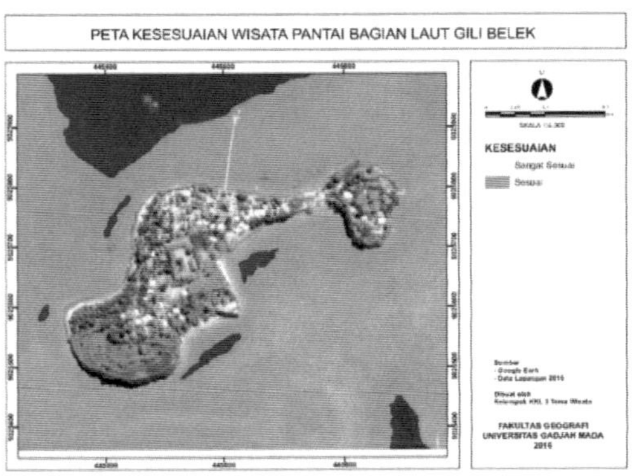

Figure 3. Marine Tourism suitability map of Gili Belek

Figure 4. Marine Tourism Suitability of Gili Maringkik

Mapping process for tourism activity at Gili Belek and Gili Maringkik divided into two objectives, namely marine tourism and beach tourism. marine toursim mapping result shows that the dyke is suitable for tourism activity expecially for sea toursim activity because due to the crystal clear waters and currents are not strong, so there is little posibility of an accident when tourists want to do snorkeling in both dyke. It is a minus value for sea tourism at Gili Maringkik and Gili Belek is absence of coral which is usually become tourist destinations for snorkeling, and no attractions that can be enjoyed besides seagrass. On the other hand, the slowly current allow tourist to swim at the waters of Gili Belek or Maringkik. Beach tourism activity show different results, which is in not all parts of the land of Gili Belek and Gili Maringkik that suitable for tourism activites. Digital analysis results from a combination of parameters such as the typology of the beach, the substrate, and so on provide results that bit of beach in Gili Maringkik which allows for tourism activites. On Gili Maringkik, because most of the dyke is a rugged coast while on Gili Belek, there are several beaches which is used as a docked and landfills.

With analysis using tiered qualitative approache, can be seen that the Region Gili Indah has the potential of marine tourism, especially for swimming and the like, but do not have beach for beach tourism. This statement is somewhat less satisfactory because of the qualitative analisys and observations, the Gili Indah also has the potential of beach tourism. This occurs because of the used method can only assess the suitability of a place to travel, but it can not assess the potential for tourism activity because tourism itself is activity that is very dynamic and is influenced by many things instead of just the physical parameters such as current speed, water clarity, substrate, and etc. There is a possibility, although a place deemed fit to travel based on tiered qualitative analysis, without the support of adequate infrastructure, it will hinder the suitability and make tourism activities to a halt.

Then there are factors which will affect the promotion and be the driving factor for tourists to go to the tourist area that is still not well known.

5. Conclusion and Recomendation

1. Conclusion
   a. Gili Belek and Gili Maringkik have marine tourism potential, instead of beach tourism, as a result of tiered qualitative analysis and output map, more regions in Gili Belek and Gili Maringkik which has high compatibility level in the tourism sector water.
   b. Determination land suitability classes at Gili Indah area is not enough just to use a tiered quantitative tiered approach only consider the physical parameters, but also requires an understanding and comprehensive analysis of the various viewpoints of science, from economics, politics, anthropology, sociology, and so forth up to the science of tourism that own.
2. Recomendation
   a. Need to do further research and study on methods to assess the tourism potential of an area, then there is distinction between tourism land, sea, mountains, and so forth.
   b. Field observation and confirmation is needed to ensure that all data can be accurately obtained and the corresponding desired target.

References

[1] Ceballos Lascuráin, Héctor 1996*Tourism, Ecotourism, and Protected Areas: The State of Nature-Based Tourism Around the World and Guidelines for Its Development*(IUCN Protected Area Programme series)

[2] Fandeli, C 2000 *Pengertian dan Konsep Dasar Ekowisata* Artikel dalam buku :*Pengusahaan Ekowisata*, editor : Chafid Fandeli. Fakultas Kehutanan UGM bekerjasama dengan Unit Konservasi Sumberdaya Alam Yogyakarta. Yogyakarta.

[3] Kresse, W dan D. M. Danko 2012*Springer Handbook of Geographic Information*(Heidelberg: Springer)

[4] Yulianda, F 2007 Ekowisata Bahari sebagai Alternatif Pemanfaatan Sumberdaya Pesisir Berbasis Konservasi *Conf. Sains* 21 February 2007 (Departemen MSP. FPIK. IPB. Bogor)

Above Ground Carbon Stock Estimates of Mangrove Forest Using Worldview-2 Imagery in Teluk Benoa, Bali

E D Candra[1*], Hartono[2], and P Wicaksono[2]

[1]Undergraduate Program of Cartography and Remote Sensing, Department of Geographic Information Science, Faculty of Geography, Universitas Gadjah Mada, Yogyakarta55281, Indonesia
[2]Department of Geographic Information Science, Faculty of Geography, Universitas Gadjah Mada, Yogyakarta 55281, Indonesia

*email: erikadwic@gmail.com

Abstract. Mangrove forests have a role as an absorbent and a carbon sink to a reduction CO2 in the atmosphere. Based on the previous studies found that mangrove forests have the ability to sequestering carbon through photosynthesis and carbon burial of sediment effectively. The value and distribution of carbon stock are important to understand through remote sensing technology. In this study, will estimate the carbon stock using WorldView-2 imagery with and without distinction mangrove species. Worldview-2 is a high resolution image with 2 meters spatial resolution and eight spectral bands. Worldview-2 potential to estimate carbon stock in detail. Vegetation indices such as DVI (Difference Vegetation Index), EVI (Enhanced Vegetation Index), and MRE-SR (Modified Red Edge-Simple Ratio) and field data were modeled to determine the best vegetation indices to estimate carbon stocks. Carbon stock estimated by allometric equation approach specific to each species of mangrove. Worldview-2 imagery to map mangrove species with an accuracy of 80.95%. Total carbon stock estimation results in the study area of 35.349,87 tons of dominant species *Rhizophora apiculata, Rhizophora mucronata* and *Sonneratia alba*.

1. Introduction

Vegetation plays an important role in maintaining the stability of the global climate. Carbon can be absorbed into the ecosystem and stored into biomass through photosynthesis of plants or soil [20]. In addition to the role of forests terrestrial / terrestrial sequestration of greenhouse gases has been known, new evidence suggests that carbon can be stored in biomass and sediment ecosystems tidal marsh vegetation such as mangroves and seagrass beds [22]. Coastal vegetation carbon stock is called blue carbon [17]. Mangrove vegetation as one that could absorb carbon has an important role in controlling the levels of CO_2 in the atmosphere. Mangrove can absorb carbon better than terrestrial ecosystems because of its ability to bury carbon in the sediments [14].

Mangrove forest biomass estimation by field surveys combined with remote sensing data is considered ideal and practical method [8]. Remote sensing model based combined with field survey data such as dbh (diameter breast high) and tree height and allometric equations for estimating biomass to develop a model of the biomass of mangrove [18, 26,21, 9]. Carbon stock assessment with remote sensing is expected to reduce the destructive method or the destruction that still preserve mangrove forests.

Estimates of biomass or carbon stock mangrove using remote sensing imagery has be done with various images such as Landsat [10, 26], RADARSAT [16], ALOS [25,27], Quickbird [9], IKONOS [23, 2], ICESat [15]. The use of different image resolutions produce different model accuracy. Mapping carbon stocks by remote sensing regardless of the species / species of mangrove vegetation can lead to errors because of different species depending on the density of the canopy which is correlated with the value of carbon on the surface [29]. As a result, in the study of mangrove carbon stocks based remote sensing, different mangrove species showed significant differences in the spectral reflection of the electromagnetic spectrum. High spatial resolution imagery have the opportunity to map carbon stocks in more detail at the level of mangrove species. The species of vegetation should be considered to estimate biomass on the surface that is accurate [18]. Based on the research of [29] also concluded that the identification of mangrove species using WorldView-2 imagery in the study area has the potential to do. Mapping carbon stock value with the image of WorldView-2 which has a spatial resolution of 1.8 meters at nadir is expected to yield higher accuracy than lower spatial resolution.

The transformation of the image of the most common and widely used for mapping and modeling mangrove is the vegetation indices. Each vegetation indices have different sensitivity to the biophysical and biochemical properties of the different vegetation such as Leaf Area Index (LAI), canopy cover, chlorophyll and nitrogen content, fragments of vegetation, biomass, and carbon stock [27]. Different types of vegetation indices have been used in a variety of mangrove mapping carbon stocks [29,26, 27, 24, 7]. Each vegetation indices can produce different level of accuracy of the mangrove carbon stocks modeling in different locations.

The purpose of this study were (1) Comparing the accuracy of the transformation of the vegetation index to map carbon stocks over the surface of the mangrove forests in Teluk Benoa, Bali using the image of WorldView-2 with the distinction of species, and (2) Mapping the value of carbon stocks and estimate the value of the total stock of carbon on the surface of the mangrove forests in Teluk Benoa, Bali.

2. Study Area and Data

The research location is included in the mangrove forest area of Taman Hutan Raya (Tahura) Ngurah Rai is located in the province of Bali. Taman Hutan Raya Ngurah Rai is an area which has been confirmed or established as the Taman Hutan Raya (Tahura), by the Minister of Forestry in 1993 with an area of 1373.5 hectares. All of the area is not used for the existing mangrove area but used for other uses, such as DAM estuary, lagoon, toll roads, and the landfill. Areas to be studied in this research is only part of Tahura Ngurah Rai. A map of the study area is shown in Figure 1.

Geographically, the study area located at 8° 43' S - 8° 44.5' S and 115° 11' E - 115°13' E. Administratively, mangrove forests that were examined included into the South Denpasar District, Denpasar in the East and the District of South Kuta, Badung Regency on the West side. Southern part of the mangrove forests is bordering with Teluk Benoa. This location is strategic located between the city of Denpasar, Kuta and Nusa Dua. In addition, the distance is very affordable from the International Airport I Gusti Ngurah Rai. These conditions make the mangrove forest areas potentially marginalized by development of the city. According Wiyanto and Fiqoh [28], mangrove species found in Teluk Benoa total of 11 species. Eleven species of mangrove is *Rhizophora mucronata* Lam., *Rhizophora stylosa* Griff., *Rhizophora apiculata* Blume, *Avicenia marina* (Forssk.)Vierh, *Avicenia officinalis* L., *Sonneratia alba* J.Sm., *Sonneratia caseolaris* (L.) Engl., *Bruguiera gymnorrhiza* (L.) Lam., *Bruguiera cylindrica* Blume, *Xylocarpus granatum* Koen., and*Ceriop tagal* C.B. Rob.

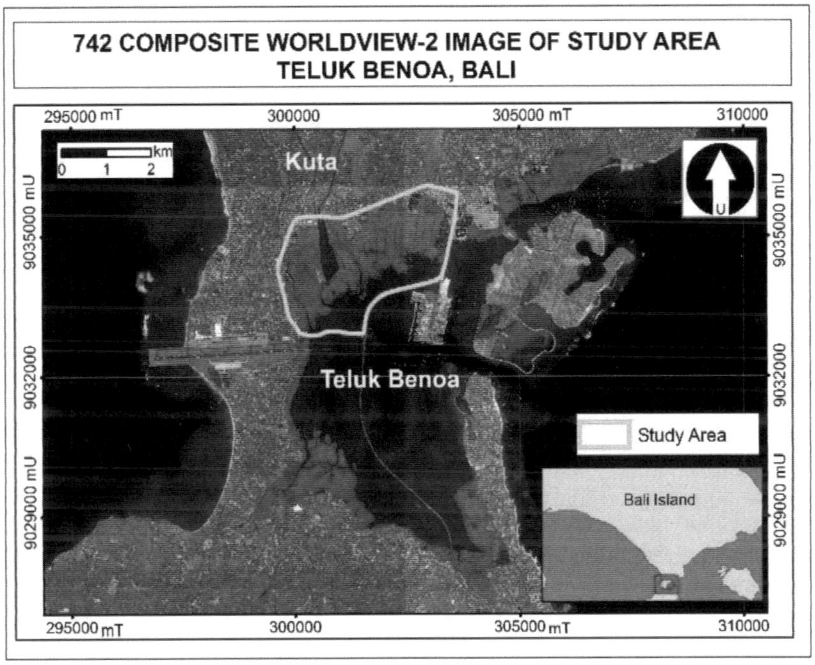

Figure 1. Worldview-2 Composite Image 752 (near infrared 1, red, blue) Study Area

Materials used in the study were WorldView-2 Imagery recording on May 20, 2015 of 16902/8820 rows / columnpart Bali Provincial. Characteristics of the sensor and the band brought in satellite Worldview-2 is shown in Table 1.

Table1. Worldview-2 Imagery Band

Band	Band Wavelength (nm)
Panchromatic	450-800
1. Coastal	400-450
2. Blue	450-510
3. Green	510-580
4. Yellow	585-625
5. Red	630-690
6. Red Edge	705-745
7. Near Infrared 1	760-895
8. Near Infrared 2	860-1040

Source: [5]

3. Method

3.1. Field survey

Field measurements conducted to obtain Diameter Breast High (DBH) data. DBH value is approximated by measuring the perimeter of the trunk at breast height of 1.3 meters. DBH value measurement performed on all individuals in plot size of 10 m x 10 m. The location is determined by the stratified sample is based on lines aligned tracking lines and lines of boats in each class of units of mangrove species. The lines have to be able to follow the tracking path that already exists in the study area, making it easier to reach the location of the current sample field survey. The sample points are given for each line that represents each class of mangrove species. Mangrove species that have a larger area given the number of more samples. There was 41 plot samples that be distinguished by model samples and accuracy test samples.

Fieldwork was conducted over seven days, which is dated May 18, 2016 until May 24, 2016. The data field is divided into two, namely the data to build the model and the data for accuracy test. The research team consists of three members, the researcher and two assistances that helped identify the species of mangrove and took measurements. Access to reach the location of the sample can be achieved in two ways, by land and water. Path within walking distance to the tracking lines that already exist and are used as travel mangrovepath. In addition to tracking path, at some locations there are dikes that can be passed. Waterways have to reach the location of sample far from land and under water is quite high. Waterway reached by boat hired from local fishermen. Accessibility is increasingly an issue of its own in this field activities.

3.2. Image correction

The image correction process is not performed geometric correction. Data obtained from Digital Globe Foundation bursts of data 2A level. At this level has been corrected geometric image globally up to 4.6 m with an accuracy of 90%. If imagery base map overlaid with Indonesia which Indonesia Topographic Map (RBI) scale of 1: 25,000, the image has been mend and not shift significantly in appearance imagery and maps. Therefore, without any geometric correction WorldView-2 images can still be used in the study with a small error. Correction radiometric imagery from WorldView-2 is done Digital Number values into spectral value at the object level surface reflectance so the image pixel value represents the value of emission spectral reflectance or the actual object. Modelling using vegetation index transformation that involves the value of each pixel should use the actual spectral values to produce an accurate model. Additionally atmospheric disturbances such as aerosol interference was minimized in the image for easy interpretation.

Worldview-2 image acquired is an image that radiometric correction has not been done to the spectral reflectance of the object. Digital Number value is converted to the value of a top-of-atmosphere spectral radiance. The radiance value is still dominated by the sun reflection values that need to be converted to the value of a top-of-atmosphere spectral reflectance. The reflectance value is still influenced by atmospheric disturbances that need to be converted into at surface reflectance to declare the value of the actual reflection of the object.Atmospheric correction to got surface reflectance used Dark Object Substaction (DOS) method with deep water as the darkest object in the image.

3.3. Mapping of Mangrove Types

The number of samples taken for the mapping of mangrove types are 60 samples. The new samples based on observations taken around the samples and during the journey to the planned sample. The entire sample was divided into two, to build a model and to test the accuracy of the resulting model of mangrove types. The field data were used to build the model with a sample size of 39. The model number of the pixel-based classification is done with supervised multispectral classification. Sample results of this field is used as a training area in the supervised classification has three classes of mangrove species.

3.4. Transformation Vegetation Index

This study uses vegetation index EVI (Enhanced Vegetation Index), DVI (Difference Vegetation Index), and MRE-SR (Modified Red Edge-Simple Ratio). EVI is one of vegetation indices that built to reduce the influence of the atmosphere and soil. EVI was developed in MODIS imagery which is a modification of the index NDVI (Normalized Difference Vegetation Index) with the background adjustment factor to correct the soil and scattering coefficients of the atmosphere. Compared with NDVI, EVI improve sensitivity in recording high biomass [11]. DVI is an basic vegetation indices using two band algorithm. DVI is a simple vegetation indices that use infrared and red bands are shared by most satellite imagery passive systems. mRE-SR is a modification of the common vegetation indices ie SR (Simple Ratio). This indices has the advantage on the edge of the red band is used. The red band edge to be more sensitive to vegetation biophysical parameters such as biomass compared to other bands [29]. Besides the red band edge has not been owned by various satellite image but it is owned by Worldview-2. Formula of the three vegetation indices are:

a. EVI

$$EVI = G\frac{\rho_{infrared} - \rho_{red}}{\rho_{infrared} + (C1x\rho_{red}) - (C2x\rho_{blue}) + L}(1+L) \quad (1)$$

G: gain factor (2,5)
C1 and C2: band correction to atmospheric aerosol scattering(6,0 and 7,5)
L: soil adjustment factor (1,0)

b. DVI

$$DVI = \rho_{infrared} - \rho_{red} \quad (2)$$

c. mRE-SR

$$mRE\text{-}SR = \frac{\rho_{rededge} - \rho_{coastal}}{\rho_{red} + \rho_{coastal}} \quad (3)$$

3.5. Modelling Biomass and Carbon Stock

The sample consisted of 41 samples of the same location as the diameter measurement sample. Total mangrove vegetation is sampled 759 mangrove stands. Mangrove stands are then calculated the value of the stock of carbon in the biomass value approach. Reference used was 0.47 from biomass is a carbon stock value according to SNI (Indonesian National Standard) 7724: 2011 [1]. The biomass value obtained by non-destructive methods, namely with methods allometry. Allometric equations were used that specific allometric every species of mangrove that aims to produce a value close to the value in the field (Table 2).

Table2. Above Ground Biomass Allometric for Mangrove Species

Mangrove Species	**Above Ground Biomass Allometric**	**Reference**
Bruguiera gymnorrhiza	$Wtop = 0{,}186 \times DBH^{2.31}$	[4]
Rhizophora apiculata	$Wtop = 0{,}235 \times DBH^{2.42}$	[19]
Rhizophora mucronata	$W_{top} = 0{,}251\ x\ \rho\ x\ D^{2{,}46}$, ρ : 1.020 kg/m^3	[12]
Sonneratia alba	$W_{top} = 0{,}251\ x\ \rho\ x\ D^{2{,}46}$, ρ : 780 kg/m^3	[12]

3.6. Statistic analysis

The statistics used for data analysis is the correlation and regression. These statistics linking measurement value in the field of value biomass is converted into carbon stock value as the dependent variable and the value of every pixel transformation of the vegetation index as independent variables. Correlation is done to determine the degree of correlation between the value of the stock of carbon in the field and the vegetation indices. This correlation test is a requirement to do regression testing. Regression tests conducted to determine the effect of the two variables and empirical equations generated is used for prediction. Empirical equation is then used to form a new image. Regression used are linear and non linear regression simple. Non-linear regression selected for both variables are assumed to have a threshold so that the relationships are not always linear so suitable to study the phenomena of vegetation.

4. Results and Discussion

4.1. Mangrove Species Classification

The results of the classification algorithm Minimum Distance to Mean produce four classes according to the given training area (Figure 2). The spatial distribution of each class was in line with the conditions on the ground. Species *Rhizophora apiculata* and *Rhizophora mucronata* dominated the artificial mangrove forests, while the mangroves that grow naturally dominated by *Sonneratia alba*. Genus *Rhizopora* especially *Rhizopora apiculata* and *Rhizophora mucronata* is found to dominate the artificial mangrove forest. It is also in accordance with the conditions of conservation of the mangrove forests by Balai Pengendali Perubahan Iklim dan Kebakaran Hutan dan Lahan of Java, Bali and Nusa Tenggara which presently houses the mangrove forest management study areas. Prior to planting mangrove conducted a study to determine the most suitable species or in accordance with the conditions of land. The result is *Rhizophora apiculata* and *Rhizophora mucronata* be selected species to be planted on the land. *Rhizophora mucronata* planted with a number of the most widely because it is considered the most suitable land and is also resistant to the disease. *Sonneratia alba* dominates the mangrove forests that grow naturally. This species is considered suitable to grow in an area close to the sea for as protector of the sea waves. Its roots were shaped roots strong breath serves to resist the wave of the sea that leads to the mainland.

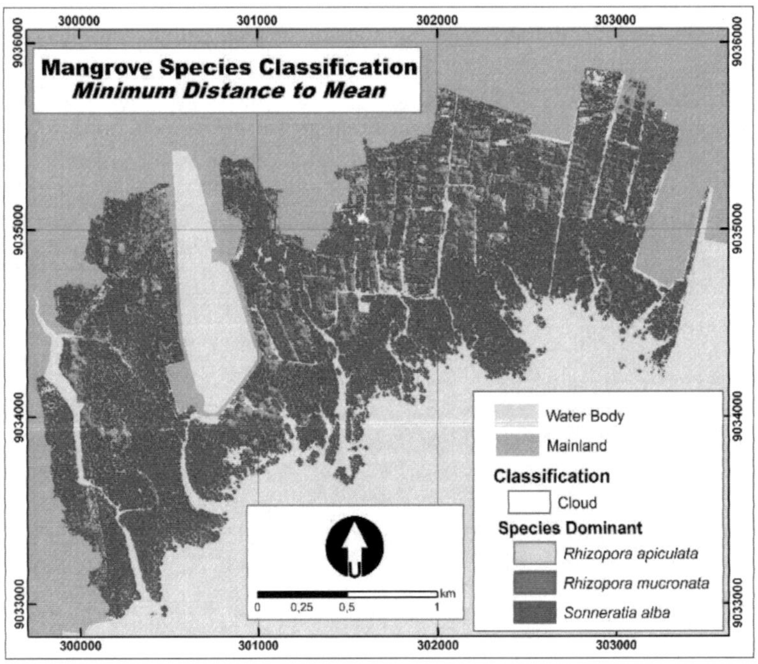

Figure 2. Results of Mangrove Species Classification Algorithm Minimum Distance to Mean

Accuracy test is done by using field samples totaling 21 samples. The method used is the error matrix involving nominal data types in this case is a class of mangrove species (Table 3).

Table 3. Error Matrix Test Accuracy Classification Minimum Distance to Mean

Classification Data	Field Data	Ra	Rm	Sa	Row Total
	Ra	4	0	0	4
	Rm	2	6	2	10
	Sa	0	0	7	7
	Cloud	0	0	0	0
	Column Total	6	6	9	21

4.2. Modeling Carbon Stock by Vegetation Indices

Modeling carbon stocks performed in each species with DVI, EVI and mRE-SR. Regression analysis is then performed to determine the influence of the independent variable is the vegetation index against the dependent variable is the carbon stocks of each species. The empirical results of regression equation is then used to build a model or predict the value of the carbon in the image. The magnitude of the effect of empirical or predictive power equation expressed by the value of determination (R^2).

Test the accuracy of carbon models each species showed that regression with mRE-SR has the highest accuracy. It can be concluded that mRE-SR is more sensitive to differences mangrove species. mRE-SR is an index that uses a modification of the Simple Ratio red wave edge. The study concluded that mRE-SR index is an index that is best able to predict the stock of carbon in each mangrove species. These results are consistent with previous studies [29] which states that the use of red tide on the edge of the vegetation index can overcome boredom in the estimated biomass of vegetation in the area of dense mangrove. The advantage of using red edge is wave property changes of vegetation will change the curve of reflection on the wave of red edge. In addition, the red wave edges are more sensitive to the calculation of biophysical parameters vegetation than other waves. Therefore, the use of mRE-SR index suitable for estimating carbon stocks at a level mangrove species in the study area.

Table 4. Highest Accuracy Model

Species	Regression	SE (ton/ha)	Accuracy
Rhizophora mucronata	Quadratic mRE-SR	12,89	85%
Rhizopora apiculata	Quadratic mRE-SR	14,52	84%
Sonneratia alba	Quadratic mRE-SR	81,15	64%

Modelling of *Rhizophora mucronata* carbon stock has a high accuracy of 85% (Table 4). Value of the stock of carbon is able to predict quite well through the vegetation index. High vegetation index value indicates the number of leaf vegetation that reflects the electromagnetic wave. The more leaves possessed the more the stock of carbon stored from the process of photosynthesis. Therefore, the carbon stocks of vegetation index predicts fairly well.Values that almost have same accuracy was shown a model of carbon stocks species *Rhizophora mucronata* and Rhizophora apiculata. This is caused by both species came from the same genus that has some characteristics of a uniform one carbon sequestration capabilities. Carbon stock value range *Rhizophora mucronata* and *Rhizophora apiculata* is not much different than the carbon stock value range of species *Sonneratia alba*.

Sonneratia alba carbon model has the lowest accuracy compared to other species models. It is influenced by several factors such as the size of the trunk diameter *Sonneratia alba* is very large when compared to other species. Big trunk size resulted in greater biomass owned. *Sonneratia alba* in the study area during recording and field survey turned out to be dominated large trunk has a characteristic size but have lower leaves cover the trunk size is not directly proportional to the number of leaves that possessed thereby affecting the amount of wave reflection. While having a diameter rods and large biomass, electromagnetic wave reflection actually heavily influenced by the object water is underneath. Therefore, the vegetation indices that highlights the characteristics of the vegetation is unable to predict with good biomass carbon stock impact on the value of the species *Sonneratia alba*.

4.3. Mapping and Carbon Stock Estimation

The total value of the stock of carbon is obtained by summing the pixel values in the form of carbon value per pixel. The total area of mangrove object in the study area were counted, 391.8488 ha. The total stock of carbon on the surface of which is kept by the mangrove vegetation can be seen in Table 5.

The total value of carbon stocks resulting from modeling based species are highly influenced by the species mapping done previously. Rated accuracy is not 100% because there was error in mapping the species so that this error also affects the suitability of modeling results with the value in the field. Therefore, to produce carbon stock value close to the value in the field is required map of mangrove species with high accuracy.

Table 5. Total Stock Carbon

Species	Regression Model	Total Carbon (tons)
Rhizophora mucronata	Quadratic mRE-SR	23.773,28
Rhizopora apiculata	Quadratic mRE-SR	7.369,35
Sonneratia alba	Quadratic mRE-SR	4.207,24
Total		35.349,87

Estimated value of carbon in the study area can be mapped to a scale of 1: 10,000 mapping with an accuracy of up to 85% (Figure 3). Estimated value of carbon in the classified into three classes of carbon values, There are: (1) 0 to 0.023 tons (2) from 0.023 to 0.035 tons, and (3) from 0.035 to 0.686 tonnes. High carbon stock value predominantly distributed in mangroves near the sea. High carbon value one of them caused by the dominance of species *Sonneratia alba* which has a stock of carbon is higher than other species. It also relates ecology, mangrove which a supply of sea and inland water that either has better growth due to supply enough water can accelerate the growth of vegetation.

Values lower carbon stocks, spread over an area of mangroves adjacent to the mainland. The dominance of species that have lower carbon stocks are species *Rhizophora apiculata* and *Rhizophora mucronata*. Low carbon stocks may be affected by human activities such as landfills flowing through the river so it accumulates in mangrove areas. Their wastes human activity is causing inhibition of growth of mangroves adjacent to the mainland due to inhibition of mangroves absorb nutrients from the soil and from pollutants trash. Mangrove growth inhibition led to the ability to store carbon stocks lower. In addition, the low carbon stocks contained in the surrounding area are covered with clouds. Low carbon stocks are influenced by the pixel values are still under the influence of clouds and cloud shadows are thin so that its value is not a reflection of the value of the actual vegetation.

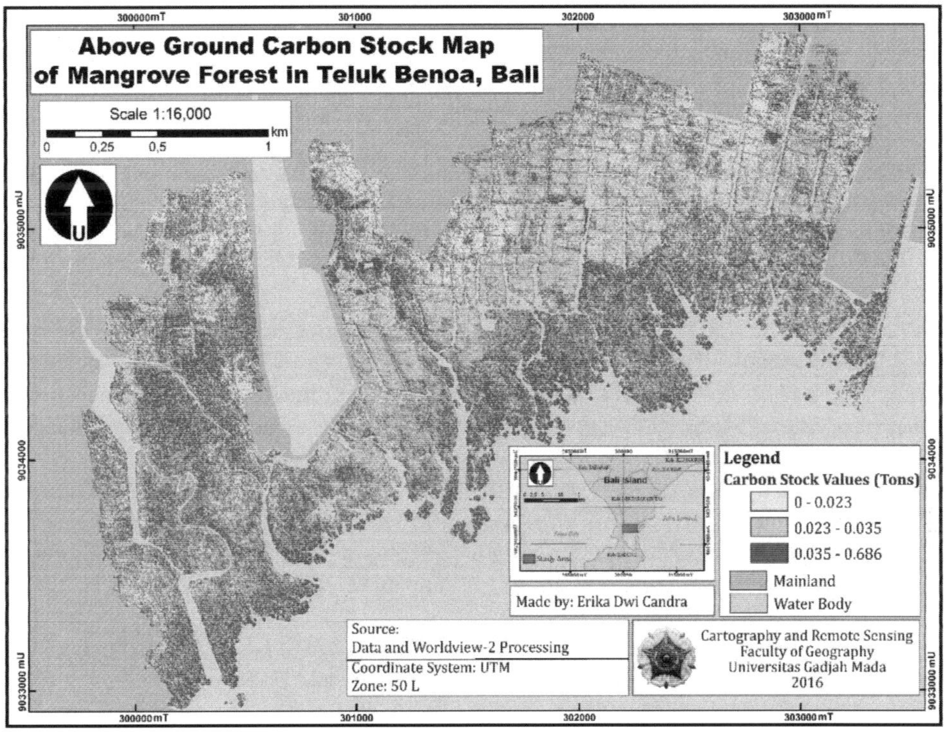

Figure 3. Above Ground Carbon Stock Mangrove Forest Map in Studies Area

5. Conclusion

Based on the research that has been done, it can be concluded that:

1. Mapping carbon stocks in the study area more accurately performed without distinction of species compared to distinguish the species. Vegetation index that has the best accuracy in mapping carbon stocks over the surface of the mangrove forests in the Teluk Benoa, Bali is an index of mRE-SR mapping with an accuracy of 85%, 84%, and 64% for each species *Rhizophora mucronata*, *Rhizophora apiculata*, and *Sonneratia alba*.
2. Estimates of mangrove forest carbon stocks in the Teluk Benoa, Bali can be mapped to a scale of 1: 10,000. With high carbon stock value dominated scattered mangrove area adjacent to the sea. Carbon stocks lower value contained in the mangrove area adjacent to the mainland. The total stock of carbon in the study area is 35349.87 tons of mapping carbon stocks of each species in an area of 392 ha.

Acknowledgments

The author would like to thank Digital Globe Foundation for Worldview-2 image granted which the research depended on. Author also would like to express our gratitude to Balai Pengendali Perubahan Iklim dan Kebakaran Hutan dan Lahan wilayah Jawa, Bali, dan Nusa Tenggara for the permit and facilities during field survey. We thank to Widyanissa Rahmayani and Wulan RT as field survey assistants.

References

[1] Badan Standardisasi Nasional 2011*SNI 7724:2011: Pengukuran dan Penghitungan Cadangan Karbon – Pengukuran Lapangan untuk Penaksiran Cadangan Karbon Hutan*(Jakarta: Badan Standardisasi Nasional)

[2] Chadwick J 2011 Integrated LiDAR and IKONOS Multispectral Imagery for Mapping

Mangrove Distribution and Physical Properties*International Journal of Remote Sensing***32**(21) 6765–6781

[3] Chen Q, Laurin G V, Battles J J, and Saah D 2012 Integration of Airborne Lidar and Vegetation Types Derived from Aerial Photography for Mapping Aboveground Live Biomass*Remote Sensing of Environment***121** 108–117

[4] Clough B F and Scott K 1989 Allometric Relationships for Estimating Aboveground Biomass in Six Mangrove Species *Forest Ecology Management***27** 117-127

[5] Digital Globe 2010a Digital Globe Data Sheet. accessedat 9 Januari 2016, from http://global.digitalglobe.com/sites/default/files/DG_WorldView2_DS_PROD.pdf

[6] Digital Globe. (2010b). Digital Globe. accessed at 20 Januari 2016, from http://global.digitalglobe.com/sites/default/files/Radiometric_Use_of_WorldView-2_Imagery%20(1).pdf

[7] Frananda H, Hartono, and Jatmiko R H 2015 Komparasi Indeks Vegetasi Untuk Estimasi Stok Karbon Hutan Mangrove Kawasan Segoro Anak Pada Kawasan Taman Nasional Alas Purwo Banyuwangi, Jawa Timur*Majalah Ilmiah Globë***17**(2) 113-123.

[8] Heumann B 2011 Satellite Remote Sensing of Mangrove Forest: Recent Advances and Future Opportunities *Progress in Physical Geography***35**(1) 87-108

[9] Hirata Y, Tabuchi R, Patanaponpaiboon P, Poungparn P, Yoneda R, and Fujioka Y 2014 Estimation of aboveground biomass in mangrove forests using high-resolution satellite data*Japanese Forest Society***19** 34–41

[10] Ishii T and Tateda Y 2004 Leaf Area Index and Biomass Estimation for Mangrove Plantation in Thailand*Geoscience and Remote Sensing Symposium***4** 2323-26

[11] Jensen J R 2004*Introductory Digital Image Processing - A Remote Sensing Perspectives* (New Jersey: Pearson Prentice Hall)

[12] Komiyama A, Poungparn S, and Kato S 2005 Common Allometric Equations for Estimating*Journal of Tropical Ecology***21**(04) 471 – 477

[13] Krisnawati H, Adinugroho W C, and Imanuddin R 2012*Monograph Allometric Models for Estimating Tree Biomass at Various Forest Ecosystem Type in Indonesia*. (Bogor: Kementerian Kehutanan)

[14] Laffoley D and Grimsditch G 2009*The Management of Natural Coastal Carbon Sinks*(Gland Switzerland: International Union for Conservation of Nature)

[15] Lefsky M A, Harding D J, Keller M, Cohen W B, Carabajal C C, Espirito-Santo F D, . . . Oliveira R 2005 Estimates of Forest Canopy Height and Aboveground Biomass using ICESat*Geophysical Research***32**, 1-4

[16] Li X, Yeh A G O, Wang S, Liu K, Liu X, Qian J, and Chen X 2007 Regression and Analytical Models for Estimating Mangrove Wetland Biomass in South China using Radarsat Images*International Journal of Remote Sensing***28**(24) 5567–82

[17] Murray B C, Pendleton L, Jenkins W A, and Sifleet S 2011*Green Payments for Blue Carbon Economic Incentives for Protecting Threatened Coastal Habitats* (Durham: Nicholas Institute for Environmental Policy Solutions)

[18] Ni-Meister W, Lee S, Strahler A H, Woodcock C E, Schaaf C, Yao T, . . . Blair J B 2010 Assessing General Relationships Between Aboveground Biomass and Vegetation Structure Parameters for Improved Carbon Estimate from Lidar Remote Sensing*Journal Of Geophysical Research***115** 1-12

[19] Ong . E, Gong W K, and Wong C H 2004 Allometry and partitioning of the mangrove, Rhizophora apiculata*Forest Ecology and Management* 395–408

[20] Patil V, Singh A, Naik N, and Unnikrishnan S 2014 Estimation of Carbon Stocks in Avicennia marina Stand Using Allometry, CHN Analysis, and GIS Methods *Wetlands***34** 379-391

[21] Patil V, Singh A, Naik N, and Unnikrishnan S 2015 Estimation of Mangrove Carbon Stocks by Applying Remote Sensing and GIS Techniques *Wetlands***35** 695-707

[22] Pendleton L, Donato D, Murray B C, Crooks S, Jenkins J, Sifleet S, . . . Baldera A 2012

Estimating Global ''Blue Carbon'' Emissions from Conversion and Degradation of Vegetated Coastal Ecosystems *Plos One*7(9) 1-7

[23] Proisy C, Couteron P, and Fromard F 2007 Predicting and Mapping Mangrove Biomass from Canopy Grain Analysis using Fourier-based Textural Ordination of IKONOS Images *Remote Sensing of Environment***109** 379–392

[24] Suyadi, Ulumudin Y I, and Vebriansyah R 2013 Indeks Vegetasi dari Citra Satelit Alos untuk Memperkirakan Cadangan Karbon Atas Permukaan di Hutan Mangrove *Berita Biologi***12**(2) 249-258

[25] Thapa R B, Watanabe M, Motohka T, and Shimada M 2015 Potential of High-resolution ALOS–PALSAR Mosaic Texture for Aboveground Forest Carbon Tracking in Tropical Region *Remote Sensing of Environment***160** 122–133

[26] Wicaksono P, Danoedoro P, Hartono, Nehren U, and Ribbe L 2011 Preliminary Work of Mangrove Ecosystem Carbon Stock Mapping in Small Island Using Remote Sensing: Above and Below Ground Carbon Stock Mapping on Medium Resolution Satellite Image*Proceedings of SPIE - The International Society For Optical Engineering* 8174 81741B-1

[27] Wicaksono P, Danoedoro P, Hartono, and Nehren, U 2016 Mangrove Biomass Carbon Stock Mapping of the Karimunjawa Islands using Multispectral Remote Sensing*International Journal of Remote Sensing***37**(1) 26-52

[28] Wiyanto D B and Faiqoh E 2015 Analisis vegetasi dan struktur komunitas Mangrove Di Teluk Benoa, Bali*Journal of Marine and Aquatic Sciences* 1-7

[29] Zhu Y, Liu K, Liu L, Wang S, and Liu H 2015 Retrieval of Mangrove Aboveground Biomass at the Individual Species Level with WorldView-2 Images *Remote Sensing*7 12192-12214

IOP Conf. Series: Earth and Environmental Science **47** (2016) 012015 doi:10.1088/1755-1315/47/1/012015

Plankton Biomass Models Based on GIS and Remote Sensing Technique for Predicting Marine Megafauna Hotspots in the Solor Waters

MIH Putra[1*], SA Lewis[2], EM Kurniasih[3], D Prabuning[1], and E Faiqoh[4]

[1]Reef Check Foundation Indonesia, Denpasar, Bali, Indonesia
[2]The Manta Trust, Dorchester, Dorset, United Kingdom
[3]Indonesian Biodiversity Research Center, Denpasar, Bali, Indonesia
[4]Marine Science Department, Udayana University, Denpasar, Bali, Indonesia

[*]Corresponding author: iherwata@reefcheck.org

Abstract. Geographic information system and remote sensing techniques can be used to assist with distribution modelling; a useful tool that helps with strategic design and management plans for MPAs. This study built a pilot model of plankton biomass and distribution in the waters off Solor and Lembata, and is the first study to identify marine megafauna foraging areas in the region. Forty-three samples of zooplankton were collected every 4 km according to the range time and station of aqua MODIS. Generalized additive model (GAM) we used to modelling zooplankton biomass response from environmental properties.Thirty one samples were used to build a model of inverse distance weighting (IDW) (cell size 0.01°) and 12 samples were used as a control to verify the models accuracy. Furthermore, Getis-Ord Gi was used to identify the significance of the hotspot and cold-spot for foraging area. The GAM models was explain 88.1% response of zooplankton biomass and *percent to full moon, phytopankton biomass*being strong predictors. The sampling design was essential in order to build highly accurate models. Our models 96% accurate for phytoplankton and 88% accurate for zooplankton. The foraging behaviour was significantly related to plankton biomass hotspots, which were two times higher compared to plankton cold-spots. In addition, extremely steep slopes of the Lamakera strait support strong upwelling with highly productive waters that affect the presence of marine megafauna. This study detects that the Lamakera strait provides the planktonic requirements for marine megafauna foraging, helping to explain why this region supports such high diversity and abundance of marine megafauna.

1. Introduction

Numerous scientific publications have reported marine megafauna population declines in recent decades, largely due to fisheries and other anthropogenic threats [1].As a result, there have been an increasing number of studies focused on establishing marine megafauna spatial models to identify key habitat and conservation priority areas. Geographic information system (GIS) and remote sensing (RS) techniques can be applied to assist with distribution or habitat modelling of highly migratorymarine megafauna[2, 3]. Moreover, GIS can be a useful tool that assistswith the strategic design and development of both terrestrial and marine conservation planning. Three components are fundamental to develop species distribution or habitat models;(i) biotic, which are related to ecological communities and the role of each species in the food web, (ii) abiotic, whichare the physical environment requirements of a species, and (iii) movement, whichisthe opportunitya species has to reach a suitable location for their abiotic and biotic requirements[4].

Firstly, biotic components are important in understanding how different species interact with each other and the resultingconsequences; for example, many marine megafauna species (e.g. cetaceans) feed on high densities of small animals such as krill and small fish [5]. As a result, understanding zooplankton biomass is critical in modelling marine megafauna distribution and behaviour [6-7], andmany studies have documented positive correlations between the occurrence of cetaceans and the distribution and abundance of zooplankton [9, 10]. These related ecological system, which are zooplankton transfer energy from primary producers toconsumers at higher trophic levels, such as fish and marine mammals [8]. Secondly, abiotic components(i.e. temperature, bathymetry,) are important in understanding theenvironmental requirements of marine megafauna species, which move actively both horizontally and vertically throughout the water column. For example, species may have specific thermal requirements needed to support their metabolic process [11]. Thirdly, it is important to understand species capability and opportunityto move and reach theirrequired biotic and abiotic conditions. This is particularly important to determine given the current climate changesituation that is altering ocean temperatures.Zooplankton are poikilothermic, which means that their physiological processesare highly sensitive to temperature[8, 12, 13], therefore changing ocean temperatures is likely to impact the distribution and composition of zooplankton, which in turn willimpact the movement and distribution of marine megafauna that prey upon them.

This study was focused in the Savu Sea off Solor Island (SI) and Lembata Island (LI) in the East Nusa Tenggara province of Indonesia.The region is known for its complex, high-energy currents, and temperature variation and is an important corridor for cetaceans, mobulids rays and other marine megafauna [14, 15]. The waters of SI and LI is critical habitat for marine megafauna, which migrate through this region to find high density prey [16]. However, the villages of Lamakera (Solor) and Lamalera (Lembata) are traditional and commercial megafauna fishing communities that primarily target mobulids, whales, and dolphins and are arguably the biggest artisanal megafauna fisheries in the world [17, 19]. Extremely high international demand for mobulids gill plates, and protein needs from marine megafauna meat has been driving these fisheries and is likely to have caused as estimated 75% decline in catch rate of mobulids in Lamakera over the last 10 years [19].In an effort to conserve vulnerable marine megafauna, the Indonesian government declared full protection for cetaceans in 1999, whale sharks (*Rhincodon typus*) in 2013, and most recently manta rays (*Manta alfredi* and *Manta birostris*) in 2014. To support the effectiveconservation and management of marine megafauna in the waters of SI and LI, identifying key habitat of marine megafauna in this region is critical. The aim of this pilot study was to develop a model fordetermining plankton biomass and distribution in order to help predict key foraging habitat for marine megafauna in SI and LI. We expected that our models would provide the following information; (i) predictions for zooplankton biomass as an ecological response, (ii) the approachesof using GIS and RS techniques to build spatial models to determine key habitat of marine megafauna.

2. Study site and sampling design

The study was conducted in the waters off SI and LI, East Nusa Tenggara province of Indonesia, including the Solor Strait and the Lamakera Strait (Figure 1). This region was chosen as the study site due to it's known abundance of marine megafauna and highly productive waters, in addition the threat from targeted megafauna fisheries makes it important from a conservation perspective.Peak season for productivity and marine megafauna occurrence is during the eastern monsoon, from June through to September, when upwelling divergence in flow transport through the straits (and Ekman pumping) is large [20]. Plankton sampling surveys were conducted over five boat tripsbetween the 5th to 18th January 2016. Zooplankton was sampled every 4km over the entire the study area (Fig. 1a).

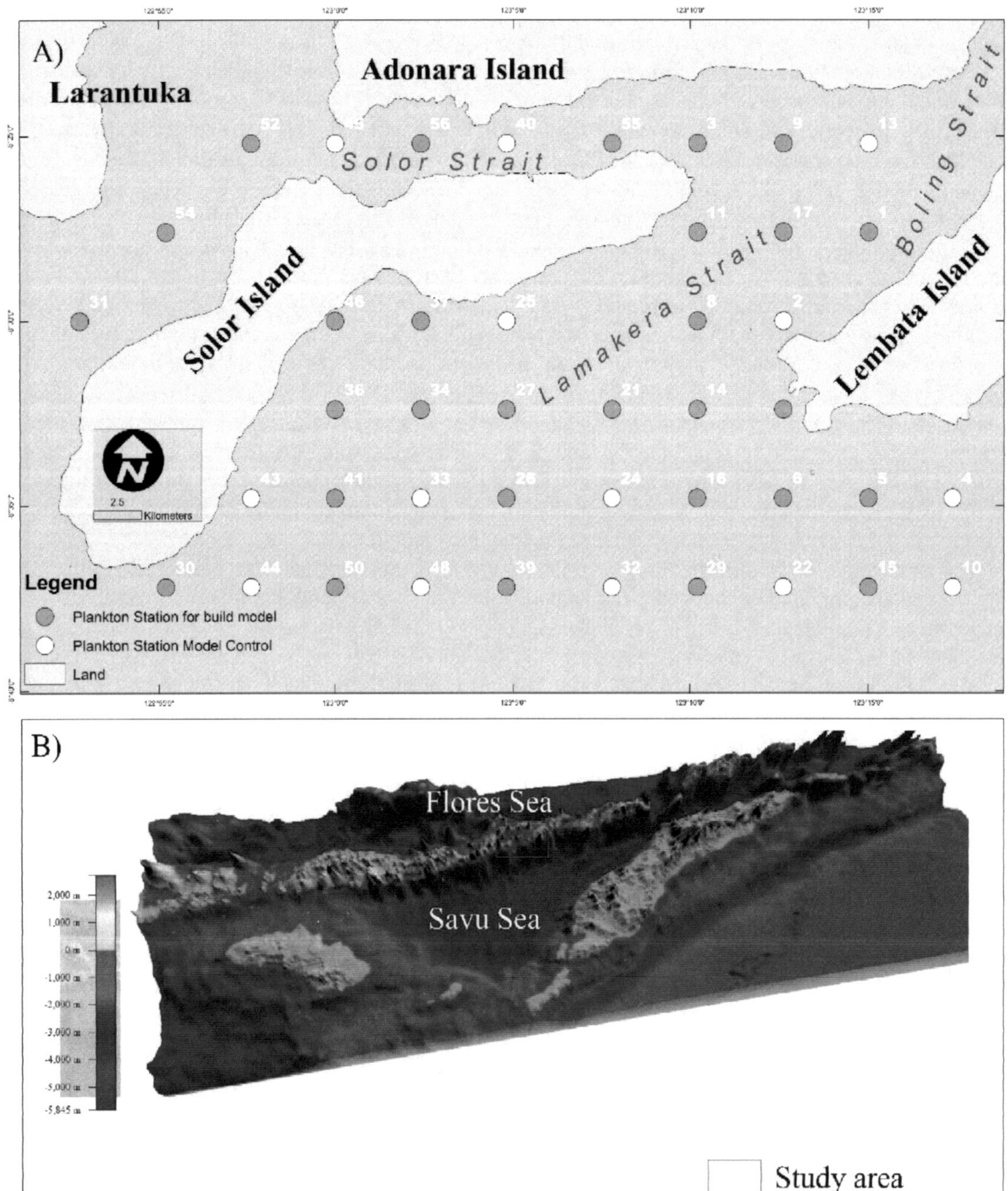

Figure 1. Maps of the study site showing a) the location of the zooplankton sampling stations (each station is 4km apart), and b) 3D bathymetric visualization of the Savu Sea.

3. Materials and Methods

3.1. Data collection

3.1.1. Zooplankton sampling

Zooplankton was collected using a 300 μm (mesh size) SEAGEAR net with a 30cm mouth diameter and 100cm length plankton net, which was trawled behind a speedboat. Trawls were conducted every 4km (Fig. 1) according to the range time and station of aqua MODIS record in the day we sampled.

During each trawl, the plankton net was towed ten meters behind the boat for 10 minutes at the surface waters. Mean distance towed was 1340 meters (± 222.5 SD), mean speed towed was 7.75 km/hours (± 1.3 SD), and mean sampling time was 9:50 AM (± 0.1 SD) such potential time for zooplankton in the surface waters. Five percent formaldehyde was used to preserve zooplankton samples until laboratory analyses, which was conducted in May 2016. The GPS position was recorded at the start and end of each trawl to determine the distance of each plankton towand thevolume of water filtered.

3.1.2. Marine megafauna observations
A Rapid ecological assessment (REA) method was used to investigate the distribution and relative abundance of marine megafauna. GPS position are planned to exploring the marine megafauna sighting during survey trip [21]. Double platform visual observation method [22]was applied, and 2 groups of observers were scanning the area using Bushnell binocular marine series with optical distance 5km during every REA. A total of five REA trips were conducted on the same day/tripthatthe plankton sampling was conducted. Data collected included; start time and finish time of survey trips, marine megafauna sighting time, peak marine megafauna encounter time, GPS location of sighting,species, behaviour, and approximate number of individuals. In addition, the weather conditions during the REA was recorded using a modified Beaufort scale[23].

3.1.3. Biophysical satellite oceanography approach
Satellite oceanography is a crucial tool used for monitoring thedynamics and productivity of water masses. Moderate Resolution Imaging Spectroradiometer (MODIS) level 3 from NASA was used to derive sea surface temperature and chlorophyll-a concentration at a 4km spatial resolution for the dates of our survey period. Chlorophyll-a concentration as a photosynthetic pigment $mg.m^{-3}$, is commonly used as a proxy for phytoplankton biomass in the surface waters (oceancolor.gsfc.nasa.gov).

3.2. Data analysis
3.2.1. Plankton biomass
In the laboratory, we filtered out 800 ml of each zooplankton sample from everysampling station/location using laboratory filter paper with a mesh size < 300 µm. The zooplankton samples were then dried in an air-conditioned room for 3 days until the dry weight of the zooplankton sampleswere constant. Analytic weight was used to measure the dry weight of zooplankton, and $mg.m^{-3}$was used as the dry weight unit of zooplankton based on the following equation:
Filtered volume

$$\text{Zooplankton biomass mg.m}^{-3} = \frac{\text{dry biomass mg}}{\text{water filterd m}^3} \quad (1)$$

3.2.2. Modelling Approach
Conventionally, linier regression has been used to determine the relationshipbetween zooplankton and phytoplankton, and in some cases there have beenstrong positive correlation, likely due toa food chain structure of zooplankton feeding upon phytoplankton. However, coastal and pelagic ecosystems are vast, dynamic and complex systems that will be more challenging for modelling linearity. We used generalized additive models (GAMs the mgcv R-package [24]) to address the issues, as they are powerful tools for modeling nonlinear relationships between zooplankton biomass and environmental properties.The final model waschosen based on the determination value (R^2) and Akaike Information Criterion (AIC) values, where the lowest AICand highest determination valueshow better performance in prediction.

3.2.3. Inverse distance weighting (IDW)
GIS modelling tools are used to conduct spatial interpolation, which estimates the values of unsampled points or data gaps in the study area. This approach is extremely useful when studying the vast marine environment, where limits such as time and money preventdata collection throughout the whole study area. Spatial interpolation is a prediction of a variable at an unmeasured location based on samples at

known locations. We used local interpolators of inverse distance weighting because the concept of computation is relevant for phytoplankton and zooplankton, where closer points are thought to be similar as result of the food web. All feature data was calculatedusing ArcGIS 10.2.2[25], using the following equation:

$$Z_j = \frac{\sum_i \frac{Z_i}{d_{ij}^n}}{\sum_i \frac{1}{d_{ij}^n}} \quad (2)$$

Where, Zjis the estimated value for the unsampled point (j), dij is the distance between the known sample (i) and the unsampled point(j), Zi is the value of the known sample (i), and n is the user-defined exponent for weighting.

3.2.4. *Model Verified*

The IDW biomass models of phytoplankton and zooplankton, which we applied,are tested to measure the accuracy prior to conducting further analyses to identify the foraging area of marine megafauna in the study region. 12 samples were used and the following equation was applied:

$$\text{RE} = \frac{(X-C)}{X} X\ 100\ \% \quad (3)$$

$$\text{MRE} = \sum_{i=0}^{n} \frac{RE}{n} \quad (4)$$

Where, X is the field data, C is the model data.Relative error (RE) is the percentage error of the model's result compared with the field data, and mean relative error (MRE)is the mean percentage error from the 12 samples that were tested.

3.2.5. *Hot spot analysis (Getis-Ord Gi*)*

We used clustering biomass models of phytoplankton and zooplankton every 0.01° (100 meters to 100 meters) to identify marine megafauna foraging sites in our study area. This analysis used the Getis-Ord Gi* [26] to identify statistically significant high values (hotspots) and low values (cold-spot) for foraging requirement of marine megafauna. Since its development in the mid-1990s the *Getis-Ord Gi** has been applied to analysing clustering in spatial data (point or area) and can be used to identify explicit areas of high use based upon specific test criteria, independent of the magnitude of abundance. The*Getis-Ord Gi**has been used substantially in recent years in various research areas [27-28]. The Getis-Ord statistic is given by the following equation:

$$G_i^*(d) = \frac{\sum_{j=1}^{n} W_{i,j} X_{i,j} - \bar{X} \sum_{j=1}^{n} W_{i,j}}{S \sqrt{\frac{(n \sum_{j=1}^{n} w_{i,j}^2 - (\sum_{j=1}^{n} w_{i,j})^2)}{n-1}}}$$

$$\bar{X} = \frac{\sum_{j=1}^{n} x_j}{n}$$

$$S = \sqrt{\frac{\sum_{j=1}^{n} x_j^2}{n}} - (\bar{X})^2 \quad (5)$$

G_i^* xi is calculated by the attribute value for feature (j), wi, j is the spatial weight between features i and j, n is equal to the total number of features and the G_i^*statistic calculated in this way is a Z-score (no further calculations are needed). Statistical significance of hot-spot and cold-spot are a measure from the Z-score and *p-value* which tell us whether or not to reject the null hypothesis, feature by feature. We applied a fixed distance band to get conceptualization of the spatial relationship and

Euclidean to measure the distance method. Statistical test result of significant hot and cold spots are showing (+/-) 3 bins reflect statistical significance with a 99 percent confidence level; (+/-) 2 bins reflect a 95 percent confidence level; (+/-) 1 bins reflect a 90 percent confidence level; and the clustering for features in bin 0 is not statistically significant. All calculation were carried out using ArcGIS 10.2.2[25].

4. Results and Discussion

4.1. Marine megafauna sighting

Marine megafauna was sighted 83% of the time during marine megafauna observation and zooplankton trawling surveys. A total of six observation survey days were conducted with a mean observation time ofnine hours per trip, from7am to 5pm.Most observation survey trips covered the whole study area around SI and LI. The majority of dolphin sightings occurred in the morning between 9am and 10am, with peak encounter time 09:20am. Tidal phase significantly affected dolphin occurrence with most sightings(61%)occurringduringan ebb tide. A similar pattern was found in Lady Elliot Island (LEI), Australia for reef manta ray (*Manta alfredi*), where feeding manta rays were observed more frequently during an ebb tide, due to a higher abundance of prey (zooplankton) accumulating around LEI during this tidal phase [29-30]. Tidal current is highly driven by the moon and during our surveys, waning crescent was the most frequentmoon phase. Results from theGAMs models (figure 2), show that the waning crescent moon phase with the percent of full moon between 2.3% - 27.8% has significant correlation with zooplankton biomass,thus drivingthe occurrence of higher tropic level species such as small fish and dolphins. Furthermore, weather appeared to affect the present or absence of marine megafauna in the waters of SI and LI. We observed that marine megafaunawere most commonly sightedwhen the weather conditionswere calmto slightwind and wave (see Table 1).

During the entire survey periodthe mean number of dolphins observed was 29 individuals (± 15 individuals SD, *n=6*), and88% of the time they were observed feeding, and were frequentlyobserved with large schools of fish and on occasion sharks were also observed hunting on the same school of fish. This usually occurred in the morning when productivity was higher.

Marine megafauna was commonly present in locations that showed higher productivity. These higher productivity locations showed a mean biomass of0.18 $mg.m^{-3}$ (± 0.03 $mg.m^{-3}$ SD) for phytoplankton, and a mean biomass of0.7 $mg.m^{-3}$(± 0.25 $mg.m^{-3}$ SD) for zooplankton. Although, significant upwelling currents did not occur during the January survey period, which is outside of the peak productivity months, our results indicate that,certain marine megafauna species such as dolphins still utilise the region, particularly higher productivity spots. Moreover,bathymetry maps of the channel between SI and LI (Lamakera Strait) are showing in figure 6a, show that the region is characterized by complex continental slope between 0^0 – 90^0, and the area in which dolphins were frequently observedwas over a slope of46^0 (± 27 SD).It has been reported that gentle continental slopessupport strong upwelling currents and thus highly productive waters that affect the presence of marine megafauna. In the Gulf of Mexico, Risso's dolphins were regularly sighted along the step of the upper continental slope,an area that provides high productivity [31]. The authors (MIH Putra and S Lewis, Persobs) have observed high diversity and abundance of marine megafauna (including, whales, dolphins, manta rays, whale sharks and ocean sunfish) using the Lamakera strait at other times of the year outside of the study period, suggesting that this areais akey habitat for marine megafauna.Both dolphins and whales have been observed resting in surface waters(logging behaviour), and also hunting on schools of small fish in the Lamakera Strait approximately 4.75 km (± 2.3 km SD) from land (figure 6), this indicates that the area is likely to be important cetacean habitat for both feeding and resting.

During our observation surveys, we were also successful in identifying hawks bill turtles, which were observed actively swimming in surface watersmost commonly in the morning. Sea turtles are grazers of macro algae and also predateon some invertebrates. Hawksbill turtles were sighted in a high productivity location(figure 6), Phytoplankton biomass in this location was 0.16 $mg.m^{-3}$ and zooplankton biomass was0.67 mg.m^{-3}. It is likely that the relative high productivity in this area is a driver in turtle occurrence at this location, as the high productivity suggests that there are greater feeding opportunities here compared to areas with lower productivity. Turtles in Ningaloo Reef, Western Australia, have also shown a similar pattern with peak sightings occurringin March, which is a time of high productivity for the region [32]

IOP Conf. Series: Earth and Environmental Science **47** (2016) 012015 doi:10.1088/1755-1315/47/1/012015

Table 1. Marine megafauna survey results with environmental condition during observation

Parameters	Explanation	Parameter range	Dolphin	Turtle
Trophic guilds[a,b]	Prey items	N/A	Fish/cephalopods feeders	Invertebrates/macro-algae feeders
Relative abundance in total (%)[b]	Number of marine megafauna abundance during survey	0 – 100	99	1
Beaufort scale[b]	Wheatear condition during encounters	1 - 5	1 (99%), 2 (1%)	2
Peak encounter time[b]	The most species encountered	7 AM – 17 PM	09:20	08:00
Feeding behavior (%)[b]	Percentage behavior of marine megafauna are indicated feeding	0 - 100	88 %	0 %
Tide condition[c]	Tide condition during encountered with marine megafauna	ebb, flood, high tide, low tide	Flood (13%), ebb(61%), high (13%), low (13%)	Ebb (100%)
Moon phases [d]	Moon phases during encountered with marine megafauna	Third quarter moon, waning crescent, new moon, waxing crescent, first quarter moon, waxing gibbous, full moon, waning gibbous	Waningcrescent (67 %), New moon (17 %), waxing gibbous (17 %)	Waningcrescent (100 %)
Phytoplankton biomass mg.$^{m-3(e)}$	Phytoplankton biomass mg.$^{m-3(d)}$ (31 d mean)	Mean 0.16 (± 0.01 SD) range 0.14 – 0.25	0.18 (± 0.03 SD)	0.16
Zooplankton biomass mg.$^{m-3(b)}$	Zooplankton biomass mg.$^{m-3(b)}$(during field survey)	Mean 0.56 (± 0.01 SD) range 0.36 – 1.45	0.7 (± 0.25 SD)	0.67
Bathymetry Slope (^{0})[f]	Slope degree of bathymetry in marine megafauna sighting	$0^o – 90^o$	46^0 (± 27 SD)	30°
Distance to the near coast (km)[b]	Distance between marine megafauna sighting and coast	N/A	4.75 (± 2.3 SD)	3

[a]Literature
[b]Field observation
[c]WXTide32
[d]NASA Horizons
[e]NASA's MODIS
[f]GEBCO

4.2. Environmental drivers on predicting zooplankton biomass response

The GAMs for zooplankton biomass and environmental predictor was tested in 2 model scenarios, which are shown in figure 2. The first model examines zooplankton biomass as a response of the predictors (*percent to full moon, distance of sampling station to the coast, sea surface temperature, and phytoplankton biomass*) and was explained as 87.7% of the deviance of zooplankton biomass in the waters of SI and LI. Results from the first model shows that moon phase (F= 3.6, edf= 5.6, p<0.05) and phytoplankton biomass (F= 9.3, edf= 7.85, p<0.05) were a more significant predictor for zooplankton biomass response than sea surface temperature (F= 4.9, edf= 1.8, p<0.05). Zooplankton biomass was not significantly related to distance of sampling station to the coast (F= 0.4, edf= 1, p>0.05). The second model shows the zooplankton biomass as the response of predictors (*percent to full moon, phytopankton biomass, sea surface temperature, and time of sampling to high tide*) was 88.8 % of the deviance of zooplankton biomass, which was higher than model 1. Model 2 identified moon phase (F= 4.9, edf= 7.6, p<0.05), phytoplankton biomass (F= 4.6, edf= 6.4, p<0.05) and sea surface temperature (F= 11.54, edf= 1, p<0.05) as the most important predictor. However,time of sampling to high tide (F= 1.8, edf= 2.2, p>0.05) which were not significant predictors.

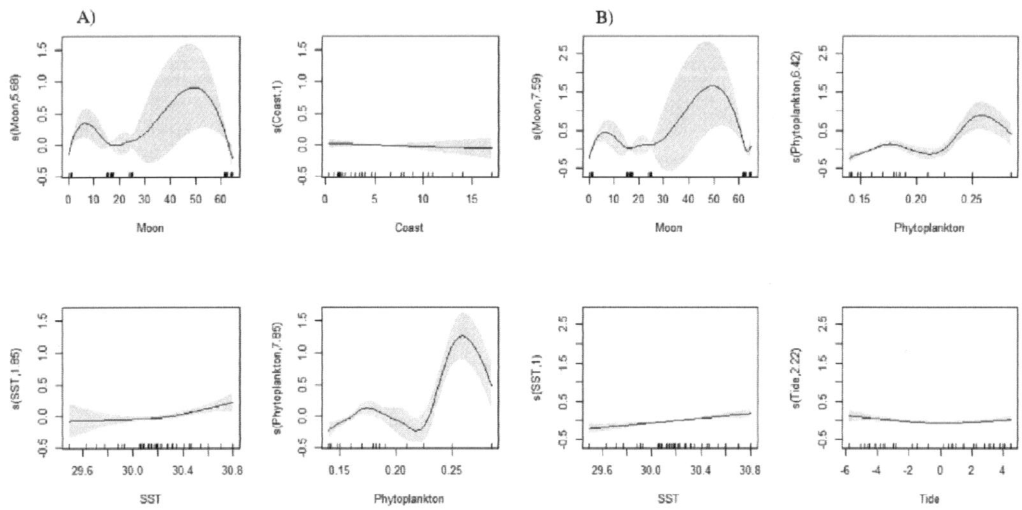

Figure 2. Generalize adaptive models final output of zooplankton biomass as the respone of a) *percent to full moon, distance to coast, sea surface temperature, and phytoplankton biomass*; b) *percent to full moon, phytopankton biomass, sea surface temperature, and time of sampling to high tide*. Zero on the y axis indicates no effect of the predictor. The magnitude reflects the importance of variable. Shade lines and error bars represent 95% confidence intervals.

The first predictor, which was the fraction of moon illumination (*percent to full moon*)was the most significant predictor for both model 1 and model 2, which showed a higher zooplankton biomass response than the other predictors. Waxing gibbous and waning gibbous are the moon phases with the highest response of zooplankton biomass (figure 2), the magnitude of zooplankton biomass significantly increased during these moon phases. A similar pattern was observed in 2015, where catch rate of the planktivore feeder, *Manta birostris*, was three times higher during waxing gibbous and waning gibbous moon phase (Misool Baseft in *unpublished data*). However, studies in Komodo National Park (KNP) and Lady Elliot Island (LEI) showed a different pattern. Manta rays were more abundant in KNP and LEI during the new moon and full moon when the tidal flux was highest [29, 33]. In contrast, the waters of SI and LI showed that during the full moon and new moon mobulids ray catch (i.e. *Manta birostris, Mobula japonica, and Mobulatarapacana*) in Lamakera were the lowest. The fraction of moon illuminated (*percent to full moon*) is a key factor that Lamakeran fisherman use to determine a potential hunting period, which reportedly starts two-four days after the new moon, i.e. a waxing crescent moon [19]. GAMs models indicate that the beginning of a waxing crescent moon (0-

10 *percent to full moon*)significantly influences the response of zooplankton biomass, and thus drives the occurrence of higher trophic level species such as manta rays, dolphins and other marine megafauna in the waters of SI and LI. The second predictor, phytoplankton biomass, also significantly correlated with zooplankton biomass in both model 1 and model 2. Since satellite-derived chlorophyll-a concentration appears to be a proxy for biomass of zooplankton, satellite data is animportant predictor for studying marine megafauna[29, 34].The third predictor, sea surface temperature,is commonly reported as being a critical predictor for modelling zooplankton.Several studies have documented that zooplankton biomass will be higher in cooler temperatures [30]. However, in this study both model 1 and model 2 found thatthere was a positive correlation between warmer SST and zooplankton biomass. However, this result does not necessarily signify that higher zooplankton biomass will be found in in warmer temperatures, this is because in the study region there is very little temperature variation in January with temperature range generally between 29°C – 30°C. Further analysis throughout the entire year is needed to determine the effect of SST on zooplankton biomass.

Table 2. Candidate generalize addaptive models for predicting zooplankton biomass as the function of various candidate explanatoryvariables, with a comparison of the model's goodness-of-fit in terms of coefficient of determination value (R^2) and Akaike Information Criterion (AIC) values. Model 2 was the better candidate model, which more lower AIC value and highest determination value.

Model	R^2	GCV Score	AIC	Deviance explained
Model 1= Zooplankton ~s(Moon)+ s(Coast)+s(SST)+s(Phytoplankton)	0.79	0.012	-71.33	87.7%
Model 2 = Zooplankton ~s(Moon)+ s(Phytoplankton)+s(SST)+s(Tide)	0.81	0.012	-73.69	88.8%

The fourth predictor, tide cycle,has been found by several studies to beanother important predictor for modelling zooplankton and thus planktivores [29-30]. Results from our GAMs output showed that tide cycle did not significantly effect zooplankton biomass in this study. This is contradictory to other studies that have shown a strong relationship between tide and zooplankton biomass. However, this is a similar case to that of reef manta rays and whale sharks in southern Mozambique that showeda weak relationship with the time to high tide [35]. Our fifth predictor, distance of sampling station to the coast, showed no significant correlationto zooplankton biomass. This result is similar to a study in Oman, which showed no strong correlation between distance to the coast and cetacean habitat.

Based on coefficient of determination value (R^2) and Akaike Information Criterion (AIC) values in Table 2. Model 2 was the most comprehensive and showed better performance in explaining zooplankton response to different oceanographic predictors in the waters of SI and LI.

4.3. Plankton spatial modelling

Spatial biomass distribution of both phytoplankton and zooplankton was completed (final) using IDW with search radius using variable method, and cell size 0.01^0 (100 meters to 100 meters). Twelve station we used to assess the accuracy ofmodels. Both models of phytoplankton biomass and zooplankton biomass was performance in high accuracy. Phytoplankton biomass of IDW models was the higher accuracy with mean relative error 4.23%than zooplankton biomass 12.14 % (see Table 3). Spatial interpolation of IDW as a part of deterministic interpolation which calculate from measured point that are closely and give most influence, than IDW assume that a distance each sampling point as critical role on build high accuracy models [36-37]. In this case, phytoplankton and zooplanktonwas sampled each 4 km according a station sample of Aqua MODIS (4km or level 3) for this region as figure 1 showed. Sampling effort which was conducted show that the plankton distribution was affected by sampling density and distance each sampling station. Although, spatial interpolation are available in several method such as kriging and spline, some researcher was found that no significant differences between method. The most significant for build a high-resolution spatial models is sampling density [38].

Spatial analysis of plankton distribution in this study was show that are both phytoplankton (0.14 mg.m^{-3}) and zooplankton (1.45 mg.m^{-3})was higher in Lamakera strait and west part of SI. The spatial patterning was get smooth on colouring transition as results fromdensity and distance station sample which showing Figure 1, 3a and 3d. Based on results of GAMs models, the phytoplankton biomass is a strong predictor for modelling zooplankton biomass and that are was approved by spatial models which was recognise that are phytoplankton biomass drives zooplankton biomass distribution. Although, in some cases there are not positive correlate as effects of role from other predictor such as tidal current which drives and accumulate zooplankton in sheltered and shallow waters location such in SI bay area (Figure 3a).Study in Kilindoni Bay off Mafia Island, Tanzania was found there are tidal currant play a critical role in accumulate zooplankton in this area and motivate whale sharks to feeding in regularly[39-40].

Table 3. Accuracy assessment of inverse distance weighting models (0.01°) of zooplankton biomass and phytoplankton biomass. (RE %) is relative error between models output and field data each station "control", and (MRE %) mean relative error is relative error for full model.

Station ID	Field data zooplankton biomass mg.m^{-3}	IDW model zooplankton biomass mg.m^{-3}	RE (%)	MRE (%)	Satellite data phytoplankton biomass mg.m-^{3}	IDW model phytoplankton biomass mg.m^{-3}	RE	MRE
2	0.78	0.65	16.59	12.14	0.17	0.17	0.00	4.23
13	0.82	0.78	4.34		0.28	0.21	26.32	
22	0.49	0.54	11.08		0.15	0.15	0.00	
24	0.66	0.51	22.57		0.15	0.15	0.00	
25	0.51	0.55	8.23		0.16	0.16	0.00	
32	0.44	0.48	9.22		0.14	0.15	5.63	
33	0.81	0.54	33.60		0.16	0.16	0.00	
40	0.52	0.55	6.41		0.18	0.18	2.70	
43	0.53	0.54	1.65		0.15	0.16	6.67	
44	0.44	0.47	7.45		0.15	0.15	0.00	
45	0.42	0.50	19.97		0.18	0.17	8.11	
48	0.45	0.47	4.57		0.14	0.15	1.35	

4.4. Detecting the foraging area of marine megafauna in the waters of Solor

Based on spatial interpolation output, 129014 station data with spatial resolution 0.01°was we test using *Getis-Ord Gi**spatial statistical analysis to generate area which identify as hotspots (high productivity) and cold-spot (low productivity) both phytoplankton and zooplankton. Four map ware generate to identify foraging area of marine megafauna in the waters of SI and LI: Figure 3c and 3f are visualize from a Z-score to determining confidence thresholds of hotspots and cold-spot both phytoplankton and zooplankton. Moreover, to identify foraging area we classify the significances hotspot of biomass threshold as *90%*, *95%*, and *99%* or *99.9%*.Statistical significant *90%* of phytoplankton biomass hotspots was (lowest *Z-score* 1.645, *p-value* 0.099, phytoplankton biomass 0.177 mg.m^{-3}; highest *Z-score* 1.957, *p-value* 0.05, phytoplankton biomass 0.188 mg.m^{-3}), *95%* of hotspots (lowest *Z-score* 1.977, *p-value* 0.047, phytoplankton biomass 0.179 mg.m^{-3}; highest *Z-score* 2.573, *p-value* 0.01, phytoplankton biomass 0.189 mg.m^{-3}), *99%* and or*99.9%* of hotspots (lowest *Z-score* 2.581, *p-value* 0.0098, phytoplankton biomass 0.194 mg.m^{-3}; highest *Z-score*13.66, *p-value* 0.00, phytoplankton biomass 0.25 mg.m^{-3}). Zooplankton biomass threshold for statistical significant of *90%* hotspots (lowest *Z-score*2.543, *p-value* 0.01, zooplankton biomass 0.662 mg.m^{-3}; highest *Z-score*

2.872, *p-value* 0.004, zooplankton biomass 0.676 mg.m^{-3}), *95%* hotspots (lowest *Z-score* 2.876, *p-value* 0.004, zooplankton biomass 0.677 mg.m^{-3}; highest *Z-score*3.504, *p-value* 0.0004, zooplankton biomass 0.754 mg.m^{-3}), *99%* and or *99.9%*hotspots (lowest *Z-score*3.51, *p-value* 0.0004, zooplankton biomass 0.755 mg.m^{-3}; highest *Z-score*23.01, *p-value* 0.00, zooplankton biomass 1.445 mg.m^{-3}).Figure 3b and 3e is spatial clustering based on Z-score and *p-value*, a highest Z-score and smallest *p-value* are indicated as hotspots for both phytoplankton biomass and zooplankton biomass. Then, a low negative Z-score and smallest *p-value* are indicated as cold-spot for both phytoplankton biomass and zooplankton biomass. Z-score near zero will classify as no clustering.

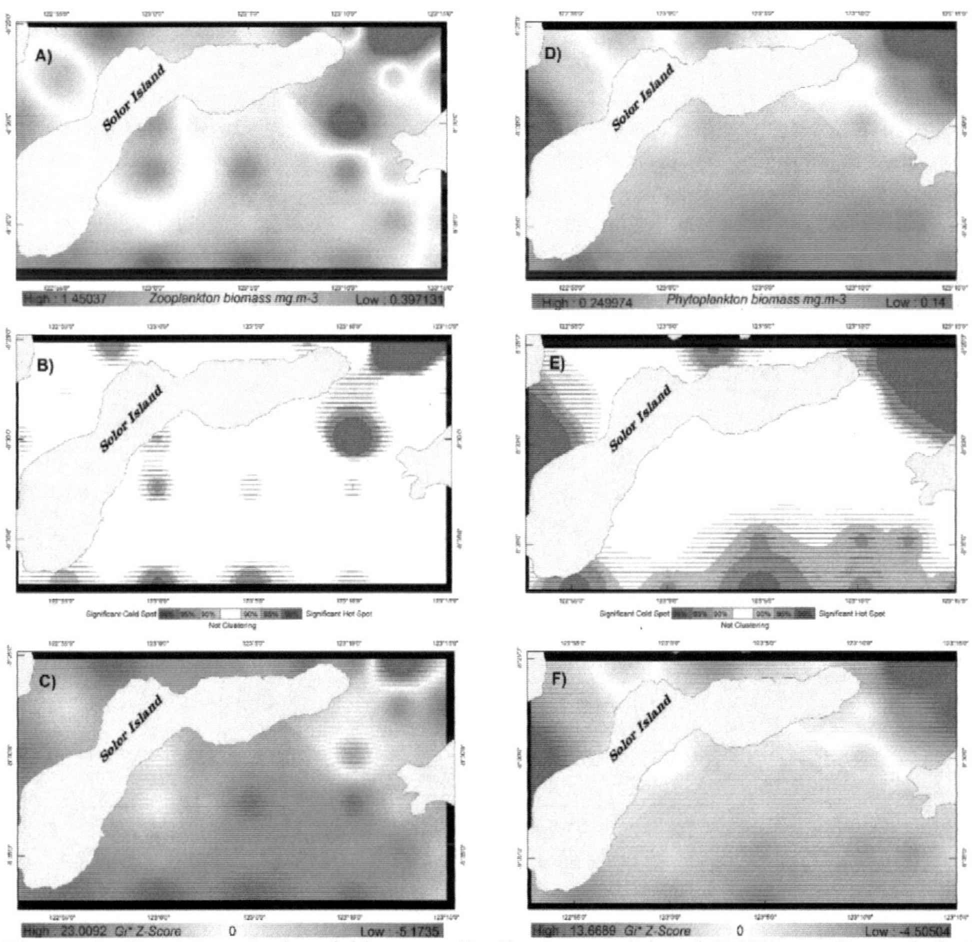

Figure 3. Spatial analysis of biomass distribution models of IDW(row 1); statistical significant hotspot and cold-spot *Getis-Ord Gi**(row 2); *Gi***Z-score* statistical significance threshold values; column right as zooplankton and column left as phytoplankton. Red colour as indicated high value (hotspots), blue colour as indicated low value cold-spot.

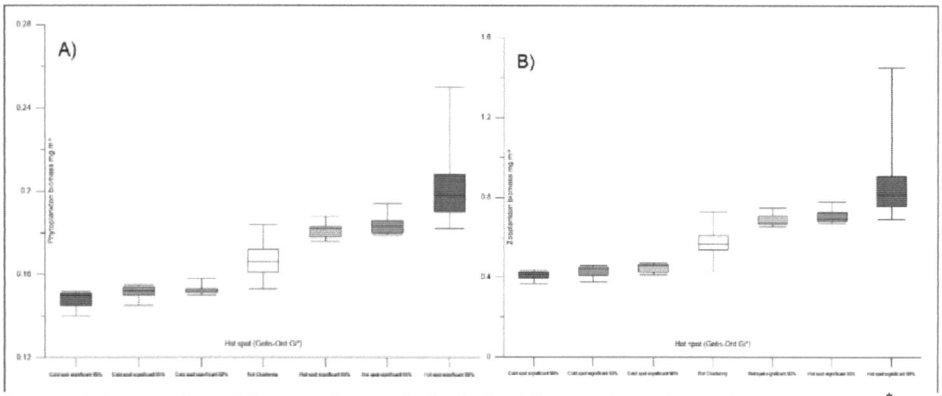

Figure 4. Comparison biomass by statistical significant clustering of *Getis-Ord Gi** a) phytoplankton mg.m^{-3}, b) zooplankton mg.m^{-3}.

A) Plankton groups showing highest productivity

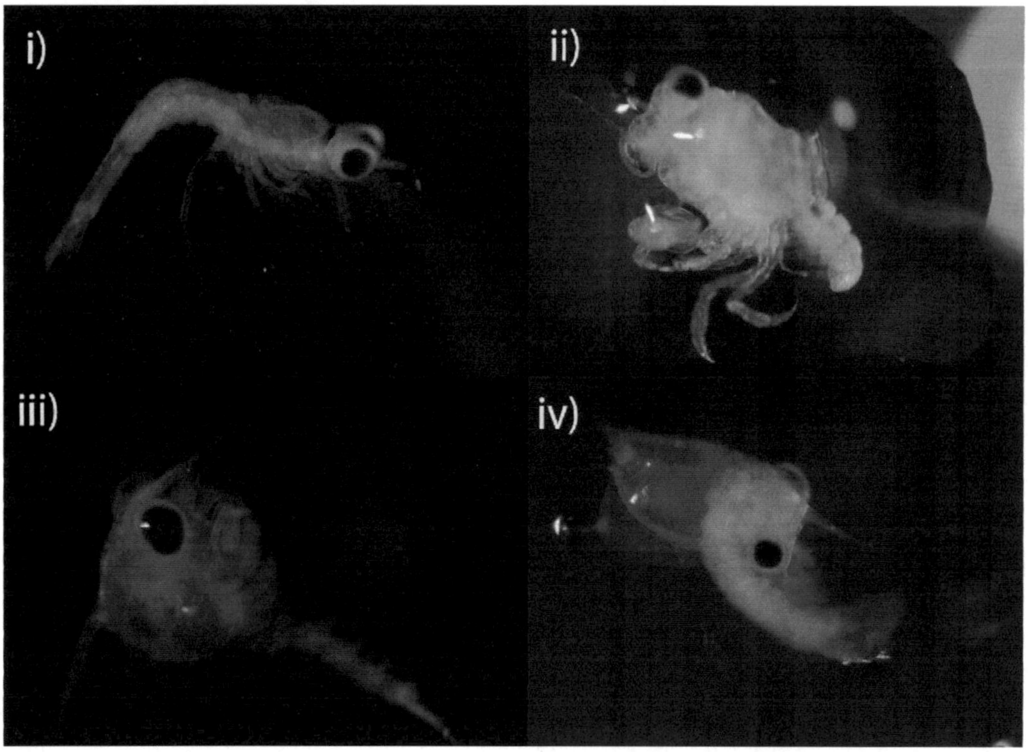

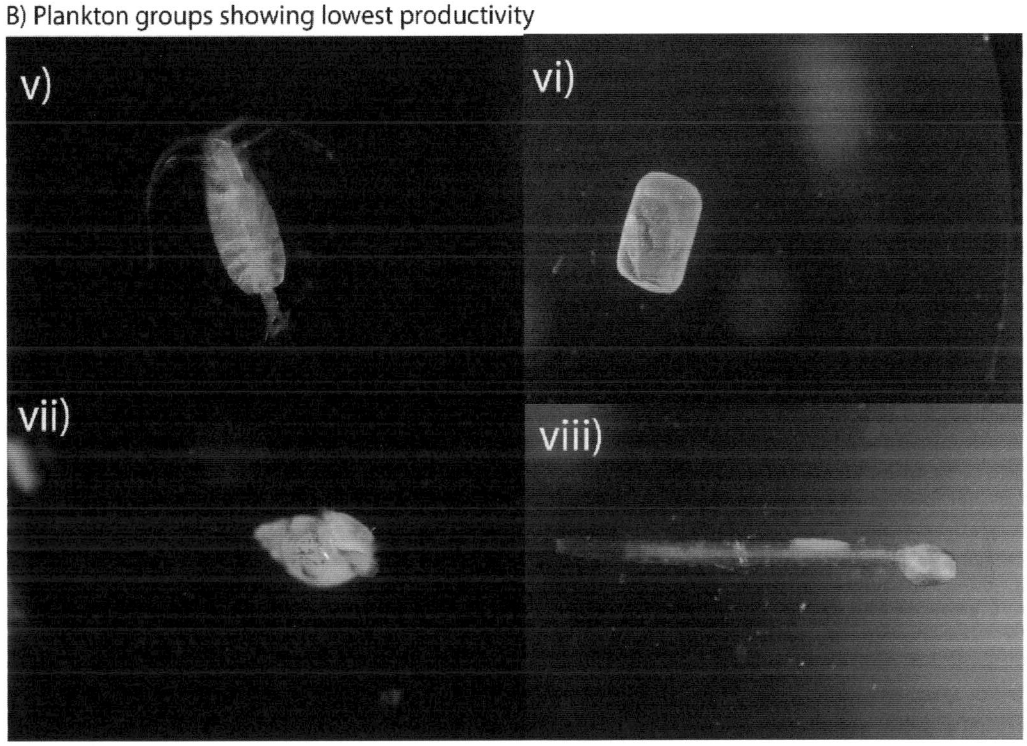

Figure 5.zooplankton group composition in different station; location with the highest productivity are dominated with *macro zooplankton* such i) shrimp, ii) crab, iii) crab, iv) crab; and location with lowest productivity are dominated withv) copepods, vi) egg vii) mollusca, vii) chaetognatha.

Overall, *Getis-Ord Gi** spatial statistical analysis was identify a potential hotspots for foraging area based on plankton biomass modelling which are distribute in high dense patch in Lamakera strait. A high dense patch of hotspots are exist in small area an approximately 15% - 30% of total waters off SI and LI. Plankton biomass was higher in specific location which has characterize by oceanographic process in this region.Combined by tidal currents surrounding straits of Solor, Lamakera, and Bolinghas driving zooplankton from various direction to the same place in Lamakera strait. Bathymetric across this channel are quite shallow in the north part and extremely slope in the south part off SI and south west of LI, which has provide a strong upwelling and drives a water masses from Savu Sea with high productivity in Lamakera strait (see Figure 6). Interestingly, that combination has resulting current turbulence and accumulate high dense patch of zooplankton biomass in this area. Shrimp and crabhas dominate the zooplankton composition in the waters with high productivity, which mean there are is main key on transferring energy to marine megafauna in trophic ecology (Figure 5) [41]. Environmental drives has strong explanatory for the dynamics of zooplankton in this area. A comparison biomass by statistical significances as Figure 4 showed we used to approvingthis theory. Statistical test was show that are 99% significant hotspot are have high range than other classify, that indicated something exceptionally unusual has happened at this location in terms of the spatialconcentration of biomass was higher in Lamakera strait as results environmental drives. Statistical one t-test we used to generalize a significances of hotspot analysis. The results was showed the foraging behavior was significantly related to plankton biomass hotspots (see Figure 6), which were two times higher compared to plankton cold-spots of phytoplankton (0.2 ± 0.01 SD vs 0.14 ± 0.003 SD mg.m^{-3}) and zooplankton (0.87 ± 0.17 SD vs 0.41 ± 0.01 SD mg.m^{-3}). Whale sharks in Kilindoni Bay, off Mafia Island, Tanzania was show same pattern that the foraging area was characterize by high dense patch zooplankton which 10 times higher (25 vs. 2.6 mg m^{-3}) [40].

Table 4. Statistical summary and sample one t-test of plankton biomass *Getis-Ord Gi**output.

Plankton	*Getis-Ord Gi**	Mean biomass mg.m^{-3}	SD	CV	*p-value*	95% confidence interval (mg.m^{-3})
Phytoplankton	Cold spot 99%	0.148	0.003	0.02	<0.05	0.1482 – 0.1484
	Cold spot 95%	0.151	0.002	0.01	<0.05	0.1516 – 0.1517
	Cold spot 90%	0.152	0.002	0.01	<0.05	0.1528 – 0.1529
	Not clustering	0.165	0.006	0.60	<0.05	0.1653 – 0.1654
	Hot spot 90%	0.180	0.002	0.01	<0.05	0.1803 – 0.1805
	Hot spot 95%	0.183	0.003	0.01	<0.05	0.1833 – 0.1835
	Hot spot 99%	0.200	0.010	0.06	<0.05	0.2004 – 0.2009
Zooplankton	Cold spot 99%	0.41	0.01	0.04	<0.05	0.408 – 0.411
	Cold spot 95%	0.43	0.02	0.05	<0.05	0.431 – 0.433
	Cold spot 90%	0.44	0.02	0.04	<0.05	0.446 – 0.448
	Not clustering	0.55	0.05	0.09	<0.05	0.554 – 0.555
	Hot spot 90%	0.68	0.02	0.03	<0.05	0.685 – 0.687
	Hot spot 95%	0.70	0.02	0.03	<0.05	0.701 – 0.703
	Hot spot 99%	0.87	0.17	0.19	<0.05	0.865 – 0.876

4.5. Applied modelling for management and conservation of marine megafauna

Our models provide fine-scale of the biomass distributionand highlight the importance of prey availability on driving a distribution of marine megafauna. The potential of foraging area was characterize by environmental properties such as phytoplankton biomass, sea topographic, sea surface temperature and tidal current, which provide their environmental requirement for feeding and thermoregulation.Three dimensional models of bathymetry in Figure 6a and 6bwas help us to understanding the distribution of marine megafauna related characteristic and complexity of sea topographic. Extremely step slope, which are characterize Lamakera strait, has support strong upwelling and thus highly productive waters that influencing the presence of marine megafauna. Even though, in several observation, dolphin has sighting in the location which are not characterize as plankton biomass hotspot, but field observation was found that are dolphin move from south part/open Ocean to Lamakera strait which are characterize by high productivity.The dolphin was move such indicationhas response the area with high productivity. Several animal movement studies has documented there are animals has move as a response to exploring food, reproduction, and climate change [42].

Pervious study was found that are this region such critical habitat for some cetacean (i.e. spinner dolphin and pantropical spotted dolphin) which can observed every year [43]. This study detects that the Lamakera strait provides the planktonic requirements for marine megafauna foraging, helping to explain why this region supports such high diversity and abundance of marine megafauna.Modelling approach are powerful to help studying marine environmental, which are large area, dynamic, and complex. Then, this study has identified that are this location being key habitat for marine megafauna and being priority location for conservation.

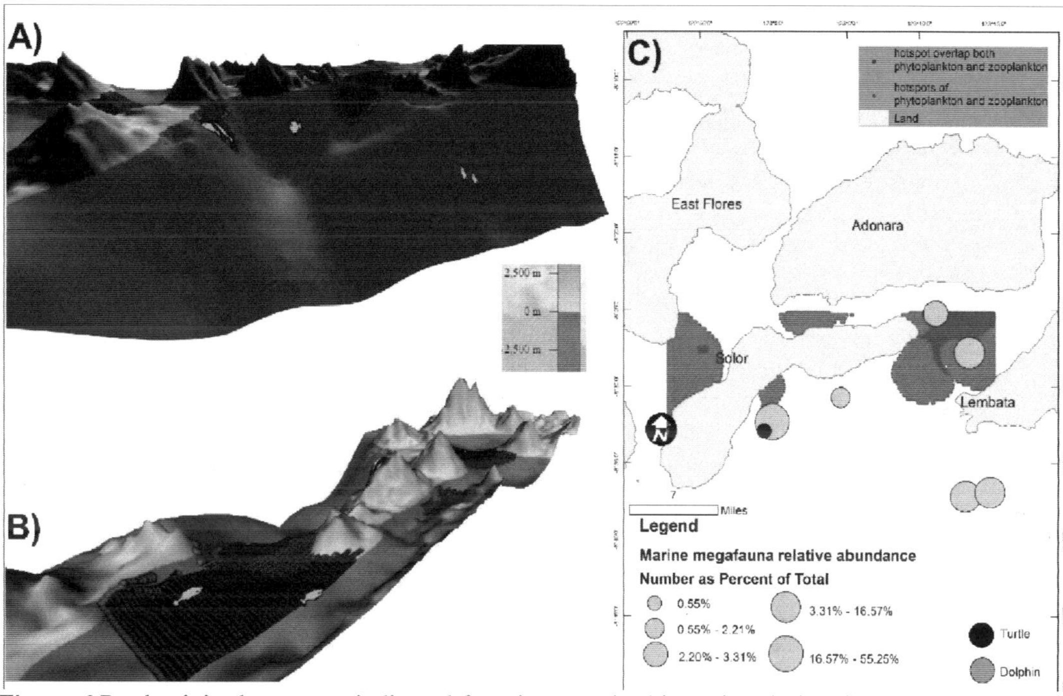

Figure 6. Productivity hotspot as indicated foraging area in this region during field survey in January 2015. 3D visualization topography models GEBCO of Solor region; a) south part of Solor and b) northeast of Solor. Red dots indicated plankton hotspot as statistically significant test of *Getis-Ord Gi** 99%, 95%, 90%. Fish symbol indicated marine megafauna sighting during study. C) 2D visualization; red color indicated hotspots of phytoplankton and zooplankton; dark red indicated hotspot overlap both phytoplankton and zooplankton.

5. Conclusion

Here we present GIS and remote sensing approach to modelling a plankton biomass hotspotswhich together triggered the marine megafauna feeding aggregation in the waters of SI and LI. Our GAMs model found that are zooplankton biomass was drives by environmental properties such as moon phase and phytoplanktonas a higher predictorfor zooplankton biomass response. Linking environmental dynamic in understanding the plankton biomass distribution is helpful to figure out the response of trophic level above such as small fish which will drives a dolphin.Our spatial statistical analysis examine that are a potential of foraging area was 2 times higher of plankton biomass than other location which identify as cold-spot. Including zooplankton biomass in predicting critical habitat of marine megafauna habitat is important to build high accuracy model, and several studies was examine that incorporating prey item was improve cetacean habitat model [2]. Moreover, spatial analysis approaches are useful tool that helps managers to recognize a critical habitat for marine megafauna especially endangered species, and further to build strategic design and management plans for MPAs.

Acknowledgment

We are grateful to the Indonesian Biodiversity Research Center (IBRC) for providing research instrument and laboratory research facilities to analyse zooplankton sample. We thank all to IBRC scientist (Andrianus Sembiring, Ni Putu Dian Pertiwi, Yuli Syamsuni, and Astria Yusmalinda) and Rafid Shidqi for laboratory work. We thank to the NASA Ocean Biology Processing Group for providing of MODIS satellite data. Lamakeran fisherman for assisting during field work on collecting zooplankton data and marine megafauna sighting. This project as part of Lamakera Manta Conservation Transition Phase-Bridge Funding Grant "*A New Day in Lamakera*" and funding has providing by Vulcan and Mac Arthur.

References

[1] Schipper J, Chanson, JS, Chiozza F, Cox NA, Hoffmann M, Katariya V, Lamoreux J, Rodrigues, AS, Stuart SN, Temple HJ and Baillie J 2008 The status of the world's land and marine mammals: diversity, threat, and knowledge*Science***322** (5899) pp.225-230

[2] Lambert C, Mannocci L, Lehodey P, Ridoux V 2014 Predicting Cetacean Habitats from Their Energetic Needs and the Distribution of Their Prey in Two Contrasted Tropical Regions. *PLoS ONE***9**(8): e105958. doi:10.1371/journal.pone.0105958

[3] Mannocci L, Catalogna M, Dore´mus G, Laran S, Lehodey P, et al 2014 Predicting cetacean and seabird habitats across a productivity gradient in the South Pacific gyre*Prog Oceanogr***120** 383–398

[4] Peterson AT 2011 *Ecological niches and geographic distributions(MPB-49) (No. 49)*. Princeton University Press

[5] Preen A R, Marsh H, Lawler, IR, Prince RIT and Shepherd R 1997 Distribution and abundance of dugongs, turtles, dolphins and other megafauna in Shark Bay, Ningaloo Reef and Exmouth Gulf,Western Australia *Wildlife Research***24** 185–208. doi:10.1071/WR95078

[6] Bradshaw JA, Higgins J, Michael K, Wotherspoon SJ and Hindell MA 2004 At-sea distribution of female southern elephant seals relative to variation in ocean surface properties*ICES Journal of Marine Science***61** 1014–1027. doi:10.1016/J.ICESJMS.2004.07.012

[7] Littaye A, Gannier A, Laran S and Wilson JPF 2004 The relationship between summer aggregation of fin whales and satellite-derived environmental conditions in the northwestern Mediterranean Sea*Remote Sensing of Environment***90** 44–52

[8] Richardson AJ 2008 In hot water: zooplankton and climate change*ICES Journal of Marine Science: Journal du Conseil***65**(3) 279–95

[9] Kideys AE, Kovalev AV, Shulman G, Gordina A and Bingel F 2000 A review of zooplankton investigations of the Black Sea over the last decade*Journal of Marine Systems***24** 355–371. doi:10.1016/S0924-7963(99)00095-0

[10] Wilson SG, Pauly T and Meekan MG 2002 Distribution of zooplankton inferred from hydroacoustic backscatter data in coastal waters off Ningaloo Reef,Western Australia. *Marine and Freshwater Research***53** 1005–1015. doi:10.1071/MF01229

[11] Sims DW 2003 Tractable models for testing theories about natural strategies: foraging behaviour and habitat selection of free-ranging sharks*Journal of Fish Biology***63** 53–73

[12] Mauchline J 1998 The biology of calanoid copepods*Advances in Marine Biology***33**710 pp

[13] McKinnon AD, Richardson AJ, Burford MA and Furnas MJ 2007*Vulnerability of Great Barrier Reef plankton to climate change. In Climate Change and the Great Barrier Reef*Great Barrier Reef Marine Park Authority. p 121–152

[14] Kahn B, Pet J 2003 Long-term visual and acoustic cetacean surveys in Komodo National Park, Indonesia 1999-2001: Management implications for large migratory marine life. In: Proceedings and Publications of the World Congress on Aquatic Protected Areas 2002. Australian Society for Fish Biology. Murdoch, WA.

[15] DeVantier LM, Turak E, Allen G 2008 Lesser Sunda ecoregional planning coral reefstratification. Reef and seascapes of Lesser Sunda ecoregion. The Nature Conservancy, Arlington, VA.

[16] Kahn B 2010*Seismic Surveys and Offshore Exploration in Highly Sensitive Marine Areas: Regulatory Guidelines and Best Practices for Contractors and Operators Working in the Indonesian Seas*." Bali: Apex Environmental. p

[17] Mustika PLK 2008*Marine mammals in the Savu Sea (Indonesia): indigenous knowledge, threat analysis and management options. PhD Thesis*(James Cook University)

[18] Setiasih N*Manta Ray of Hope Indonesia Fishery Investigation*. Shark Saver and WildAid.

[19] Lewis SA, Setiasih N, O'Malley MP, Campbell SJ, Yusuf M and Sianipar AB 2015 Assessing Indonesian manta and devil ray populations through historical landings and fishing community interviews (No. e1642). *PeerJ PrePrints*.

[20] Potemra JT, Sprintall J, Hautala SL and Pandoe W 2003 Observed estimates of convergence in the Savu Sea, Indonesia *J. Geophys. Res***108**(C1), 3001, doi:10.1029/2002JC001507.

[21] Khan B 2003*Solor-Alor visual and acoustic cetacean survey: interim report April-May 2003*

survey period. The Nature Conservancy South East Asia Center for Marine Protected Area. Bali. Indonesia.

[22] WWF-Hongkong 2003*Marine Mammals. In Marine Mammals Survey Technique. Proceeding of the workshop on cetacean species identification, Marine Mammals survey Technique and Review of Regional Knowladge, 29-30 Januari 2003, Indonesia.* Departemen Kelautan dan Perikanan dan WWF-Indonesia, Jakarta, Indonesia.

[23] Khan B 2001. *Komodo National Park: A Rapid Ecological Assessment of Cetacean Diversity, Abundance, and Distribution: Monitoring Report April 2001. 1999-2000 Synopsis.* TNC Indonesia Program. Coastal and Marine Conservation Center. Bali, Indonesia.

[24] R Core Team 2012 R: a language and environment for statistical computing. Vienna, Austria: R Foundation or Statistical Computing. Available: http://www.R-project.org/.

[25] ESRI 2011*ArcGIS Desktop: Release 10.* Redlands, CA: Environmental Systems Research Institute.

[26] Ord JK, Getis A 1995 Local spatial autocorrelation statistics: distributional issues and an application*Geogr. Anal***27** (4), 286–306.

[27] Rud R, Shoshany M, Alchanatis V 2013 Spatial–spectral processing strategies for detection of salinity effects in cauliflower, aubergine and kohlrabi*Biosyst. Eng***114** (4), 384–396.

[28] Kuletz KJ, Ferguson MC, Hurley B, Gall AE, Labunski EA and Morgan TC 2015 Seasonal spatial patterns in seabird and marine mammal distribution in the eastern Chukchi and western Beaufort seas: Identifying biologically important pelagic areas*Prog Oceanogr***136** (2015) 175–200. http://dx.doi.org/10.1016/j.pocean.2015.05.012

[29] Jaine FRA, Couturier LIE, Weeks SJ, Townsend KA, Bennett MB, Fiora K, Richardson AJ 2012 When Giants Turn Up: Sighting Trends, Environmental Influences and Habitat Use of the Manta Ray Manta alfredi at a Coral Reef*PLoS ONE***7**(10): e46170. doi:10.1371/journal.pone.0046170

[30] Armstrong AO, Armstrong AJ, Jaine FRA, Couturier LIE, Fiora K, Uribe-Palomino J, Weeks SJ, Townsend KA, Bennett MB, Richardson AJ 2016 Prey Density Threshold and Tidal Influence on Reef Manta Ray Foraging at an Aggregation Site on the Great Barrier Reef*PLoS ONE***11**(5): e0153393. doi:10.1371/journal.pone.0153393

[31] Baumgartner MF 1997 The distribution of Risso's dolphin (Grampus griseus) with respect to the physiography of the northern Gulf of Mexico*Mar Mamm Sci***13**(4):614–638

[32] Sleeman JC, Meekan MG, Wilson SG, Jenner CK, Jenner MN, Boggs GS, Steinberg CC and Bradshaw CJ 2007 Biophysical correlates of relative abundances of marine megafauna at Ningaloo Reef, Western Australia. *Marine and Freshwater Research***58**(7), pp.608-623.

[33] Dewar H, Mous P, Domeier M, Muljadi A, Pet J, Whitty J 2008 Movements and site fidelity of the giant manta ray, Manta birostris, in the Komodo marine park, Indonesia*Mar Biol***155** 121–133.

[34] Richardson AJ 2004 Climate impact on plankton ecosystems in the northeast Atlantic *Science***305**1609–1612. doi:10.1126/science.1100958.

[35] Rohner CA, Pierce SJ, Marshall AD, Weeks SJ, Bennett MB and Richardson AJ 2013 Trends in sightings and environmental influences on a coastal aggregation of manta rays and whale sharks*Marine Ecology Progress Series***482** pp.153-168.

[36] Apaydin H, Sonmez FK and Yildirim YE 2004 Spatial interpolation techniques for climate data in the GAP region in Turkey*Climate Research***28**(1) pp.31-40.

[37] Earls J and Dixon B 2007 Spatial interpolation of rainfall data using ArcGIS: A comparative study *In Proceedings of the 27th Annual ESRI International User Conference***31**

[38] Dirks KN, Hay JE, Stow, CD, Harris D 1998 High-resolution studies of rainfall on Norfolk Island. Part II: Interpolation of rainfall data*J Hydrol***208**(3-4):187–193

[39] Thums M, Meekan M, Stevens J, Wilson S and Polovina J 2012 Evidence for behavioural thermoregulation by the world's largest fish*Journal of The Royal Society Interface*p.rsif20120477.

[40] Rohner CA, Armstrong AJ, Pierce SJ, Prebble CE, Cagua EF, Cochran JE, Berumen ML and Richardson AJ 2015 Whale sharks target dense prey patches of sergestid shrimp off Tanzania. *Journal of plankton research*, p.fbv010.

[41] McClellan CM, Brereton T, Dell'Amico F, Johns DG, Cucknell A-C, Patrick SC, Penrose R, Ridoux V, Solandt JL, Stephan E, Voiter SC, Williams R, Godley BJ 2014 Understanding the Distribution of Marine Megafauna in the English Channel Region: Identifying Key Habitats for Conservation within the Busiest Seaway on Earth*PLoS ONE***9**(2): e89720. doi:10.1371/journal.pone.0089720

[42] Jaine, FAD 2013*The movement ecology of the reef manta ray Manta alfredi in eastern Australia PhD Thesis*School of Geography, Planning and Environmental Management, The University of Queensland.

Distribution of Glass Eel by the Water Surface Salinity Using Landsat TM at Pelabuhan Ratu Bay, West Java

D S Irianto, Supriatna,and TjiongGiok Pin

Department of Geography, Faculty Mathematics and Natural Sciences, Universitas Indonesia

E-mail: desaptr@gmail.com;ysupris@yahoo.com; tjiong.giok@sci.ui.ac.id.

Abstract. Eel (*Anguilla spp.*) is consumed fish that has an important economic value, either for local or international market. Pelabuhanratu Bay is an area with big potential for supplying eel seed. One of important factor, which affect an eel existence, is salinity, because eel migrate from fresh water, brackish, and sea naturally although the otherwise so that need ways to describe the distribution of glass eel by the salinity. To find out the percentage of salinity, it obtained from Landsat 8 Imagery in year 2015 using salinity prediction of Algorithm Cimandiri. The research has been conducted at Cimandiri Estuary, Citepus Estuary, and Cimaja Estuary based on wet and dry months. The existence of glass eel which is obtained from the catch was occurs on dry month when the most catch was occurs at the edge of estuary. The catch is reduced if it's farther from the edge of estuary, at the beach towards the sea and the inside of the river mouth with the percentage of salinity towards the sea is increase while the percentage of salinity towards the river is decrease.

1. Introduction

Eel (*Anguilla spp.*) is the consumption of fish have economic value is important both for the local market and abroad. Market demand will be very high discharge fish reaches 500,000 tons per year, mainly from Japan and Korea, the main suppliers of eel is China and Taiwan [3]. In Indonesia there are seven known species of eel that its spread is almost evenly across the waters that empties into the deep sea.This fish spawning places suspected in the Indian Ocean, namely West Sumatra [5] and [8] and later the larvae are carried by currents and tides in the direction of the coastal. One of the coastal of the south coast of Java that is directly related to the Indian Ocean was Pelabuhanratu. Therefore, these waters have a large potential for the provision of seed eel.

One of the factors that influence the life of the eel is salinity, because naturally fish eel migrate from freshwater, brackish, and marine or vice versa. Measurement of salinity directly is difficult because it requires a huge time and cost. However, [11]have made the prediction equation salinity through sea color sensing or prediction equation salinity through sea color sensing or CDOM (Colored Dissolved Organic Matter) by utilizing multiple-temporal image of Landsat-7 ETM+. The higher the CDOM then the lower salinity and low CDOM then high salinity tend. Salinity basically has no color signal directly, the value of salinity can be estimated from the colors signals are dominant elements contained in waters, such as CDOM.This research, the distribution of salinity values obtained from the equation, by utilizing Landsat 8 OLI (the operational land imager) throughout the year 2015.

According to [8] in the waters of Indonesia there are seven species of eel. *Anguilla celebensis* and *Anguilla borneensisis* an endemic animal in Kalimantan and Sulawesi. *Anguilla interioris* and Anguilla obscura located in waters north of Papua Island. Anguillabicolorpasifica can be found in the northen part of Indonesia's waters (Pacific Ocean). Anguilla bicolor is around the Indian Ocean (to the west of island of Sumatera and the southern island of Java). Anguilla marmorata is a type of eel with cosmopolitan distribution across the tropical waters.

Pelabuhanratu Bay is one of the southern of the island of Java which hasa great potential in providing seed eel.Locations where eel fishing is at the mouth of the river.During the temporary stay in the brackish waters of eel fish in the PelabuhanratuBay many communities to a outcome was later arrested for cultivated or sold directly. Many of the at least catches obtained by eel catchers, depends on the good or bad the salinity in the estuary, as salinity is influenced by tides, sea currents, precipitation in catchment area, evaporation, precipitation and bathymetry of the marine. The condition which aspects influenced the writers to do research in the Pelabuhanratu Bay.

2. Methods

Eel Larvaemigrate from freshwater brackish, and marine or vice versa. Once mature, the fish discharge to the sea in order to reproduce and then die, then the larvae are carried by the current to spawning results beach and towards freshwater through the river. Freshwater has low salinity due to its concentration of less than 0.05 ppt, while the salinity seawater tends to be high because the content of more than 5 part per thousand (ppt) concentration [2]. Before entering the river from the sea, eel larvae are catched by people in the Pelabuhanratu Bay. The results of the arrest depend on factors of ups and downs of tidal, rainfall and other can affect the uneven salinity mixture. Value distribution of salinity can be known from the results of the interpretation of the image by using the equation of the salinity of the Cimandiri algorithm [9].

Making a map of the distribution of salinity with Landsat imagery by 2015 is done using the equation to the algorithm Cimandiri [9], as follows:

$$\text{Salinity (ppt)} = 29{,}983 + 165{,}047 \times (B2) - 260{,}277 \times (B3) + 2{,}609 \times (B4) \quad (1)$$

Where:

B2 = *Blue waves*
B3 = *Green waves*
B4 = *Red waves*

Further data processing is performed after the observation field. Map of the spread of salinity results in data processing data processing early will be merged with the larval fish catching point location where the dots are derived from information obtained from the results of the interview fishermen who catch f eel larvae('impun'). Other data processing are as follows:

1. Data processing precipitation Rainfall data were not presented in the form of a map but in tabular form for then analyzed with salinity
2. Processing of data forecasting tides seawater surface Data about the ups and downs of sea water is not presented in the form of a map but in tabular form for later analyzed by the salinity of the surface and the distribution of eel larvae (impun). To simplify the analysis will be made of graphs forecasting tides per hour

The processing of data of sea surface currents (direction and speed of the flow) average data speed and direction of the surface current remains daily tabular data are presented as the source of the data. The current sea level at each time of observation are also presented for analysis.

3. Result and Discussion

In this research on the measurement of salinity surface waters is done based on a wet month (March 24, and April 25, 2015) and the dry month (August 15 and September 16, 2015). For more details can be seen in Figure 1, 2, and 3.

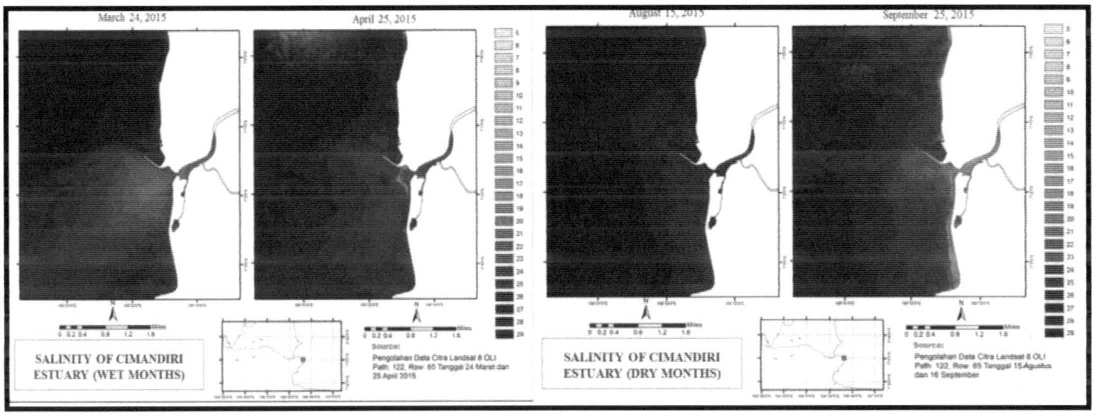

Figure 1. Salinity Surface Waters of Cimandiri Estuary in wet and dry months

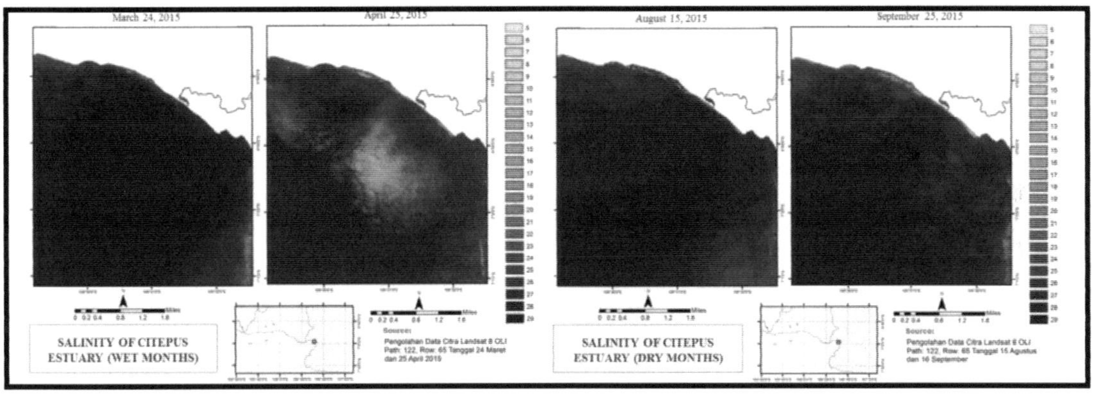

Figure 2. Salinity Surface Waters of Citepus Estuary in wet and dry months

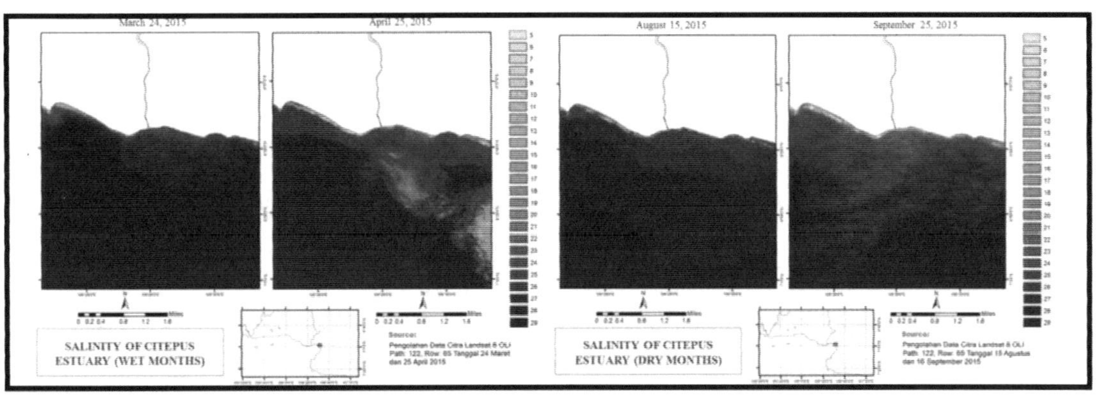

Figure 3. Salinity Surface Waters of Cimaja Estuary in wet and dry months

Based on the results of the measurements of salinity surface waters using Landsat ETM using Cimandiri Algorithm [9] based on known CDOM:

1. The whole area of study is the estuary in the Pelabuhanratu Bay which has a minimum and maximum salinity values that be different. The value of the salinity in the waters of the Pelabuhanratu Bay has the minimum salinity values 5,02683 ppt on April 25, 2015 and the maximum value is 32,2708 ppt on August 15, 2015.The Cimandiri Estuary of the minimum salinity value 5,0268 ppt and 29,7146 ppt maximum value on April 25, 2015.The Cimandiri Estuary the minimum salinity values is 5,5553 ppt and 29,7146 ppt maximum value on April 25, 2015 and estuary salinity value Cimaja Estuary minimum 5,4249 ppt on August 15, 2015 and the maximum value 29,7146 ppt on April 25, 2015.
2. The region with salinity 0.5-5 ppt of the results from Algorithm salinity model was not found in the area of research in the region of the Pelabuhanratu Bay (Cimandiri, Citepus and Cimaja Estuary).
3. Areas with salinity value 5-18 ppt of the results from Algorithm salinity model can be found on the estuary and part of the Pelabuhanratu Bay.
4. Areas with salinity value 18–30 ppt of the results from Algorithm salinity model can be found on most of the estuary and the Pelabuhanratu Bay.

Estuary area have salinity values varied between 0.5-30 ppt [6].The value of the salinity of the Pelabuhanratu Bay based on the estuary of the Cimandiri, Citepus and Cimaja obtained from the results of the prediction at the Cimandiri Algorithm upon salinity CDOM made by (Supriatna et al., 2015) averaged just consists of two zones are Mexo-mesohaline Mexo-and polyhaline. Based on table 1, the zoning of the Cimandiri estuary, Cimaja estuary and Citepus Estuary averages has 2 zones are Mexo-mesohaline (5-18ppt) and Mexo-polyhaline(18-30 ppt)

Table 1. The zone of Surface Salinity estuaries Citepus Cimaja and Cimandiri

Location	Salinity (ppt) by 2015				Zone
	March 24	April 24	August 15	Sept 16	
Cimandiri	28.8009	29.7146	28.1239	28.3983	Mexo-mesohaline and Mexo-polyhanline
	5.04429	5.02683	5.58921	5.07053	
Citepus	29.5044	29.7146	28.1374	28.3983	Mexo-mesohaline and Mexo-polyhanline
	6.33286	5.55539	6.88039	5.81782	
Cimaja	29.5044	29.7146	27.5934	28.3983	Mexo-mesohaline and Mexo-polyhanline
	5.96557	5.78921	5.42493	5.68135	

Source: Data processing (2016)

Here are the waters of zoning in the Cimandiri, Citepus and Cimaja Estuary upon salinity surface waters are seen based on the wet and dry month:

1. In the wet month at the Cimandiri Estuary on March 24, and April 25, 2015, the water has a salinity of 5-18 ppt (Mexo-mesohaline) which is the water of the river enough to dominate the water because of zoning extends approximately 2 kms from the mouth of the river which took place on March 24, while on April 25, 2015. The water of the river extends more or less is a 450 m from the mouth of the river. It can occur due to high rainfall which resulted in a flow of water from the River to the great sea and the mouth of the estuary is Cimandiri River is great river compared Cimaja and Citepus River.
2. In dry in the month of the Cimandiri Estuary on August 15, and September 16, 2015, Mexo-polyhaline zone (18-30ppt) dominating the waters of the estuary. Salinity tend to increase due to low rainfall and sea water brought to the estuary with the help of the tides tend to be much.
3. In the wet and the dry months in the Citepus estuary is Mexo-polyhaline zone, which dominated (18-30 ppt) that is because different estuary type compared to the Cimandiri

Estuary. The mouth of the estuary sand covered Citepus and including small when compared to the estuary mouth of the Cimandiri which resulted in the distribution of water to the sea slightly so that more sea water entering the mouth of the estuary.

4. In the wet and the dry months in the Cimaja estuary is Mexo-zone dominated polyhaline (18-30 ppt). In the wet zone Mexo-mesohaline (5-18 ppt) very small only approximately 70 m from the mouth of the river. That is because the mouth of the Cimaja is not great and water flow its not heavy because it is the type of rocky river.

In the wet months on March 24, and April 25, 2015, have the precipitation resulting in high salinity surface waters estuary at a point sample belongs are range between 17-19 ppt caused the presence of input water is rain water that neutral. On the waters of the Pelabuhanratu Bay itself, the highest salinity on March 24, and April 25, 2015 has a high value is 29.5044 ppt and 32.1454 ppt.

In the dry months on August 15, and September 16, 2015, salinity or high tend to be larger than in the wet.It is also caused by low rainfall even on August 15, and September 16, 2015, no rains in the Pelabuhanratu Bay. On the waters of the Pelabuhanratu Bay itself, the highest salinity has a value of 32.2708 ppt and 29.1798 ppt.

Tides sea water becomes an important factor of mixing fresh water with salt water in the estuary area.The tidal Indonesia into four different,Java's south coast have tidal type mixture leaning daily doubles (mixed tide, prevailing semi diurnal).This is a tide that occurs twice and twice the tide receded during the day but sometimes one plug and one time recede by having different time and altitude. In this study the tidal data used is tide forecasting data that was in the fishing port of Pelabuhanratu. Forecasting data that are used based on the wet (24 March and 25 April 2015) and the dry months (August 15 and September 16, 2015).The adjusted data recording on satellite images Landsat, where time is observed by image is 09.00-10.00.

Condition of the tidal all the time by waters are experiencing image coverage.Where on March 24, 2015, high advance water of 76.26 cm, dated April 25, 2015, level high water of 38.49 cm, August 15, high water of 35.91 cm face and on September 16, 2015, the level of the waters of 40.29 cm. On March 24, 2015 is the highest water level. Tidal is the external power that evokes the movement of a mass of water and affect change high water at the mouth of the face [2], because the densities of fresh water lower than sea water, the mass of the freshwater will tend to be at the top of the mass of sea water. To be able to happen then the mass mixing sea water must be greater than the mass of the freshwater and there should be a driving force that resulted in the throes of water. The higher the face of the water, the greater the mass of the water that goes into the river, so that the mixing of salinity that occurs around the mouth of the greater. The greater mixing of salinity that occurs, the value of increased salinity estuaries.In contrast, low to high sea indicates at least the mass of sea water coming into the estuary, allowing the mixing of salinity that occurs in the estuary into a small.
The results processing of imagery, shows that at the time the highest pairs is 76.26 cm on March 24, 2015, salinity in the estuary of the Cimandiri tend to moderate that is contained in the Mexo-mesohaline salinity value of 15-19 ppt, while in the estuary of the Citepus and Cimaja tend high are ranges of 20-24 ppt. On April 25, 2015 high face of water were in 38.49 cm where the salinity at the mouth of the Cimandiri tend to moderate range of 16-19 ppt, while in the estuary of the Citepus and Cimaja tend high are 20-23 ppt.On August 15, 2015 level of water were in 35.91 cm where the salinity at the mouth of Cimandiri, Citepus and Cimaja tend high are 21- 23.5 ppt. On September 16, 2015 face high water was at 40.29 cm where the salinity at the mouth of Cimandiri, Citepus and Cimaja tend high are 21-23.6 ppt.

It proves that the higher the face of the water, the greater the mass of the water that goes into the river, so that the mixing of salinity that occurs around the mouth ofthe greater.The greater mixing of salinity that occurs, the value of increased salinity estuaries. In contrast, low to high sea indicates at least the mass of sea water coming into the estuary, allowing the mixing of salinity that occurs in the estuary became small is not always evident, because the tidal a variable cannot stand on its own without the influence of other variables.

The speed and direction of the currents of the ocean surface is one of the factors that influence the spread of salinity in the sea.Sea surface current velocity can determine how fast sea water with high salinity can be mixed with river water that has low salinity. The faster the speed of the current sea level then the faster mixing between seawater and river water salinity variation will happen so that the higher. If the speed of the current sea level slowly, then mixing that occurs will be less and less, so the low salinity variation.

Judging from the geographical location, the Pelabuhanratu Bay only got the direction of currents that come from the west and south. The direction of flow on the surface of the waters in the Pelabuhanratu Bay in wet (March 24, and April 25, 2015) and the dry months (August 15, and September 16, 2015). In the wet (rainy season) sea current is moving from the west to the east or south while in dry (dry season) move from south to north or west.

The speed and direction of the flow of surface salinity in the estuary waters. In the wet months on March 24, 2015 at current speed 0.29 m/s with the direction of flow from east to west, the salinity of the surface waters in the vicinity of the mouth of the Cimandiri tend low, due to the location of the mouth of the Cimandiri leading towards the west with the flow direction. At the mouth of the Citepusand Cimaja Estuary, the salinity of the surface are likely to be high because of the location of the mouth of the estuary to the south so no transfers at a rate of flow.

In the month of March 25, 2015, the current speed 0.21 m/s in a direction from north to south flow of the salinity in the estuary of the Cimandiri increase and at the mouth of the Citepus and Cimaja tends to fall because the direction of the mouth of the estuary at Citepus and Cimaja southwards where the direction of the flow on March 25, 2015 from north to south.

In the dry month of August 15, 2015, the speed flow 0.27m/s with the direction of flow from west to east the salinity at the mouth of the Cimandiri likely to moderate because on that date the rain does not occur and the direction of ocean currents towards the mouth of the estuary so that sea water entering the mouth of the estuary of more of the water that comes from the river and at the mouth of the Citepus and Cimaja higher compared with Cimandiri estuary ranged from 20.0-23.5 ppt.

In the dry months on September 16, 2015, speed flow 0.13m/s in a direction from north to south flow of the salinity in the estuary of Cimandiri, Citepus and Cimaja, tend high are ranges between 20-24 ppt.

Changes of salinity on free waters is relatively small when compared with that occurs in coastal regions.As it known to many coastal waters to enter freshwater from the mouth of estuaries, especially at a time of much rain. Closely related to the presence of salinity adjustments osmotic pressure between the cytoplasm of cells in the body of the fish with the circumstances surrounding salinity. In addition to the closely related to the osmotic pressure adjustment, then the salinity also determine the buoyancy of the pelagic eggs of its nature. Besides changes in the salinity of the water mass changes often show and stability conditions.

Eel larvae (impun) do the spawning at a depth of 400-6,000 metres under the sea with a temperature of 16°-18°C and the parent discharge will die after spawning process.After that eel larvae will be heading to waters with low salinity to mature before returning to sea for breed. Based on Figure 4, the following relationship with salinity results catch fish eel larvae (‘impun’) at the mouth of Cimandiri, Citepus and Cimaja in wet and dry month:

1. On the wet months (March 24, and April 25, 2015) are not carried out the arrest, because there isno larvae. In the wet at the mouth of the surface salinity waters of Cimandiri estuary on March 24, 2015, ranges between 15-18 ppt and on April 25, 2015, ranged between 14.3-18 ppt in the Mexo-mesohaline zone. At the mouth of the surface salinity Citepus Estuary waters on March 25, and April 25, 2015, ranges between 20-24 ppt in the Mexo-polyhaline zone. At the mouth of the surface salinity waters Cimaja estuary on March 24, 2015, ranges 20-24,7 ppt and on April 25, 2015 ranged from 20-23,4 ppt in the Mexo-polyhaline zone.

2. In the dry months (August 15, and September 16, 2015) catches eel larvae ('impun') reached 179 kg, where the mouth of the Cimandiri with the largest capture results is 102 kg, the mouth of Citepuswith results is 26 kg and the mouth of the Cimaja with the results is 51 kg.
3. On the dry months in the estuary salinity surface waters of Cimandiri estuary on August 15, 2015 range between 21-23 ppt with the results captured 56 kg and on September 16, 2015, ranged from 21-24 ppt with the results captured 46 kg in the Mexo-polyhaline zone. At the mouth of the surface salinity Citepus estuary waters on August 15, 2015, range between 21-24 ppt with the results captured September 16, 2015, and 15 kg range between 20-22 ppt with the results captured 11 kg in the Mexo-polyhaline zone. At the mouth of the surface salinity waters Cimaja estuary on August 15, 2015, range 22-24 ppt with the results captured 27 kg and on September 16, 2015, range between 20-23 ppt with the results captured 24 kg in the Mexo-polyhaline zone.
4. On August 15, 2015, the most catches in the estuary mouth is in the Cimandiri estuary that is on point samples 1 and 2 with catches of 25kg and 13kg with the highest salinity 21.67 ppt and 21.43 ppt. Getting away from the mouth of the estuary, the catch is reduced as was the case in the sample points 3, 4 and 5 with a value of salinity increases towards the sea and the salinity value towards river decreases with salinity value 20.93 ppt, 22.27 ppt and 22.81 ppt. On September 16, 2015, the catch at point 1 and 2 samples with catches of 20 kg and 12 kg with the highest salinity 21.96 ppt and 20.81 ppt. Same as of on August 15, 2015, on September 16, 2015, also happens that is increasingly far from the mouth of the estuary, the catch is reduced as was the case in the sample points 3, 4 and 5 with a value of salinity increases towards the sea and the salinity value towards river decreases with salinity value 20.56 ppt, 23.57 ppt and 23.64 ppt.
5. On the Citepus estuary on August 15, 2015, catches at each sample point location is relatively the same amount of the catch 15 kg with an average value of salinity in the sample point 23 ppt. The number of catches at the mouth of the estuary more compared with catches at the point of the other samples. As of August 15, 2015, the date of September 16, 2015, the number of catches at each sample point location relative amount equal to 11kg with an average value of salinity in each sample point is 21 ppt and the number of catches at the mouth of the estuary more compared with catches at the point of the other samples.

On August 15, 2015, the most catches in the Cimaja estuary is at the mouth of the estuary at the point samples 1 and 2 with catches of 8 kg with the highest salinity 23.12 ppt and 22.74 ppt. Getting away from the mouth of the estuary, the catch is reduced as was the case in the sample points 3, 4 and 5 with a value of salinity increases towards the sea and the salinity value towards river decreases with salinity value 23.58 ppt, 22.59 ppt and 21.81 ppt. On September 16, 2015, the catch at point samples 1 and 2 has a value of catches and the salinity value equal to catches of 8 kg with the highest salinity 21.70 ppt. Same as of August 15, 2015, on September 16, 2015, also happens that is increasingly far from the mouth of the estuary. The catch is reduced as was the case in the sample points 3, 4 and 5 with a value of salinity increases towards the sea and the salinity value towards river decreases with salinity value 22.78 ppt, 21.56 ppt and 20.13 ppt.

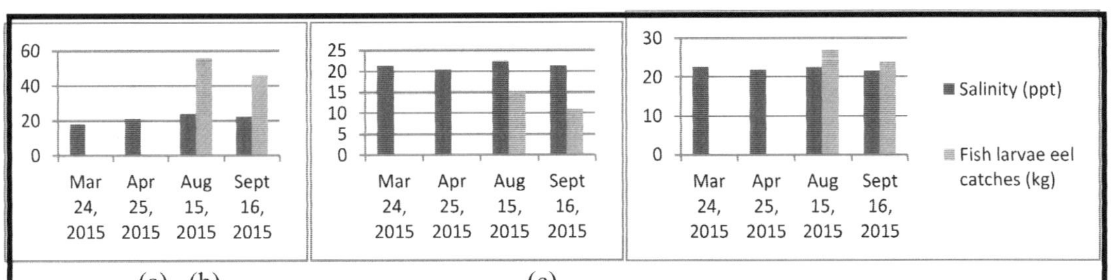

Figure 4. The graph of the relationship of the level of salinity with the results of taking eel fish larvae

4. Conclusion

Based on the results of analysis and discussion, then it can be inferred that the fish larvae eel distribution at the mouth of Cimandiri River, Citepus River, and Cimaja River obtain from the results of the drilling program at the fisherman's catch and salinity occurred during to the dry months where most catches occurred at mouth of the estuary. The catch is reduced if farther from the mouth of the estuary is on the coast towards the sea and the inside of the mouth of the river with the highest salinity towards the sea is increasing while the value of salinity towards the river reduced. Variations of salinity in the estuary area are determined by the big nothingness of mixing that is affected by rainfall, tidal and the sea current conditions. High salinity occurs when rainfall is low, the state of the tide at the speed of sea current that high move horizontally. Low salinity occurs when high rainfall, the tidal is state ebbs and the current speed of moving horizontally.

References

[1] Amalia N 2014 *Estuarine zone at Muaragembong Estuary Based on Salinity by Remote sensing* (Depok: Dept. of Geography, Universitas Indonesia)

[2] Budioyo R 2013*Effect of Salinity on Growth Phase Eel Fish Glass Eel Fish Farming Technology as an Alternative eel*(Surakarta:UniversitasSebelasMaret)

[3] Haryono 2008Eel, eel-eared: Potential and cultivation aspects. Field of Zoology, Biology-LIPI Research Center*Fauna Indonesia***8**(1): 22-26

[4] IOCCG 2000Remote Sensing of Ocean Colour in Coastal, and Other Optically Complex, Waters. Sathyendranath, S. (ed.)*Report of the Internasional Ocean-Colour Coordinating Group*No. 3, IOCCG (Daetmouth, Canada)

[5] Kottelat2013 The fishes of the inland waters of southeast Asia: a catalogue and core bibliography of the fishes known to occur in fresh water, mangroves and estuaries*The Raffles Bulletin of Zoology*, Suplemen No. 27. ISBN 978-2-8399-1344-7

[6] Ladya D C 2015Cimandiri estuary zones *based onsalinity*(Depok: Dept. of Geography, Universitas Indonesia)

[7] Nybakken J W 1992*Marine Biology: An Ecological Approach*(Jakarta: PT. Gramedia)

[8] Sarwono B 1993*Cultivation of Eel*. Ed. 7(Jakarta: PT. PenebarSwadaya)

[9] SupriatnaL, Supriatna J, Koestoer R H, Takarina N D 2016 Algorithm model for determination of Cimandiri Estuarine boundary using remote sensing*AIP Conference Proceedings***1729** 020079.

[10] Venice System 1958*Symposium on the classification of brackish water* Venice. Archivio di Oceanografia e Limnologia.

[11] Wouthuyzen S, Sediadi A, S Tarigan, and H Indarto 2008PengukuransalinitaspermukaanTeluk Jakarta melaluipenginderaanwarnalautmenggunakan data multi-temporal Citra satelit Landsat-7 ETM+.*Proced. PIT MAPIN XVII* Bandung 10-12-2008.

Monitoring The Land Accretion Development at Coastal Area of Blanakan, Subang Indonesia

Nandi, Ginna Meriana, Lili Somantri

Department of Geography Education, Universitas Pendidikan Indonesia
Jl. Dr.Setiabudi No. 229 Bandung, 40154

corresponding author's e-mail: nandi@upi.edu

Abstract. A land accretion is formed by deposition in estuaries. Recently, a land development in Subang coastal area has raised an increase. Beside its potential, coastal areas are also threatened with damage including abrasion, accretion, loss of mangrove forests, and sea water intrusion. One of the coastal areas that have been arising in very extensive land is Blanakan coastal in Subang Regency. This study aims to monitor the development of a land accretion that have been arise during the period of 1990 to 2015 and also to examine the use of a land accretion and analyze the impact of a land accretion to the social and economic conditions in the Blanakan Coastal Areas. The method used in this research was descriptive quantitative method. In this research, The Landsat imageries were overlaid came from 1990, 2000, 2010, and 2015 to determine the development of a land accretion. Based on the results of Landsat imagery overlaid over the period 1990-2015. Overall, during the period 1990-2015, accreted land formed was an area of 782.9 hectares and abrasion area of 73.3 hectares with changes in the most far reaching 1580.3 m. The use of land accretion in the Blanakan Coastal mostly used for a fishpond with a key commodity is Milkfish and Bago shrimps. The impact of land accretion to the social and economic conditions was reflected through the five indicators such as livelihoods, income, education, health, and ownership of assets.

Keywords: coastal area, monitoring, land accretion, Blanakan, Subang Indonesia

1. Introduction

Coastland is an area of encounter between land and sea, in which a mutual relation still takes place[1]. As an archipelago state with the coastline reaching 99.093 km, the potential of coastline area in Indonesia is very vast and advantageous for various economic activities[2]. One of the provinces possessing potential along the coastline being currently developed is the province of West Java [3].The length of coastline in the province of West Java extending in the northern part from Cirebon district to Bekasi district is approximately 365 km long and in the southern part from Ciamis district to Sukabumi district is around 355 km.

Subang district is geographically located in the northern part of the province of West Java. Of all existing sub-districts, there are four coastal districts, namely Blanakan, Sukasari, LegonKulon, and Pusakanagara sub-districts. The length of coastline in the coast of Subang district reaches the number of 68 km [4].

The utilization of coastal areas to sustain people's economic activities frequently causes destructions, such as abrasion, accretion, seawater intrusion, and the decreasing number of mangrove forests and the damage of coral reefs [5, 6]. Several beaches in Subang district have been objected to abrasion as in Muara Karang – Legon Wetan until Pondok Bali, as well as along the area of Patimban. Sedimentation happened in the estuary indicated with deltas throughout the river as the aftermath of land accretion along the coastline. BPLHD reported that the furthest abrasion occurred in Indramayu sub-district of 48.57 km, whereas the longest accretion ensued along Subang Beach reaching the length of 50.44 km[7].

Land accretion is a land formed by sedimentation at river mouth. In Subang district, the widest land accretion was formed in Blanakan sub-district with the average of 686.53 m, while the land accretion formed in Pusakanagara sub-district had the average 159.32 – 538.50 m [7,8]. Up to the present time, the land accretion in Blanakan coast is only used as an embankment. However, the utilization has not been maximized because it is hampered by weather or high tides that it causes crop failure. The measurement of the width of the land accretion was lastly conducted by BPLHD West Java in 2004 which mentioned that the land accretion in Subang district was 6000 Ha in 2002. Furthermore, there is no government agency that conducts the land accretion measurement. The measurement is usually performed by companies having business in the area[9]. Based on the facts mentioned, the land accretion measurement is required so it can be utilized optimally to bolster the community's economic needs and local revenue. Moreover, this research will analyze the usage of land accretion by the community. The effects of the existence of the land accretion on social and economic condition of people in Blanakan Coast will also be identified.

2. Methods

The method used in this research is the descriptive quantitative method. The population in this study was Blanakan sub-district. The sample in this study was a village with land accretion consisting of 7 villages, such as Cilamaya Girang, Rawameneng, Jayamukti, Blanakan, Langensari, Muara, and Tanjungtiga. As for the number of inhabitant sample was counted using purposive sampling with Slovin formulary; that is of 99 people.

The data recollection technique in this research was study of literature, study of documentation, interview, and field observation. The data processing technique was divided into 2; the first one was Landsat imagery interpretation covering RBG transformation, image enhancement, and cropping. Meanwhile the second one was social data processing performed by conducting interviews, tabulation, and creating tables and graphics.

The data analysis technique in this study was separated into 2, which were to analyze the development of the land accretion using perpendicular grid to measure the distance of development of the land accretion throughout 1990 – 2015. The illustration of the usage of perpendicular grid can be seen in picture 2. As for the utilization of the land accretion area and the effects of existence of the land accretion on socioeconomic condition was conducted using cross-tabulations with parameters of income, livelihood, education, and ownership of assets/facilities.

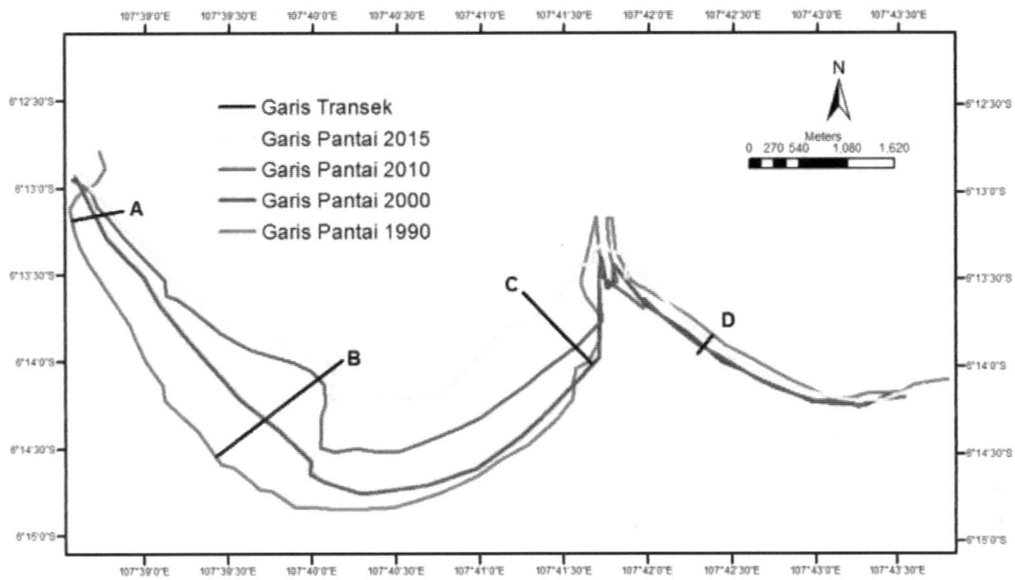

Figure 1. Transect measurements lines

3. Results and Discussions

3.1 The development of the Land Accretion in Blanakan Coast Subang District throughout 1990-2015

To identify the land accretion development in Blanakan Coast, a Landsat imagery overlay was conducted from the year 1990, 2000, 2010, and 2015. Each landsat coastline was delineated and then overlaid. The outcome of the Landsat imagery overlay throughout 1990-2015 showed transformations in the coastal area of Blanakan sub-district caused by land accretion as well as abrasion. These two coastal dynamics take turns in transforming the visual of Blanakan coast. Below is the table of the development of the land accretion in area and distance throughout the period of 1990 – 2015 visualized in table 1 and table 2.

Table 1. The Development of Accretion and Abrasion in Blanakan Coast throughout 1999-2015.

Year	Area (Ha)	
	Accretion	Abrasion
1990-2000	242.6	69
2000-2010	266.9	8.2
2010-2015	277.3	0
total	786.8	77.2

Table 2. The Development of Width of Accretion and Abrasion in Blanakan Coast throughout 1990 – 2015

Sample	Length (m)			
	1990- 2000	**2000- 2010**	**2010- 2015**	***Total***
A	255.4	100.6	89.1	445.1
B	684.7	721.9	173.7	1580.3
C	-64.5	270,3	509	444.5
D	-165.3	18	60	-87.3
Total	**177,6**	**277,7**	**208**	**2382.6**

Overall, the land accretion formed in the period of 1990-2015, the land accretion is of 786.8 Ha and the abrasion is of 77.2 Ha. For the past 25 years, the land of Blanakan Coast has undergone the furthest changing reaching 1,580.3 m.

The most significant accretion is seen in the coastal area in the western of Ciasem River, covering CilamayaGirang, Rawameneng, Jayamukti, Blanakan, Langensari, and Muara villages. The area of the land accretion forms throughout 1990 – 2015 is 739.6 Ha. Besides, abrasion happened during 1990-2000 was 18.6 Ha wide. Meanwhile, at the eastern side of Ciasem River, there is only Tanjungtiga village and the most dominant dynamic was the abrasion throughout 1990-2010 that was 51,9 Ha in. Apart from abrasion, in the period of 2010-2015, land accretion is also formed but its area is not as enormous as it is in the coastal area in the western part of Ciasem River, of 47.2 Ha. Averagely, throughout 25 years, there has been land accretion formed with the total in width 786.8 Ha; roughly 31.5 Ha of land accretion is formed each year.

The land accretion formed in Blanakan coast is caused by over sediment runoff from Ciasem River which is then pushed by the stream towards the western part, proven by the wind rose resulted in data recordings from 1996 until 2010 in northern Subang beach which indicated dominant wind facing to west and southwest. Moreover, because the effects of wind, the sedimentation happening in the study areas in Blanakan coast is also influenced by the fact that the western side of Subang district coast is more slope compared to the coast in the eastern part. The sediment is then trapped in Blanakan sub-district bay [7]. Sedimentation is also influenced by the changing of coastline shape. Moreover, the coastal water is also affected by the dynamics of interaction between water inputs from the sea and freshwater from the river [10].

The western part of Blanakan Coast (bordered by Ciasem River) is an area with the widest area of land accretion. This is because its morphology which is shaped as a bay (protruded inside) and slope. This protruded inside and slope shape is what causes sediment from Patimban Beach, Pamanukan is easily trapped in the area and formed a land accretion. Apart from the morphology, there are also rivers with small estuary that all sediments coming from various directions will be collected and deposited there. As for the eastern part of Blanakan, there is only Tanjungtiga village which for the period of 1990 – 2010 had undergone fairly great abrasion. Its morphological shape which is not too protruded (relatively even) to the sea eases the strong sea stream and high tidal wave to move pass it, eroding the land. Furthermore, the decreasing of mangrove forests density has reduced its function as a wave retainer that the abrasion was difficult to withstand [7].

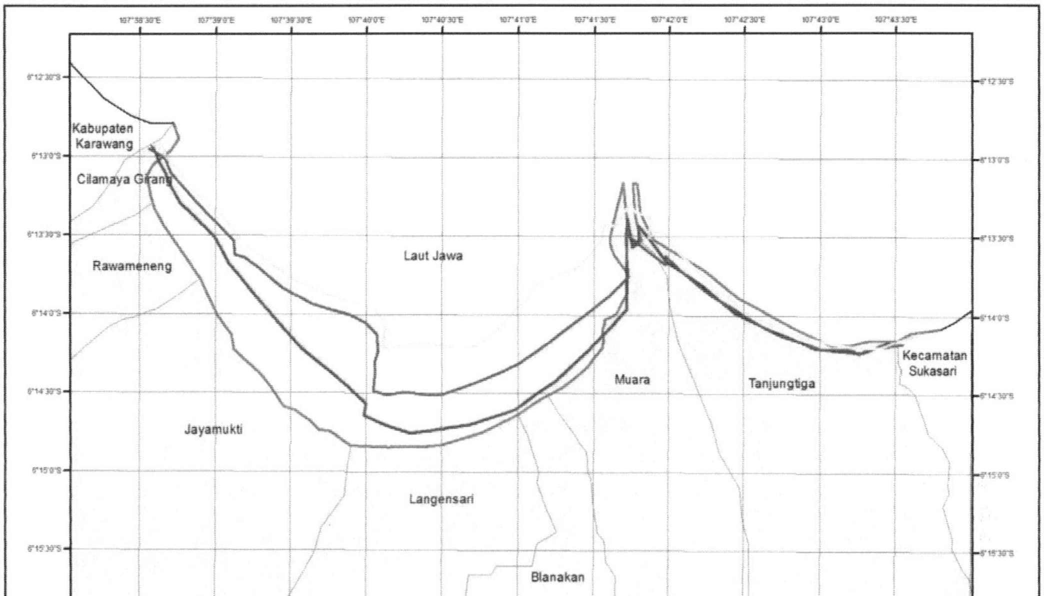

Figure 2. The map of the development of land accretion throughout 1990 – 2016

3.2 The Utilization of the Land Accretion Area in Blanakan Coast in Blanakan Coast

The existence of land accretion is utilized by coastal communities to open an embankment to provide their economic needs. All land accretion areas in Blanakan Coast is used as embankments. According to juridical law, land accretions belong to Perhutani. However, in reality, Perhutani allows the people to work on them under the circumstances that they have to care about the sustainability of ecosystems there. To run the land accretion, people need to administer licensing letters by paying around IDR 500.000 – IDR 2.000.000 to Perhutani but through village head. By paying the money, people normally acquire 2 Ha of land for each. Since as time passes the land accretion formed is getting many, the number of people interested in managing them is increased, too. Therefore, Perhutani cooperates with local community to set up an Institution of Forest Villagers/Lembaga Masyarakat Desa Hutan (LMDH) in every village. LMDH functions as a bridge between Perhutani and the society.

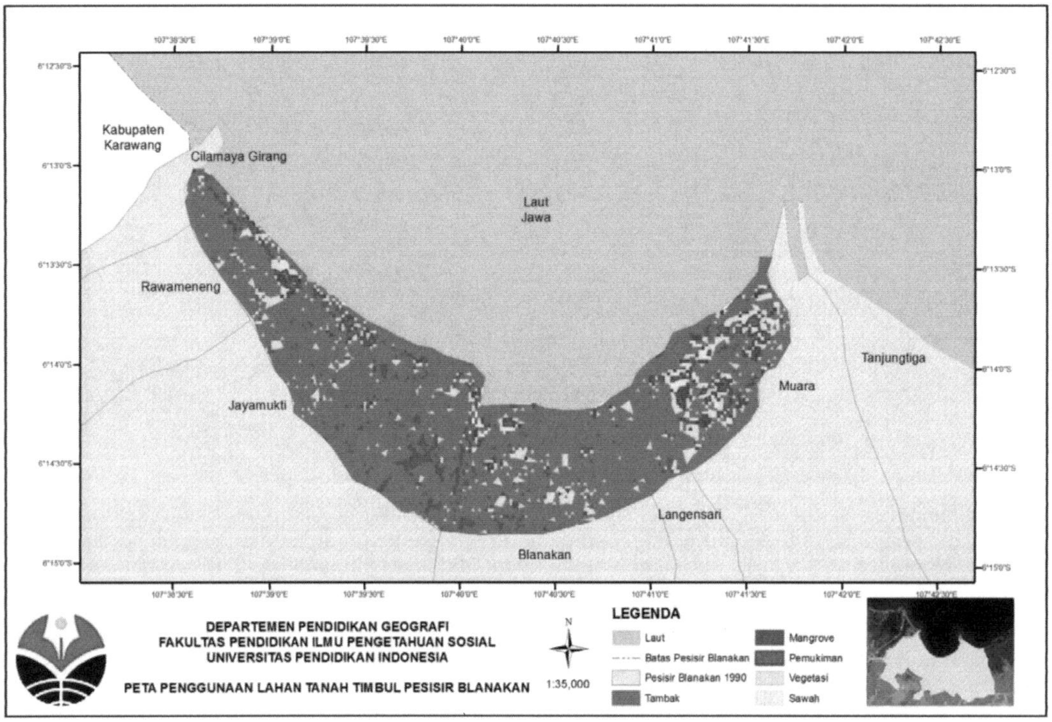

Figure 3. The map of the utilization of the land accretion in Blanakan Coast

3.3The Impact of Land Accretion on Social and Economic Condition of Blanakan Coast Society

The existence of land accretion has provided several impacts on social and economic condition of Blanakan Coast Society, Subang District. Accreted land helps people in the society to satisfy their primary needs. The socioeconomic condition can be identified using 5 indicators, i.e. livelihood, income, education, health, and ownership of assets.

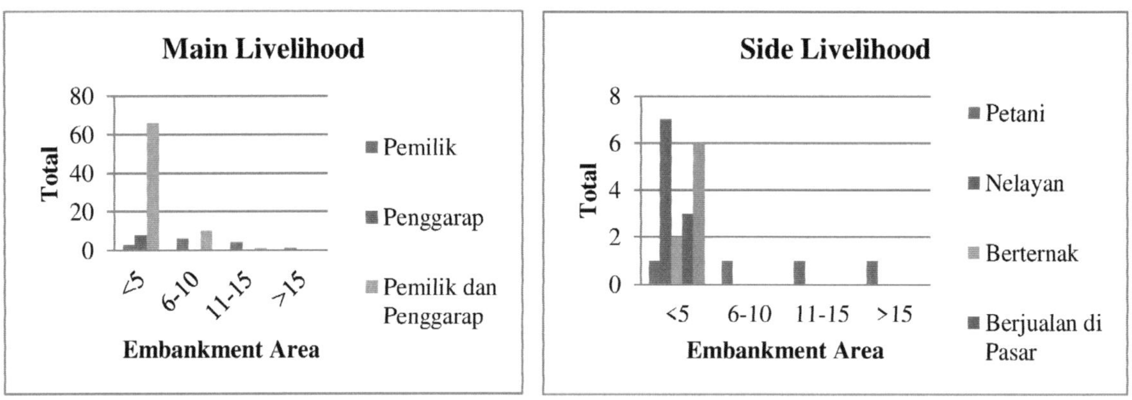

Figure 4. The embankmentarea with main and side livelihood

Majority of people who live in Blanakan Coast are the owners of as well as the workers at embankments with varied embankment area ranging around 1-5 Ha.

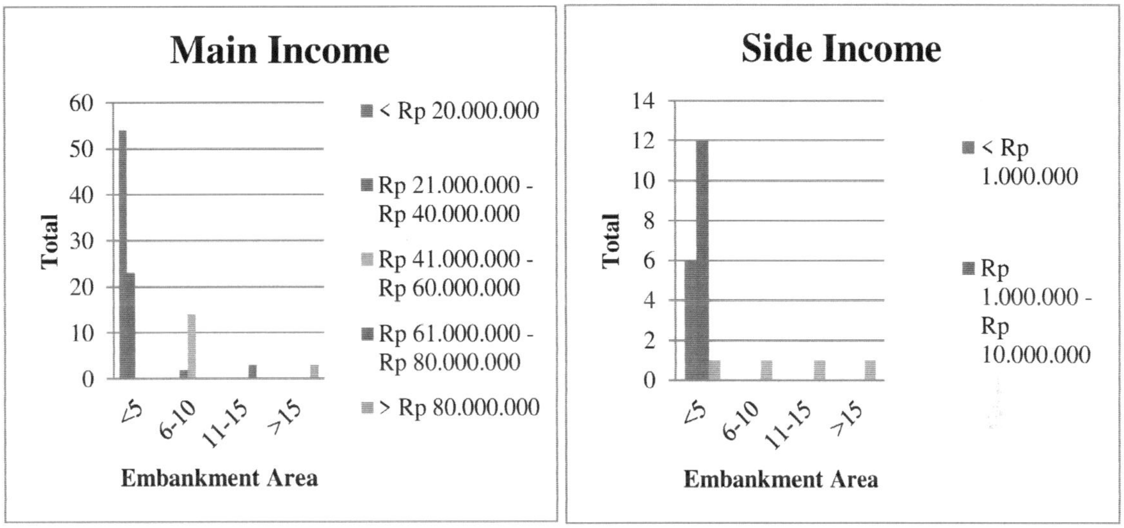

Figure 5. The embankmentarea with main and side income

People who own embankments with the area of 1-5 Ha commonly have side livelihood to meet their necessities, such as farmers, fishermen, farmers, sellers at markets, and owners of small shops. On the other hand, the owners of embankments with the area of above 5 Ha usually own rice fields as side livelihood. The wider the embankments one has, the higher their income is. The main income of < IDR 20,000,000 places first belonging to embankment owners with an area of 1-5 Ha. The most side income is found at the group of farmers whose embankments are of 1-5 Ha, as what has been discussed earlier that it is the group with most side livelihoods. Embankment farmers whose income is < IDR 20.000.000 have more schooled compared to other income groups. Found in this group are children with level of education ranging from Primary School, Junior High School, Senior High School, and University. All respondents involved in this research state that they wish their children should be educated until attending the university so they can get better professions and income compared to their parents.

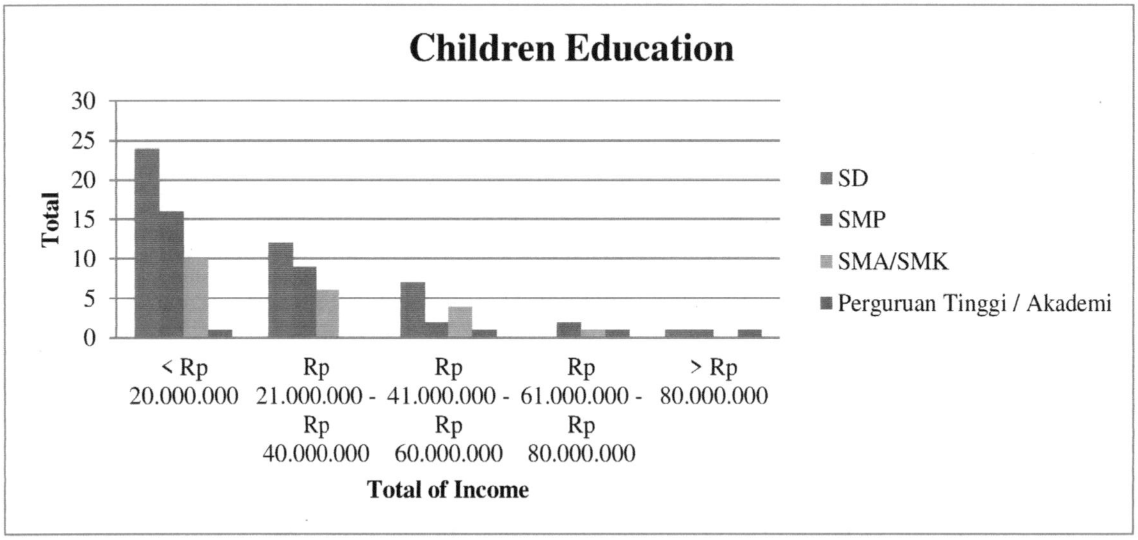

Figure 6. Total of Income with Children Education

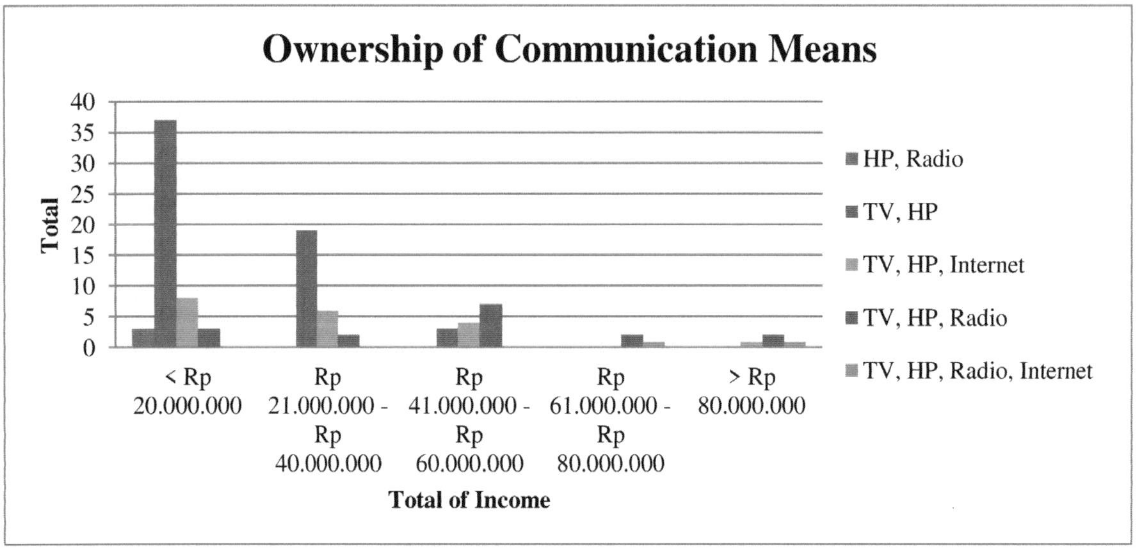

Figure 7. Total of Income with Ownership of Communication Means

As for the indicator of health, all respondents state that they have had their own lavatory even though the condition is still varied, from the ones that are feasible to use and the less feasible ones. The awareness to live healthily has been decent. The last socioeconomic which is the ownership of assets, are separated into the ownership of residence, ownership of communication means, and ownership of transportations. Of all respondents, only two of them rent their house, typically newcomers who come as embankment workers. For the ownership of communication means, the lesser the farmers' income is, the more limited the communication means they have. Most of them only have televisions and cellphones. On the other hand, the group of embankment farmers whose income is >IDR 80,000,000 owns more varies.

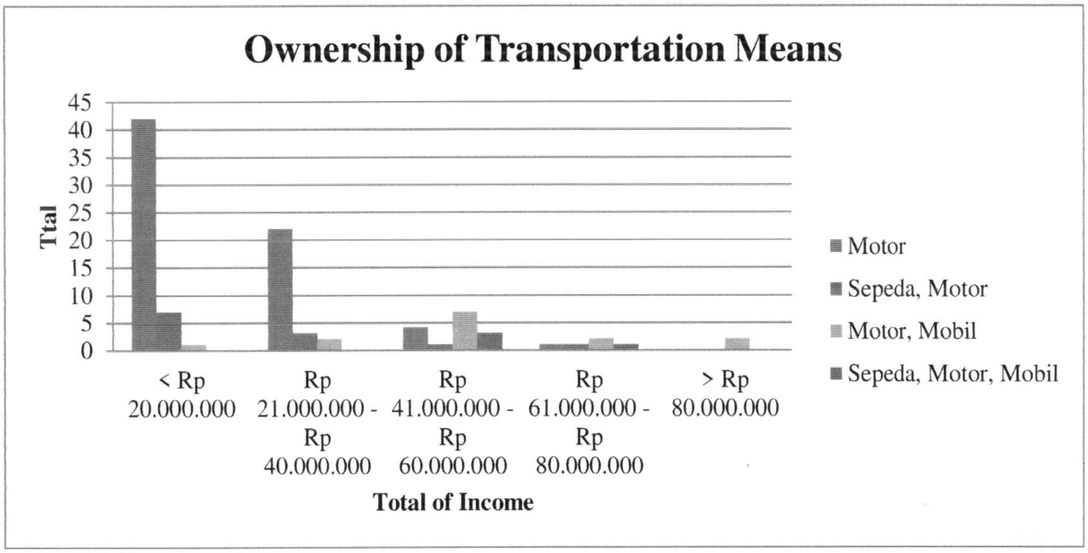

Figure 8. Total of income with ownership of transportation means

Similar to the ownership of communication means, the lesser the embankment farmers' income is, the fewer the vehicles they have. Nearly all of the respondents ride motorbikes and in the group with the income of > IDR 20,000,000, the highest ownership of transportation means belongs to motorbikes group. Meanwhile the most varied ownership of transportation means belong to the group of embankment farmers with the income of IDR 41.000.000 – IDR 60.000.000, that are bicycles, motorbikes, and cars.

4. Conclusion

Based on the landsat imagery overlay throughout 1990-2015, overall, during the period, the land accreted is 782.9 Ha and the abrasion is of 73.3 Ha with furthest changing reaching 1.580,3 m. The dynamics in Blanakan Coast, Subang District can be divided into 2. The western part (Cilamaya Girang, Rawameneng, Jayamukti, Blanakan, Langensari, and Muara Village) is the area with more land accretion, which is 741.2 Ha with the furthest accretion of 1,580.3 m. Meanwhile, for the eastern part, there is only Tanjungtiga village, the dominant dynamic is abrasion. During 1990-2010, abrasion has occurred spreading to 75.3 Ha abrade the coast as far as 147.3 m. However, since 2010-2015, land accretion was formed as wide as 41.3 Ha and the coast undergoes an addition up to 60 m. Based on the result of the landsat imagery interpretation, it is known that the utilization of land accretion in Blanakan Coast is as follows. From the total of area of 786.8 Ha, the land accretion formed from 1990-2015, an area of 639.9 Ha is used by the community there as embankments. Besides, there are also mangroves of 43.5 Ha, vegetation of 71.6 Ha, rice fields of 5.4 Ha, and residency of 26.4 Ha. The existence of embankments in Blanakan Coast helps the economy of local community. With these embankments in accreted land, the livelihood becomes more varied. The more the embankments owned, the higher their income, the ownership of assets and their welfare regarding education and health. For the government of Subang Regency, this research can be considered as a recommendation on the protection of coastal areas, especially in deciding zones of land utilization such as embankments and mangrove. This is to avoid conflicts which may arise when the area is managed by local community.

References

[1] Triatmojo B 1999. *Teknik Pantai*(Beta Offset: Yogyakarta)
[2] Eric C F Bird and Otto S R Ongkosongo 1980*Environmental changes on the coasts of Indonesia*(The United Nations University's Programme on the Use and Management of Natural Resources. Tokyo, Japan)
[3] * * *, *BAPPEDA Jawa Barat* 2004 Bandung
[4] * * *, Geospatial Information Agency*2012 Hasil survey geografi dan toponimi, Badan Informasi Geospasial* http://www.bakosurtanal.go.id/ (accessed at 20.12.2013)
[5] PrasetyaG 2007The role of coastal forests and trees in protecting against coastal erosion. Coastal protection in the aftermath of the Indian Ocean tsunami: What role for forests and trees? Proceedings of the Regional Technical Workshop, Khao Lak, Thailand, 28–31 August 2006*Compiled and edited by Susan Braatz, Serena Fortuna, Jeremy Broadhead and Robin Leslie. Food and Agriculture Organization (FAO) of the United Nations, Regional Office for Asia and the Pacific* 39 Phra Atit Road, Bangkok, Thailand
[6] Nandi2014 Coastal conservation policies and integrated coastal zone management (ICZM) in indonesia *International Journal of Conservation Science***5**(3)pp 387-396
[7] Taufiqurohman A, and Ismail M F 201. Analisis Spasial Perubahan Garis Pantai di Pesisir Kabupaten Subang Jawa Barat*Jurnal Ilmu dan Teknologi Kelautan Tropis***4**(2)pp 280-289.
[8] Kasim F 2012Pendekatan Beberapa Metode dalam Monitoring Perubahan Garis Pantai Menggunakan Dataset Penginderaan Jauh Landsat dan SIG*Jurnal Ilmiah Agropolitan***5**pp 620-635
[9] * * *, BPLHD Jawa Barat2004 Bandung
[10] Nandi 2015*The Ecological Dynamics of a Coastal Lagoon: A Study of the Morphodynamics, Land Use Change, Community Participation, and Spatial Planning of the Segara Anakan Lagoon in Indonesia [Disertation]*(Universitat Leipzig, Germany)

IOP Conf. Series: Earth and Environmental Science **47** (2016) 012018 doi:10.1088/1755-1315/47/1/012018

Evaluating the performance of Lyzenga's water column correction in case-1 coral reef water using a simulated Wolrdview-2 imagery

Masita Dwi Mandini Manessa[1, 2*)], Muhammad Haidar[3], Syarif Budhiman[4], Gatot Winarso[4], Ariyo Kanno[5], Tatsuyuki Sagawa[6], Masahiko Sekine[5]

[1] Graduate School of Science and Engineering, Yamaguchi University, Japan
[2] Center for Remote Sensing and Ocean Science (CReSOS), Udayana University, Denpasar, Bali, Indonesia
[3] Thematic Mapping Division, Geospatial Information Agency of Indonesia
[4] Remote Sensing Application Center, National Institute of Aeronautics and Space (LAPAN)
[5] Graduate School of Sciences and Technology for Innovation, Yamaguchi University, Ube, Japan
[6] Remote Sensing Technology Center of Japan, Tokyo, Japan

*) e-mail: masita@yucivil.onmicrosoft.com, mdm.manessa@gmail.com.

Abstract. This study identified the performance of Lyzenga water column correction under controlled condition of noise, water quality, and bottom type diversity. We build a simulated multispectral imagery, represent the subsurface reflectance of coral reef environment that receives by Worldview-2 satellite. The evaluation was performed to two different subsurface reflectance: Lee et al.'s and Maritonia et al.'s radiative transfer model. As a result, in the case of simple Maritonia et al.'s reflectance model, it shows that the effect of the water column is removable. However, if the actual just above surface reflectance is as more complex as Lee et al.'s model, then the Lyzenga's water column method could reduce the effect of water attenuation but not removed. In fact, based on the mathematical presentation of Lyzenga's model on simulation reflectance formula the effect of depth and water attenuation are in multiplication function. Thus, even after the correction, the water column attenuation effect stillremain.

1. Introduction

The variation of bottom type, water depth, and water attenuation (scattering and absorption in the water column) generated a complicated problem in coral reef mapping. The water attenuation became one of the issues in creating an accurate map. In order to answer this problem Lyzenga[1] proposed an image based correction method to reduce the effect of water attenuation, well known as Lyzenga's water column correction. Since then this correction method has widely applied due to the simplicity. The past publication [2, 3, 4, 5, 6] shows the superiority of this approach with varied accuracy result. Unsurprisingly, many researchers doubt the performance [7, 8]. This opinion built on a fact of Lyzenga's water column correction rely on a lack assumptions; in particular, it assumes no substantial changes in water quality or less diverse bottom types. For the past decade, several studies has developed a more accurate correction combine the empirical and analytical method such as Purkis [7] and Mishra et al. [8]. However, the availability of optical water parameter used in the analytical method is rare, especially in developing country such as Indonesia. Hence, Lyzenga's method still the best option to be applied in the area with lack water optical data.

As mention before the performance of Lyzenga's water column correction was controlled by the water quality, bottom type diversity, and image noise. A method evaluation using a real imagery was risky because the possibility of error caused by instrument, the time gap of image recording and measurement date, or image noise. Moreover, the actual performance and each parameter effect could not describe precisely. Then this study evaluated the Lyzenga's water column correction under controlled variable, noise, and error. We build a simulated imagery that represents the coral reef environment by combining the measurement and model data.

The usage of simulated imagery to evaluate the performance of water column correction was not for the first time. Zoffoli et al. [9] had performed a similar analysis method to test the performance of an analytical water column correction. Moreover, our simulation set up bears a close resemblance to Zoffoli et al. [9]. However, this study was performed the analysis on the empirical water column correction of Lyzenga [1].

Since there are many radiative transfer model [i.e. 10, 11, 12, 13] to simulated the reflectance received by the satellite. This study used two major models that well known for modelling the reflectance of water surface. First is the simple reflectance model of Maritonia et al. [11] latter be mention as Maritonia et al.'s model. Second is more sophisticated reflectance model of Lee et al. [13] that latter be mention as Lee et al.'s model.

Moreover, our simulation dataset was modeled in hyperspectral range then we convert into the wavelength signature of Worldview-2 Multispectral images as the basis of our study. The Worldview-2 image was intentionally used because it has large visible spectral range than the common multispectral image (SPOT-6, Quickbird, RapidEye, etc.). Especially the coastal band that located in shorter visible wavelength that claims to be useful and sensitive for coastal application such us Coral reef classification mapping or bathymetry.

2. Concept of Lyzenga's Water Column Correction

Lyzenga's proposed the water column correction that based on two assuption: the depth play significant rule in changing the reflectance of object; and water quality is contant all over image. Thus, the radiance or reflectance received by satellite explain that bottom reflectance is exponential relation to water depth and downwelling attenuation. The natural logarithm function of the reflectance value (ρ) was added to linearize the attenuation effect with respect to depth. Thus, a transformed reflectance (X)can be built. In this step of linearization, the Lyzenga method also engages noise correction using the average value of deep-water pixel radiance[9]. The equation of transformed reflectance can be expressed as:

$$X_i = \log(\rho_i - \rho_{\infty i}) \tag{1}$$

Whereρ_i is observed spectral reflectance and$\rho_{\infty i}$represents the water depth-averaged reflectance at band i.

Transformed radiance (X) bands i and j are a linear function of water depth (Equation 6) and are linearly relate each other, for a given bottom types (Lyzenga, 1981). Then If the relationship between reflectance and depth is linear and the bottom type is constant, the scatter plot of transformed reflectance (X) bands i and j will fall ideally in a straight line and expressed as:

$$X_i = mX_j + y \tag{2}$$

The slope (m) of this straight line represents the relative attenuation ($\frac{k_i}{k_j}$) in each combination of bands. The different bottom types represented in a scatter plot should create a similar line, the variation which indicates changes in depth. The gradient of each line would be identical because the ratio of the attenuation coefficients ($\frac{k_i}{k_j}$) is independent of bottom type.

The y-intercept of each line is subsequently used, independent of free of depth, as an index of bottom type; this is known as the depth-invariance index (Yij), which can be written as:

$$Y_{ij} = X_i - \frac{k_i}{k_j} X_j \quad (3)$$

where X_i and X_j represent the transformed radiance at bands i and j, respectively; and ki/kj is the irradiance attenuation coefficient of water in bands i and j. The the ratio of attenuation coefficient (ki/kj) only holds when pixels of the same object at different depths are distributed linearly for the paired bands (Lyzenga, 1978).

3. Simulated Worldview-2 Image

3.1. Overview

Simulated imagery is a simulated the subsurface reflectance based on radiative transfer model for the coral reef environment. The simulated imagery is build base on several assumptions that refer to the condition of the coral reef ecosystem. First, water is very clear and less suspended matter. Second, chlorophyll-α concentration is around 0.02 - 10 mg/m3. Third, clear sky where no significant noises such us cloud, haze, and smokes. Fourth, due to a high spatial imagery then pixel mixing spectral is less. Last, the water surface is calm with no significant white caps.

3.2. Maritonea et al.'s Model

Maritonea et al.'s [11] proposed the surface reflectance function that simplifying assumptions to derive analytical formulae expressing the reflectance of shallow waters as a function of observation depth and bottom depth and albedo. The bottom reflectance assumed as an approximately linear function of the bottom reflectance and an exponential function of the water depth as the same that explain by Lyzenga [1]. Then the just above surface reflectance is written as follows

$$\rho_{rs} = \left(\rho_{\infty} + \left((p_b - \rho_{\infty})e^{-h2K}\right)\right) \quad (4)$$

where ρ_{∞} is the in-water volume scattering reflectance at infinite depth, ρ_b is the bottom reflectance, K is the effective attenuation coefficient, and h is water depth. Note that the coefficient $2K$ in Equation (4) is downweliing and upwelling diffuse attenuation coefficient that assumes equal to the vertical diffuse attenuation coefficient for upward flux.

The calculation of K or attenuation coefficient and ρ_{∞}or reflectance at infinite depth was refer to Morel and Maritonia [11] as follows

$$K(z,\lambda) = K_w(\lambda) + K_{bio}(z,\lambda) \quad (5)$$

$$K_{bio}(z,\lambda) = X(\lambda)(Chl(z))^{e(\lambda)} \quad (6)$$

$$K_w(\lambda) = a_w(\lambda) + (1/2)b_w(\lambda) \quad (7)$$

$$\rho_{\infty} = \frac{b_b(\lambda,z)}{a_w(\lambda)} \quad (8)$$

$$b_b(\lambda,z) = (1/2)b_w(\lambda) + b_{pb}(\lambda,z) \quad (9)$$

$$b_{pb}(\lambda,z) = \left(0.002 + 0.02(0.50 - 0.25\, log_{10}(Chl(z)(550/\lambda))\right) \cdot (0.30(Chl(z)^{0.62}) - b_{w\,550}) \quad (10)$$

where a_w and b_w stand for the absorption and scattering coefficients of optically pure sea water, respectively. a_w , b_w , X and e for each specifiec wavelenght is refer to Table 2 in Morel and Maritonia (2001).

3.3 Lee et al.'s Model

The subsurface reflectance model by Lee et al. [13] was the basis in the Hydrolight Software and had been widely used. The remote sensing the just above surface reflectance for nadir-viewing as follow:

$$\rho_{rs} \approx \frac{0.5 r_{rs}}{1 - 1.5\, r_{rs}} \tag{11}$$

$$r_{rs} = \left(r_{rs}^{dp} \left(1 - e^{-\left[\frac{1}{\cos(\theta_s)} + D_u^C\right] kH} \right) + \frac{\rho_b}{\pi} e^{-\left[\frac{1}{\cos(\theta_s)} + D_u^B\right] kH} \right) \tag{12}$$

with r_{rs}^{dp} is the subsurface reflectivity of the deep water; ρ_b is the bottom reflectance; θ_s is subsurface solar zenith angle; D_u^c and D_u^B are the optical path-elongation factors for scattered photons from the water column and the bottom, respectively; k is the attenuation coefficient; and H is the depth.

The radiative transfer equation parameters depend greatly on the optical properties of the water expressed by

$$r_{rs}^{dp} \approx \left(0.084 + 0.17 u(z,\lambda)\right) u(z,\lambda) \tag{13}$$

$$u(z,\lambda) = \frac{b_b(z,\lambda)}{a(z,\lambda) + b_b(\lambda z,\lambda)} \tag{14}$$

$$k(z,\lambda) = a(z,\lambda) b_b(z,\lambda) \tag{15}$$

$$D_u^c(z,\lambda) \approx \sqrt{1.03(1 + 2.4u(z,\lambda))} \tag{16}$$

$$D_u^b(z,\lambda) \approx \sqrt{1.04(1 + 5.4u(z,\lambda))} \tag{17}$$

$$b_b(z,\lambda) = b_w(\lambda) + b_{bp}(z,\lambda) \tag{18}$$

$$a(z,\lambda) = a_w(\lambda) + a_p(z,\lambda) + a_y(z,\lambda) \tag{19}$$

with u is the backscattering diffuse attenuation ratio;a_w , a_p, a_y are the attenuation coefficients for pure water, chlorophyll-bearing particles, and covarying yellow matter (CDOM); b_w, b_{bp} are the backscattering coefficient for water and particles. z is chollorophil-a consentration (mg/m3). λ is wavelength (nm).

3.4 Input Parameters

The most abundant bottom types group in shallow coral reef ecosystem are divided as a substrate, life coral, dead coral, and algae. Although these features have different spectral characteristic and can separate as homogenous pixels, in reality, there can be a significant amount of intermixing between these features. This research is focusing on evaluating the homogenous feature and neglected the possibility of various features in each pixel. Moreover, each bottom type was set to have the same distribution of depth and IOP value (Figure 1). Then the inputs value to calculate the just above surface reflectance in equation a and b is shown in Figure 1 and details as follow:

1. TOA solar zenith angle (θ_s) set as 30 then it convert to subsurface solar zenith angle (θs) using Fresnel equation.
2. Depth (H) is 35 random number in normalizing distribution from 0.5 – 19 meter
3. Inherent optical properties (IOP) of D_u^c, D_u^B , V, k, and Kare modeled based on Chlorophyll concentration (z), then we set the Chlorophyll concentration (z) around 0.2 – 0.25 mg m^{-3}

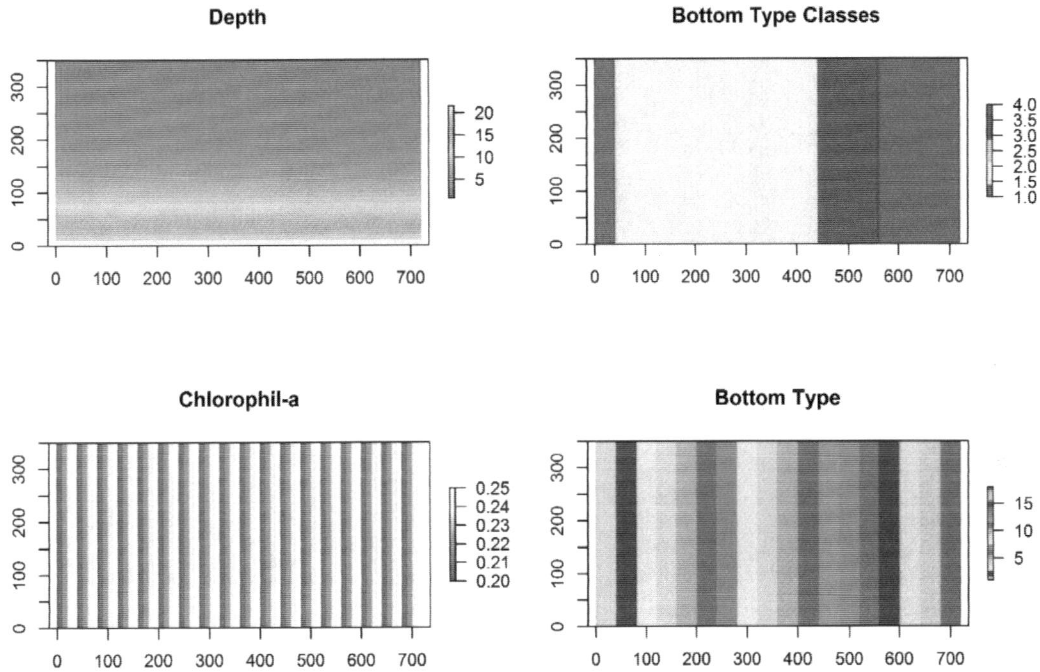

Figure 1. Input data to build the artificial imagery (a) Depth distribution, (b) five bottom type classes (from left to right: class 1: Substrate, class 2: Life coral, class 3: Dead coral, class 4: Algae), (c) Chlorophyll-a distribution, (d) 18 bottom Types distribution (from left to right: Sand, Cyanobacteria, *Acropora sp.*, *Porites sp.*, *Galaxea sp.*, *Heliopora*, *Sinularia sp.*, Coral, Xenia, Softcoral, *Porites sp.*, Dead Coral Tuft Algae, Dead Coral Coralline Algae, Dead Coral Bleaching, Coralline algae)

4. Bottom reflectance (ρ_b) is 18 type of bottom type of one sand, ten coral reefs, four dead corals, and three algae that measured in Derawan (Marlina, 2004)

Furthermore, the full spectral of simulated imagery were converted to each three multispectral bandwidth Rrs (*i.e.*, blue band, green band, and red band) with the following equation [14]:

$$R_{rs}^{wv}(B_i) \approx \frac{\int_{400}^{700} R_{rs}(\lambda) F_{Bi}^{wv}(\lambda) d}{\int_{400}^{700} F_{Bi}^{wv}(\lambda) d} \tag{20}$$

With F_{Bi}^{wv} as the Multispectral bands response function for band i.

3.5 Simulated Image of Coral Reef Environment
Figure 2 shows the simulated image from Maritonia et al.'s model. In the shorter wavelength band (Coastal – Blue bands) some dark bottom type such as algae were unable to model because the object bottom reflectance have smaller reflectance value than deep water ($\rho_\infty > p_b$) and shows as blank white color in Figure 2(a) and 2(b). Based on Maritonia et al.'s based simulated image result, two points can be marked as a note. First, even in the real multispectral image data, all coral reef bottom type can show value, but it not contains a bottom reflectance information. In the case of a dark object, there are opportunities that reflectance received by satellite is just the noise from the water column, water surface, and atmosphere. Second, significant removal of the bottom type in shorter wavelength shows opposite result from the positive claim of Coastal band in Worldview-2.

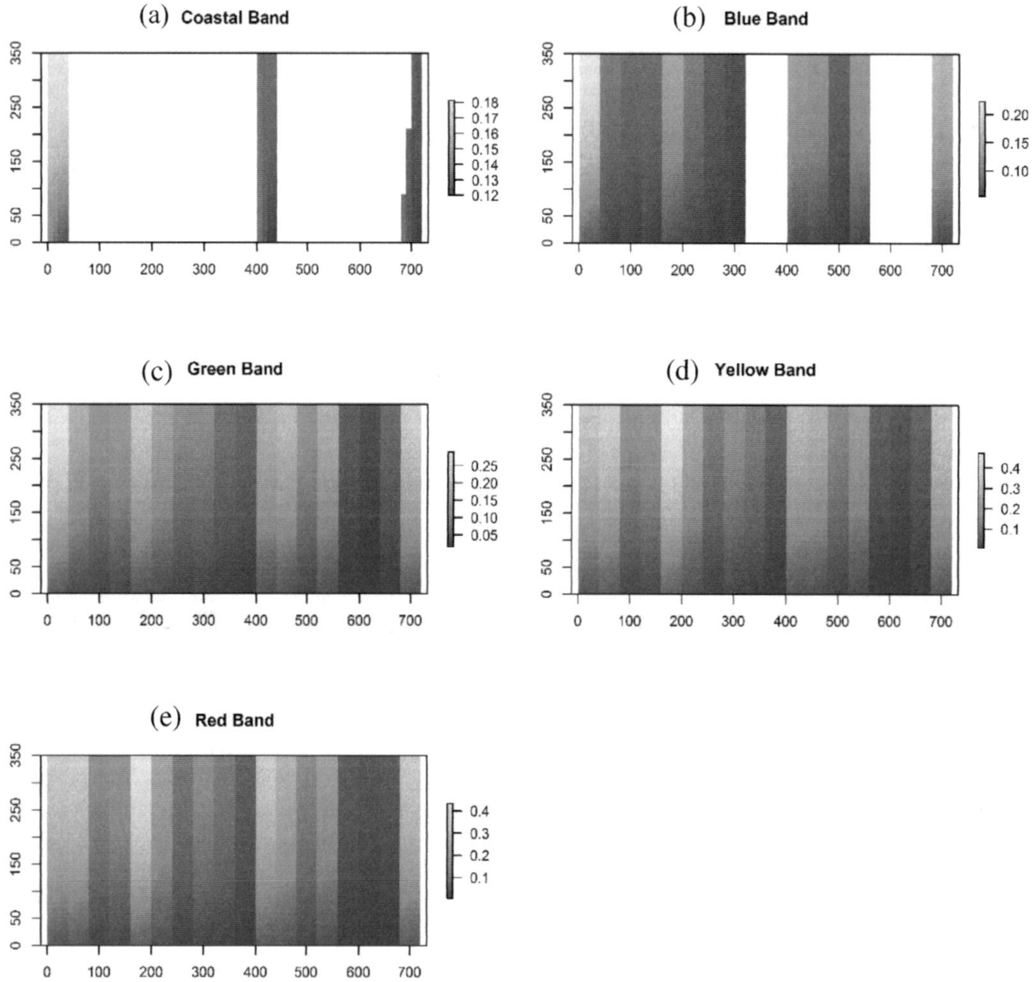

Figure 2. Simulated image of Maritonia et al.'s model:(a) Grey shaded of Coastal band, (b) Grey shaded of Blue band, (c) Grey shaded of Green band, (d) Grey shaded of Yellow band and (e) Grey shaded of Red band.

While the simulated image from Lee et al.'s model was shown in Figure 3. In the case of Lee et al.'s based simulated image, the two marked point from Maritonia et al.'s based simulated image was not applied. The dark object removal did not appear in any bands, and the coastal bands still give valuable information for all bottom types. Additionally, the attenuation effect of water depth was shown more intensively affected the just above surface reflectance especially in longer wavelength bands (Yellow and Red band).

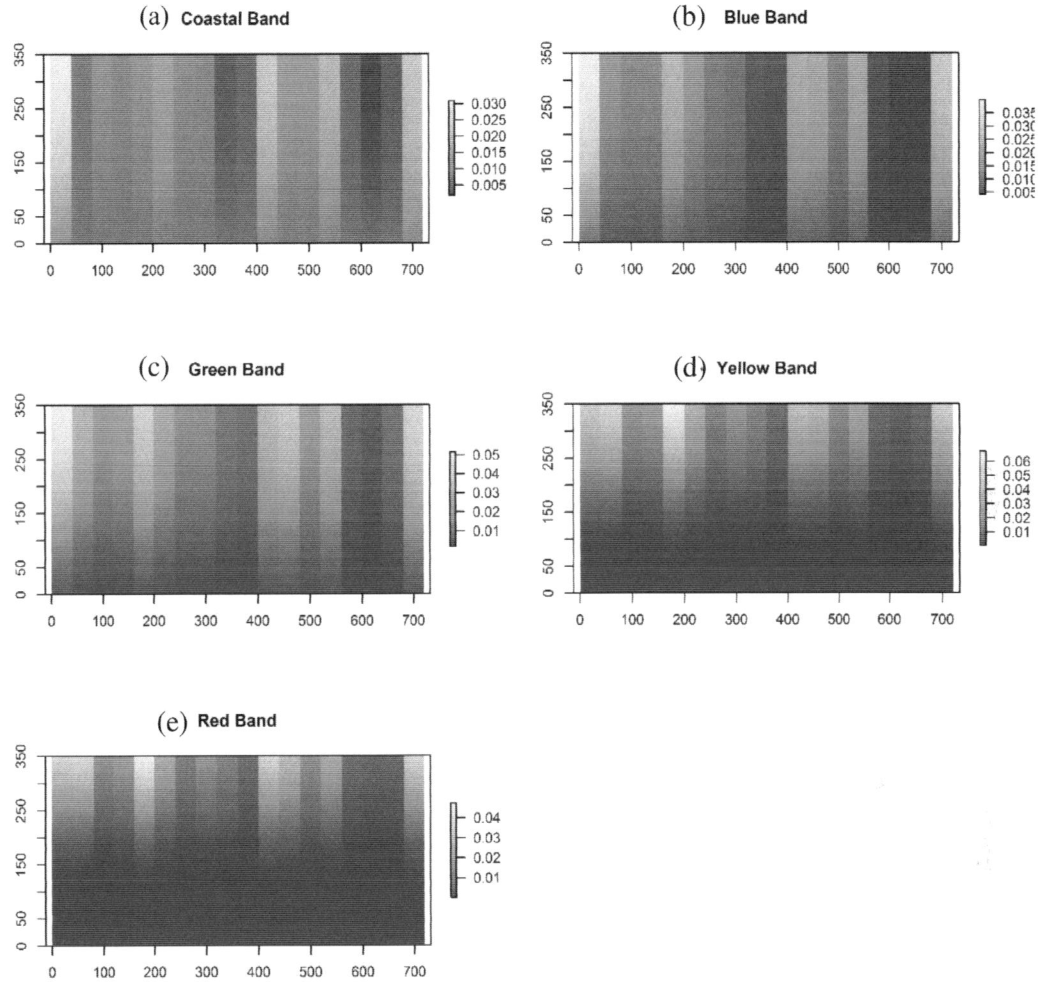

Figure 3. Simulated image of Lee et al.'s model:(a) Grey shaded of Coastal band, (b) Grey shaded of Blue band, (c) Grey shaded of Green band, (d) Grey shaded of Yellow band, and (e) Grey shaded of Red band

4. Performance of Lyzenga's Water Column Correction

The reflectance of visible bands (Figure 4.a. and 4.b) decreases follow with increasing depth, due to the exponential function of depth and attenuation coefficient (Equation 4 and 12). The effect of water attenuation for both models was different. In the case of Lee et al.'s model, the longer visible band wavelength (i.e., red band) shows a more dramatically effect of water attenuation. This result is consistent with Kirk [15], for wavelength higher than 550 nm the attenuation coefficient is increasing rapidly with depth, occur due to a high ratio of scattering and absorption.

Figure 4.b and 5.b shows the linearized reflectance (X_i) of Maritonia et al.'s and Lee et al.'s based simulated images, respectively. Theoretically, a straight line form is created if we plotted the linearized reflectance with depth, as the Maritonia et al.'s model performed (Figure 4.b). Since the Lee et al.'s based simulated images have more complex's relation between reflectance and depth the plotted linearized reflectance with a depth not show more curvy line form (Figure 5.b).

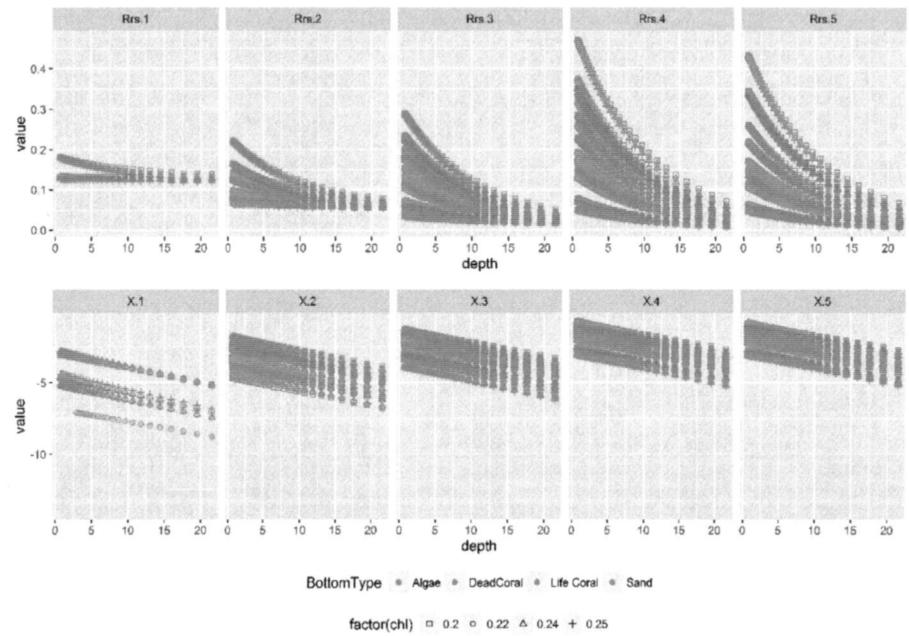

Figure 4. The scatterplot of each bottom type classes shows the effect of depth for Maritonia et al.'s model (a) reflectance value and depth, (b) linearized reflectance and depth

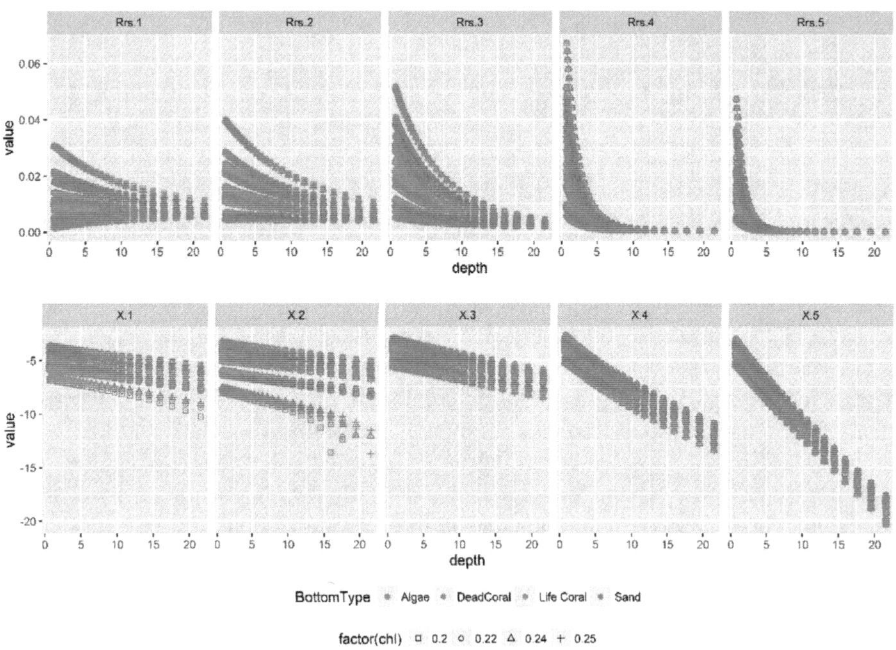

Figure 5. The scatterplot of each bottom type classes shows the effect of depth for Lee et al.'s model: (a) reflectance value and depth, (b) linearized reflectance and depth

4.1 The Removal of Water Attenuation Effect

The performance of Lyzenga's water column correction (Figure 6.a and 6.b) proved that the effect of scattering and absorption in the water column are reduced able. The performance of Lyzenga water column correction on Maritonia et al.'s based simulated images work as Lyzenga explains in the paper that depth invariance index is free of water attenuation effect, but the Lee et al.'s model result shows slightly different.

If the actual surface reflectance is more complicated thenMaritonia et al.'s reflectance model (Eq.4), as we used in the Lee et al.'s based simulated imagery. Then the Lyzenga water column correction could not remove the water attenuation effect but could reduce the effect. Another remaining problem in the case of Maritonia et al.'s model, where the slope ($\frac{k_i}{k_j}$) of each bottom types for some band pairs was sometimes not equal for each bottom types. This occurred due to the water quality variation. Even this study set the water quality in the range of 0.2 – 0.25 mg m-3 which in small difference. But the effect was significantly affecting an object with dark reflectances, such as algae.

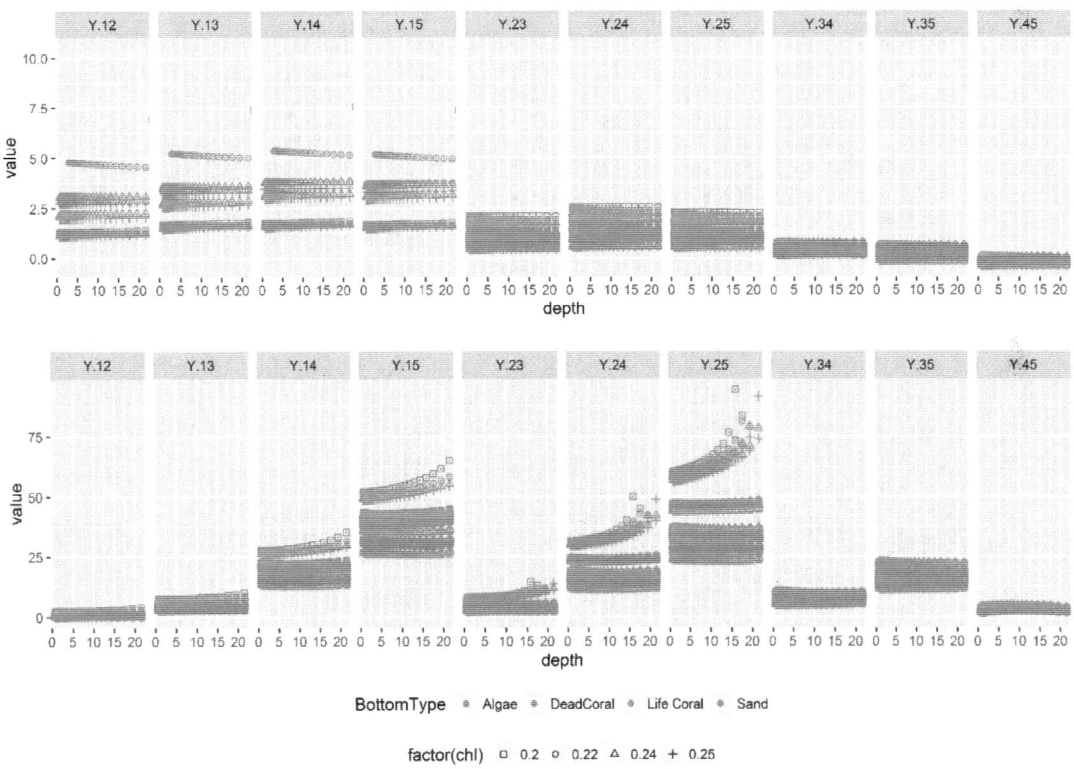

Figure 6. The depth invariance index for (a) Maritonia et al.'s and (b) Lee et al.'s model versus depth

5. Mathematical Explanation of Simulation Results

By using a mathematical expression, we check the performance of Lyzenga water column correction on simulated reflectance data. First for the Maritonia et al.'s model, the surface reflectance (Equation 13) that is an exponential function of depth are converted as a linear function of depth using Equation 1 and shows in detail as follows:

$$X_i = \log\left(\left(V_i + \left((\rho_{b_i} - V_i)e^{-hk_i}\right)\right) - \left(V_i + \left((\rho_{b_i} - V_i)e^{-\infty k_i}\right)\right)\right) \quad (21)$$

$$= \log\left(\left(V_i + \left((p_{b_i} - V_i)e^{-hk_i}\right)\right) - (V_i)\right)$$

$$= \log\left((p_{b_i} - V_i)e^{-hk_i}\right)$$

$$= \log(p_{b_i} - V_i) - hk_i$$

Then$m$ or the slope of two pair data for Maritonia et al.'s model is written as:

$$m == \frac{\left(log(p_{b_i} - V_i) - hk_i\right)_{max} - \left(log(p_{b_i} - V_i) - hk_i\right)_{min}}{\left(log\left(p_{b_j} - V_j\right) - hk_j\right)_{max} - \left(log\left(p_{b_j} - V_j\right) - hk_j\right)_{min}} \quad (22)$$

To calculate m, Lyzenga's method assumed that the water quality is homogeneus and a single bottom type is used, then p_b, V and $k$value are the same all over depth range, thus the calculation of m simplyfied is as follows

$$m = \frac{k_i(h_{max} - h_{min})}{k_j(h_{max} - h_{min})} = \frac{k_i}{k_j} \quad (23)$$

Equation 23 shows that the slope of pair between two linearized reflectance of i and j bands is equal with $\frac{k_i}{k_j}$, then equation 3 could written as

$$Y_{ij} = \log(p_{b_i} - V_i) - hk_i - \frac{k_i}{k_j}\left(\log\left(p_{b_j} - V_j\right) - hk_j\right)$$

$$= \log(p_{b_i} - V_i) - \frac{k_i}{k_j}\log\left(p_{b_j} - V_j\right) \quad (24)$$

Then equation 24 shows that's no depth variable in the final product of Lyzenga's Water Column Correction.

Second for the Lee et al.'s model that shows water attenuation could not be removed. Then we evaluated the performance mathematically. Begin with substitute the Lee's surface reflectance function (Eq.2) to transformed reflectance equation (Eq.5) as follows

$$X_i = \log\left(\left(r_{rs\,i}^{dp}\left(1 - e^{-\left[\frac{1}{\cos(\theta_s)}+D_{u\,i}^C\right]k_iH}\right) + \frac{\rho_i}{\pi}e^{-\left[\frac{1}{\cos(\theta_s)}+D_{u\,i}^B\right]k_iH}\right) - \left(r_{rs\,i}^{dp}\right)\right)$$

$$= \log\left(\frac{\rho_i}{\pi}e^{-\left[\frac{1}{\cos(\theta_s)}+D_{u\,i}^B\right]k_iH} - r_{rs\,i}^{dp}e^{-\left[\frac{1}{\cos(\theta_s)}+D_{u\,i}^C\right]k_iH}\right) \quad (25)$$

In order to solve equation 2, $D_{u\,i}^B = 1.3786D_{u\,i}^C$. Then the equation is simplified as

$$X_i = \log\left(\frac{\rho}{\pi}e^{-\left[\frac{1}{\cos(\theta_s)}+D_u^B\right]kH} - r_{rs}^{dp}e^{-\left[\frac{1}{\cos(\theta_s)}+D_u^C\right]kH}\right)$$

$$= \log\left(\frac{\rho}{\pi}e^{-\left[\frac{1}{\cos(\theta_s)}+1.3786D_u^C\right]kH} - r_{rs}^{dp}e^{-\left[\frac{1}{\cos(\theta_s)}+D_u^C\right]kH}\right)$$

$$= \log\left(\left(\frac{\rho}{\pi}e^{-\left[\frac{1}{\cos(\theta_s)}+0.3786D_u^C\right]kH} - r_{rs}^{dp}\right)e^{-\left[\frac{1}{\cos(\theta_s)}+D_u^C\right]kH}\right) \quad (26)$$

$$= -\left(\frac{1}{\cos(\theta_s)} + D_{u_i}^C\right)k_iH + \log\left(\frac{\rho_{b,i}}{\pi}e^{-\left[\frac{1}{\cos(\theta_s)}+0.3786D_{u_i}^C\right]k_iH} - r_{rs\,i}^{dp}\right)$$

$$= -\left(\sec(\theta_s) + D_{u_i}^C\right)k_iH + \log\left(\frac{\rho_{b,i}}{\pi}e^{-\left[\sec(\theta_s)+0.3786D_{u_i}^C\right]k_iH} - r_{rs\,i}^{dp}\right)$$

The transformed reflectance X_i for Lee's reflectance is shown in Equation 26. Then the slope of pair X_i and X_j for Lee et al.'s model is was follow

$$m = \frac{\left(-(\sec(\theta_s) + D^C_{u\,i})k_iH + log\left(\frac{\rho_{b,i}}{\pi}e^{-|\sec(\theta_s) + 0.3786D^C_{u\,i}|k_iH} - r^{dp}_{rs\ i}\right)\right)_{max} - \left(-(\sec(\theta_s) + D^C_{u\,i})k_iH + \log\left(\frac{\rho_{b,i}}{\pi}e^{-|\sec(\theta_s) + 0.3786D^C_{u\,i}|k_iH} - r^{dp}_{rs\ i}\right)\right)_{min}}{\left(-\left(\sec(\theta_s) + D^C_{u\,j}\right)k_jH + log\left(\frac{\rho_{b,j}}{\pi}e^{-|\sec(\theta_s) + 0.3786D^C_{u\,j}|k_jH} - r^{dp}_{rs\ j}\right)\right)_{max} - \left(-\left(\sec(\theta_s) + D^C_{u\,j}\right)k_jH + log\left(\frac{\rho_{b,j}}{\pi}e^{-|\sec(\theta_s) + 0.3786D^C_{u\,j}|k_jH} - r^{dp}_{rs\ j}\right)\right)_{min}} \quad (27)$$

As explain in subchapter 2.2, Lyzenga's method assumed that the water quality is homogeneus and a single bottom type is used, then p_b, V and $k$value are the same all over depth range, thus the calculation of m is written as follows

$$m = \frac{-(\sec(\theta_s) + D^C_{u\,i})k_iH_{max} + (\sec(\theta_s) + D^C_{u\,i})k_iH_{min} + \log\left(\frac{\rho_{b,i}}{\pi}e^{-[\sec(\theta_s) + 0.3786D^C_{u\,i}]k_iH_{max}} - r^{dp}_{rs\ i}\right) - \log\left(\frac{\rho_{b,i}}{\pi}e^{-[\sec(\theta_s) +}\right.}{-(\sec(\theta_s) + D^C_{u\,i})k_iH_{max} + (\sec(\theta_s) + D^C_{u\,i})k_iH_{min} + \log\left(\frac{\rho_{b,i}}{\pi}e^{-[\sec(\theta_s) + 0.3786D^C_{u\,i}]k_iH_{max}} - r^{dp}_{rs\ i}\right) - \log\left(\frac{\rho_{b,i}}{\pi}e^{-[\sec(\theta_s) +}\right.}$$

$$= \frac{(\sec(\theta_s) + D^C_{u\,i})k_i(H_{min} - H_{max}) + \log\left(\frac{\frac{\rho_{b,i}}{\pi}e^{-[\sec(\theta_s) + 0.3786D^C_{u\,i}]k_iH_{max}} - r^{dp}_{rs\ i}}{\frac{\rho_{b,i}}{\pi}e^{-[\sec(\theta_s) + 0.3786D^C_{u\,i}]k_iH_{min}} - r^{dp}_{rs\ i}}\right)}{\left(\sec(\theta_s) + D^C_{u\,j}\right)k_j(H_{min} - H_{max}) + \log\left(\frac{\frac{\rho_{b,j}}{\pi}e^{-[\sec(\theta_s) + 0.3786D^C_{u\,j}]k_jH_{max}} - r^{dp}_{rs\ j}}{\frac{\rho_{b,j}}{\pi}e^{-[\sec(\theta_s) + 0.3786D^C_{u\,j}]k_jH_{min}} - r^{dp}_{rs\ j}}\right)} \quad (28)$$

$$= \frac{\left(\sec(\theta_s) + D^C_{ui}\right)k_i}{\left(\sec(\theta_s) + D^C_{uj}\right)k_j}$$

Assumed that $\log\left(\frac{\frac{\rho_{b,i}}{\pi}e^{-[\sec(\theta_s) + 0.3786D^C_{u\,i}]k_iH_{max}} - r^{dp}_{rs\ i}}{\frac{\rho_{b,i}}{\pi}e^{-[\sec(\theta_s) + 0.3786D^C_{u\,i}]k_iH_{min}} - r^{dp}_{rs\ i}}\right)$ and $\log\left(\frac{\frac{\rho_{b,j}}{\pi}e^{-[\sec(\theta_s) + 0.3786D^C_{u\,j}]k_jH_{max}} - r^{dp}_{rs\ j}}{\frac{\rho_{b,j}}{\pi}e^{-[\sec(\theta_s) + 0.3786D^C_{u\,j}]k_jH_{min}} - r^{dp}_{rs\ j}}\right)$ will give a very small value than could be neglected then Equation 28 shows that the slope of a pair between band i and j in the case of Lee et al.'s model was not only the attenuation coefficient ratio but include the and solar zenith value. Finally, the Depth Invariance Index equation (Eq. 3) for Lee's model can be simplified and expressed as

$$Y_{ij} = \left(-\left(\sec(\theta_s) + D^C_{ui}\right)k_iH + \log\left(\frac{\rho_{b,i}}{\pi}e^{-[\sec(\theta_s) + 0.3786D^C_{ui}]k_iH} - r_{rs}\right)\right) - \frac{\left(\sec(\theta_s) + D^C_{ui}\right)k_i}{\left(\sec(\theta_s) + D^C_{uj}\right)k_j}\left(-\left(\sec(\theta_s) + D^C_{uj}\right)k_jH + \log\left(\frac{\rho_{b,j}}{\pi}e^{-[\sec(\theta_s) + 0.3786D^C_{uj}]k_jH} - r_{rs}\right)\right) \quad (29)$$

$$= \log\left(\frac{\rho_{b,i}}{\pi}e^{-0.378633D^C_{ui}k_iH} - r^{dp}_{rs\,i}\right) - \log\left(\frac{\rho_{b,j}}{\pi}e^{-0.378633D^C_{uj}k_jH} - r^{dp}_{rs\,j}\right)\left(\frac{\left(\sec(\theta_s) + D^C_{ui}\right)k_i}{\left(\sec(\theta_s) + D^C_{uj}\right)k_j}\right)$$

Above equation shows the Lyzenga water column method could not remove the effect of depth and water attenuation. In fact, the effect of depth and water attenuation are in multiplication function. Thus, even after the correction, the water column attenuation effect will remain. In the case of IOP with chlorophyll-a between 0.2 – 0.25 mg m-3, if the attenuation coefficient is small such us blue band (kblue ≈ 0.04) or green band (kgreen ≈ 0.06) then the depth invariance index have less water column effect. More over, a failure of correction will occur for the pair of a longer visible band wavelength, namely red band, due to the high attenuation coefficient (kred ≈ 0.4) that reach 9 and 6 times more higher than in the blue and green band, respectively.

One downside regarding this study is that each bottom type had been set to have the same opportunity to appear in the same number and same depth, which unnaturally thing happen in the coral reef ecosystem. Unfortunately, there is no scientific information about the representative distribution of each bottom types. It became the limitation to made more natural simulated imagery with representative distribution in the coral reef environment. However, the initial simulated imagery already answers the main aim of identifying the performance of Lyzenga's water column correction.

6. Conclusion

The simulated imagery provides powerful tools for imagery or methodology evaluation. Our investigation on simulated Worldview-2 imagery over coral reef environment shows that the Lyzenga's water column correction plays differently in both Maritonia et al.'s and Lee et al.'s surface reflectance model. In the case of Maritonia et al.'s based simulated image, the effect of attenuation coefficient was almost removable in any pair of six visible bands. While, for the Lee et al.'s based simulated image the Lyzenga water column correction tecnique was only able to reduce the effect of water attenuation. Supprisingly, the assumption that ratio of attenuation coefficient is equal at any bottom types was hardly to proven. In our simulation, a slight variation in water quality (0.2 - 0.25 mg m-3) variation shows an significant effect on the performance of Lyzenga's water column correction.

The present study has only examined the effect of water column correction under zero noise condition, which gave a positive result. Therefore, the other factor that is influencing the just aboved surface reflectance should be investigated, such as noise, bottom types diversity, intermixing pixels, water quality, and number of representative sample. Research into solving several of this problems is already underway.

7. Acknowledgement

The first author acknowledges support from the Indonesia Endowment Fund for Education (LPDP) by the Ministry of Finance. We are very grateful to Marlina Nurlidiasariforgave permission to use her spectral measurement data.

References

[1] Lyzenga, D.R 1978 Passive remote sensing techniques for mapping water depth and bottom features *Applied Optic***17**(3) pp 379–383.

[2] Pulliza, DT 2004 A Multi-Sensor Comparison For Coral Reef Habitat Mapping: A Case Study Using A Tropical Patch Reef Environment In Biscayne National Park, Florida (Doctoral Dissertation, University OF Puerto Rico Mayaguez Campus).

[3] Pahlevan, N.; Valadanzouj, M.; Alimohamadi A 2006 A Quantitative Comparison to Water Column Correction Reference 1Techniques for Benthic Mapping Using High Spatial Resolution Data. Paper presented at the ISPRS Commission VII Mid-term Symposium *Remote Sensing, From Pixels to Processes*(Enschede, The Netherlands)

[4] Nurlidiasari, M., Budiman, S 2005 Mapping coral reef habitat with and without water column correction using Quickbird image *International Journal of Remote Sensing and Earth Science***2**pp 45–56.

[5] Sagawa, T., Mikami, A., Aoki, M.N., and Komatsu, T 2012 Mapping seaweed forests with IKONOS image based on bottom surface reflectance *Proc. SPIE* **8525**: 85250Q. doi: 10.1117/12.975678

[6] Budhiman, S., Parwati, E., and Emiyati 2012 The effect of the extent of coral reef area on uniform bottom reflectance determination for water column correction using Landsat ETM *International Journal of Remote Sensing and Earth Sciences (IJReSES)* **9**(2) pp 88-99.

[7] Purkis, SJ and Pasterkamp, R 2004 Integrating in situ reef-top reflectance spectra with Landsat TM imagery to aid shallow-tropical benthic habitat mapping *Coral reef.* **23**(1) pp 5-20

[8] Mishra, D.R., Narumalani, S., Rundquist, D., Lawson M 2005 High-resolution ocean color remote sensing of benthic habitats: a case study at the Roatan Island, Honduras *IEEE Geoscience and Remote Sensing Letters***43**(7) pp 1592--1604

[9] Zoffoli, M. L., Frouin, R., and Kampel M 2014 Water column correction for coral reef studies by remote sensing *Sensors***14**(9), 16881-16931.

[10] Gordon, HR, Brown, OB, and Ando Jacobs, MM 1975 Computed Relationships Between The Inherent And Apparent Optical Properties Of A Flat Homogeneous Ocean *Applied Optics***14** (2), pp 417-427.

[11] Maritorena, S., Morel, A., and Gentili, B 1994 Diffuse reflectance of oceanic shallow waters: Influence of water depth and bottom albedo *Limnology and oceanography***39**(7), pp 1689-1703.

[12] Lafon, V., Froidefond, J. M., Lahet, F., and Castaing P 2002 SPOT shallow water bathymetry of a moderately turbid tidal inlet based on field measurements *Remote sensing of Environment***81**(1), pp 136-148.

[13] Lee, Z.P., and K.L. Carder 1999 Effect of spectral band numbers on the retrieval of water column and bottom properties from ocean color data *Applied Optic***41**(12) pp 2191-2201.

[14] Lee, Z.P., A. Weidemann, and R. Arnone 2013 Combined effect of reduced band number and increased bandwidth on shallow water remote sensing: the case of Worldview-2 *IEEE Trans. Geoscience and Remote Sensing***51** (5) pp 2577-2586.

[15] Kirk, J.T.O 1981Estimation of the scattering coefficient of natural waters using underwater irradiance measurements *Australia Journal of marine Freshwater Research***32** pp 533-539.

The Relationship between the Health of Coral Reefs to The Damages of Jelateng Sub Watersheds in the Northern Coast of Sekotong, West Nusa Tenggara

A Zaenuri, A N Saputra, D A Kudiastuti, I Fajria, L AWijayanti, and P A Aryaguna
Masters Program of Remote Sensing, Faculty of Geography, Universitas Gadjah Mada

*email: pramaardhaaryaguna23@gmail.com

Abstract. Coral reefs are unique ecosystem and most of them are found in tropical areas. Remote sensing technology has the ability to present the information about the existence of marine resources, such as coral reefs. Remote sensing technology also provide the spatial information on wider area. This study is aimed to see how significant the effect of watersheds damage toward the condition of coral reefs because of erosion which is occurred in the upstream and sedimentation which is occurred in the downstream of the watershed. Acquisition of the data in this study used remote sensing image Landsat 7 ETM+ and Lansat 8 OLI in multi temporal, the data collection of this study used photo-quadrate technique and taking sample in the field. Spectral analysis and transformation of CTA (Classification Tree Analysis) was used to generate the level of the health of coral reefs. The method of RSI (Red Silt Index) was used to find out the degree of turbidity of the water. Analysis of the changes was done to find out the changes which were occurred on coral reefs based on the changes of turbidity that were occurred between 2000 till 2015. Based on the result, it shows that there is no effect on the changes of turbidity value toward the health condition of coral reefs in the northern coast of Sekotong Sub District 2000 till 2015. This is due to the use of RSI index or turbidity index that cannot be used in the condition of shallow water.

Keywords: Coral reefs, Landsat, Sekotong

1. Introduction

Coral reefs are a landform formed by the activity of organisms in the marine environment. The shape and the formation of coral reefs is influenced by some factors such as temperature, salinity, depth of sea, ocean circulation, supplies nutrients, turbulence and turbidity. To achieve maximum growth, coral reefs require clear water, with the water temperature relatively warm or hot (23^0-25^0C), big wave motion and smooth water circulation, also protected from sedimentation process. Secondary term for the existence of coral reefs is constant salinity of seawater. Coral reefs in Indonesia estimates occupy an area of 7,500 km^2 with the richness biological type of coral reef in the world [8]. Only 6,2% of those coral reefs are in good condition. This damage of coral reef is caused by some factor such as:

1. Human greed
2. Ignorance and indifference
3. Weak law enforcement

The appropriate way to help research the detection of the distribution of coral reefs is by using Remote Sensing. Remote Sensing more efficient in the use of time and cost, also data image can be analyzed visually and digital [6].

Landsat TM imagery commonly used for observation and management of coral reefs, because Landsat TM imagery has a high enough spatial and spectral resolution [11]. Landsat TM has 2 bands with spectral resolution which able to identify objects under the water. The ability of electromagnetic waves to penetrate to the water is very low because water will absorb some of the electromagnetic energy. It makes the ability to identify object under the water surface limited. Object can be observed or detected only at the depth of 15 to 30 meters. The condition of a watershed can be seen by the condition of the watershed downstream. Many downstream conditions contained sediment can be associated that the conditions experienced erosion. Watershed conditions which are decrease will certainly affect to the condition of coral reefs in the coastal area [4]. We need study to know how the erosion and watershed damage affected to the coral reef condition.

This study has some purposes:

1. Monitoring and mapping of the coral reef health conditions, especially on the northern coast of Sekotong some part of Jelatang watershed.
2. Monitoring and mapping the water conditions, especially the turbidity level of the northern coast of Sekotong some part of Jelatang watershed.
3. Analyze the relationship effect of pollution on the coral reef health condition in multi-temporal.

2. Study Area

The area in this study is northern coast of Sekotong, West Nusa Tenggara. Sekotong in West Lombok is well known with its sea ecosystem diversity. Pearl, seaweed, and fish are some comodities from Lombok. Unfortunately, the condition of coral reef in Sekotong are not in good condition. Only 10% of the coral reef are life coral and the other are death coral and smashed coral [4]. The high level of sedimentation is suspected as the main factor of the coral reef damage in Sekotong. Because of the condition of the coral reef and its diversity of sea ecosystem, Sekotong are choosen as the study area.

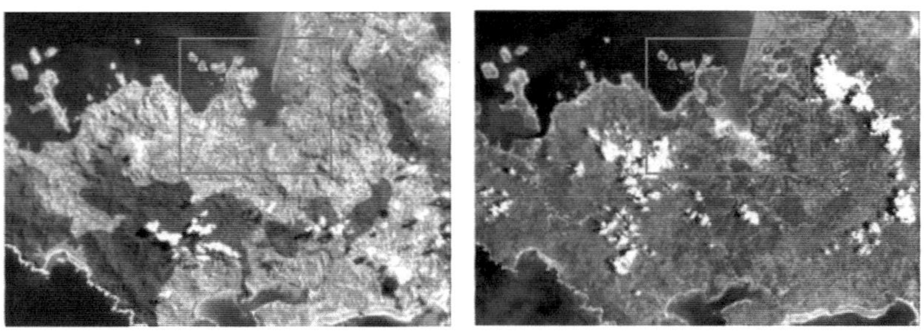

Figure 1. The changes of sedimentation at Northern coast of Sekotong.(a) Landsat 7 ETM+ true color composit image recorded in 2000.(b) Landsat 8 OLI true color composit image recorded in 2015.

3. Method

Spectral correction at level *radiance at sensor* and *reflectance at surface* are applied to Landsat 7 ETM+ and Landsat 8 OLI. Corrected images at the level of *reflectance at surface* used to build make coral reef mullet map and corrected images at the level of *radiance at sensor* used to build *Red Silt Iindex (RSI)* to estimated the level of sedimentation. The value of sedimentation from RSI index compared to the filed condition using regresion analysis[7]. To get the RSI value we can use the formula[5] as shown bellow.

$$\mathrm{RSI} = \frac{V_{\lambda 2} - V_{\infty}}{V_{\lambda 1} - V_{\infty}} \tag{1}$$

where V_2 : wave length of band 2 in image
V_1 : wave length of band 1 in image
$V\infty$: darkest pixel value in sea area

RSI value are between 0 - 2. 0 value are located in sea far from the land (low sedimentation) and 2 are located in sea near the land (high sedimentation). Figure 3 shown the map as result of RSI value ehich represent the sedimentation condition

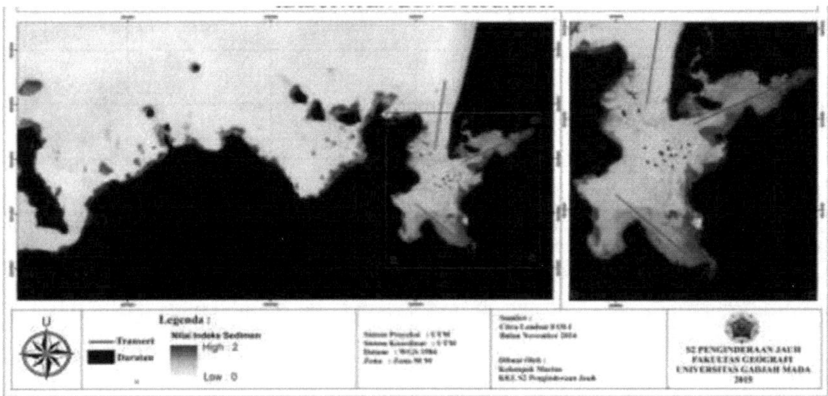

Figure 2. *Red Silt Deposition Index* (RSI) Map Northern coastal of Sekotong West Lombok 2015

Radiometric correction process to get the *reflectance at surface* is using FLAASH (*Fast Line Of Sight Atmospheric Analysis of Hypercubes)*[10]. The *sun glint* correction also applied to the images to reduce the disturbance causing by the water coloum in the sea that will change the pixel value[2]. *Sun glint* correction[9] are applied by using this formula

$$R_i' = R_i - b_i \times (R_{NIR} - Min_{NIR}) \quad (1)$$

Where R_i : reflectance value in visible band
b_i : regression value
R_{NIR} : reflectance value in near infra red band
Min_{NIR} : minimum reflectance value in near infra red band

After radiometric correction is done at level of *at surface reflectance* we assume that there is no atmospheric disturbace in the image so the specral value in images are appropriate with the spectral reflectance curve. Radiometric correction are applied to image recorded in 2000 and 2015 so we can assume that pixel value in 2000 and 2015 are represent the same condition. The coral reef maping are build using unsupervised classification with iso-data algorithm. This method classified the coral reef into 8 classes based on the value of sepectral reflectance. Figure 6 is the map represent the ditribution of coral reef as the resul of unsupervised classification.

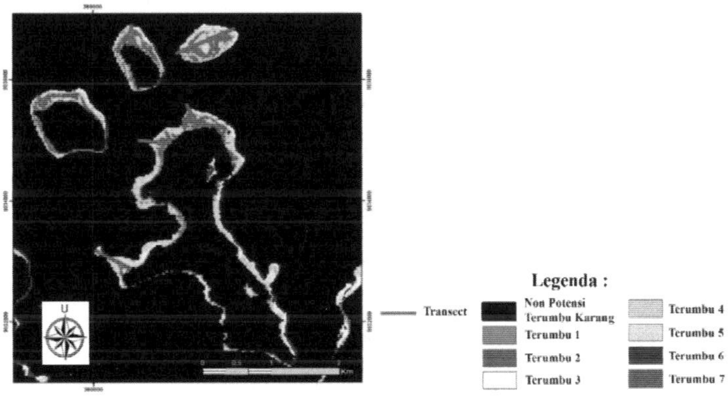

Figure 3. *Coral reef* Map Northern coastal of Sekotong West Lombok 2015

Sample for the field check are obtained using transect line. Transect line are made by notice the variation of the condition in area study. Figure 7 are show the transect line of the sample.

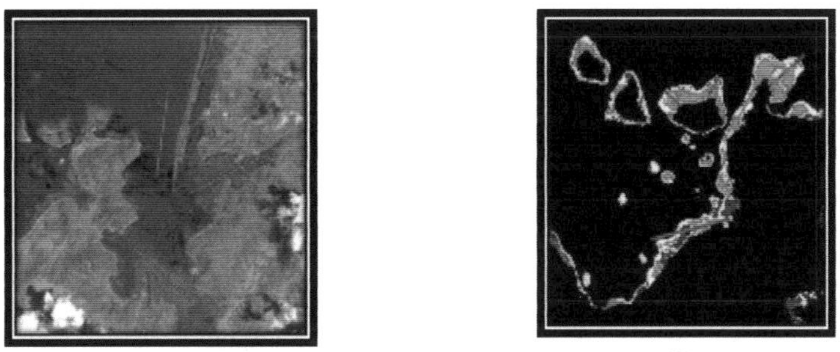

(a) (b)

Figure 4. Transect line field sample on both of map unit, (a) Transect Sample RSI; (b) Transect sample coral reef condition

4. Discussion

Classification Tree Analysis method is used for benthic habitat mapping. This method is a part of image mining (data mining) methods as a variation of the decision tree or regression decision tree method. The advantage of this method is there is no minimum limit on the training sample so that the sample still can be processed, although only one pixel. Samples were gained 1 pixel only since the limitation of sampling in the field.

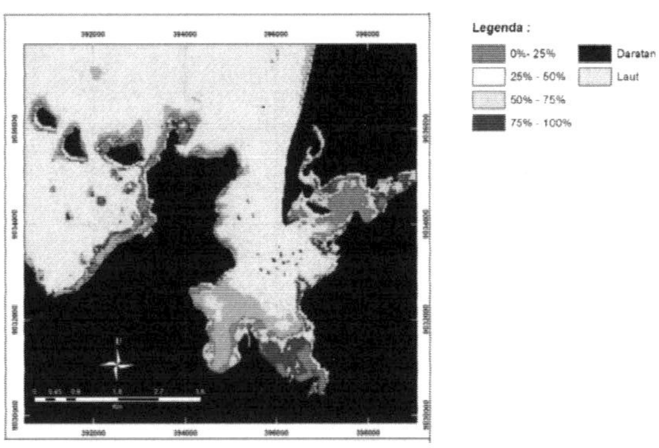

Figure 5. Map of Coral reef habitat in 2015

After getting the data benthic habitat, the next method is extracting a class that has a limit reef mapping to map the health of coral reefs.

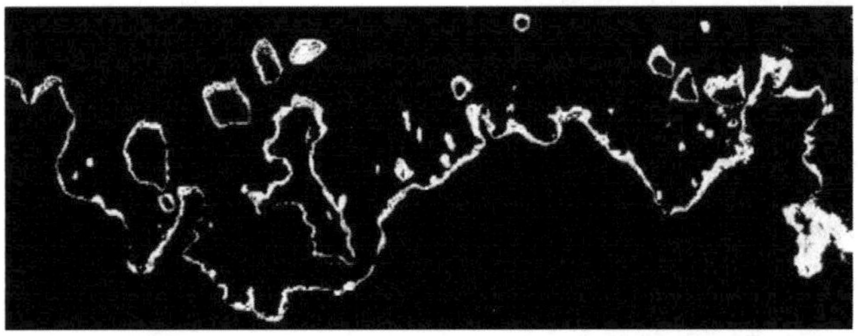

Figure 6. Masking of sea boundary and land

The field survey is aimed to determine the composition of benthic habitat with photoquadrat methods [3]. However, in the practice there were various limitations in the field survey. The photo shoot was done without the use of frames or quadrant assistance as restrictions. The result which was obtained in the Sekotong Sub District showed that there are eight categories of benthic habitat, they are coral, coral rubble, coral sand, seagrass sand, coral sand, seagrass sand reefs, coral rubble sand, seagrass and sand which can be seen in Figure 3.

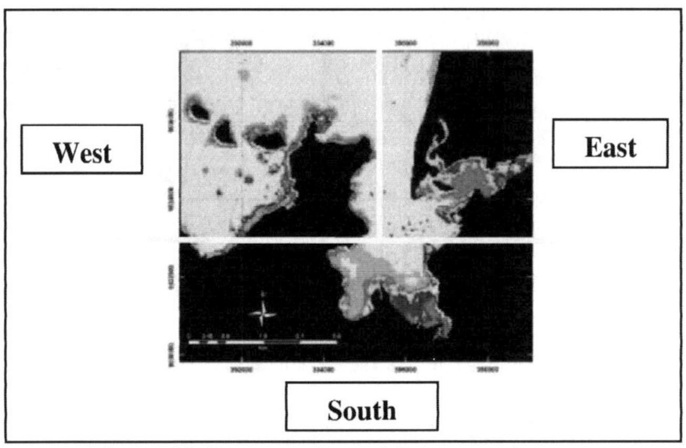

Figure 7. Segmentation of Sekotong Areas

4.1. Mapping of coral reefs health

Mapping of the health of coral reefs used empirical method in which this method used regression value of 3 visible bands with the percentage of coral reef health on the field which was taken by transect photos.

Table 1. The sample coral reef health toward pixel value.

Koordinat X	Koordinat Y	B1	B2	B3	Mean Percent Cover (%)	Percent cover (%)
391815	9035665	263	389	115	33.3	25 - 50
391785	9035695	267	463	187	68.3	50 - 75
392685	9034765	352	496	139	40.0	25 - 50
392715	9034795	363	566	215	83.5	75 - 100
392745	9034795	369	550	238	39.0	25 - 50
392715	9034765	438	627	225	97.5	75 - 100
392175	9035515	515	745	237	70.0	50 - 75

Regression Statistics	
Multiple R	0.691558
R Square	0.478252
Adjusted R Square	-0.0435
Standard Error	25.23169

Table 2. Statistic value from the regression result of pixel value and percentage of coral reef health.

	Coefficients	*Standard Error*	*t Stat*	*P-value*	*Lower 95%*	*Upper 95%*	*Lower 95.0%*	*Upper 95.0%*
ntercept	-26.856	56.281	-0.477	0.666	-205.966	152.255	-205.966	152.255
B1	-0.480	0.636	-0.755	0.505	-2.503	1.543	-2.503	1.543
B2	0.492	0.591	0.833	0.466	-1.388	2.372	-1.388	2.372
B3	-0.026	0.447	-0.059	0.957	-1.450	1.397	-1.450	1.397

The result of equation in the regression analysis can be seen as follow:

$$y=-26.856-(0.48*B1)+(0.492*B2)-(0.02*B3) \quad (2)$$

Where: B1 = Blue band
B2 = Green band
B3 = Red band

Then, the equation was applied to Landsat 8 OLI 2015 that had been corrected by surface reflectance and masking of coral reef.Previously, delimitation of sea boundary derived from the spectral values manually using PCA bands PC 2, which PC 2 gives a clear boundary between deep water and shallow water.

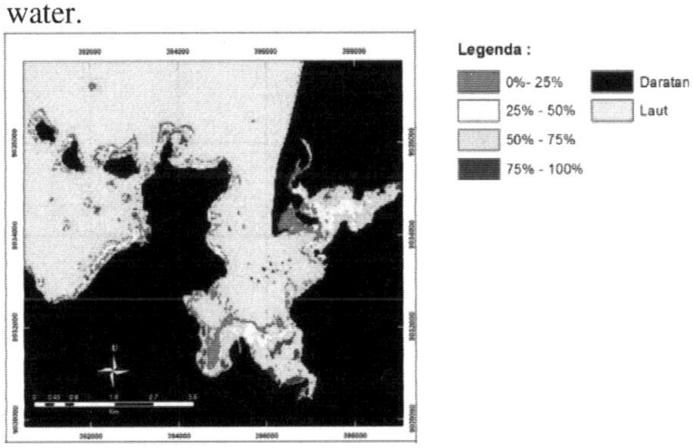

Figure 8. Map of Coral Reef health in 2015

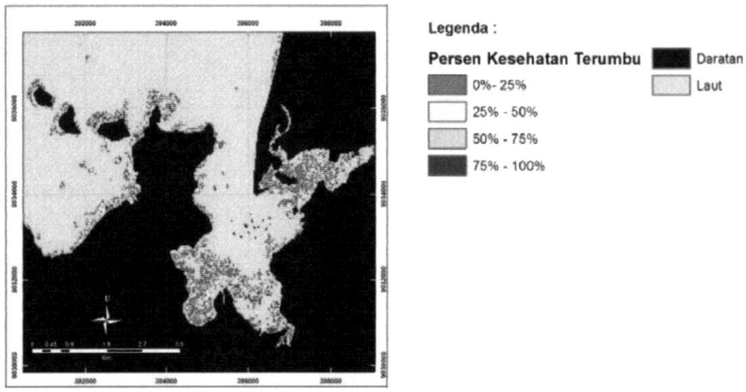

Figure 9. Map of coral reef health in 2000

The equation was also applied to the image in 2000 with the assumption that the pixel values in 2000 and 2015 represented the same reef conditions which were clear from atmospheric disturbances because it has already corrected till surface reflectance.

North coast of Sekotong in some parts of DAS Jelatang is generally swifting water area, because of the large ship traffic from outside Lombok Island, tourist boats and or fishing boats. From 4 classes of existing coral reef health, in 2000 the health of coral reefs were dominated by a class of 25-50% health, covering to 1.7424 million m^2 from the total covering area of Coral reef amounted to 5.8824 million m^2 on the north coast Sekotong. In 2015, the health condition was dominated by 25-50% class health with 2.5101 million m^2 width from 7.9713 million m^2 of total covering area of coral reefs. Coral reef area with class of 0-25%, which is still in the same condition to 2015, is 486 900 m^2 area. Coral reef area with class of 25-50%, which still in the same condition to 2015, is 550 800 m^2 area. Coral reef area with class of 50-75%, which still in the same condition to 2015, is 520 200 m^2 area. Coral reef area, with class of 75-100%, which still in the same condition to 2015, is 249 300 m^2 area.

4.2. Monitoring and mapping the condition of water turbidity level

The first stage to do in the process of mapping the level of turbidity is to make linear regression analysis between the actual turbidity value in the field and the value of RSI index which were obtained from the previous work. Values of correlation (R^2) from regression between the pixel values of the RSI index and turbidity index shows the value is 0,87 which shows that there is a match between the pixel values and field of turbidity data.

$$y = (246{,}69 * b1) - 109{,}62 \tag{3}$$

Note: b1 is pixel value from RSI index since the regression turbidity value in the field does not use pixel value from the real imagery.

Table 4. Regression of pixel value of RSI index with the value of turbidity as the results of laboratory in 2015.

Coordinate X (E)	Coordinate Y (N)	Pixel value	Turbidity
394305	9032395	0.4635	1.53
394275	9032605	0.5080	12.96
394605	9031735	0.4524	7.35
394515	9031765	0.5267	23.05
395415	9031465	0.7031	66
395385	9031555	0.6336	55
396645	9031915	0.4857	21.66
396555	9032065	0.4628	12.68
397275	9033775	0.4761	4.91
397245	9033685	0.4639	3.87
398565	9034645	0.6864	64
398385	9034375	0.6575	41.73
391005	9035965	0.5133	0.84
391695	9036085	0.4870	0.95

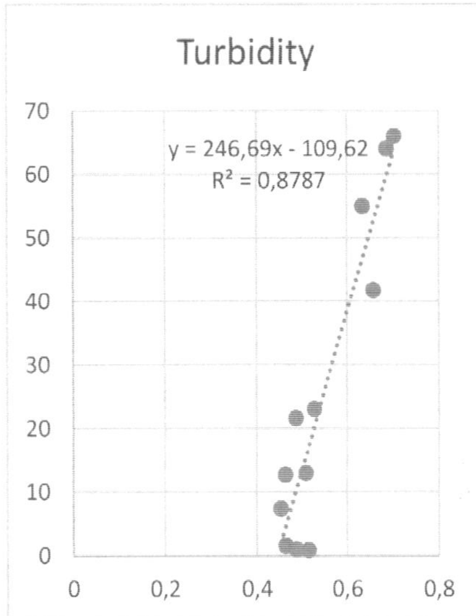

Map of modeling result of turbidity level in 2015 and 2000 can be seen in Figure 10 and Figure 11 as follows:

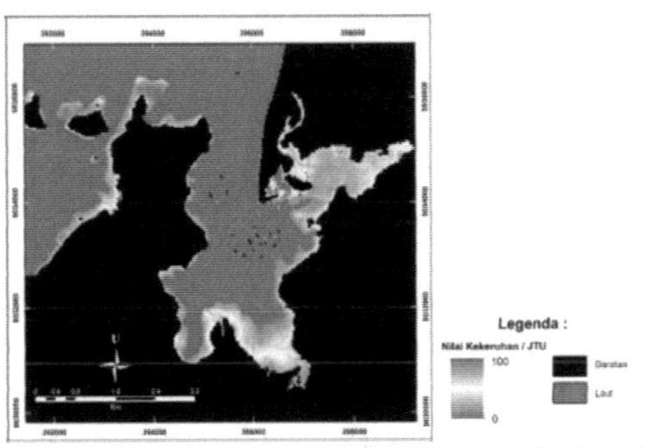

Figure 10. Map of Turbidity level in 2015

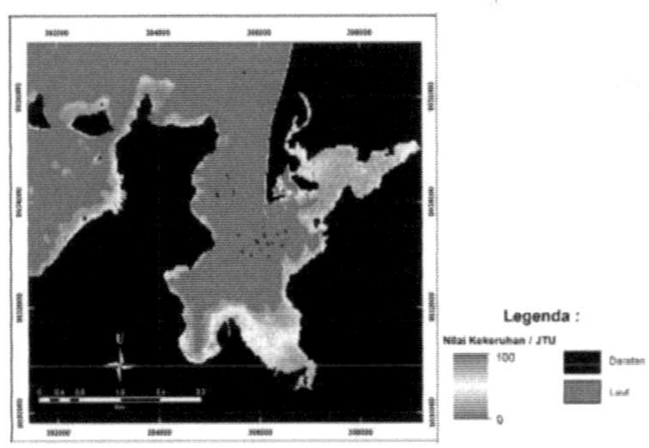

Figure 11. Map of turbidity level in 2000

Table 5. Comparison between turbidity value and condition of coral reef health in 2000-2015.

Coordinate X	Coordinate Y	Turbidity value in 2000 (FTU)	% Coral reef health in 2000 (%)	Turbidity value in 2015 (FTU)	% Coral reef health in 2015 (%)
394305	9032395	17,49	0-25	62,30	25-50
394275	9032605	34,96	50-75	36,37	50-75
394605	9031735	15,73	0-25	19,81	0-25
394515	9031765	2,06	25-50	39,70	50-75
395415	9031465	43,19	50-75	63,44	75-100
395385	9031555	52,85	25-50	64,12	75-100
396645	9031915	60,46	75-100	36,57	50-75
396555	9032065	50,87	50-75	48,87	75-100

397275	9033775	46,39	50-75	15,88	25-50
397245	9033685	26,35	25-50	29,30	25-50
398565	9034645	25,93	25-50	31,41	50-75
398385	9034375	44,67	25-50	48,10	25-50
391005	9035965	56,83	75-100	19,90	50-75
391695	9036085	26,62	0-25	19,06	25-50

From 14 existing sample points, most of them have higher turbidity value in 2015. This shows that the process of sedimentation during last 15 years in the northern coastal areas of Sub District Sekotong increased, which caused the turbidity levels in the water also increased. The condition indicates the mutual influential relationship between the increasing of turbidity and the degradation of coral reef health. This happens because the increasing of turbidity can reduce the photosynthesis ability of coral reef organisms which causes the degradation of coral reef condition and eventually the death of coral reef.

Based on the values in the table above, it can be seen that the turbidity value is not related to health condition of coral reef. The inversely patterns of both conditions are not found. Indeed, the pattern changes value which indicate the relation from both condition are not found. The reason of the inverse pattern is bias in reading the pixel value. Bias in the reading the value is due to the use of algorithm for modeling the health of coral reefs, reading the value of sediment which is very high as a coral reef, as the result, it can be seen on the table that the sample number 6 of turbidity is high but the value for the percentage of healthy reef in the sample is also high. Error in reading the value of coral reefs health also occurs in other samples. These results indicate that the hypothesis used in this study was rejected. It is due to the value of turbidity resulting from RSI index using Landsat 7 and Landsat 8 can not be directly used to estimate the health condition of coral reefs.

5. Conclusions

1. Monitoring of the health condition of coral reefs in the northern coast of the Sekotong Sub District shows that within 15 years, from 2000-2015 there has been a deterioration of the health of coral reefs, especially in coastal area because it has a lot of land use activities, such as land use change into residence and tourism place, while for the more remote areas of the coastal reef health condition is better.
2. Monitoring of waters condition, especially the level of turbidity in the north coast of the Sekotong Sub District shows that within 15 years, from 2000-2015there has been turbidity increasing/improvement, especially in the Lembar bay area because of the activities on loading and unloading ships and pond.
3. Based on the result of this research, there is no effect of changes in turbidity value on the health condition of coral reefs in the northern coast of the District Sekotong in 2000-2015. This is due to the use of the index RSI cannot be used for shallow water.
4. Assessment of the relationship between turbidity waters and coral reef health conditions should not be done using the RSI index. It should be done by using the interpolation method which does not depend on the value of the pixel.

Reference

[1] Karsidi, A. dkk., 2012 *Informasi Geospasial Terumbu Karang Indonesia* (Cibinong :Pusat Pemetaan dan Integrasi Tematik BIG)

[2] Kay, S., Hedley, D.K., and Lavender S 2009 Sun Glint Correction of High and Low Spatial Resolution Images of Aquatic Scenes: a Review of Methods for Visible and Near-Infrared Wavelengths *Molecular Diversity Preservation International. Remote Sensing.* **1** pp 697-730.

[3] Preskitt, L.B, Vroom, P.S., and Smith C.M 2004 A Rapid Ecological Assessment (REA) Quantitative Survey Method for benthic Algae Using Photoquadrats with Scuba *Pasific Science* **58**. No. 1 pp 201-209.

[4] Richmond, H. R. *et al.* 2007Watershed and Coral Reefs: Conservation Science, Policy, and Implementation *ProQuest Research Library*. Bioscience; Jul/Aug 2007 : 57

[5] Sakuno, Y. and Oki, K., 2015Relationship between Turbid Water and Coral Damage Distribution Using ALOS AVNIR-2 Images and Diving Survey Data Immediately after the Heavy Rain Disaster of the Amami-Oshima Island, Japan *Scientific Research Publishing. Advanced Remote Sensing*. **4**, pp 25-34.

[6] Siswandono 1987*Aplikasi Penginderaan JauhuntukPemetaan Terumbu Karang Kepulauan Seribu dengan software microBrian.Tesis*. Yogyakarta:UniversitasGadjahMada.

[7] Sugiyono 1999 *Statistika Untuk Penelitian*. Alfebeta. Bandung

[8] Supriharyono, M.S 2000*Pelestarian dan Pengelolaan Sumberdaya Alam di Wilayah Pesisisr Tropis* (P.T. Gramedia Pustaka Utama. Jakarta)

[9] Steinmetz, F., Deschamps, P., and Ramon, D 2011 *Atmospheric correction in presence of sun glint: application to MERIS* (Optical Society of America)

[10] USGS 2015 *Landsat- Earth Observation Sattelites* Fact Sheet.

[11] Winarso,G.,Suprapto, T., Asriningrum, W., Hastuti, M 2001 Inventarisasi Sumberdaya Kelautan Pulau Biak Menggunakan Data Inderaja. *Prosiding MAPIN Pertemuan Ilmiah Tahunan X* 11 – 12 Juli 2001, Mataram

The Mapping of Temporary Evacuation Site (TES) and Tsunami Evacuation Route in North Pagai Island, Mentawai Islands Regency – Indonesia

Aprizon Putra[1], Herdiana Mutmainah[2]

[1]Research Institute for Coastal Resources and Vulnerability
Agency of Research and Development of Marine and Fisheries
[2]Ministry of Maritime Affairs and Fisheries
Padang- Painan km 16 Padang – West Sumatra 25245, Indonesia

Abstract: Mentawai Islands Regency, especially north Pagai island, suffered two earthquakes on April 15 2016. The local government in cooperation with the relevant parties had tried to minimize casualties before the disaster or during the disaster by making an evacuation route to the TES. The purpose of this study was to the mapping of TES and tsunami evacuation route using the approach of Geographic Information Systems (GIS) for disaster mitigation of tsunami-potential earthquake in north Pagai island.The research was conducted at 3 locations in the coast of Sikakap village, 4 locations in the coast of Taikako village, 3 locations in the coast of Silabu village, 7 locations in the coast of Saumanganya village, and 3 locations in the coast of Matobe village. The effort taken was to evacuate people to the TES with an average distance of 372.62 m from the beach. The results showed that the feasibility of TES that could accommodate residents, among others, were in the hills behind Sikakap Port, Taikako Silaoinan hills, the hills near the Mapinang Silabu village chief's office and Mapinang hills, Gulukguluk Saumanganya and Panatarat Matobe hills.

Keywords: TES, Evacuation route, Tsunami, SIG, North Pagai Mentawai.

1. Introduction

Indonesia is subscription area of natural disasters. Various regions in Indonesia is a disaster-prone, especially the earthquake disaster, tsunami, flood and volcanic eruptions. Indonesian territory surrounded by Eurasian plate, Indo-Australian plate and Pacific plate. Where at one time these plates will move and cause a earthquake potential tsunami [1], as happened in Aceh in 2004 and the Mentawai islands, in particular north Pagai island, in 2010.

Historically, the waters of western Sumatra have a very high level of seismicity which can be seen from the distribution of earthquake centers in the region with the depth of epicenter deeper eastward. The quakes are generally associated with Sunda subduction system that indicates the relative movement between the Indo-Australian plate and the Eurasian plate. Because the system has been Sunda subduction, earthquakes with magnitude above 7.0 Mw have partially produced a sizeable tsunami and damaging [2].

[3],[4],[5],[6][7] add based on the data catalogs that major earthquakes which have occurred and caused a tsunami in the waters of western Sumatra among others (**Fig. 1**) were in 1797 (M~8,4); 1833

(M~9,0); 1861 (M~8,5); 1935 (7.7 Mw); 2000 (7.9 Mw); 2002 (7.2 Mw) 2004 (Mw 9,15); 2005 (Mw 8,5); 2007 (8.4 Mw) (7,9 Mw); 2008 (7,3 Mw) dan 2010 (7,7 Mw) (7.8 Mw) (7.2 Mw).

The tsunami that hit North Pagai Island on October 25, 2010 with 7.7 Mw magnitudes occurred at coordinates 3.29 south latitude; 100.7 east longitudes with 20.6 km depth had caused more than 449 fatalities, 270 serious injuries and 142 minor injuries, while the refugees numbered 14.983 residents [8]. Furthermore, on April 15 2016, North Pagai Island suffered two earthquakes more i.e at 17:24 pm with the strength of 5.0 Mw located at coordinate 3.58 south latitude; 100.45 east longitude in 52 km depth, and the aftershocks occurred at 17:56 pm with the strength of 4.5 Mw at coordinates 3.74 south latitude; 100.51 east longitude in 10 km depth [9]. The earthquake which occurred was close to the location of the earthquake that triggered the tsunami on October 25, 2010.

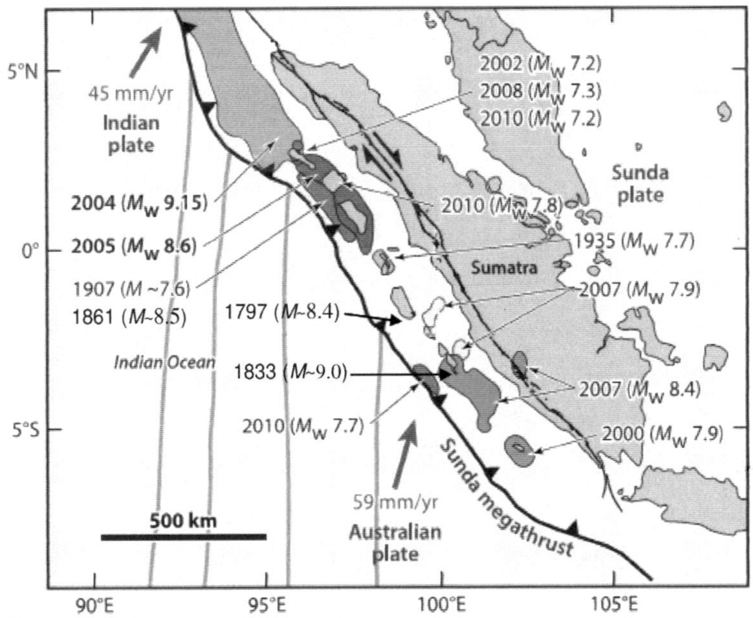

Figure 1. Map of the History of earthquakes and tsunamis in the waters of western Sumatra from[3], Konca et al. (2008)[4], Shearer and Burgmann (2010)[5], Hill et al. (2012)[6], and Meltzner et al. (2015)[7].

The tsunami occurred in north Pagai Island, especially in the coastal villages such as the village of Saumanganya, Silabu, Matobe, Betumonga, and Taikako. According to the records of the Agency for Disaster Management (BPBD) of West Sumatra province, there were 282 people killed, 411 people missing and unaccounted, 77 people severely injured, 25 people slightly injured, and more than 100 houses damaged. Victims who evacuated were 150 residents of Sikakap and 150 residents of Silabu.All community activities in north Pagai island are concentrated in the coastal villages[10].The villages of Saumanganya, Taikako and Sikakap are coastal villages which have the most residents in North Pagai Island having 9767 people in 2015[11].

Because of the tsunami disaster that would harm the society, the society then should be alert and vigilant to face any threat of disaster. However, this attitude must be accompanied by a correct understanding including the understanding of the procedures of how to save themselves because a false understanding would endanger them. One of the ways is with the mitigation of tsunami-potential earthquake disaster in the form of TES and evacuation route before or during the disaster. The community needs the TES and the evacuation route so that if the disaster happens people can go to the TES quickly through the correct route in accordance with the evacuation route to reduce the risk of casualties when tsunami occurs. The purpose of this study was to determine the mapping of TES and tsunami evacuation route using GIS approach to mitigate tsunami-potential earthquake disaster in north Pagai Island, Mentawai islands.

2. Research Methodology

The research was conducted in the hamlet of Sibaibai, central Sikakap and eastern village of Sikakap; the hamlet of Pasibuat, Muara Taikako, Bulakmonga, and Kaute of Taikako village; West Silabu , Maguiruk and Tumalei of Silabu village; Mapinang, Pinairik, Mabulau Buggei, Patutukat Sibau, and Pasapuat of Saumanganya village; Matobe Tunang, Polaga, Mangaungau of Matobe village (**Fig. 2**).The research was held in March 2016 (before the April 15 2016 earthquake).

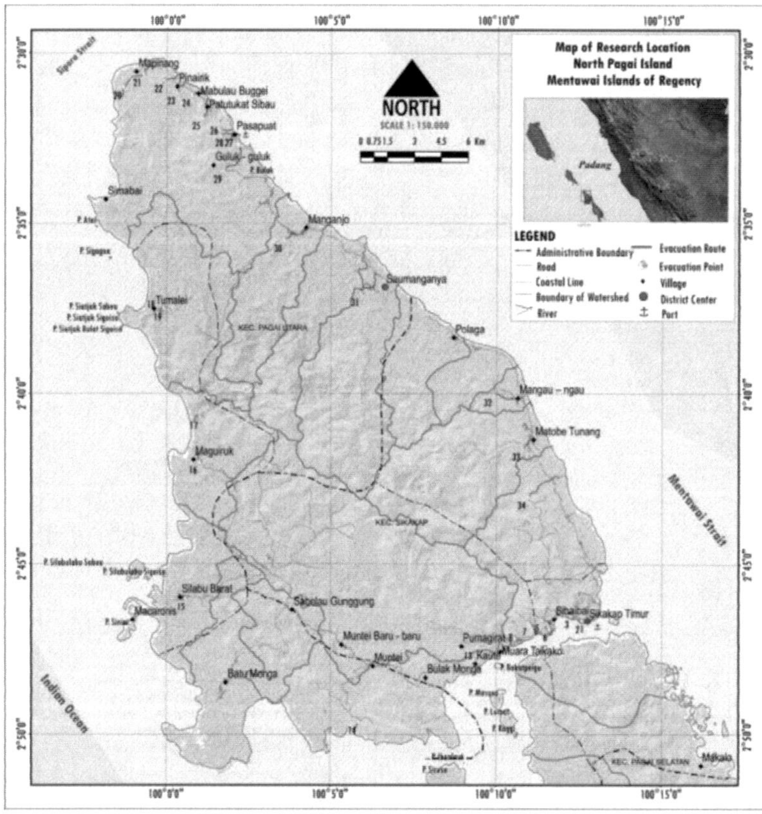

Figure 2. Map of the research sites.

The research method used was mapping survey with GIS approach using network analysis. The step of determining the TES location was by identifying the distribution of residential settlements [12].The elements of residential settlements distribution was used to determine residential areas which were vulnerable to tsunamis. The next step was verifying the results of map interpretation by conducting field surveys. The result of this verification was used to update the settlement data, while updating the road data was done based on field surveys using the Global Information System (GPS) in accordance with the road network map ofnorth Pagai island. The road network data was needed in the analysis process to determine evacuation routes to the TES. The results of the analysis were in the form of a coverage polygon with location buffer as far as 270 m from the central location of TES.

The method used to determine the distance to TES referred to the Institute of Fire Safety and Disaster Preparedness Japan (IFSDPJ)[13]explains that the evacuation speed is 0,751 m / sec (the speed walk of the elderly). The time spent is <10 min. This study used a 6-minute time spent because it was assumed that was the time citizens needed to go to the nearest TES. The processing time of evacuation was 6 minutes = 6 x 60 seconds = 360 seconds, so the distance of TES was 360 seconds x 0.751 m / sec = 270.36 m = 270 m (**Fig. 3**). After determining the distance as far as 270 m in 6 minutes, then the network analysis process was carried out using the road network data and the distribution of residential settlements. From the process, the distance to the TES required at each study site could be determined. The result of the coverage analysis as far as 270 m was then overlaid with the distribution of residential settlement data. The results of the overlaid could determine the reachable and the unreachable TES from residential cluster.

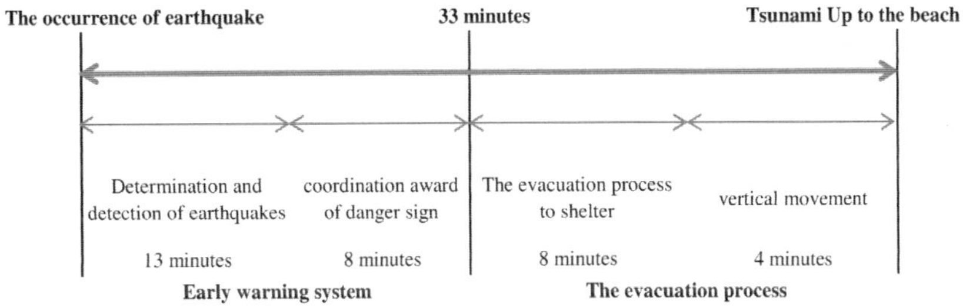

Figure 3. Time of the earthquake until the tsunami reached the shore [13].

3. Results and Discussion

3.1 The Altitude of TES

The TES altitudes in north Pagai island are very diverse with different percentages (Table 1).The map of location altitudes was used to determine the location of TES to be reached by residents before tsunami occurs located at the height of> 15 metres above mean sea level (MAMSL) which was a safe evacuation site [14] states that the classification of tsunami hazard zone can be determined from the contour of the area. Based on the height of area, the risk zones which are prone to tsunamis are classified as very high risk at the height of 0-5 meters at, high risk at 5-10 meters, temporary secure at 10-15 meters, and secure at> 15 meters (**Fig. 4**).

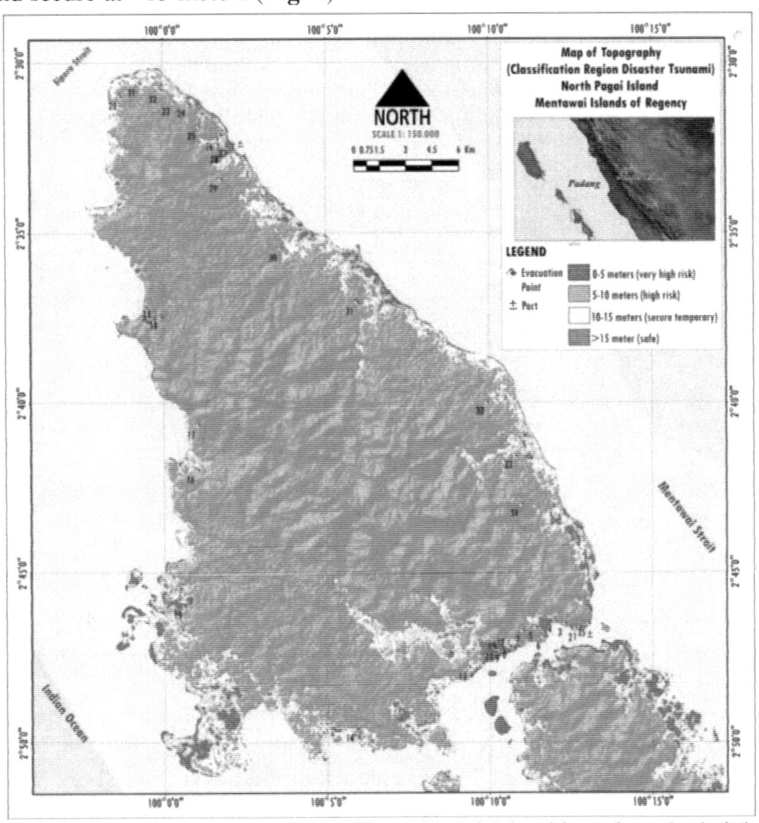

Figure 4. Map of tsunami-prone zones in North Pagai Island based on the height of TES.

Table 1. The average height of TES in north Pagai Island.

Village	Hamlet	Evacuation Site	Average (MAMLS)	Morphology of area
Sikakap	East Sikakap	3	21.5	Hillsides
	Central Sikakap	3	23.6	Hillsides
	Sibaibai	3	48.8	Highland
Taikako	Pasibuat	2	15.7	Hillsides
	Muaro Taikako	4	10.3	Hillsides
	Kaute	1	17.5	Highland
	Bulakmonga	1	33.1	Highland
Silabu	West Silabu	2	25.7	Hillsides
	Tumalei	2	45.2	Hillsides
	Maguiruk	2	47.2	Hillsides
Saumanganya	Mapinang	3	24.2	Highland
	Mabulau Buggei	2	47.7	Hillsides
	Patutukat Sibau	2	32.1	Hillsides
	Pinairik	1	20.5	Hillsides
	Pasapuat	3	19.1	Hillsides
	East Saumanganya	1	29.5	Hillsides
	Manganjo	1	22.5	Hillsides
Matobe	Matobe Tunang	1	23.3	Hillsides
	Mangau –ngau	1	25.5	Hillsides
	Polaga	1	28.4	Hillsides

Table 2. The population of the coastal area of north Pagai Island.

Hamlet	Village	Sub-district	Population
Central Sikakap	Sikakap	Sikakap	673
East Sikakap	Sikakap	Sikakap	1039
Sibaibai	Sikakap	Sikakap	660
Pasibbuat	Taikako	Sikakap	191
Muara Taikako	Taikako	Taikako	418
Kaute	Taikako	Taikako	258
Bulakmonga	Taikako	Taikako	271
West Silabu	Silabu	North Pagai	181
Maguiruk	Silabu	North Pagai	283
Tumalei	Silabu	North Pagai	172
Mapinang	Saumanganya	North Pagai	201
Pinairik	Saumanganya	North Pagai	119
Mabulau Buggei	Saumanganya	North Pagai	252
Patutukat Sibau	Saumanganya	North Pagai	105
Pasapuat	Saumanganya	North Pagai	530
Manganjo	Saumanganya	North Pagai	460
East Saumanganya	Saumanganya	North Pagai	348
Matobe Tunang	Matobe	Sikakap	411
Mangau –ngau	Matobe	Sikakap	85
Polaga	Matobe	Sikakap	135
	Jumlah		6792

Source: BPS of Mentawai Islands Regency, 2016

3.2 The hamlet's monographs, road network and evacuation route selection

The hamlet's monographs data were used as a parameter to determine the number of people who would be evacuated to the predetermined TES (**Table 2**). In this study, not all the population of north Pagai Island was evacuated, because a small proportion of the population had been in a place at the height of > 15 MAMSL so that the evacuation was not necessarily done. The assumption was that the point that became the starting point of the evacuation of the entire population identified as the tsunami victims was determined by analyzing areas which height were less than 15 MAMSL.

Whilst, the road network became the main parameter because in the road network data attributes were the road class, road width and road length (**Table 3**) in which the making of the evacuation route map would consider the length of the route to be taken by the people in the coastal village of North Pagai. The following is the parameter of evacuation routes to be taken by residents to get to the TES which had been selected.

Table 3. Parameter of north Pagai island road network.

Evacuation route	Road Class`	Average width (m)	Average length (m)
I	Main road (PT Minas Pagai Lumber road)	18	619
II	Village/hamlet road	3.32	376.46
III	Footpath/ground	1.5	389.745

According to (**Table 2** and **Table 3**) above, the number of people estimated to be the victims of the tsunami disaster was 6792 residents. The tsunami disaster mitigation was based on the parameters used for the determination of TES and evacuation routes namely: altitude, road networks and the hamlet monographs. The road network data were then analyzed with GIS approach using network analysis in which could be known the information of the travel time to the TES through the 38 pre-determined evacuation routes (**Fig. 5** and **Table 4**) in north Pagai island coastal areas which were vulnerably affected by the tsunami.

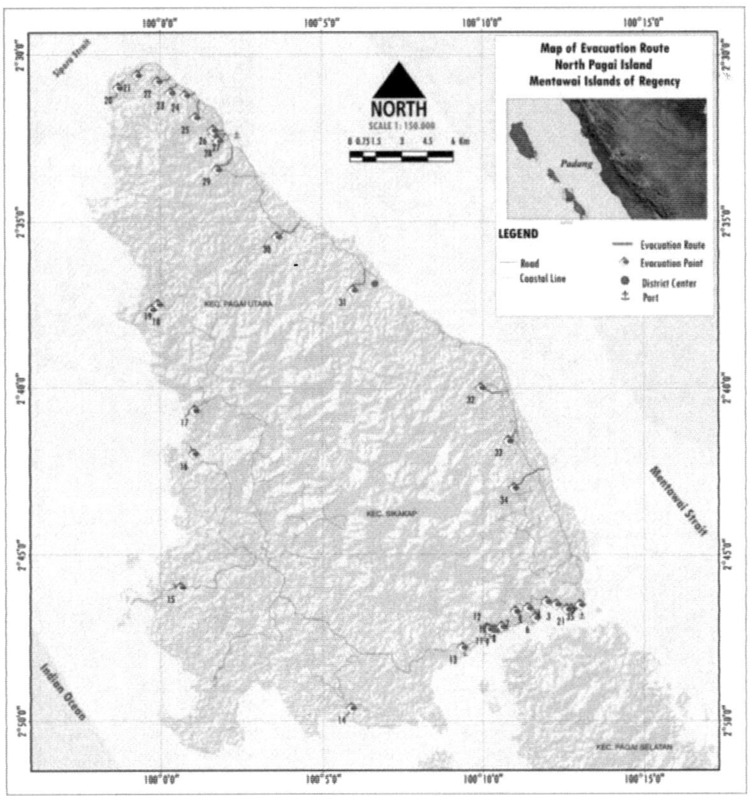

Figure 5. The map of TES and tsunami evacuation routes in north Pagai island.

IOP Conf. Series: Earth and Environmental Science **47** (2016) 012020 doi:10.1088/1755-1315/47/1/012020

Figure 6. The location of evacuation route in central sikakap.

Figure 7. The location of evacuation route in SMP N 1 Sikakap.

Figure 8. The location of evacuation route in the gate of Pasibuat.

Figure 9. The plan of evacuation shelter in the port of PT. Minas Pagai.

Figure 10. The location of TES and evacuation routes in west Silabu.

Figure 11. The condition during tide.

Figure 12. The location of evacuation route in east Saumanganya.

Figure 13. The condition of bridge as an evacuation route in the Pasapuat Port.

Figure 14. The gate of Matobe village in Matobe Tunang which was a tsunami evacuation route.

Figure 15. The condition Tsunami evacuation route in Polaga.

Table 4. The characteristics of tsunami evacuation routes in north Pagai island.

Location	Length (m)	Width(m)	Roadtype	Lane	Condition	time (minute)
E_01 Central Sikakap	250	3.5	concrete	village	well	± 3
E_02 Central Sikakap	96	3.5	concrete	village	well	± 1
E_03 East Sikakap	140	3.5	concrete	village	well	± 3
E_04 East Sikakap	415	3.5	concrete	village	well	± 3
E_05 East Sikakap	140	3.5	concrete	village	well	± 2
E_06 Sibaibai	607	3.5	concrete	village	damaged	± 7
E_07 Sibaibai	188	3.5	concrete	village	well	± 2
E_08 Sibaibai	937	3.5	concrete	village	damaged	± 13
E_09 Pasibbuat	365	3	concrete	village	damaged	± 3
E_10 Pasibbuat	313	3.5	concrete	village	well	± 3
E_11 Muara Taikako	145	3.5	concrete	village	well	± 2
E_12 Muara Taikako	305	3.5	concrete	village	well	± 3
E_13 Muara Taikako	500	3.5	concrete	village	well	± 5
E_14 Muara Taikako	664	3.5	concrete	village	well	± 7
E_15 Kaute	577	18	ground	company	well	± 5
E_16 Bulakmonga	619	1.5	ground	village	basic	± 7
E_17 West Silabu	530	3	concrete	village	damaged	± 5
E_18 West Silabu	428	3	concrete	village	damaged	± 4
E_19 Tumalei	502	1.5	ground	village	basic	± 5
E_20 Tumalei	778	1.5	ground	footpath	basic	± 8
E_21 Maguiruk	150	1.5	concrete	village	damaged	± 2
E_22 Maguiruk	215	1.5	concrete	village	damaged	± 2
E_23 Mapinang	1.49	1.5	soil	footpath	basic	± 13
E_24 Mapinang	720	3.5	concrete	village	well	± 8
E_25 Pinairik	879	3.5	concrete	village	damaged	± 9
E_26 Mabulau Buggei	825	3.5	concrete	village	damaged	± 9
E_27 Patutukat Sibau	611	3.5	concrete	village	damaged	± 7
E_28 Patutukat Sibau	1.21	3.5	concrete	village	damaged	± 13
E_29 Pasapuat	579	3.5	concrete	village	damaged	± 7
E_30 Pasapuat	650	3.5	concrete	village	damaged	± 7
E_31 Pasapuat	676	3.5	concrete	village	damaged	± 7
E_32 Mabulau Buggei	1.854	5	concrete	village	damaged	± 14
E_33 Manganjo	1.825	3.5	concrete	Village	damaged	± 14
E_34 East Saumanganya	1.79	3.5	concrete	Village	damaged	± 14
E_35 Matobe Tunang	1.429	3.5	concrete	Village	damaged	± 13
E_36 Matobe Tunang	581	3.5	concrete	Village	well	± 7
E_37 Mangau -ngau	2.195	3.5	concrete	Village	damaged	± 15
E_38 Central Sikakap	136	3.5	concrete	Village	well	± 2

According to [15], a bunch of people running together can cover a distance of 3225.81 m within 20 minutes. Considering the strength differences (based on age and sex), the length of the evacuation route should be less than 3225.81 m. The length of the evacuation route E-01 to E38 had met the criteria of the tsunami evacuation route length with an average length of 382.44 m which was less than the maximum distance of a bunch of people running together with maximum distance of 3225.81 m. It can be concluded that the evacuation routes that had already existed in North Pagai Island were very appropriate for evacuation when a tsunami potential-earthquake hit.

3.3 Determining TES

3.3.1 Sikakap

Residential settlements located in the coast of Sikakap village included in the tsunami-prone zones were western and central Sikakap, and Sibaibai.Based on field surveys, it was found out that, in general, residential settlements in the eastern and central Sikakap and Sibaibai were within ± 10 m from the shoreline with a population of 2372 inhabitants[11]. The coast of Sikakap village was not severely affected by the tsunami although based on the factsin the field as well as the interviews of local people there were some buildings severely damaged by the earthquake before the tsunami. Geographically, Sikakap village is close to hills as an alternative of TES when tsunami hits. The position of Sikakap village as the center of economy and administration, as well as the main gate of north Pagai island has resulted in a lot of buildings that have more than 2 floors as evacuation shelters.

To determine the location of TES in the village of Sikakap **(Fig.16)**, it was necessary to know the location altitude of TES region **(Table 1).** In the study site could be seen 8 evacuation routes to TES. There were 3 routes in eastern Sikakap with the height of 21.5 MAMSL, 2 routes in central Sikakap with the height of 23.6 MAMSL, and 3 routes in Sibaibai with a height of 48.8 MAMSLto TES. The average time needed to TES was ± 4 minutes with the furthest route of 415 m from central Sikakap.

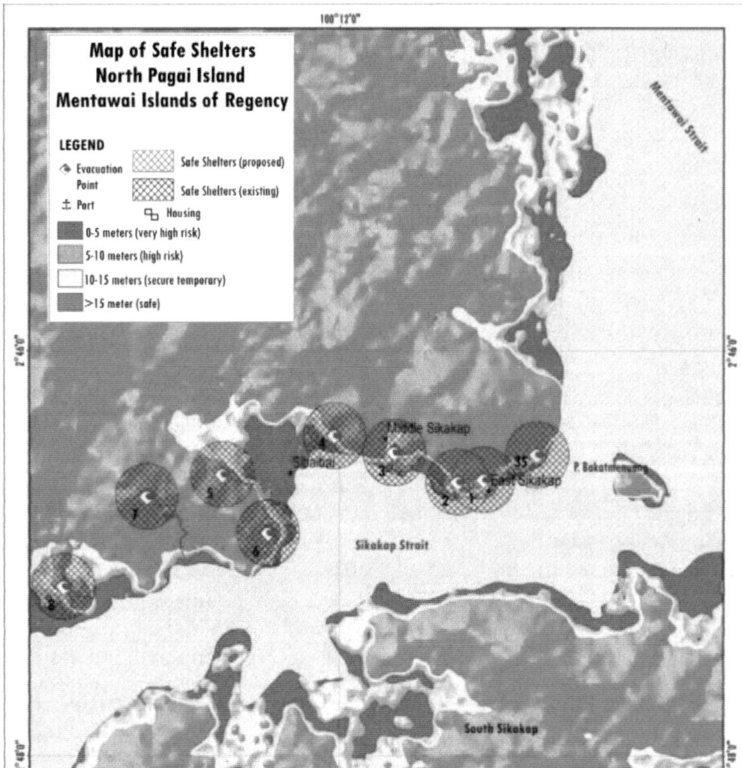

Figure 16. The map of TES in Sikakap village scale 1:30.000.

3.3.2 Taikako

Residential settlements in the coast of Taikako village included in tsunami-prone zones were the hamlets of Pasibuat, Muaro Taikako, Kaute dan Bulakmonga. According to field surveys, the residential settlements in Pasibuat which were ± 68 m from the shoreline with a population of 191 inhabitants, in Muaro Taikako which were± 40 m from the shoreline with total population of 418 inhabitants, in Kaute which were ± 310 m from the shoreline with total population of 258 inhabitants, and in Bulakmonga which were ± 1.6 km from the shoreline with total population of ± 271 inhabitants had not yet received permanent housing assistance planned to be built of 80 units[16].According to BPS data (2016), the number of residents in Taikako village was 1138. The people in Taikako village proposed permanent housing on the land of PT. Minas Pagai at an altitude of 120 MAMSL and on Silaoinan hills with the height of 46 MAMSL.The coastal area of Taikako village was not so badly affected by the tsunami on October 25, 2010. The suggested evacuation shelter in the village was located in Kaute within ± 90 m towards the south of the port of PT. Minas Pagai consisting of a 3-storey building.

To determine the location of TES in the village of Taikako **(Fig.17),** it was necessary to know the location altitude of TES region **(Table 1).** In the study site could be seen 6 evacuation routes to TES. There were 2 routes in Pasibuat with the height of 15.7 MAMSL, 2 routes in Muaro Taikako with the height of 10.3 MAMSL, and 1 route in Kaute with a height of 17.5 MAMSL, and 1 route in Bulakmonga with a height of 33.1 MAMSL to TES. The average time needed to TES was ± 4,87 minutes with the furthest route of 664 m from Muaro Taikako.

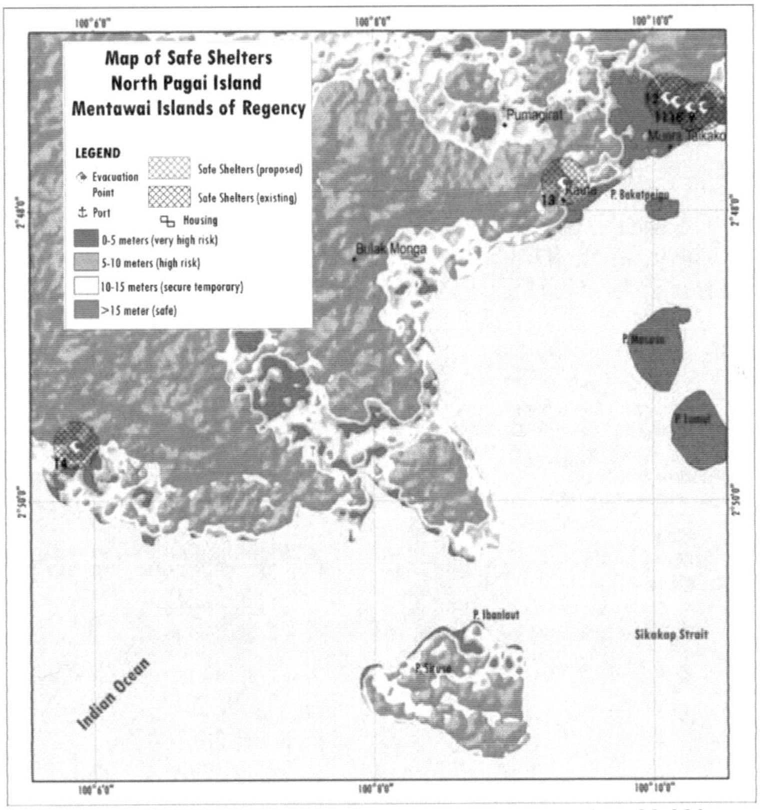

Figure 17. The map of TES in Taikako village scale 1:30.000.

3.3.3 Silabu

Residential settlements located in the coast of Silabu village included in tsunami-prone zones were the western Silabu, Maguiruk, and Tumalei. According to field surveys, the residential settlements in west Silabu which were ± 1,2 km from the shoreline with a population of 181 inhabitants, in Maguiruk which were ± 532 m from the shoreline with total population of 238 inhabitants, and in Tumalei which were 851 m from the shoreline with total population of ± 172 inhabitants had not yet received permanent housing assistance planned to be built of 87 units[16]. Based on the facts in the field supported by interview with the local community, western Silabu was also prone to flood (rob).According to[11] data, the population of the coast of Silabu village was 636 inhabitants. Local people asked for permanent housing on the high land near the office of Silabu village chief to anticipate tsunami dan rob in Western Silabu. The coastal areas of Silabu village were badly hit by the tsunami, one of them was Macaronis resort located on the island of Sinaia. In 2010 (**Fig. 18**), the resort was hit by 3 m high tsunami waves that damaged buildings and mangroves which protect the coastal area [17].

Figure 18. Macaronis before and after tsunami 2010.

To determine the location of TES in the village of Silabu **(Figure 19),** it was necessary to know the location altitude of TES region **(Table 1).** In the study site could be seen 6 evacuation routes to TES. There were 2 routes in western Silabu with the height of 25.7 MAMSL, 2 routes in Maguiruk with the height of 47.2 MAMSL (the suggested TES in the north area), and 2 routes in Tumailei with a height of 45.2 MAMSLto TES which did not yet have evacuation shelter. The average time needed to TES was ± 4,66 minutes with the furthest route of 778 m from Tumalei.

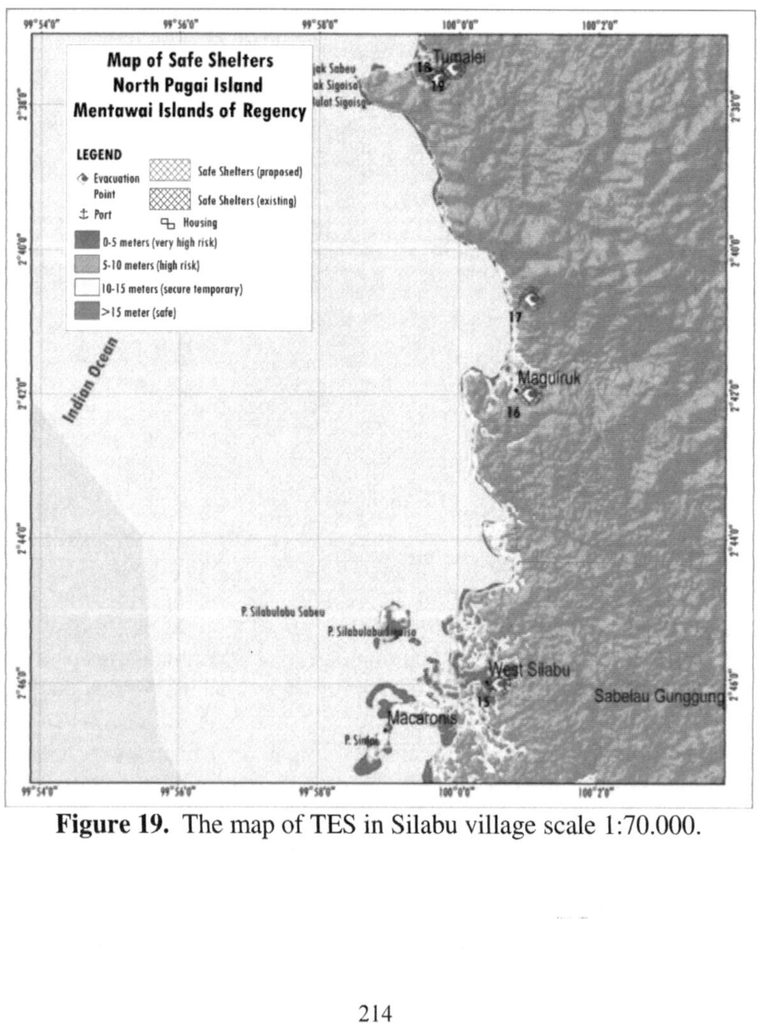

Figure 19. The map of TES in Silabu village scale 1:70.000.

3.3.4 Saumanganya

Residential settlements located in the coast of Saumanganya village included in tsunami-prone zones were Mapinang, Pinairik, Mabulau Buggei, Patutukat Sibau, Pasapuat, Manganjo and east Saumanganya. According to field surveys, the residential settlements in Pasapuat which were ± 110 m from the shoreline with a population of 530 inhabitants, in Manganjo which were ± 745 m from the shoreline with total population of 460 inhabitants, in Pinairik which were ± 656 m from the shoreline with a population of 119 inhabitants, in east Saumanganya which were ± 550,8 m from the shoreline with total population of 348 inhabitants, in Mabulau Buggei which were ± 230 m from the shoreline with total population of 242 inhabitants, and in Patutukat Sibau which were 321 m from the shoreline with total population of ± 105 inhabitants had not yet had permanent housing planned to be built of 88 units[16]. Based on the facts in the field supported by interview with the local community, western Silabu was also prone to flood (rob).According to BPS (2016) data, the number of population of the coast of Saumanganya village was 2025 inhabitants. Similar to Taikako village, the people living in the coastal areas asked for permanent housing on the land of PT. Minas Pagai Lumber at the height of 100 MAMSL, and in Gulukguluk hills at the height of130 MAMSL.Whilst the evacuation shelter proposed in Pasapuat was ± 200 m towards the west at the construction site of the pier port which planned to be the 2ndport after Sikakap port consisting of 3-storey building. Pasapuat was the area in the eastern part of North Pagai Island which was the worst affected by the tsunami, beside the western area of Batumonga village such as Muntei Barubaru with 98 fatalities and Sabeugugung with 75 fatalities. Those areas had been given an assistance of permanent housing by related parties [18].

To determine the location of TES in the village of Saumanganya **(Fig.20),** it was necessary to know the location altitude of TES region **(Table 1).** In the study site could be seen 12 evacuation routes to TES. There were 3locations in Mapinang, one of them was the suggested TES in western Mapinang with average height of 24.2 MAMSL, 1location in Pinairik with the height of 20.5MAMSL,3 locations in Pasapuat with the height of 19.1 MAMSL, 1locations in Manganjo with the height of 22.5 MAMSL, 1 location in Mabulau Buggei with the height of 47.7MAMSL, 2 locations in Patutukat Sibau with the height of 32.1 MAMSLand 1 location in east Saumanganya with a height of 29.5MAMSLwhich did not yet have evacuation shelter. The average time needed to TES was ± 9,07 minutes with the furthest route of 1.823kmfromMabulau Buggei, Manganjo and east Saumanganya.

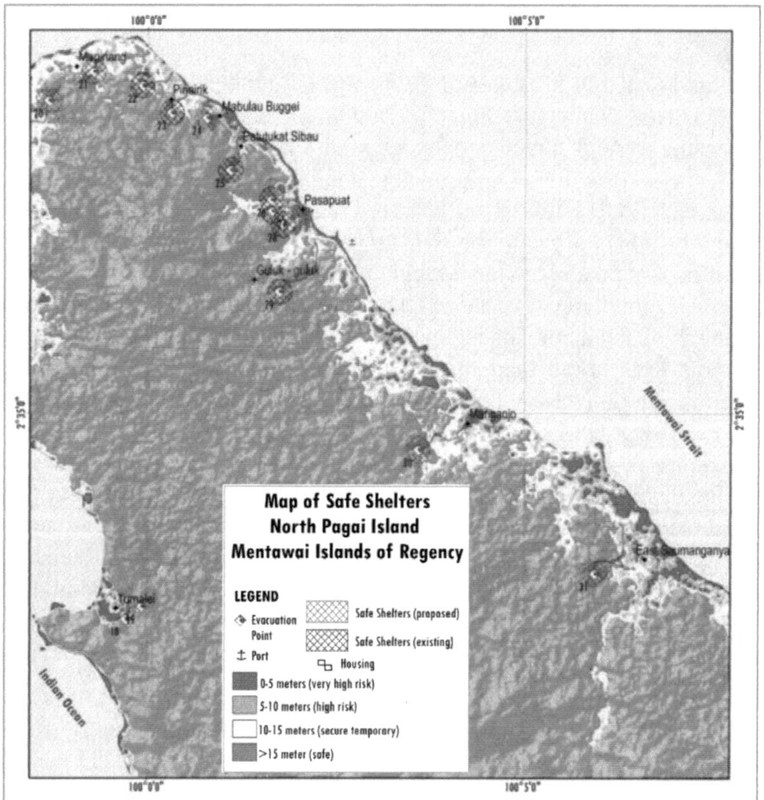

Figure 20. The map of TES in Saumanganya village scale 1:65.000.

3.3.5 Matobe

Residential settlements located in the coast of Matobe village included in tsunami-prone zones were Matobe Tunang, Polaga and Mangaungau. According to field surveys, the residential settlements in Matobe Tunang were ± 643 m from the shoreline with a population of 411 inhabitants, in Polaga were ± 182 m from the shoreline with total population of 135 inhabitants, and in Mangaungau were 325 m from the shoreline with total population of 85 inhabitants. According to BPS (2016) data, the number of population of the coast of Matobe village was 631 inhabitants. Local residents asked for permanent housing which located on Panatarat hillsat the height of 85above sea level within ± 2 km from Matobe Tunang, and ± 2,5 km from Mangaungau, while the suggested permanent housing was located in Polaga at a church near the Posyandu/Polmades of Polaga. Based on the information from the interview conducted to local residents, in the locations was also planned to be built a 3-floor building which was the allocation of aid for trans-area settlement. The coast of Matobe village was not badly affected by the tsunami, though according to information from local people there were several buildings severely damaged because of it. However, there were not any fatalities when the tsunami hit north Pagai Island on October 25 2010.

To determine the location of TES in the village of Matobe **(Fig.21)**, it was necessary to know the location altitude of TES region **(Table 1).**In the study site could be seen 2 evacuation routes to TES. The TESin Mangaungau with the height of 25.5 MAMLS, in Matobe Tunang with the height of 23.3 MAMLS and in Polaga did not have evacuation route signs yet, and were the suggested location of TES. Based on the field observation, one of the routes to the TES location with the height of 28.4 MAMLS was the area of Panatarat hills plantation, while the evacuation shelter in Matobe village was not yet available. The average time needed to the TESwas ± 9.25 minutes with the furthest route of± 2.195 kmfromMangaungau.

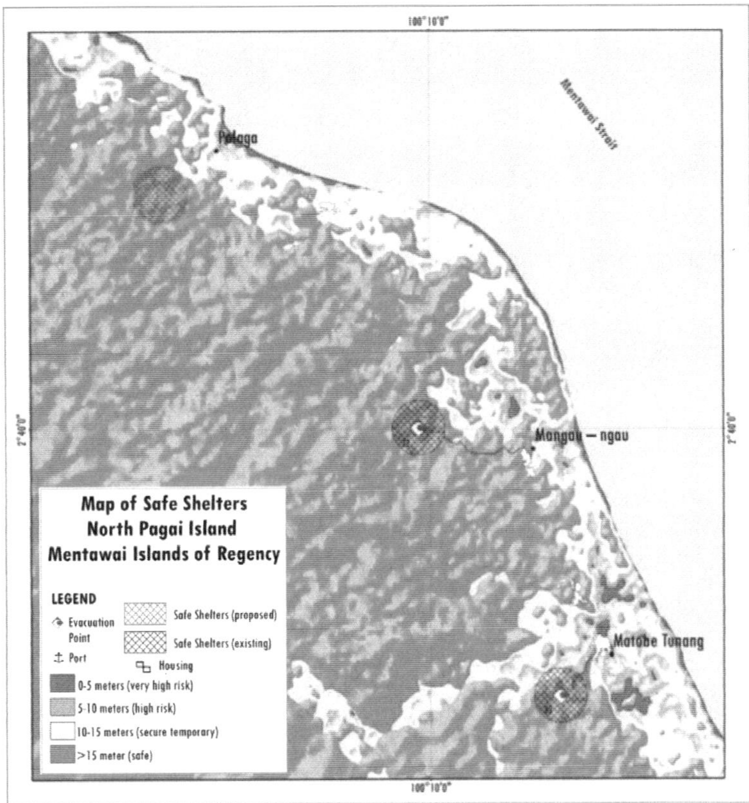

Figure 21. The map of TES in Matobe village scale 1:30.000.

4. Conclusions and Recommendations

4.1 Conclusions

In general, the residential settlements in the village of Sikakap were within ± 10 m from the coastline with a population of 2372 inhabitants. The average time needed to get to TES was ± 4 minutes with the furthest route of 415 m from Central Sikakap. The residential settlements in Taikako village were in average within ± 504 m from the coastline with a population of 1138 inhabitants. The people of the coast of Taikako village asked for permanent housing on Silaoinan hillsat the height of 46 MAMLS. The average time needed to get to TES was ± 4,87minuteswith the furthest route of 664 m from Muaro Taikako. The residential settlements in Silabu village were in average within ± 218,17 m from the coastline with a population of 707 inhabitants. Local residents asked for permanent housing on the hill near the office of Silabu village chief at the height of 25,7 MAMLS. The average time needed to get to TES was ± 4,66minutes with the furthest route of 778 m from Tumalei. The residential settlements in Saumanganya village were in average within ± 435.33 m from the coastline with a population of 1804 inhabitants. The residents of the coast of Saumanganya village asked for permanent housing on the hill of Gulukguluk at the height of130 MAMLS. The average time needed to get to TES was ± 9,07minuteswith the furthest route of ± 1.823 km fromMabulau Buggei, Manganjo and east Saumanganya. The residential settlements in Matobe were in average ± 383.33 m from the shoreline with a population of 631 inhabitants. Local residents insisted on permanent housing which located on Panatarat hill sat the height of 85m above sea level within ± 2 km from Matobe Tunang, and ± 2,5 km from Mangaungau, while the suggested permanent housing was located in Polaga at a church near the Posyandu/Polmades of PolagaThe average time needed to the TESwas ± 9.25 minutes with the furthest route of ± 2.195 km from Mangaungau.

4.2 Recommendations
Intensive socialization needs to be done for each coastal village in North Pagai Island because historically these villages had been hit by tsunami. The socialization is in the form of an introduction to the evacuation route map so that if the calamity occurs, residents are more directed in the action to save themselves. The evacuation routes need to be provided with signs to show the evacuation direction so that residents can easily identify the evacuation routes. Maintenances of the road and the access of the evacuation routes of north Pagai Island coast needs to be done.

5. Acknowledgements
Gratitude is expressed to the Head of Research Institute for Coastal Resources and Vulnerability, who has given full support to writers in conducting research using DIPA APBN 2016. Gratitude is also expressed to Mr.Harfiandri Damanhuri, S.Pi, M.Sc, a lecturer of Bung Hatta University Padang, who has given input related to the environmental condition of North Pagai Island as well as accompanying writers while conducting the survey. Writers are also thankful to those who have helped the administration process and in the field.

References
[1] Natawidjaja, D.H., Sieh, K., Chlieh, M., Galetzka, J., Suwargadi, B.W., Cheng, H.,Edwards, R.L., Avouac, J.-P., Ward, S.N. 2006 Source parameters of the great Sumatran megathrust earthquakes of 1797 and 1833 inferred from coral microatolls *J. Geophys. Res.* **111** B06403. http://dx.doi.org/10.1029/2005JB004025.

[2] Sieh, K., Natawidjaja, D.H., Meltzner, A.J., Shen, C.-C., Cheng, H., Li, K.-S., Suwargadi, B.W., Galetzka, J., Philibosian, B., Edwards, R.L 2008 Earthquake supercycles inferred from sea-level changes recorded in the corals of West Sumatra *Science***322** 1674 1678. http://dx.doi.org/10.1126/science.1163589.

[3] Briggs, R.W., Sieh, K., Meltzner, A.J., Natawidjaja, D., Galetzka, J., Suwargadi, B., Hsu, Y.-j, Simons, M., Hananto, N., Suprihanto, I., Prayudi, D., Avouac, J.-P., Prawirodirdjo, L., Bock, Y 2006 Deformation and slip along the Sunda mega-thrust in the great 2005 Nias-Simeulue earthquake *Science***311** 1897-1901. http://dx.doi.org/10.1126/science.1122602.

[4] Konca, A.O., Avouac, J.-P., Sladen, A., Meltzner, A.J., Sieh, K., Fang, P., Li, Z., Galetzka, J., Genrich, J., Chlieh, M., Natawidjaja, D.H., Bock, Y., Fielding, E.J., Ji, C., Helmberger, D.V 2008 Partial rupture of a locked patch of the Sumatra megathrust during the 2007 earthquake sequence*Nature***456** 631 - 635. http://dx.doi.org/10.1038/nature.07572.

[5] Shearer, P., Bürgmann, R 2010 Lessons learned from the 2004 SumatraeAndaman megathrust rupture*Annu. Rev. Earth Planet. Sci.* **38** 103-131. http://dx.doi.org/10.1146/annurev-earth-040809-152537.

[6] Hill, E.M., Borrero, J.C., Huang, Z., Qiu, Q., Banerjee, P., Natawidjaja, D.H., Elosegui, P., Fritz, H.M., Suwargadi, B.W., Pranantyo, I.R., Li, L., Macpherson, K.A., Skanavis, V., Synolakis, C.E., Sieh, K 2012 The 2010 MW 7.8 Mentawai earthquake: very shallow source of a rare tsunami earthquake determined from tsunami field survey and near-field GPS data*J. Geophys. Res.***117** B06402. http://dx.doi.org/10.1029/2012JB009159.

[7] Meltzner et al. 2015

[8] Satake, K., Nishimura, Y., Putra, P.S., Gusman, A. R., Suhendar, H, Fuji, Y., Tanioka, Y., Latief, H., Yulianto, E 2013 Tsunami Source of the 2010 Mentawai, Indonesia Earthquake Inferred from Tsunami Field Survey and Waveform Modeling*Pure Appl. Geophys***170**1567–1582, doi:10.1007/s00024-012-0536-y.

[9] Akbar, R 2016 Dua Gempa Beruntun Goyang Kepulauan Mentawai*Sindo News* [Online], http://daerah.sindonews.com/read/1101415/174/dua-gempa-beruntun-goyang-kepulauan-mentawai-1460722845. Last accessed 24.06.16.(2016, 15 April)

[10] Harismanto 2010 Korban Tewas Jadi 31 dan 166 Hilang Diterjang Tsunami*Tribun News* [Online]. http://www.tribunnews.com/regional/2010/10/26/korban-tewas-jadi-31-dan-166-hilang-akibat-tsunami. Last acceseed 24.06.16.(2010, 26 October).

[11] [BPS] Badan Pusat Statistik 2016*Kabupaten Kepulauan Mentawai Dalam Angka Tahun 2016*(Publisher. Badan Pusat Statistik Kabupaten Kepulauan Mentawai, Tua Pejat)

[12] Laurini, R 1996 Spatio-Temporal Databases: From Moving to Active Geographic Objects*CISM International Centre for Mechanical Sciences***365** 0254-1971. doi:10.1007/978-3-7091-2684-4_30.

[13] Budiarjo A *2006 Evacuation Shelter Building Planning For Tsunami-Prone Area; A Case Study of Meulaboh City, Indonesia. Master Thesis unpublished.* International Institute for Geo-Information Science and Earth Observation, Enschede, 112 pp.

[14] Permana H 2007*Pedoman Pembuatan Peta Jalur Evakuasi Bencana Tsunami*(Kementerian Negara Riset dan Teknologi, Jakarta)

[15] Rahmadhani N 2014*Analisis Aksesibilitas Shelter Evakuasi Tsunami di Kota Padang Berbasis Sistem Informasi Geografis. Master Thesis not published.* Diponegoro University. Semarang.

[16] Pambudi A 2016 Pembangunan 385 rumah khusus di Mentawai ditarget selesai tahun ini. *Antara News* [Online]. http://www.antaranews.com/berita/557934/pembangunan-385-rumah-khusus-di-mentawai-ditarget-selesai-tahun-ini. Last acceseed 24.06.16 (2016,29April)

[17] Loughran 2010 Survivor releases amateur video of Mentawai tsunami. *Observers France24* [Online]. http://observers.france24.com/en/20101029-survivor-releases-video-mentawai-tsunami-indonesia-macaronis-surf-resort. Last acceseed 24.06.16(2010, 29 October)

[18] [BPBD] Badan Penanggulangan Bencana Daerah 2012*Rencana Kontinjensi Menghadapi Bencana Tsunami Provinsi Sumatera Barat*(Badan Penanggulangan Bencana Daerah Provinsi Sumatera Barat, Padang)

Observing Deformation at Mt. Raung East Java Based on PALSAR-2 Imagery by Using Interferometric SAR

Arliandy P. Arbad[1], Achmad Ardy[2], and Ridwan A. Ashari[3]

[1]Institute of Industrial Science, The University of Tokyo, Tokyo 153-8505, Japan.
[2]Faculty of Agriculture, Dept. of Agronomic-Tech, Lampung University, Indonesia.
[3]Dept. of Geodesy and Geomatic, Universitas Gadjah Mada, Yogyakarta, Indonesia.

E-mail : arbad@iis.u-tokyo.ac.jp

Abstract. In August 2015, Indonesia Center of Volcanology and Geological Hazard Mitigation (CVGHM) recorded of tectonic activities at Mt. Raung with maximum amplitude 2-32 mm and continuing the tremor quakes until the beginning of the 2016 eruption period. Mt. Raung is located at East Java Province, one of most active stratovolcano in Indonesia, typically erupt with explosive eruptions and another deadly hazards such as pyroclastic flow, lahar and volcanic gases. Radar imagery consequently proposes of value device for mapping and assessing of volcano oppurtunities. By this study, we propose InSAR method to observe deformation in Mt. Raung. Interferometric SAR derives the phase difference based on two images of PALSAR-2 observations taken in January 2015 and January 2016. According to the processing of interferometric SAR, those images must be coregistered into a stack, and we selected 2015 imagery as master and the other imagery as slave. We estimate the interferogram result to know the line-of-sight then be flattened by removing the topographic phase an inflating volcano (or any other landform) produces a pattern of concentric fringes in a radar interferogram from which the ffects of viewing geometry and topography have been removed. Finally, we expect the result ofInSAR processing technique to investigate ground deformation of Mt. Raung. It would be a capable and cost-effective way of enhancing the techniques normally used in geodetic monitoring to assess the next eruptive events.

1. Introduction

As we know Indonesia is a part of ring of fire, some of hazard disasters caused tectonic activates and volcanic activities are frequently occurred. Most of interesting evidence, around 13 % of the world's active volcanoes are located in Indonesia. Tectonically, the active volcanoes are the result of a collision between Indian-Australian, Eurasian, and Philippine Plates [10] Large number of people living close to volcanoes site because these areas usually contain some of the most mineral rich soils, which provide perfect conditions for farming. Lava and material from pyroclastic flows are weathered to form nutrient rich soils which can be cultivated to produce healthy crops and prosperous harvests. Volcanic eruptions produce disaster materials such as the lava, pyroclastic fall, pyroclastic flows, pyroclastic surges, lateral blast, debris avalanche, volcanic tsunamis, mud, flooding and harmful gases [9]. A large number of volcanoes are situated in the (so-called) ring of fire area. Analysis from geographic and topographic data suggests that Indonesia is the most volcanically active in the world, with numerous eruptions each year and millions of people living on the flanks of the volcanoes [4].

The objective of this study is to observe the characteristics of mount Raung activities by using InSAR, the most commonly technique is used remote sensing methods, remote sensing has been largely implemented to characterize of volcano eruptions, both of optical remote sensing and SAR (Synthetic Aperture Radar)[5], SAR data involve the reflection and radiation of electromagnetic waves. In this research, we propose to use the InSAR analysis method to process SAR (Synthetic Aperture Radar) images taken from ALOS22/Palsar2 data. Characteristic of SAR image is an inherent characteristic of coherent imaging, including SAR imaging is the presence of speckle noise, which is random, deterministic, interference pattern in an image formed with coherent radiation of a medium containing many sub-resolution scatters. Mapping active volcano using SAR (e.g[7]) in mount Merapi to classify the pyroclastic deposits has applied to observe active volcano.

2. Methodology

Mount Raung is located in East Java Province, one of the most active volcanoes in Indonesia which short eruptions cycle among 1 year to 5 years. According to reported from Centre of Volcanology and Geological Hazard Mitigation Indonesia, of tectonic activities at Mt. Raung with maximum amplitude 2-32 mm and continuing the tremor quakes until the beginning of the 2016 eruption period.

2.1. Data Processing

For analyzing the eruption events in June to July 2015 we used SAR data derived from ALOS2/Palsar-2 Images which L-band frequency characteristic onboard from Advanced Land Observing Satellite (ALOS2) with active microwave sensor. Table below show details data are used for this study.

Table 1. Data information

Date	Sensor	Processing Type	B_{perp}[m]	Stack	Polarization	Orbit
20Jan 2015	ALOS2/Palsar2	L.1.1	0.0	Master	HH	Ascending
19Jan2016	ALOS2/Palsar2	L.1.1	237.3	Slave	HH	Ascending

PALSAR-2 is L-band frequency (active microwave remote sensing) which characterize enable to penetrate cloud (cloud-free) and day-and-night observation. Images below are used for interferometric SAR Processing. The synthetic aperture radar signal processing could be implemented simply as follow below.

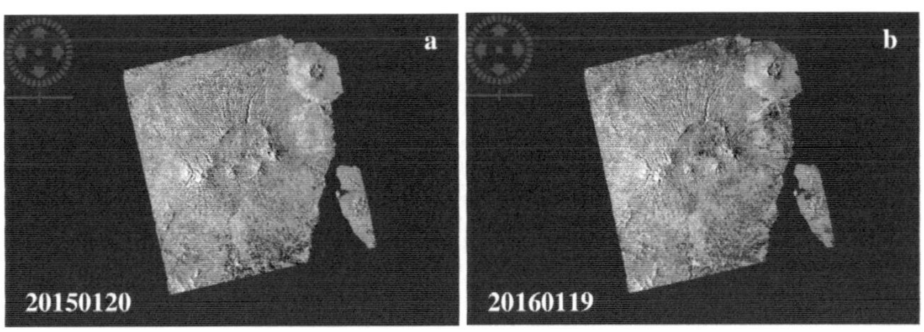

Figure 1. image (**a**) is show Master image which recorded in January 20, 2015 and image (**b**) is show slave image which recorded in January 19, 2016.

2.2. Flow Chart

In this research we proposed InSAR method in the conventional single-interferogram approach, derives from two radar images of the same area acquired at different times to measure ground displacement. The technique uses the phase difference of backscattered signals from the two acquisitions to measure differential motion in the Line Of Sight (LOS) direction include vertical and horizontal components. The InSAR deformation image produced from two SAR images that associated to the eruption in the beginning of 2016.

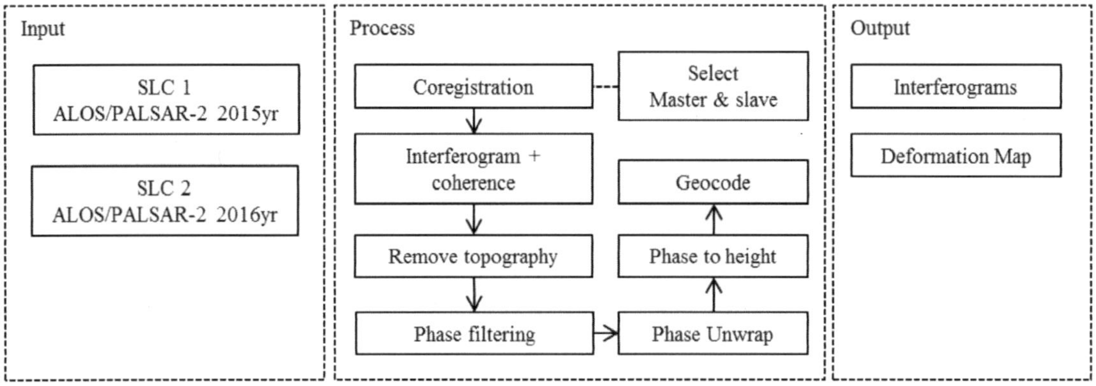

Figure 2. Flowchart study

We presented the interferometric analysis using snap software, a common architecture for all Sentinel Toolboxes is being jointly developed by Brockmann Consult, Array Systems computing and C-S called Sentinel Application Platform by ESA, (2016) to get interferogram and line of sight displacements, the interferometric phase was unwrapped with the SNAPHU program. To show ground deformation with high spatial resolution and accuracy on the large areas observation we used DInSAR (Differential Synthetic Aperture Radar Interferometry) explained by Massonnet(2008)[11]; Gabriel *et al*, (1989) [3] which phase can be subtracted from the SAR interferogram to remove topography. DInSAR used to analysis the line of sight displacement for monitoring and detecting change volcano in long-term and short-term [6] and also for studying fault mechanism [1].

3. Results and Discussion

Mount Raung is located in East Java Province, one of the most active volcanoes in Indonesia which short eruptions. We observed mount Raung based on activities since June 21st 2015 and continuing until July 4th 2015 which magma and volcanic materials. Figure below as shown the ALOS2/PALSAR images from different time. As basic study, some changes happened inside the calderas due to volcanic activities during the eruptions period.

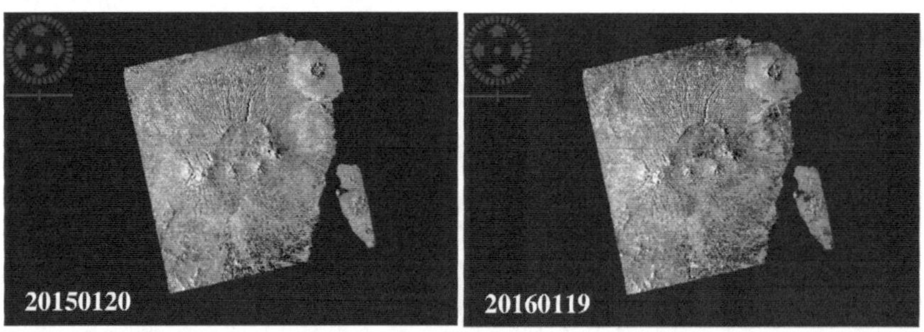

Figure 3. ALOS2/PALSAR Images of Mt. Raung taken Pre-eruptive and Post-eruptive.

Regarding to Centre of Volcanology and Geological Hazard Mitigationone of the most active volcanoes in Indonesia, which short eruptions period, mount Raung activity currently, located at the bottom of the caldera. In February 1902, cone of central caldera appearedaround 90 m. Character of mount Raung is explosive eruptions (Strombolian) as happened in 1586, 1597, 1638, 1890, 1953 and 1956, the resulting ash was thrown up into the air and never happens hot clouds glide blanketed much of the body in 1953. The main danger eruption of mount Raung is the danger of a direct result of the eruption such as hot clouds glide and throw pyroclastics. The period of eruption. Based on the history of eruptive activity shortest period between two eruptions is 1 year and the longest 90 years.

3.1. Coregistration and Coherence Image

Following the SLC images, we can confirm by coregistered images into each eruption period, detail coregistration images as shown below.

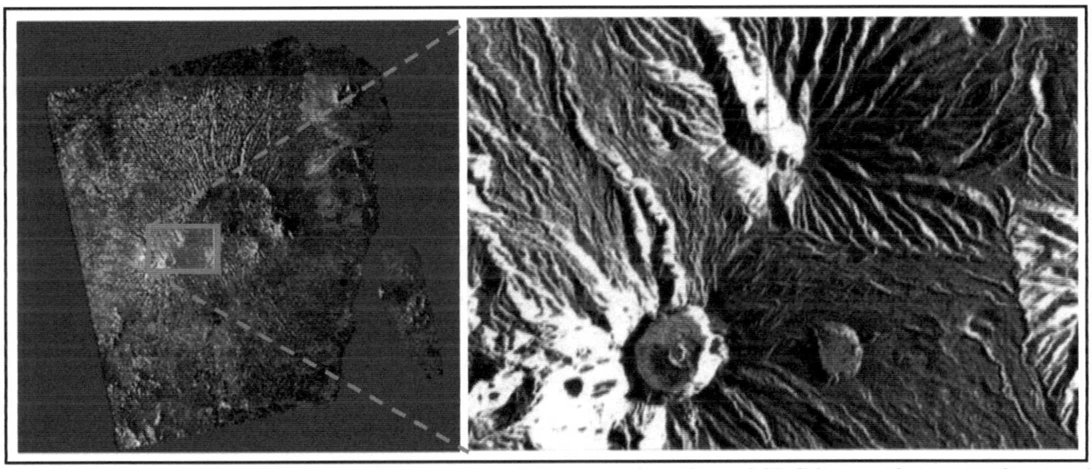

Figure 4. Images from the shape of area are made by coregistration of SLC images between slave and master images, taken in different time.

For The RGB View can be useful for amplitude change detection. Regarding to ESA SNAP (2016) those images, we will see things that have changed in red or green and things that have not changed in yellow. It is also a visual indication that the coregistration has properly aligned both images. The resulting of RGB view should look mostly yellow. Poor registrations will have badly lined up terrain. The first result of interferograms as RGB could be allowed us to study interferometric SAR in volcano disaster in particular landslide case.

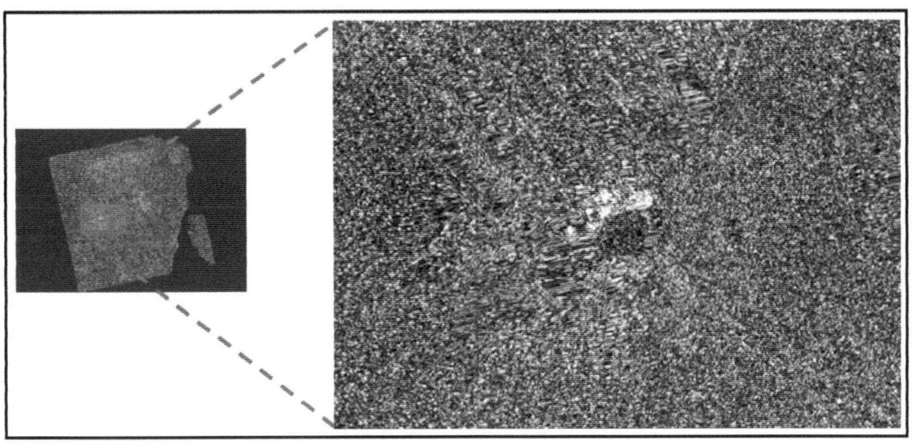

Figure 5. Images from the shape of coherence estimation of Mt. Raung obtained from interferogram between master and slave image.

The coherence band shows how similar each pixel is between the slave and master images in a scale from 0 to 1. Areas of high coherence will appear bright. Areas with poor coherence will be dark. In the image, vegetation is shown as having poor coherence and buildings have very high coherence.

3.2. Deformation Analysis

We processed SAR images from two different datasets from PALSAR2 (Table 1.) by using the SNAP procedure to generate LOS deformation rate maps of the mount Raungarea for the following observation periods: 2015.01.20 (PALSAR-2 track 131-7020; Figure **1a**), 2016.01.19 (PALSAR-2 track 131-7020; Figure **1b**). All the tracks proceed into SNAPHU developed at Stanford university, processing after removed topographic phase, phase unwrapping is a key step during the signal processing of interferometric synthetic aperture radar (InSAR) data. The precision and efficiency are two key problems of phase unwrapping.

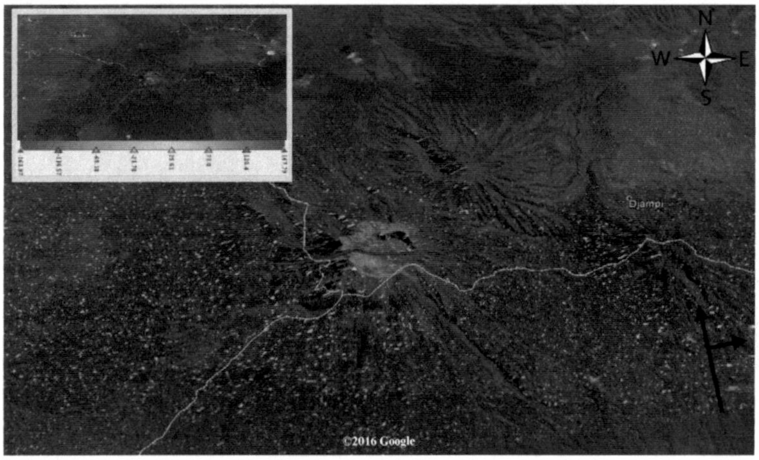

Figure 6. Land surface changes occurred in the Mt. Raung, which the result of geocoded processing overlaid with Google Earth.

Following the interferograms shown in **Figure 6**, from January 2015 to January 2016, a small amount of subsidence persisted near the caldera of mount Raung, but the northern of mount Raung began to rise slightly. Uplift spread throughout the central part of the caldera and southern part of mount Raung indicated subsidence. Northeast of mount Raung, is located mount Ijen in the same caldera, here mount Ijen activities also could be assessed in land surface changes. The color progression inside the caldera is reversed between uplift and either subsidence. The surface changes profile plot are outlined below.

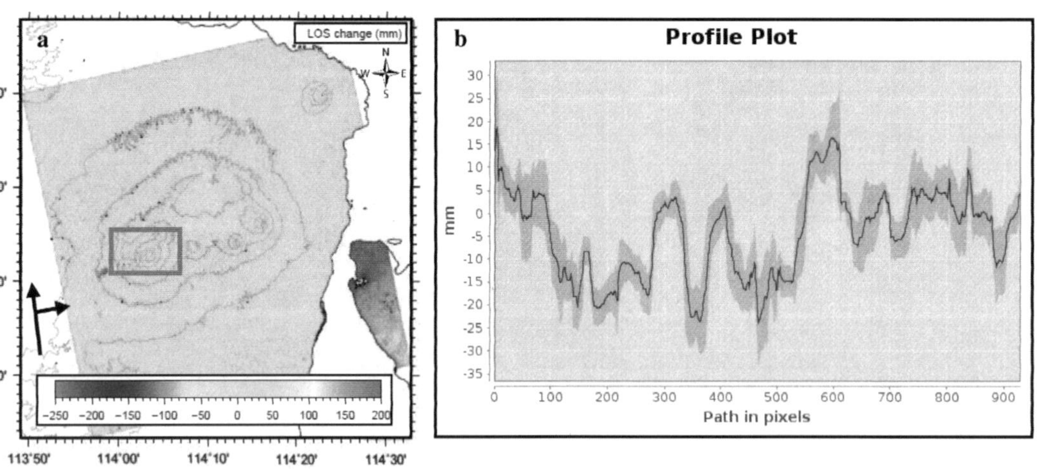

Fsigure 7. Image (a) is a complete scene processing of ALOS2/PalSAR-2 and image (b) is a profile plot of land surface changes related to **Figure 6.** the blue line is representative of path in pixel value.

Moreover, the result indicates that mount Raung started inflating soon after the end of June to the beginning of July (**Figure 6**). Related to CVGHM report, during the period August 2015 - January 19, 2016 earthquake recorded in the Post PGA Raung. The path-following phase unwrapping algorithm usually has high computational efficiency, but low coherence areas are prone to have accumulated errors. The result indicate that mount Raung started inflating soon after the end of June to the beginning of July (**Figure 7b**). Related to CVGHM report, during the period August 2015 - January 19, 2016 earthquake recorded in the Post PGA Raung, consisting of: Earthquake Tremor, volcanic earthquake, Tectonic earthquake and Blowing (which began recorded on August 15, 2015). Since August 5, 2015 RSAM value is fluctuating decline and relatively stable. Interferograms color shown in (**Figure 7a**) it could allowed us to interpret a wider area, including mount Ijen located in northeast of mount Raung.

As the results of InSAR processing, some volcanic eruptions have been preceded or accompanied by extremely large water level variations that were observed without special equipment. The simplicity of these observations is appealing and unequivocally demonstrates that groundwater levels change in response to volcanic deformation. Much more can be learned, however, from high-resolution water level measurements in well-characterized aquifers (Dzurisin, D., 2006). As an example; we could do assessment at Songgon districts position on the slopes of the mount Raung cause these districts traversed by several rivers that disgorge in the mount Raung. Binauand Badeng rivers with a length of 12 km, the Manggaran river 10 km and 8 km flow at Kumbo River [14].

For the human life assessments (risky and vulnerability) the results of these processing is allowed us to assess from geographic and topographic data suggests that Indonesia is the most volcanically active in the world, with numerous eruptions each year and millions of people living close to the volcanoes.Located in 3 regions mount Raung activities have been spreading threat for human life.

3.3. Future Study

The result given the new sight of volcanic activities monitoring and implementing of InSAR to observe volcano. Based on our study, we plan to check ground truth (field survey) analysis by surveying surrounding the volcano to understand of volcanic activities associated to the fault and vegetation surrounding the mount Raung. And find the relationship between land surface changes through the ground water level changes also we can find correlation ground water flow effect associated to deformations in mount Raung.

4. Conclusion

PALSAR-2 data can be used to assess subsidence and volumetric changes in the areas of volcanoes. By this study we claimed that using InSAR method it is possible to calculate large mass movements in the two observations during the different time observations to get clues for further study we can use same technique to quantify volumetric changes in the summit areas of volcanoes dome and the utility of InSAR for measuring the complex geophysical signals at mount Raung is evident given the variety of deformation measurements revealing both eruptive and non-eruptive behaviour, following the interseismic deformation over the Java fault, our study could supports to understand the surface movements along the fault, which agrees well with previous geological studies. For future study we will do risk and vulnerability assessment including this study to create disaster prone-areas due to volcanic activities and InSAR time-series analysis for monitoring sense of motion in East Java volcanic system and fault along the South Semarang to East Java about 300 km.

Acknowledgments

Thank you for LPDP (Indonesia Endowment Fund for Education) Scholarship funding, Indonesia Ministry of Finance. JAXA (Japan Aerospace Exploration Agency) to provide ALOS22/PALSAR-2 images. Remote Sensing of Environment and Disaster Laboratory, Industrial Institute of Technology University of Tokyo and also Earthquake Research Institute for sharing knowledge on volcano sense.

References

[1] Currenti, G., Bonaccorso, A., Del Negro, C., Guglielmino, F., Scandura, D. and Boschi, E 2010FEM-based inversion for heterogeneous fault mechanisms: application at Etna volcano by DInSAR data *Geophysical Journal International* **183**(2) pp.765-773.

[2] Dzurisin, D 2006 *Volcano deformation: new geodetic monitoring techniques*(Springer Science & Business Media)

[3] Gabriel, A.K., Goldstein, R.M. and Zebker, H.A 1989 Mapping small elevation changes over large areas: differential radar interferometry *Journal of Geophysical Research: Solid Earth* **94**(B7), pp.9183-9191.

[4] Loughlin, S., Sparks, S., Brown, S., Jenkins, S. and Vye-Brown, C. eds 2015*Global Volcanic Hazards and Risk*(Cambridge University Press)

[5] Lu, Z., Kwoun, O. and Rykhus, R 2007 Interferometric synthetic aperture radar (InSAR): its past, present and future *Photogrammetric engineering and remote sensing* **73**(3), p.217.

[6] Papageorgiou, E., Foumelis, M. and Parcharidis, I 2012 Long-and short-term deformation monitoring of Santorini Volcano: unrest evidence by DInSAR analysis *IEEE Journal of Selected Topics in Applied Earth Observations and Remote Sensing* **5**(5), pp.1531-1537.

[7] Solikhin, A., Pinel, V., Vandemeulebrouck, J., Thouret, J.C. and Hendrasto, M 2015 Mapping the 2010 Merapi pyroclastic deposits using dual-polarization Synthetic Aperture Radar (SAR) data*Remote Sensing of Environment***158** pp.180-192.

[8] Solikhin, A., Thouret, J. C., Gupta, A., Harris, A. J., and Liew, S. C 2012 Geology, tectonics, and the 2002–2003 eruption of the Semeru volcano, Indonesia: Interpreted from high-spatial resolution satellite imagery *Geomorphology***138**(1), 364-379.

[9] Tilling, R.I 1989 Volcanic hazards and their mitigation: progress and problems *Reviews of Geophysics***27**(2), pp.237-269.

[10] Zaennudin, A 2010 The characteristic of eruption of Indonesian active volcanoes in the last four decades*J. Lingkungan dan Bencana Geologi***1** pp.113-129.

[11] Massonnet, D. and Souyris, J.C 2008 *Imaging with synthetic aperture radar*(CRC Press)

[12] Snap Processing and Tutorials 2016 from http://step.esa.int/main/doc/tutorials/

[13] Mt. Raung Report taken from http://www.vsi.esdm.go.id/

[14] BPS Report taken from https://bondowosokab.bps.go.id/website/pdf_publikasi/Statistik-Daerah-Kecamatan-Sumber-Wringin-2015.pdf

IOP Conf. Series: Earth and Environmental Science **47** (2016) 012022 doi:10.1088/1755-1315/47/1/012022

Measuring Deformation in Jakarta through Long Term Synthetic Aperture Radar (SAR) Data Analysis

Agustan[1], Albertus Sulaiman[1], Takeo Ito[2]

[1]Center of Technology for Regional Resources Development (PTPSW) – BPPT, Jakarta, Indonesia
[2] Research Center for Seismology, Volcanology and Disaster Mitigation – Nagoya University, Japan

Corresponding author: agustan@bppt.go.id

Abstract. Jakarta as a home for more than 10 millions habitant facing complex environmental problems due to physical development that cause physical deformation. Physical deformation issues such as decreasing environmental carrying capacity, land cover changes and land subsidence have occurred. Recent studies shows that the long of shoreline changes in a span of 13 years from 2002 to 2015 around 14 km due to land reclamation in Jakarta bay. Previous studies also concluded that Jakarta suffer a sinking phenomena due to its rapid subsidence rate, approximately 260 mm/year in northern part of Jakarta. During the 2007 to 2011, the land subsidence phenomena in Jakarta was observed by InSAR based on ALOS-PALSAR data and found that the subsided areas only occurred in certain areas, mainly in Pluit and Cengkareng regions, with a subsidence of approximately 70 cm for 4 years. Land subsidence is generally related to geological subsidence i.e. sediment consolidation due to its own weight and tectonic movements; or related to human activities such as withdrawal of ground water and geothermal fluid, oil and gas extraction from underground reservoirs, and collapse of underground mines. The amount of subsidence or uplift can be estimated from the number of concentric fringes that appear in the interferogram. This research utilizes Synthetic Aperture Radar (SAR) data observed from ALOS-2 (L-band) and Sentinel-1 (C-band) satellites. By interfering two single look complex (SLC) images from different observation epoch, it is found that the subsided area that has been identified before continues to subside. This occurs especially in Pluit region and has been revealed by interfering ALOS-2 data up to year 2016. The deformation in this area is approximately 12 cm from November 2015 to September 2016. The process of land reclamation also clearly identified by Sentinel-1 image by series data processing in Sentinels Application Platform (SNAP) software.

1. Introduction

The Earth's shape changes from time to time and can be categorized into periodic and un-periodic change. The Earth body tide is one example of the periodic change, whereas ground deformation is an example for the un-periodic change and it is irreversible. Ground deformation may be related to seismo-tectonic processes such as earthquakes, faulting, volcanism, landslide; anthropogenic processes such as ground water pumping and mining; or other environmental phenomena such as glaciation and deglaciation.

Land subsidence that is defined as the motion of surface shifts downward relative to height datum such as mean sea level, is also classified into ground deformation. Land subsidence is generally related to geological subsidence i.e. sediment consolidation due to its own weight and tectonic movements; or related to human activities such as withdrawal of ground water and geothermal fluid, oil and gas extraction from underground reservoirs, and collapse of underground mines [1], [2].

Jakarta, as the capital of Indonesia, also has an urban development that consequence on extensive land use and land cover change and increase in groundwater extraction [3]. Therefore, land subsidence phenomena is believed that have been occurred and correlated to disaster such as inundation[4] due to high rainfall in some areas in Jakarta [5].Previous studies shown that subsidence rate in Jakarta are approximately 260 mm/year in northern part of Jakarta [6], [7], [8] and [9], whereas Pluit and Cengkareng regions, subsided of approximately 70 cm from 2007 to 2011 [5].

This article discusses the ground deformation in Jakarta revealed by Synthetic Aperture Radar (SAR) data in long-term period, as a continuation of previous study [5]. Each pixel of SAR data contains a complex number that carries amplitude and phase information about the microwave field backscattered by all objects in corresponding resolution cell projected on the ground. These kinds of information are stored in complex format and therefore it is also known as single look complex (SLC) that is composed of a regular grid with complex values or phasors [10].For that reason, ground deformation in Jakarta is conducted through amplitude image analysis and interferometric analysis. Amplitude image analysis is utilized to identify land surface changes, whereas interferometric analysis to obtain land subsidence phenomena.

2. Data and Methods

Long-term ground deformation observation in Jakarta by SAR data analysis utilizes SAR data that were observed by C-band (wavelength 5.6 cm) Sentinel-1A, and L-band (wavelength 24 cm) ALOS-2 (Advanced Land Observing Satellite-2) satellite systems. It is important to assess the ability of these two systems for ground deformation studies in tropical region that influenced by high precipitation. Previous studies show that C-band radar is more sensitive to temporal decorrelation caused by meteorological events, whereas L-band SAR offers better performance in terms of coherence, ionospheric perturbations are smaller at C-band [11] and [12]

The Sentinel-1 satellite system is a constellation of two satellites (namely Sentinel-1A and Sentinel-1B) equipped with a C-band SAR sensor developed by European Space Agency (ESA). The Sentinel-1A was launched on April 3rd 2014, with temporal resolution is 12 days and supports four different operational imaging modes that are providing different resolution and coverage: Interferometric Wide swath (IW) mode, Extra Wide swath (EW) mode, Strip-Map (SM) mode, and Wave (WV) mode [13], [14], [15]. Three Sentinel-1A data in Single polarization IW mode for November 12th 2015, April 4th 2016 and June 15th 2016 are downloaded from Sentinels Scientific Data Hub and processed with SNAP – S1TBX (Sentinel Application Platform – Sentinel 1 Toolbox) software [16].

The ALOS-2 satellite system developed by Japan Aerospace Agency (JAXA) and was launched on May 24th 2014, with temporal resolution is 14 days and support mainly two modes for worldwide observation. One is the 10 m resolution for 70 km swath for Stripmap mode and the other is the 350 km swath for ScanSAR mode. However, there is another mode namely high-resolution spotlight mode with 1 to 3 m resolution [17]. Three ALOS-2 data in Stripmap mode for November 17th 2014, January 25th 2016 and September 5th2016 are downloaded through ALOS-2 User Interface Gateway (AUIG) website: https://auig2.jaxa.jp/ips/home under ALOS-2 Principle Investigator scheme. These data are processed by using GAMMA SAR Software [18] under Nagoya University licence. The summary of data sets is listed in Table 1.

The first part, both data sets are processed to obtain the image for Jakarta region by exploiting the amplitude components. The steps include: orbit file correction, radiometric calibration, speckle reduction, and geometric correction. The second part is interferometric data analysis to identify the land subsidence. There are three interferograms resulted from both data sets through coregistration, interferogram calculation, topographic phase removal, interferogram filtering and geometric correction.

Table 1. Summary of Data Sets

	Epoch-1	Epoch-2	Epoch-3	Interferogram	Perpendicular Baseline (m)	Time differences (days)
Sentinel-1A	12/11/2015	04/04/2016	15/06/2016	1511_1604 1511_1606 1604_1606	111 75 35	144 216 72
ALOS-2	17/11/2014	25/01/2016	05/09/2016	1411_1601 1411_1609 1601_1609	57 94 37	434 658 224

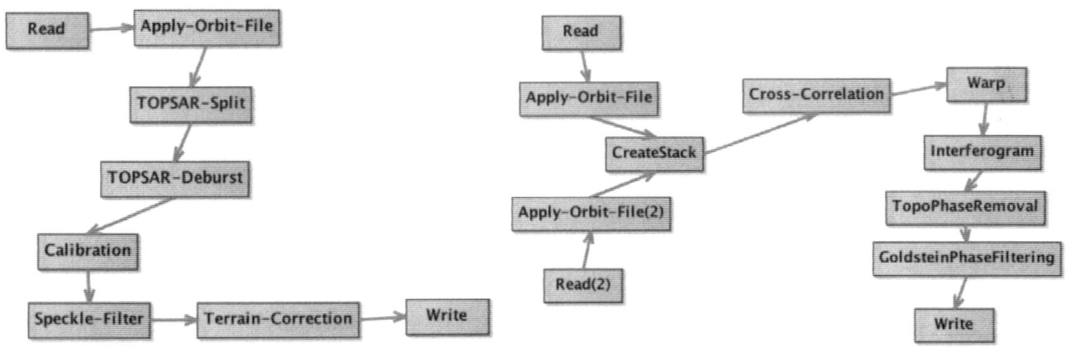

Figure 1. Data Processing Flow at S1TBX for Image (left) and Interferogram (right) generation

3. Result and Discussion
The results of SAR data analysis are presented into image analysis for land surface changes and interferogram analysis for subsidence identification.

3.1. Image Data Analysis
Series of SAR data observed by Sentinel-1A and ALOS-2 and processed by S1TBX are shown in Figure 2 and Figure 3.

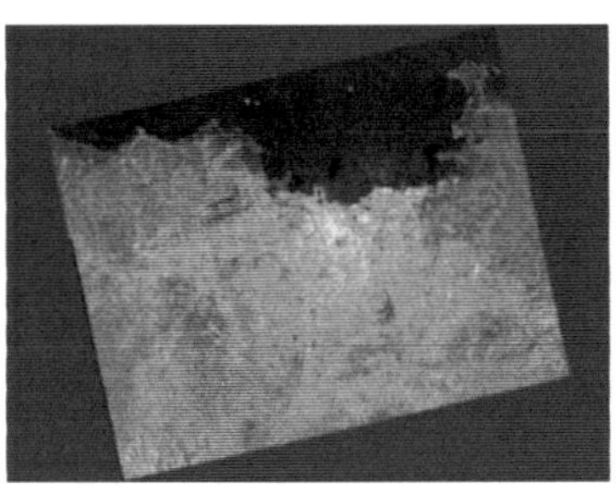

Sentinel-1A, VV Polarization
12/11/2015

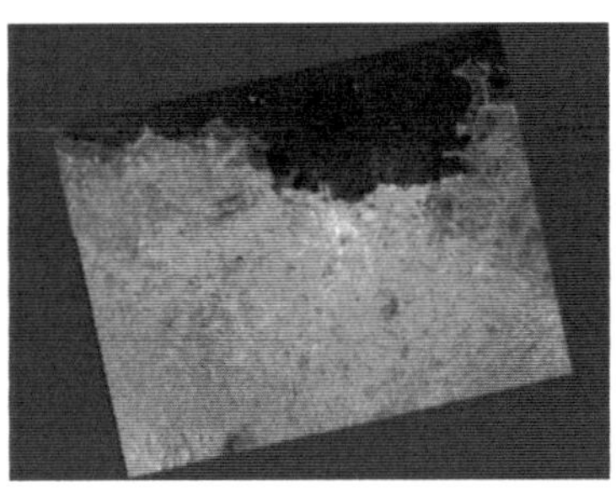

Sentinel-1A, VV Polarization
04/04/2016

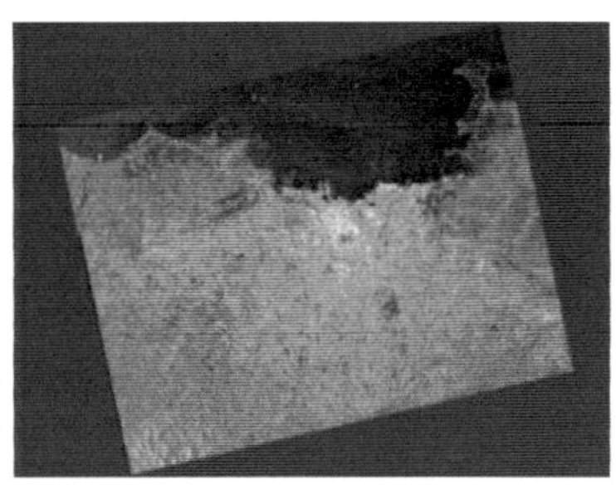

Sentinel-1A, VV Polarization
15/06/2016

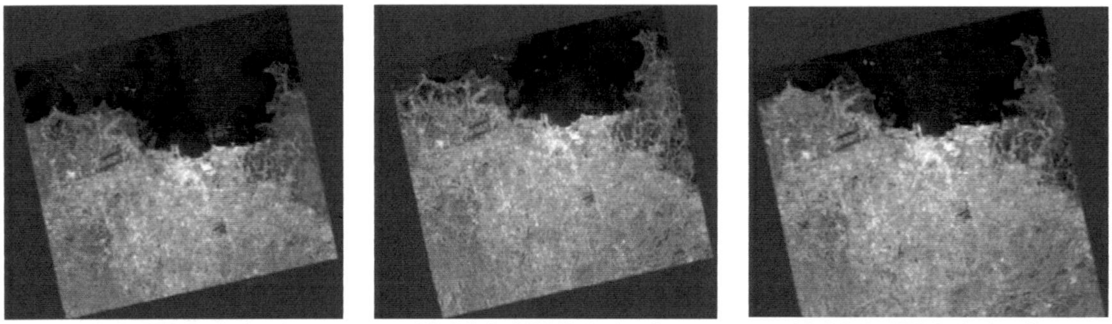

ALOS-2, HH Polarization 17/11/2014 | ALOS-2, HH Polarization 25/01/2016 | ALOS-2, HH Polarization 05/09/2016

Figure 2. SAR Images for Jakarta observed by Sentinel-1A and ALOS-2

ALOS-2, HH Polarization 17/11/2014 | Sentinel-1A, VV Polarization 12/11/2015 | ALOS-2, HH Polarization 25/01/2016

Sentinel-1A, VV Polarization 04/04/2016 | Sentinel-1A, VV Polarization 15/06/2016 | ALOS-2, HH Polarization 05/09/2016

Figure 3. Deformed Area observed by Sentinel-1A and ALOS-2

It is shown that the land surface deformation is spotted on the coastal area and had been started on November 2014 and related to land reclamation activity. Dual polarization of ALOS-2 data can be explored further with color composite technique. S1TBX provides three default color composites namely dual polarization ratio, dual polarization multiple and dual polarization difference.

Dual Polarization Ratio Color Composite | Dual Polarization Multiple Color Composite | Dual Polarization Difference Color Composite

Figure 4. Color Composite Technique for Deformed Area

3.2. Interferometric Data Analysis

For interferometric data analysis to identify land subsidence, this experiment found that the S1TBX can generate interferogram based on Sentinel-1A. However, the results are not good since the atmospheric noise affects much for C-band radar data. For ALOS-2 data, it is found that the S1TBX cannot generate a proper interferogram since there is no facility to apply orbital file and therefore coregistration result not good and affects the interferogram formation. For that reason, ALOS-2 data are processed by using GAMMA SAR Software.

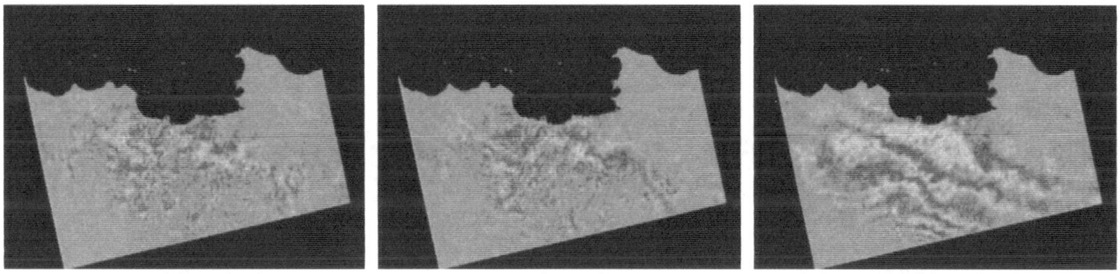

DInSAR of Sentinel-1A	DInSAR of Sentinel-1A	DInSAR of Sentinel-1A
12/11/2015_04/04/2016	12/11/2015_15/06/2016	04/04/2016_15/06/2016
Bperp 111m	Bperp 75m	Bperp 35m
144 days	216 days	72 days

Figure 5. Interferogram based on Sentinel-1A processed by S1TBX

Figure 5 shows the interferogram from Sentinel-1A data analysis by using S1TBX. All the interferograms have a small baseline, therefore the topographic effect can be neglected. However, the signal appears mainly from atmospheric path and orbital error and cannot be analysed.

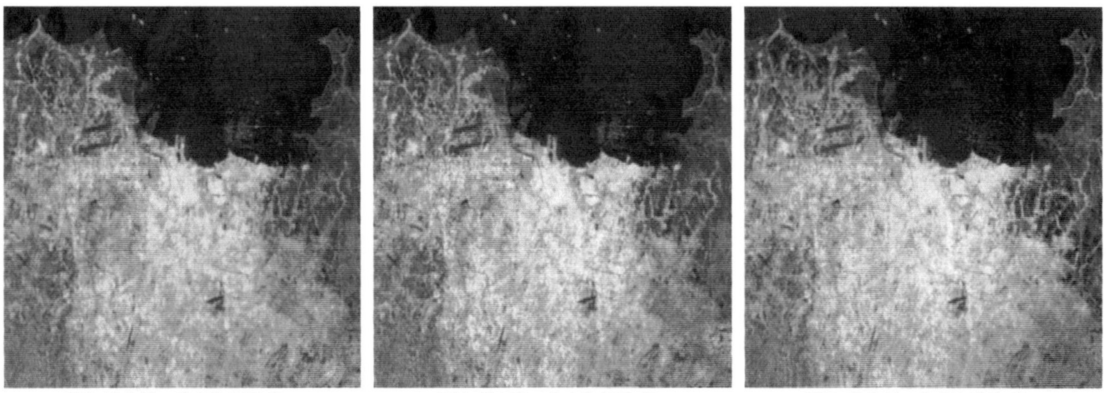

DInSAR of ALOS-2	DInSAR of ALOS-2	DInSAR of ALOS-2
17/11/2014_25/01/2016	17/11/2014_05/09/2016	25/01/2016_05/09/2016
Bperp 57m	Bperp 94m	Bperp 37m
434 days	658 days	224 days

Figure 6. Interferogram based on ALOS-2 processed by GAMMA SAR Software

Figure 6 shows the interferogram from ALOS-2 data analysis by using GAMMA SAR Software. All the interferograms have a small baseline, therefore the topographic effect can be neglected. Some subsided areasare spotted based on differential interferometric SAR that indicated by color fringes(mainly by purple) and always found in all interferograms. It is spotted in Pluit region on the north of Jakarta and in Bekasi region. These results are combined with previous study [5] that identify some subsided area from 2007 to 2011. It is found that the Pluit area is still affected by subsidence phenomena that are shown in following figures.

Figure 7. Identification of subsided areas in Jakarta based on DInSAR technique. (a) Interferogram from 2007-2011 based on ALOS-PALSAR data, (b) subsided area in Pluit region from 2007-2011 or 1472 days based on ALOS-PALSAR data, (c) subsided area in Pluit area from 2014 -2016 or 658 days based on ALOS-2 data, image not rectified.

4. Conclusions

This research shows the ability of SAR data to identify ground deformation in Jakarta area by analysing the amplitude and phase components. Sentinel-1A data and S1TBX software are useful to obtain the land surface changes based on amplitude analysis. This research also found that atmospheric phase affects much to C-band SAR data as already identify by previous studies [11] and [12]. Subsidence phenomena in Jakarta area can be identified by ALOS-2 data based on interferometric analysis. It is found that the Pluit area is still affected by subsidence since 2007.Recent data analysis, from November 2014 to September 2016 or in 658 days, the Pluit area subside for almost 12 cm.

Acknowledgments

Authors would like to thank to Japan Aerospace Agency (JAXA) through Principle Investigator of scheme that provide ALOS-2 data; and also to ESA for providing Sentinel-1A and SNAP-S1TBX software.

References

[1] Raucoules D, Colesanti C, and Carnec C2007 Use of SAR interferometry for detectingand assessing ground subsidence *Comptes Rendus Geosciences* **339**(5) pp 289–302 http://dx.doi.org/10.1016/j.crte.2007.02.002.

[2] Yuill B, Lavoie D, and Reed D 2009 Understanding subsidence processes in coastal Louisiana *Journal of Coastal Research* pp 23–36

[3] Hudalah D and Firman T 2012 Beyond property: Industrial estates and post-suburban transformation in Jakarta Metropolitan Region *Cities* **29** pp 40–48

[4] Murakami S, Kawase M and Komine H 2012 Land subsidence in Mekong delta by using InSAR and future development for vulnerability assessment in consideration of global climate change *International Workshop on Geo-engineering for responding to climate change and sustainable development of infrastructure (HUEGEO 2012)* 17-18 Dec 2012

[5] Agustan, Sanjaya H and Ito T2013 Jakarta Land Subsidence And Inundation Vulnerability Based On SAR Data *Proceedings of the 34th Asian Conference of Remote Sensing***1**pp 187-194.

[6] Abidin HZ, Andreas H., Djaja R, Darmawa D, Gamal M 2008 Land subsidence characteristics of Jakarta between 1997 and 2005, as estimated using GPS surveys *GPS Solutions*.**12** (1), pp 23–32

[7] Abidin, H., Andreas, H., Gumilar, I., Fukuda, Y., Pohan, Y. E., and Deguchi, T2011 Land subsidence of Jakarta (Indonesia) and its relation with urban development *Natural Hazards* **59**(3), pp 1753–1771, http://dx.doi.org/10.1007/s11069-011-9866-9.

[8] Ng, A. H.-M., Ge, L., Li, X., Abidin, H. Z., Andreas, H., & Zhang, K2012 Mapping land subsidence in Jakarta, Indonesia using persistent scatterer interferometry (PSI) technique with ALOS PALSAR *International Journal of Applied Earth Observation and Geoinformation* **18**(0), pp 232-242. doi: http://dx.doi.org/10.1016/j.jag.2012.01.018

[9] Chaussard, E., Amelung, F., Abidin, H., & Hong, S.-H 2013 Sinking cities in Indonesia: ALOS PALSAR detects rapid subsidence due to groundwater and gas extraction *Remote Sensing of Environment* **128**(0), pp 150-161. doi: http://dx.doi.org/10.1016/j.rse.2012.10.015

[10] Hanssen, R.F. 2001. Radar Interferometry: Data Interpretation and Error Analysis. *Kluwer Academic Publishers*, Dordrecht, pp 328.

[11] Rignot, E.; Mouginot, J2012 Ice flow in Greenland for the international polar year 2008–2009 *Geophys. Res. Lett.* **39**, L11501.

[12] Mouginot, J.; Scheuchl, B.; Rignot, E2012Mapping of ice motion in Antarctica using synthetic-aperture radar data *Remote Sens* **4** pp 2753–2767.

[13] Potin, P.; Rosich, B.; Roeder, J.; Bargellini, P2014Sentinel-1 mission operations concept In *Proceedings of the IEEE IGARSS*, Québec, QC, Canada, 13–18 July 2014; pp. 1465–1468.

[14] Torres, R.; Snoeij, P.; Geudtner, D.; Bibby, D.; Davidson, M.; Attema, E.; Potin, P.; Rommen, B.; Floury, N.; Brown, M2012GMES Sentinel-1 mission *Remote Sens. Environ.* **120**, pp 9–24.

[15] Geudtner, D.; Torres, R.; Snoeij, P.; Davidson, M.; Rommen, B 2014 Sentinel-1 system capabilities and applications In *Proceedings of the IEEE IGARSS*, Québec, QC, Canada, 13–18 July 2014; pp 1457–1460.

[16] Veci, L., Lu, J., Prats-Iraola, P., Scheiber, R., Collard, F., Fomferra, N. and Engdahl, M 2014 The Sentinel-1 Toolbox In *Proceedings of the IEEE International Geoscience and Remote Sensing Symposium (IGARSS)*, pp. 1-3. IEEE.

[17] Arikawa, Y, Saruwatari H, Hatooka Y, Suzuki S2014 PALSAR-2 launch and early orbit operation result In *Proceedings of International Geoscience and Remote Sensing Symposium (IGARSS)* 2014, pp 3406–3409.

Drought Vulnerability Mapping with Geomorphological Approach in Yogyakarta Special Region (DIY) and Central Java

Sudaryatno

Department of Geographic Information Science Faculty of Geography, Universitas Gadjah Mada, Yogyakarta, Indonesia

sudaryatno@ugm.ac.id

Abstract. This study aims to determine the level of vulnerability of the geomorphologic drought that occurred in Central Java and Yogyakarta Special Region. This study examines geomorphologic drought. Parameters used were slope, drainage, Available Water Capacity (AWC), permeability, landform, and land use. Landsat 8 and SRTM data were used for the extraction of physical parameters, such as slope, drainage, landform, and land use. The method used in this study is scoring and weighting. Query results were used for data classification by overlaying drought geomorphologic parameters. The expected outcome of this research is to map the geomorphologic drought vulnerability on Central Java and Yogyakarta Special Region. Drought vulnerability was divided into wet, normal and dry classes. Distribution of the dry class is frequent. Some of the dry classes are distributed on the steep till extremely steep slope region and on the structural and karsts landform. This was related to AWC value where region with high AWC contributed to the poor drainage of the soil, such as at Kulonprogo, Purworejo, Kebumen, Blora, Wonogiri, Purbalingga, Pekalongan, Jepara and Kudus regency. Normal classes are distributed on the sloping till steep slope, have moderate till well-drained soil and low AWC, such as at Gunung Kidul, Pati, Temanggung regency, and Magelang city. Wet classes are distributed on the flat or almost flat and sloping region. Most of the wet classes are distributed on volcanic hills and coastal area. Those regions are well-drained and the land uses are mostly for settlement and farming, such as at Sleman, Yogyakarta city, Klaten, Bantul, and Wonosobo regency.

Keywords: Geomorphology, Drought, Mapping, Landsat 8, SRTM

1. Preface

The recent change in global climate has become a new problem in Indonesia. The effect of that change in global climate also affects the micro climate, like the increase of drought locations. The drought areas in Java Island, especially in Central Java and DIY (Daerah Istimewa Yogyakarta – Yogyakarta Special Region) always increase every year. Both areas have static drought vulnerability areas per year, but the drought area increases by 30% in average, every year. As a result, the increase of drought area might aggravate the water crisis in those areas.

Drought might be caused by meteorological or geomorphological factor. In Indonesia, meteorological drought happens more often than geomorphological drought. Since researches about geomorphological drought are scarce, this research is trying to assess the geomorphological drought that is caused by physical factors, such as topography and non-water-absorbing rocks [6]. Geomorphological approach used in this research studies the slope, geology (stone), types of soil, hydrology, and vegetation cover.

The advancement of remote sensing technology today made various researches with wide coverage easier. One of the easiness is the availability of Landsat 8 Image / LDCM (Landsat Data Continuity Mission) which is available for free. The supremacy of Landsat 8 can be utilized for various studies and applications, such as drought assessment. In the other hand, Landsat 8 Image also available multitemporal, so it is highly possible to perform an analysis with sequential time order. Physical parameters and geomorphological information which support the drought vulnerability research could also be extracted from the image. Afterwards, the mapping of drought vulnerability zone is done with Geographic Information System. Reinterpretation and research input are then presented according to Cartographical Rule.

2. Formulation of the problem

The formulation of the problem in this research is:

- How is the level of geomorphological drought vulnerability in DIY and Mid-Java Province?

3. Objective

The objective of this research is to:

- Analyse the vulnerability of geomorphological drought in Yogyakarta Special Region and Central Java Province.

4. Benefits of research

The benefits of this research are:

a) Providing information about the distribution of drought areas in Yogyakarta Special Region and Central Java Province.
b) Development of Geographical Information System and Remote Sensing science and technology in case study field applications.
c) As a reference for land regulation, farming, and vegetation choosing, which are appropriate to the land geomorphology.

5. Research methodology

This research is done in several stages. The data gathering and selection stage includes literature study, material procurement, and data selection. Afterwards, the selected data are converted to digital form.

The next stage is the digital image processing. In this stage, there are geometric and radiometric image correction, composite forming, multispectral classification, and parameter extraction which are used to determine the drought vulnerability form Landsat 8 image. An Image Processing Software is used to aid the processing of remote sensing satellite image data.

The derivative data which is used in this research is processed after geometric corrections. The function of the correction is to produce data which has appropriate special position with the coordinates on the field. Geomorphological data processing is done by scoring and weighting for classification. The scoring and weighting for each parameter isn't equal. The dominant drought parameters are given wider range of values then minor parameters. First class rating shows that the parameter affects drought highly, whereas the last rating shows that it is has less effect on drought. The used parameters in research are slope, drainage condition, Available Water Capacity (AWC), permeability, landform and land usage. The landform classification is appropriate with the [7] classification and the result of land usage classification is according to Landsat 8 interpretation. The interpretation is done with seven interpretation elements.

The value of drought vulnerability can be collected from drought index calculation. The formula used to calculate the drought index is:

$\Sigma\{(RixBi)G\}$

Note:

Ri: : Rating
Wi : Weight
G : Number of Parameter

The used parameters, classes, and dignity in research are shown at Table 1:

Table 1. Drought Vulnerability Parameter Class and Weight

Parameter	Class	Rating (Ri)	Weight (Bi)
Slope (R1 x B1) Weight = 2	• Flat • Fairly Sloping • Sloping • Fairly Steep • Steep • Very Steep • Sheer	1 2 3 4 5 6 7	2 4 6 8 10 12 14
Drainage (R2 x B2) Weight = 2	• Good • Average • Bad • Very Bad	1 2 3 4	2 4 6 8
Available Water Capacity (AWC) (R3 x B3) Weight = 4	• 100 mm/m • 150 mm/m • 200 mm/m • 250mm/m • 300 mm/m	1 2,5 4 5,5 7	4 10 16 22 28
Permeability (R4 x B4) Weight = 4	• Medium Grain Sandstone • Limestone • Schist • Tuff • Weathered Gabbro • Weathered Granite • Clay • Basalt • Slate Stone	7 7 7 7 7 7 7 7 4	21 21 21 21 21 21 21 21 12
Permeability (R4 x B4) Weight = 3	• Fine Sand • Dust • Fine Grain Sandstone • Rough Gravel • Medium Gravel • Fine Gravel • Rough Sand • Medium Sand • Sand Dunes • Turf	4 4 1 1 1 1 1 1 1 1	12 12 3 3 3 3 3 3 3 3
Landform (R5 x B5) Weight = 2	• Volcanic (V) - V1 Volcanic Mountains - V2 Volcanic Hills	 4 3	 8 6

	- V3 Volcanic Plains	2	4
	• Denudasional (D)		
	- D2 Denudasional Hills	7	14
	• Fluvial (F)		
	- F Fluvial Plains	1	2
	• Karst		
	- K1 Southern Java Karst Hills	6	12
	- K2 Northern Java Karst Hills	6	12
	• Marin		
	- M1 North Beach Plains	5	10
	- M2 South Beach Plains	4	8
	• Structural		
	- S1 Mountains	5	10
	- S2 Hills	5	10
	- S3 Structural Plains	5	10
Land Usage (R6 x B6) Weight = 2	• Body of Water • Forest, Mixed Farms, Plantations, Fishpond • Settlement, Bushes • Dry Land Farming, Moor, Field • Open Land, Built Land	1 4 6 5 7	2 4 6 8 10

Source: [1, 2, 3, 4, 5]

The calculation of scoring result is done using the Geographic Information System Query. In this method, there are tabular and spatial data calculations. Data is processed digitally, so it is done faster and easier with many parameters. The resulting drought index is then classified into three classes; Dry, Normal, and Wet.

6. Research outcome

The geomorphological drought in August had wider coverage than it is during wet and normal conditions. These conditions are triggered by landform and the soil ability to store or let water flow. Geomorphological Drought Map in Yogyakarta Special Region and Central Java in 2013 is provided in Figure 1.

The research shows that the distribution of geomorphological drought vulnerability with similar dry class in every regent within Yogyakarta Special Region and Central Java. The drought vulnerability level is the result of interaction of slope, drainage condition, Available Water Capacity (AWC), permeability, landform and land usage factors in an area.

Table 2. Land Drought Vulnerability in Yogyakarta Special Region and Central Java (in Hectare)

Regency	Low	Normal	High	Area (Ha)
Banjarnegara	13.046	40.761	61.138	115.464
Bantul	25.273	16.952	9.863	54.003
Banyumas	798	61.250	76.813	139.423
Batang	21.687	20.216	45.304	87.636
Blora	131	32.555	159.468	195.904
Boyolali	30.328	36.543	45.734	113.171
Brebes	29.034	68.427	77.623	177.438

Cilacap	30.962	107.875	90.004	241.174
Demak	8.311	31.896	59.352	100.309
Grobogan	1.080	97.344	102.523	202.052
Gunungkidul	225	76.261	71.748	149.090
Jepara	1.894	6.700	88.371	97.782
Karanganyar	3.695	11.399	62.936	79.006
Kebumen	8.362	31.826	92.044	134.093
Kendal	15.570	54.756	29.953	100.911
Klaten	63.130	2.430	4.374	70.284
Kodya Magelang	-	1.683	163	45.166
Kodya Pekalongan	284	3.335	1.052	57.782
Kodya Salatiga	3.924	852		111.756
Kodya Semarang	10.469	13.346	14.661	1.855
Kodya Surakarta	1.723	311	2.676	159.044
Kodya Tegal	638	2.366	697	87.834
Kodya Yogyakarta	3.287			4.693
Kudus	62	16.968	27.907	113.984
Kulon Progo	5.968	22.611	28.908	81.160
Magelang	36.811	25.264	49.138	10.7864
Pati	3.931	65.919	88.387	10.3842
Pekalongan	5007	32.082	50.350	4.801
Pemalang	16.322	44.490	52.681	102.354
Purbalingga	622	20.426	59.764	38.844
Purworejo	3.937	47.346	55.283	5.7847
Rembang	2.553	31.570	69.141	99.952
Semarang	36.424	31.135	34.286	50.973
Sleman	47.465	2.509	7.588	4.734
Sragen	206	37.341	61.133	98.747
Sukoharjo	9.635	15.693	25.388	3.715
Tegal	35.007	34.968	28.373	87.488
Temanggung	12.688	25.817	48.560	194.719
Wonogiri	1369	51.531	138.189	101.203
Wonosobo	43.867	21.940	34.922	3303
Total	535.725	124.6691	195.6497	3.781.402

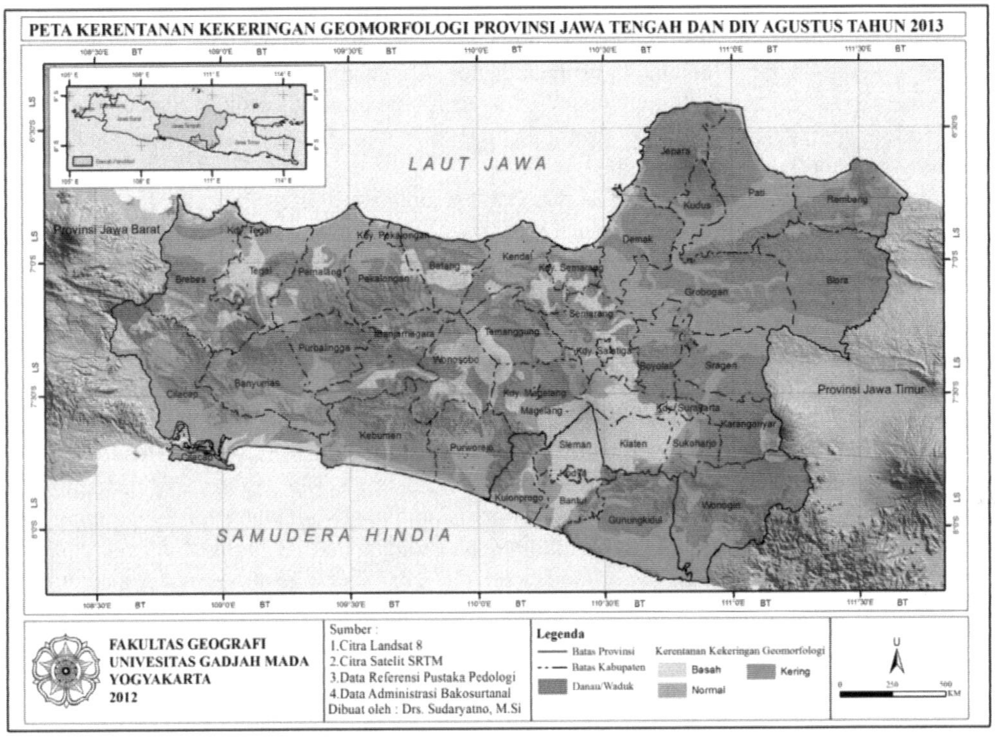

Figure 1: Geomorphological Drought Vulnerability Map in DIY and Central of Java in August 2013

Wet classes lie on some areas within Sleman, Klaten, Boyolali, Bantul, Magelang, Wonosobo, Kendal, Temanggung, Batang, Tegal, Kodya Semarang, Demak coast, and small parts of Pemalang Regency. Normal classes are distributed on the southern part of Mid-Java, which are some of the areas of Gunungkidul regency, Cilacap, Wonogiri, Kulonprogo, and Purworejo Regency, and northern parts of Mid-Java, which covers some of Pati regency, Kendal, Rembang, Pekalongan, Tegal, Pemalang, Brebes, Kodya Tegal, Kodya Semarang, and Kodya Pekalongan.

Normal classes are also available on some parts of Blora regency, Grobogan, Purbalingga, Sragen, Boyolali, Temanggung, Magelang, Banyumas, Banjarnegara, and Wonosobo.

Regencies that mostly included in the dry class are Wonogiri, Karanganyar, Sragen, Blora, Rembang, Pati, Jepara, Kudus, Grobogan, Temanggung, Magelang, Kulonprogo, Kebumen, Purbalingga, Demak, Pemalang, Banjarnegara, Brebes, Banyumas, Cilacap, and Pekalongan.

Wet area distributions are narrower than dry and normal areas. This condition is caused by different permeability conditions. The permeability on wet areas is composed of fine sand, rough sand, and sand dunes. The permeability of sand is very good since it allows water flows well. Dry areas are mostly composed of materials which let water flows poorly, such as clay, which let water flows slowly. Clay also tends to absorb water to its pores, which cause water to be saturated. On wet conditions, soil on clay areas would be clayey and harder to pass. However on dry season, the condition tilted; water evaporate so the soil contracts and cracking. This condition is what called as drought.

The most affective factor from geomorphological drought is the soil condition. This soil data is then derived to AWC data, permeability, and drainage. That soil condition, added with topography and landform factors, resulting in the identifier of soil condition which determine its vulnerability to water. For example, regosol soil which consists of clay in plains would absorb water. In different condition such as escarpment, water would flow downwards, slowly.

Slope variation causes the variation of the surface quantity and runoff speed; land with steeper slope will have greater runoff speed than those with less steep slope, causing non-maximum rainwater infiltration and surface water supply.

Land texture and structure are the primary factors in determining the value of Available Water Capacity. Land with clayey texture has a very high level of AWC, while sandy land has a very low AWC level. Rooting depth is also a significant factor which also determines the value of AWC. Research result shows that the high level of AWC on each type of soil is proportional to the high level of geomorphological drought in Mid-Java and DIY, indicating that AWC is also a dominant parameter in determining geomorphological drought level.

Good permeability exists on land used as rice field, settlement, dryland farming, moor, and bushes. If associated with slope factor, good permeability exists on volcanic hills, volcanic plains, alluvial plains, and marin plains. There also exist spring belts on the buckling hillsides. This indicates a very good concentration of land water. Flat to fairly sloping slopes exist on areas with fluvial plains landform.

The studied research areas had good, average, bad, and very bad drainage classes, and are affected by the landform of an area. In the other hand, fluvial landform also has good drainage, because it is close to water source, so that water can flow and not saturate and sediment in an area

Another affective parameter in drought vulnerability is the land use. Dry class areas vulnerable to geomorphological drought had land utilized as forest, mixed farms, plantations, and fishpond. This is related to the land ability which the land and water crisis able to absorb water for a long time. Normal and wet class areas vulnerable to geomorphological drought mostly consist of body of water, settlement, bushes, dryland farming, moor, and rice field. This is related to the water availability and wet land condition, hence the soil tend to be fertile.

7. Conclusion

a) The distribution of dry class is more prevalent than other classes. Most of the dry class are distributed in steep to sheer slopes and have structural and karst landform. This is related to the type of land on those areas which affect high level of AWC, so that the drainage became bad. The examples are Kulonprogo, Blora, Purworejo, Kebumen, Wonogiri, Purbalingga, Pekalongan, Jepara, and Kudus regencies.
b) Normal classes are distributed at scarps to gentle slope. This class had average drainage to good, so that the AWC level is low. The examples are Gunung Kidul, Pati, Temanggung regencies, and Magelang city.
c) Wet classes are distributed at gentle slope to plains. Mostly distributed at middle and down slopes to coast, this class had good drainage so that most of the land uses are settlement and farming. The examples are Sleman, Klaten, Bantul, Wonosobo regencies and Yogyakarta city.

References

[1] Adi P and Subhakti 2011 *Agihan Tingkat Kerentanan Kekeringan di Provinsi Daerah Istimewa Yogyakarta Menggunakan Parameter - parameter Geomorfologi dan Meteorologi (Skripsi)* (Yogyakarta Fakultas Geografi UGM)

[2] Anisa T and Miranti 2011 *Identifikasi Lahan Pertanian Rawan Kekeringan dengan Metode Sistem Informasi Geografi Studi Kasus Kabupaten Indramayu Provinsi Jawa Barat. Karya Tulis* (Bogor IPB)

[3] Darmawijaya M Isa 1997 *Klasifikasi Tanah Dasar Teori Bagi Peneliti Tanah dan Pelaksana Pertanian di Indonesia* (Gadjah Mada University Press Fakultas Pertanian UGM)

[4] Dudal R and M Soepraptohardjo 1957 *Soil Classification in Indonesia* (Bogor Pusat Penelitian Tanah dan Agroklimatologi)

[5] Soepraptohardjo, M 1976 *Jenis Tanah di Indonesia. Seri 3C. Klasifikasi Tanah. Training Pemetaan Tanah* (Bogor)

[6] Suyono 2007 *Kajian Kekeringan dan Banjir dengan Pendekatan Geografi di satuan Wilayah Sungai Pemali - Comal Provinsi Jawa Tengah. Laporan Penelitian* (Yogyakarta Fakultas Geografi UGM)

[7] Zuidam, Van a Zuidam-Cancelado 1979 *Aerial photo - interpretation in terrain analysis and geomorphologic mappin* (The International Institute for Aerospace Survey and Earth Sciences (ITC))

IOP Conf. Series: Earth and Environmental Science **47** (2016) 012024 doi:10.1088/1755-1315/47/1/012024

Modelling tropical cyclone hazards under climate change scenario using geospatial techniques

M A Hoque[1,2], S Phinn[1], C Roelfsema[1], and I Childs[3]

[1]Remote Sensing Research Centre, School of Geography, Planning and Environmental Management, The University of Queensland, Brisbane, QLD 4072, Australia
[2] Department of Geography and Environment, Jagannath University, Dhaka-1100, Bangladesh
[3]Queensland Centre for Population Research, School of Geography, Planning and Environmental Management, the University of Queensland, Brisbane, QLD 4072, Australia

m.hoque2@uq.edu.au

Abstract. Tropical cyclones are a common and devastating natural disaster in many coastal areas of the world. As the intensity and frequency of cyclones will increase under the most likely future climate change scenarios, appropriate approaches at local scales (1-5 km) are essential for producing sufficiently detailed hazard models. These models are used to develop mitigation plans and strategies for reducing the impacts of cyclones. This study developed and tested a hazard modelling approach for cyclone impacts in Sarankhola upazila, a 151 km^2 local government area in coastal Bangladesh. The study integrated remote sensing, spatial analysis and field data to model cyclone generated hazards under a climate change scenario at local scales covering < 1000 km^2. A storm surge model integrating historical cyclone data and Digital Elevation Model (DEM) was used to generate the cyclone hazard maps for different cyclone return periods. Frequency analysis was carried out using historical cyclone data (1960-2015) to calculate the storm surge heights of 5, 10, 20, 50 and 100 year return periods of cyclones. Local sea level rise scenario of 0.34 m for the year 2050 was simulated with 20 and 50 years return periods. Our results showed that cyclone affected areas increased with the increase of return periods. Around 63% of study area was located in the moderate to very high hazard zones for 50 year return period, while it was 70% for 100 year return period. The climate change scenarios increased the cyclone impact area by 6-10 % in every return period. Our findings indicate this approach has potential to model the cyclone hazards for developing mitigation plans and strategies to reduce the future impacts of cyclones.

1. Introduction

Tropical cyclones are devastating natural disasters, typically generating high winds, intense rainfall, terrestrial flooding and storm surges [1]. Globally, many coastal areas are regularly affected by these disasters. By the several studies, it is expected and predicted that under most climate change scenarios, the intensity and frequency of tropical cyclones will increase significantly [2, 3]. The overall impacts of tropical cyclones were very high in many coastal areas of the world. These disasters often cause significant loss of life, large scale property and environmental damage [4]. Cyclone hazard modelling is an effective management process under preparedness stage of cyclone disaster management which can provide realistic scenarios identifying areas that may be affected in the future [5]. These models could be used by policymakers and administration to develop mitigation plans and strategies. Thus, hazard modelling procedure can reduce the impacts of cyclones on people, property and environment.

There are various advanced mathematical storm surge modelling approaches are available [6, 7, 8]. However, a simple modelling approach at local scales (1-5 km) is essential for producing detailed hazard models. Geospatial techniques using remote sensing and spatial analysis are a potentially effective tool for modelling tropical cyclone hazards in simple manure integrating local knowledge [3, 9]. Moreover, sea lever rise scenarios can be simulated easily using geospatial approach to assess the climate change impact on tropical cyclones. Currently, very few studies are available on cyclone hazard modelling using storm surge model under climate change scenario based on geospatial approach at the local scale.

The aim of this study was to develop and test a hazard modelling approach under climate change scenario using geospatial techniques for estimating tropical cyclone impacts at the local scale covering <1000 km^2 in Sarankhola upazila, Bangladesh.

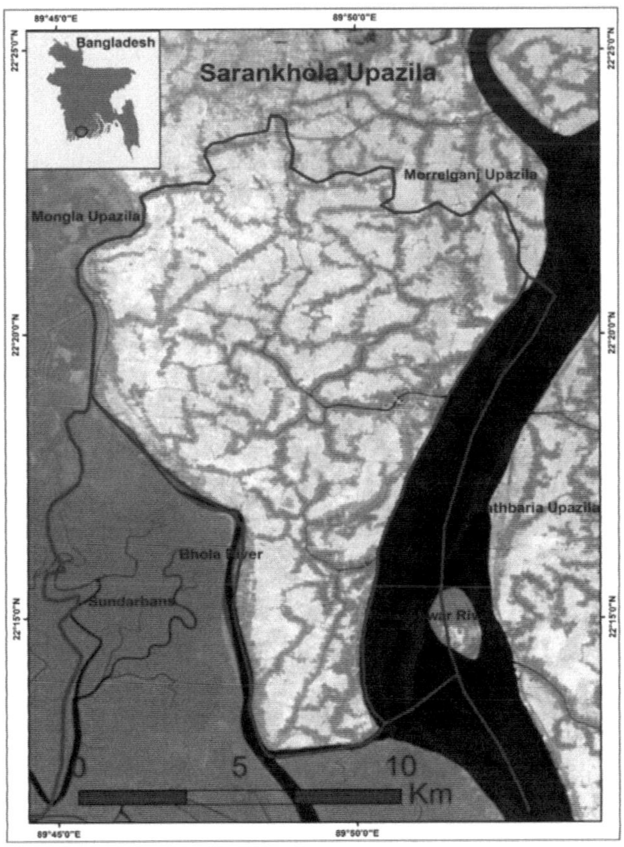

Figure 1. Study area, Sarankhola upazila under Bagerhat district in Bangladesh drawn on Landsat 8 OLI image of 15/04/2014

2. Study area and dataset

The study area is Sarankhola upazila, a local government area (about 151.24 sq. km), under Bagerhat district in Bangladesh (figure 1). The area is geographically located between 22°13´- 22°24´ N latitude and 89°46´- 89°54´E longitude in the western coastal area of Bangladesh. The area is open to Bay of Bengal by the coastal Baleshwari and Bhola rivers and is highly vulnerable to tropical cyclones induced storm surges, strong winds and flooding [10]. This coastal upazila is highly populated and was severely affected by tropical cyclone Sidr in 2007 and Alia in 2009.

The historical cyclone data (1960-2015) of western coastal region of Bangladesh was used in this study for frequency analysis to calculate different return period surge heights. These data were acquired from the Bangladesh Meteoroidal Department (BMD). The SRTM1 data at 30 m resolution were downloaded from United States Geological Survey (USGS) Earth explorer website and used as a basic input in the storm surge model. We have further processed this SRTM 1 data to remove vegetation artefacts by canopy height estimation model for producing bare earth DEM. The local sea level rise scenario data was acquired from the published study [11]. The study used the local tide gauge data which are closed to our study area for projecting sea lever rise scenario.

3. Methods

Geospatial techniques were used to develop the surge models. The local sea level sea rise scenario was integrated with these surge models to assess the climate change impacts. The resultant surge models were used for hazard analysis. The processing flows used in this study are discussed in the following sections and outlined in the figure 2.

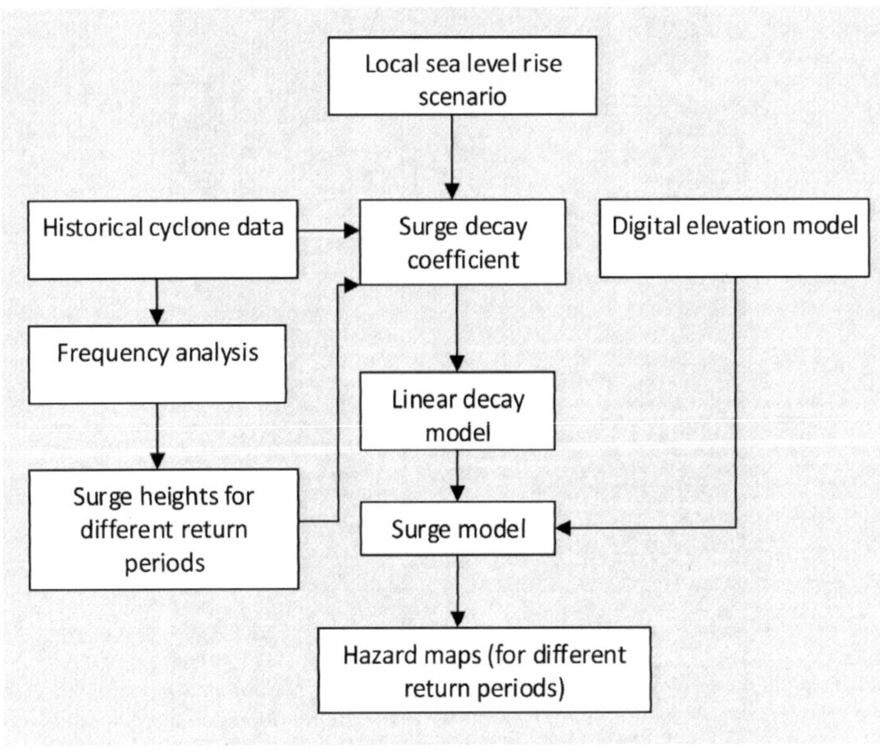

Figure 2. Flowchart of the hazard modelling approach used in this study

3.1 Frequency analysis and surge heights

Historical cyclone data (1960-2015) were used for frequency analysis to calculate different return period surge heights. The coastal area of Bangladesh is divided into three zones. The study area is located in the western coastal zone. Therefore, western coastal region cyclone data were only considered in the frequency analysis. Based on Gumbel Distribution, storm surge heights were calculated using the frequency analysis for return periods 5, 10, 20, 50 and 100 were 4.65, 5.39, 6.10, 7.01 and 7.70 meters respectively.

3.2 Climate change impact and tropical cyclone

The local sea level rise scenario was used to assess the consequences of global climate change on tropical cyclone storm surges. The local sea level rise scenario of 0.34 m for the year 2050 was projected based on the local gauge station data which are closed to study area [11]. This scenario was simulated with 20 and 50 years return periods surge models.

3.3 Surge model and surge decay coefficient

Storm surge flooding depth is considered as linearly decay towards inland. Based on this concept, the storm surge model was developed in this study. We used the ArcGIS spatial analyst tool for developing the storm surge model. In this process, surge decay coefficient (SDC) was calculated for each surge height. SDC is a function of friction caused by surface forms and land use which helps to find out how the surge depth decreases inland gradually (figure 3). The SDC is calculated using the equation 1 and illustrated in figure 3. SDC values were calculated using field data of tropical cyclone Sidr (2007). Using the SDC values and bare-earth DEM, various raster calculator formulas were used in ArcGIS Software environment to develop storm surge models for different return periods.

$$SDC = \frac{Surge\ height - Average\ elevation\ at\ the\ end\ of\ surge}{Total\ inundation\ width - Width\ of\ constant\ surge} \tag{1}$$

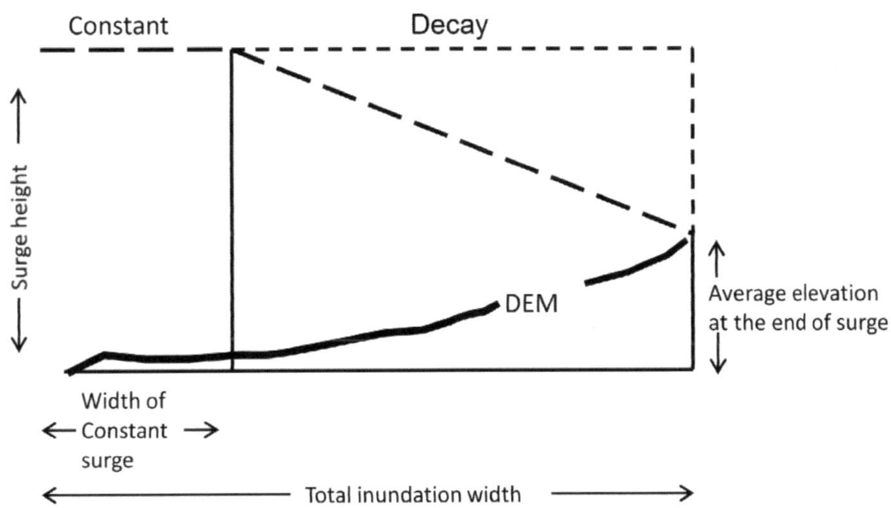

Figure 3. Schematic diagram outlining the linear surge decay model

3.4 Development of hazard map at present and future climate

Hazard maps were produced based on inundation depths derived from surge models for different return periods under present and future climate. Five levels of hazard classes were assigned for every hazard map by inundation depths ranges according to expert opinions. The assigned hazard classes are very low hazard (< 0.5 m), low hazard (0.5-1.5 m), moderate hazard (1.5-2.5 m), high hazard (2.5-3.5 m) and very high hazard (>3.5 m).

4. Results and discussion

4.1 Hazard map at present climate

Hazard maps (figure 4) were prepared by categorizing the inundation depths from storm surge models into five classes under present climatic conditions. The produced hazard maps show that hazard prone areas increase with the increase of return period and surge depth. In 10 years return period, 31% of the study area is located from moderate to very high hazard zone, while it will be 51% in 20 years return period.

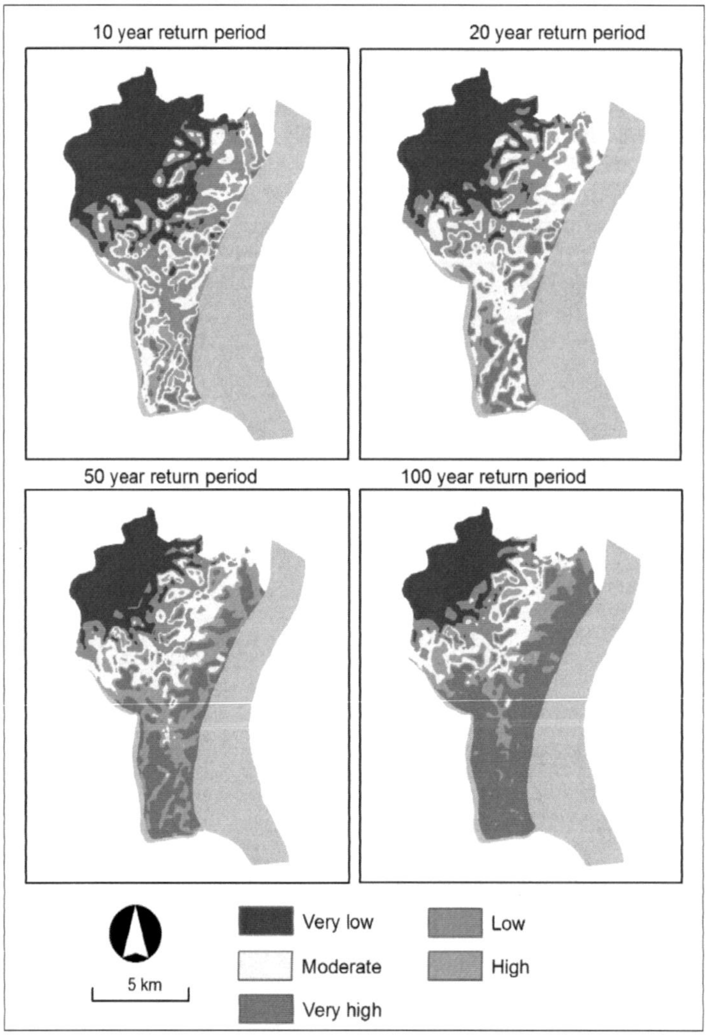

Figure 4. Hazard maps under 10, 20, 50 and 100 year return periods at present climate

On the other hand, for the 50 years return period, 63% of the study area was located from moderate to very high hazard zone, while it will be 70 % in 100 years return period. Moreover, 80% of the area will be inundated above 0.5 surge height in 100 return periods.

4.2 Hazard map under future climate

Hazard maps under future climate change impact were prepared by integrating 0.34 m local sea level rise scenario with existing storm surge models of 20 and 50 years return periods and these models then classified into five hazard classes. Figure 5 presents the hazard maps in both climate change and normal scenario for the year 2050. The study found that the area in the moderate to high hazard zones will increase by 10% in 2050 for 20 year return period under climate change scenario, while it will be 9% for 50 years return period.

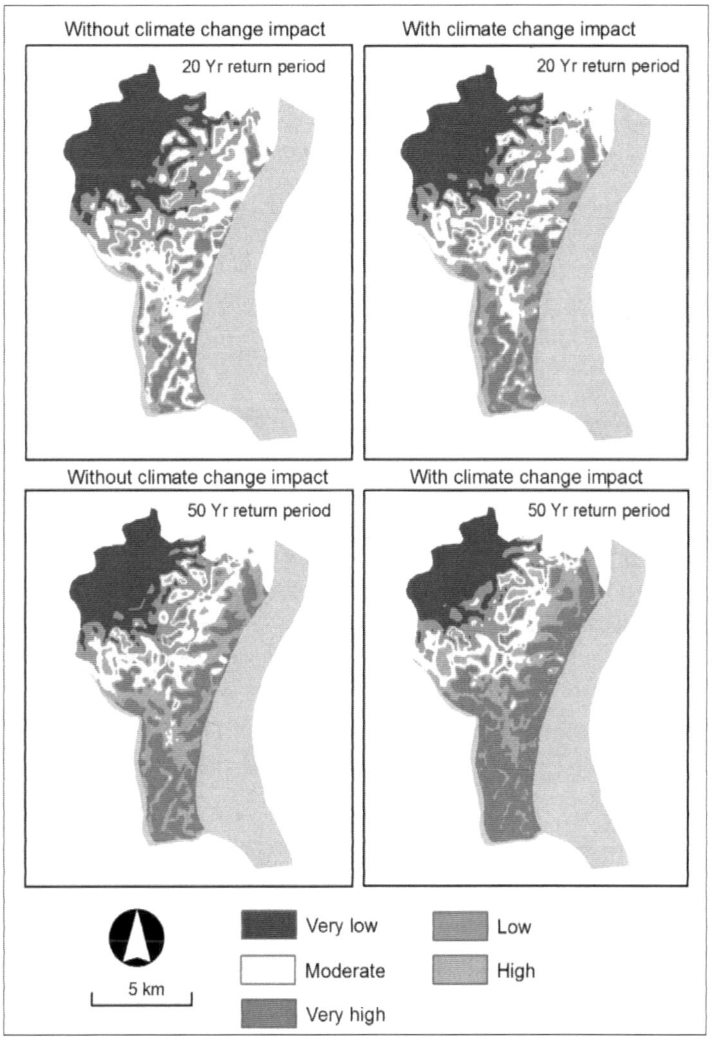

Figure 5. Hazard maps at 0.34 m sea level rise scenario under 20 and 50 year return periods for the year of 2050

5. Conclusions

The study demonstrates the capability of geospatial approach for modelling tropical cyclone hazards by integrating local knowledge with other environmental data representing conditions under climate change scenario at the local scale.

Acknowledgment

This paper is a part of Hoque's PhD project on cyclone disaster mapping, monitoring and management using satellite remote sensing and spatial analysis. The authors would like to thank the University of Queensland (UQ) for providing financial supports, Remote Sensing Research Centre at UQ for providing research facilities, and School of GPEM UQ for funding in fieldwork.

References

[1] Li K and Li GS 2013 *Risk assessment on storm surges in the coastal area of Guangdong Province* Nat. Hazards 1-11

[2] Krishnamohan K, Mohanakumar K and Joseph P 2014 *Climate change in tropical cyclones and monsoon depressions of North Indian Ocean Monitoring and Prediction of Tropical Cyclones in the Indian Ocean and Climate Change* (Netherlands: Springer) pp 33-39

[3] Yin J, Yin Z and Xu S 2013 *Composite risk assessment of typhoon-induced disaster for China's coastal area* Nat. Hazards 69 pp1423-1434

[4] Knutson TR et al. 2010 *Tropical cyclones and climate change* Nat. Geosci. **3** pp 157-163

[5] Joyce KE, Belliss SE, Samsonov SV, McNeill SJ and Glassey PJ 2009 *A review of the status of satellite remote sensing and image processing techniques for mapping natural hazards and disasters* Prog. Phys. Geogr.

[6] Arthur W, Schofield A, Cechet R, Sanabria L 2008 Return period cyclonic wind hazard in the Australian region *28th AMS Conf. on Hurricanes and Tropical Meteorology*

[7] Middlemann MH and Middelmann M 2007 *Natural hazards in Australia: identifying risk analysis requirements* Geoscience Australia

[8] Puotinen M 2007 *Modelling the risk of cyclone wave damage to coral reefs using GIS: a case study of the Great Barrier Reef, 1969–2003* Int. J. Geogr. Inf. Sci. 21 pp 97-120

[9] Rana M, Gunasekara K, Hazarika M, Samarakoon L and Siddiquee M 2010 *Application of remote sensing and GIS for cyclone disaster mangement in coastal area: a case study at barguna district, Bangladesh* Int. Archives of the Photogrammetry, Remote Sensing and Spatial Information Science vol XXXVIII, Part 8 pp122-126

[10] Nadiruzzaman M and Paul BK 2013 *Post-Sidr public housing assistance in Bangladesh: a case study* Environ. Hazards **12** pp 166-179

[11] Lee HS 2013 *Estimation of extreme sea levels along the Bangladesh coast due to storm surge and sea level rise using EEMD and EVA* J. of Geophys. Res- Oceans 118 pp 4273-4285

Modeling of Erosion on Jelateng Watershed Using USLE Method, Associated with an Illegal Mining Activities (PETI)

I N Ananda[1], F V Aswari[1], D A Narmaningrum[1], A S A Nugraha[1], M A A Asidiqi[1], and Y Setiawan[1]

[1] Postgraduate of Remote Sensing Programme, Faculty of Geography, Gadjah Mada University, Indonesia

Email: ananda.irvankpj08@gmail.com, firnavestriaswari@gmail.com, dyah.arshinta.n@gmail.com, sediyoadi07@gmail.com, ugm_abdulaziz@yahoo.com, djoeliandsetiawan@gmail.com

Abstract. The Indonesian archipelago has abundant mineral resources, and it causes many mining activities. Mineral resource is natural based resource which cannot be renewable. An abandon mining pit makes a hole in land surface and it increase the erosion severity level on the rainy season. This erosion would brought sediment to the sea, and it causes damage the ecosystem of the coastal. Erosion modeling in Jelateng watershed performed temporally using remote sensing image data, which consist of LANDSAT-5 (1995), and LANDSAT-8 (2015), and supported by field data as well. The parameters for modeling of erosion through rasterization process as input from erosion USLE models to IDRISI software. The results shown that in 1995, the majority of the area has a low level of erosion. The low erosion rate is less than 183.67 tons/hectare/year and high erosion rate is 408.34 up to 633 tons/hectare/year. Compare with in 2015, erosion models shown that erosion is most prevalent on the upstream area of Jelateng watershed, with low erosion rate is less than 432.2 tons/hectare/year and high erosion rate is 615.64 up to 1448.31 tons/hectare/year.

Keywords: Erosion, Modelling, USLE, Remote Sensing Imagery

1. Introduction

Nature resources exploitation causes excessive damage to nature and it one of the causes of civil conflict. Many mining activities in the Lombok Island found at Sekotong District, West Lombok Regency. From the exploration activities known that the District Sekotong has many mountainous areas which contains minerals, such as gold, copper, and silver. Gold mining found more among other mining activities, since gold mining is the most wanted due to its promising prospect. Over the time, the area that is considered to have the prospect of gold, silver and copper is crowded by gold hunters who most of them are illegal (miners without permission). Illegal gold mining (PETI)'s operation is already extends nearly 1000 hectares into an open, which includes a protected forest area, and also in land owned by citizens. This condition bring in environmental damage to the mining site due to most of illegal miners make holes measuring approximately 1 x 1.5 m2 with a depth of up to 3-7 m to get the minerals, and when it is not productive, then it will be abandoned as it is, without any backfilled (Ngurah Ardha, 2010). As a result of this kind of mining operations, mined land cannot function normally back, the land will be exposed due to dredged, and it will be eroded prone in case of heavy rainfall.

2. Method

This study's objective is to analyze LANDSAT OLI and TM imagery for extracting erosion parameters and modeling erosion rate between 1995 and 2015 years in Jelateng watershed. Methodology used in this study can be seen in Figure.1.

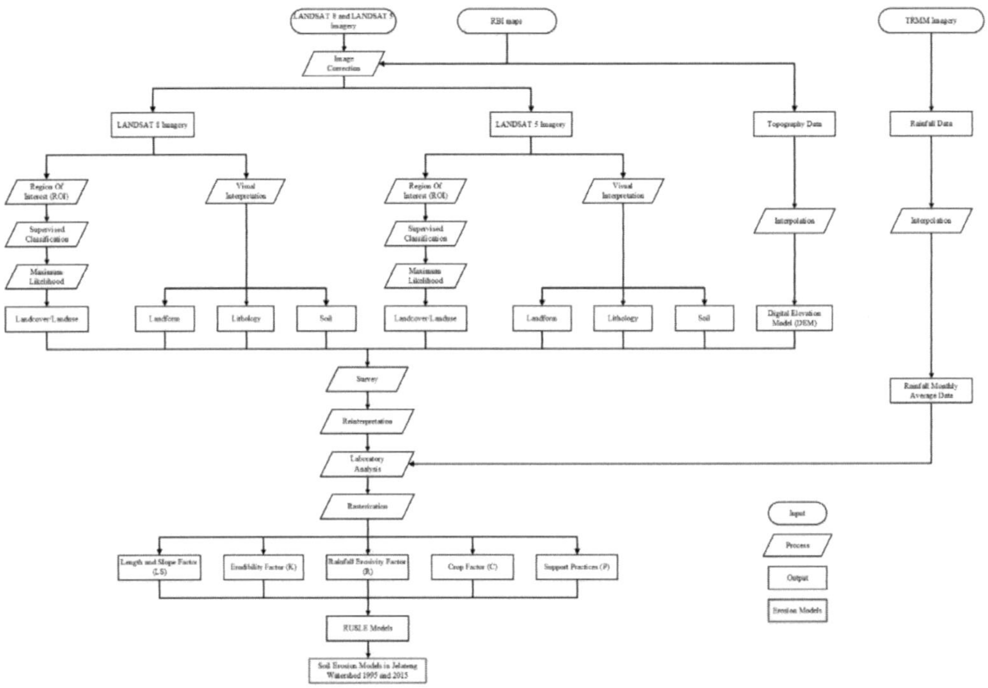

***Figure 1**. Methodology used in this study*

2.1. Erosion Factor

Climate and geology structure are the main factors that influence the process of soil erosion. Besides the characteristics of the soil and vegetation, which both depend on two previous factors and influence each other. Excluding these factors, human activities also gave a significant influence on the change rate of soil erosion (Suripin, 2004). Hardiyatmo (2006), which states that climate change such as hot-cold cycles, wet-dry resulted in the break rocks into smaller particles and weak bond between the particles. In general, the factors that cause soil erosion is consist of:

1. Climate
2. Soil conditions
3. Topography
4. Land cover/land use
5. Soil disturbance due to human activities

2.2. Universal Soil Loss Equation (USLE)

An understanding of the processes of erosion can be obtained more easily by arranging a model. Soil erosion models can be classified into three: a) an empirical model, b) physical models, and c) conceptual model. The empirical model is based on the key variables obtained from research and observation during the process of erosion. One common example of the empirical models used to estimate the magnitude of surface erosion is USLE (Universal Soil Loss Equation), USLE method developed by Wischmeier and Smith (1978) is the most common method used to predict the magnitude of erosion. The term "universal" or "general" this indicates that the equation or the method can be used to predict the magnitude of erosion for a wide variety of land use conditions and different climatic conditions (Asdak, 1995). Universal Soil Loss Equation is used to determine the weight of soil loss due to erosion. According to Smith and Wischmeier (1978) in Hardiyatmo (2006) the amount of the lost soil is influenced by six factors:

1. The length of the slope
2. The Slope itself
3. Land cover/Land use
4. Soil management
5. Kind of soil
6. Rainfall

USLE allows planners predict an average rate of erosion of certain land on a slope with a particular rainfall pattern for any kinds of soil and the implementation of land management (land conversion). USLE is designed to predict the long-term erosion of sheet erosion (sheet erosion) with erosion groove under certain conditions (Suripin, 2004). To determine the weight of the soil on the surface that lost due to rain, which soil at the surface also contains vegetation. Thus it can be generated using USLE equation (Hardiyatmo, 2006). Based on a statistical analysis of the data of more than 10,000 years of erosion and run off, and management of the physical parameters are grouped into five main variables whose values for each point can be expressed numerically (Suripin, 2004). And based on Forestry Ministry, USLE Formula can be expressed as follows:

$$A = R \, x \, K \, x \, LS \, x \, C \, x \, P$$

Where:

A	=	Total soil lost (ton/hectare/year)
R	=	Erosivity annual rainfall average (usually expressed as the energy impact of rainfall (MJ/h) x maximum intensity for 30 minutes (mm/h)
K	=	soil erodibility index (tonnes x ha x jam) divided by (ha x mega joule x mm)
LS	=	Length of slope and slope angle index
C	=	Plant management index
P	=	Soil conservation effort index

3. Results and Discussion

3.1. Rainfall Erosivity Factor (R)

Rainfall data used from rainfall monthly average data. Rainfall data used has a range of between 5.482 cm to 6.128 cm. Rainfall can be quite low because basically Lombok is a dry area. Although it does not dry as North Lombok which is a shadow-rain area, Jelateng watershed does not have high rainfall variation because only a slight difference of precipitation that falls within Jelateng watershed. Rainfall which comes into the Jelateng watershed can be said spreading evenly. From the calculation rainfall erosivity, showed that the rainfall erosivity index in Jelateng watershed has a value of 22.3537 to 26.0085 cm (Figure.2 and Figure.3), that values are small because erosivity value will directly proportional with rainfall in that area.

***Figure 2**. Erosivity factor of Jelateng watershed in 1995*

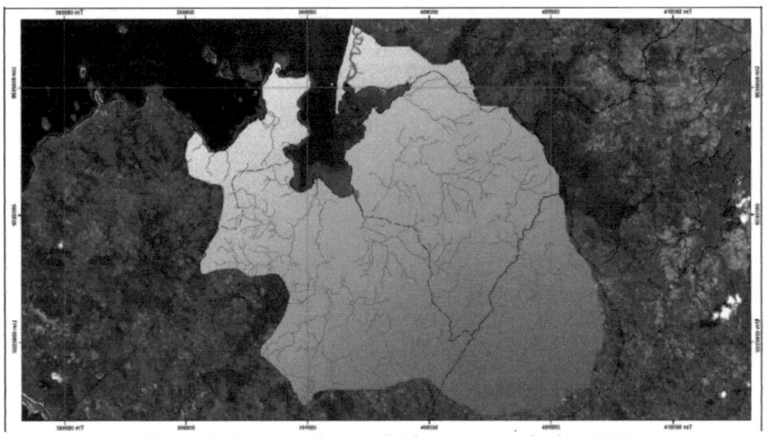

***Figure 3**. Erosivity factor of Jelateng watershed in 2015*

3.2. Erodibility Factor (K)

Erodibility calculations performed highest value obtained on the 6th sample points (Table 1), where the value of erodibility (K) is 0.312. The high erodibility scale is because the soil has low organic material (2.08%). Organic material act as preventing from soil erosion because the organic material acts as a binding agent of soil particles, thus preventing release of particles that cause erosion, besides organic materials also acts to bind water so as to reduce the rate of water can cause erosion. Land at 6th point also has a permeability class incompressible, meaning that the ground is very hard to pass the surface water into the soil so that water will have a strong runoff and cause erosion of the soil surface by the flow of water so that the erosion will be higher. The lowest erodibility occurred in 10th soil sample which has erodibility value is 0.086. Low erodibility value means the land at that point has a high resistance to erosion. Although it has a low organic matter (2.48%) but the ground has a very quick permeability (41.01 cm / h) so that the water infiltrated into the soil so that it will reduce the runoff that will cause erosion. Soil erodibility also influenced by various land use in the area, land use will result in compression of the soil.

Table 1. Erodibility calculations from laboratory analysis

Sample	Organic Material %	Permeabilitas (cm/hour) Value	Class	Lp %	Db %	Ps %	Texture
1.	2.85	1.94	Low	15.09	25.84	59.07	Sandy loam
2.	2.31	2.11	Moderate	35.15	23.33	41.52	Silty loam
3.	2.41	24.74	High	13.79	20.16	66.05	Sandy loam
4.	5.66		-	19.89	23.33	56.77	Sandy loam
5.	3.7	0.63	Low	25.31	27.23	47.47	Sandy loam
6.	2.08		High	12.37	29.6	58.03	Sandy loam
7.	2.8	4.64	Moderate	14.13	19.07	66.8	Sandy loam
8.	1.26	0.14	Low	7.95	22.59	69.47	Sandy loam
9.	6.44	5.41	Moderate	15.12	19.81	65.07	Sandy loam
10.	2.48	41.01	Very High	11.73	22.11	66.16	Sandy loam

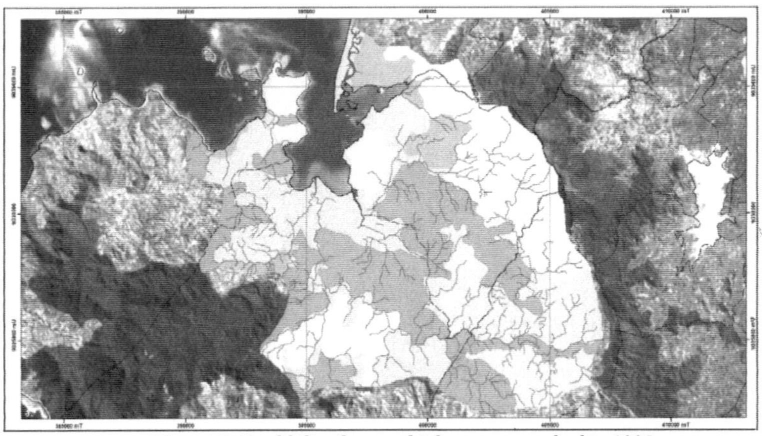

Figure 4. *Erodibility factor of Jelateng watershed in 1995*

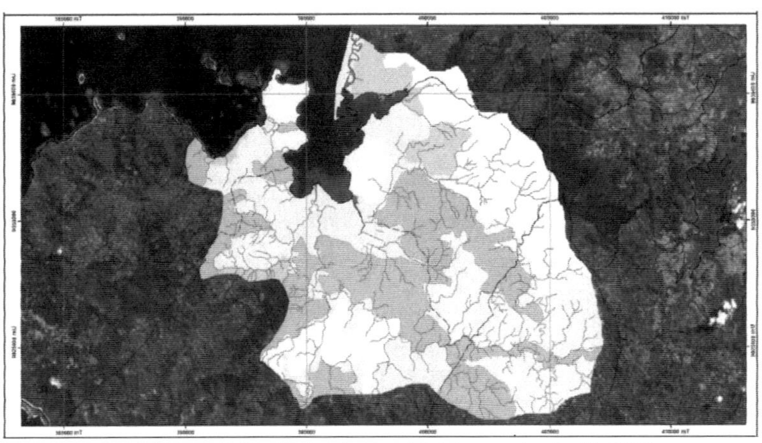

Figure 5. *Erodibility factor of Jelateng watershed in 2015*

3.3. Length and Slope Factor (LS)

Length and slope factor (LS) measurements obtained by the shooting range and slope using Abney level. The field data of 11 samples showed variations in slope, so as to represent the topographical conditions in the Jelateng watershed. Third sample point with slope of 68° and with a short slope length is approximately 8 meters. Lowest degree slope in the sample 2.2 and 3.2 which is 38° with a slope length of approximately 6 meters. Topography-related erosion is not very influential, although the slope on a sample of 2.2, 3 and 3.2 relatively steep to very steep but long slope on the short sample. Unlike the samples of 2.1, 2.2, 2.3, and 2.4 which has a relatively steep slope up to very steep slopes with a length of 15-30 meters. The results of this sample measurement can indirectly illustrate the erosion rate is high as a result of the steep slope and slope length long power lead scraper that water has more power to transport material and runoff will be more. Length and slope factor can provide conditions in the field due to these factors affect the erosion.

Table 2. Slope length and stepness factor

Sample	Slope (Degree)	Slope (Percent)	Slope Length	Factors (L)	Factors (LS)
1	40	83.85	4.25	0.44	13.09
2	54	137.50	6.5	0.54	25.11
2.1	52	127.87	20	0.95	41.88
2.2	38	78.07	6	0.52	14.34
2.2	65	214.13	15	0.82	47.45
2.3	55	142.67	20	0.95	45.11
2.4	60	172.99	30	1.17	61.48
3	68	247.08	8	0.60	36.21
3.1	42	89.97	8	0.60	19.39
3.2	38	78.07	6	0.52	14.34
3.3	50	119.07	5	0.48	19.84

Measurement slope and length data then becomes a factor used to calculate the LS factor. LS factor calculation results show the influence of the length and slope erosion modeling related. The higher value of the length and slope the erosion will be higher it shown with LS factor is also high. Table calculation can be seen in Table 2. LS factor from the results of field measurements indicate the range of values from 13.09 to 61.48.

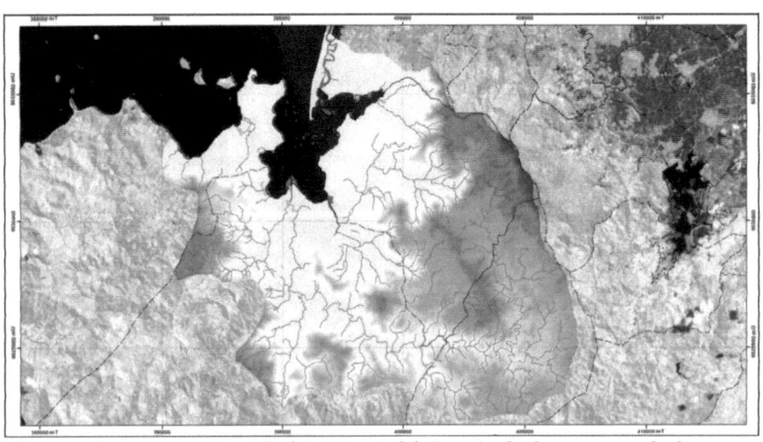

Figure 6*. Digital Elevation Model (DEM) of Jelateng watershed*

3.4. Crop Factor (C)

Factor value of plants used in this study are derived based on land use and direct observations in the field. This is done so that the value obtained is actual or current conditions at the site, so the expected value of the estimate of erosion have a high level of accuracy.

***Figure** 7. Crop factor of Jelateng watershed in 1995*

The parameter value crop factor of 1995 in the Jelateng watershed have a range between 0 to 0.45. When viewed as a whole, the dominant factor is the factor of high density vegetation classes and medium with value 0.01. Parameter value crop factor of 2015 in the Jelateng watershed have a range between 0-0.45. When viewed as a whole, the dominant factor is the factor of medium density vegetation classes (0.01) and low density vegetation classes (0.05). If you look at these two years can be said to occur significant changes in land cover between 1995 and 2015 can be seen with decreasing vegetation density and rising bare land.

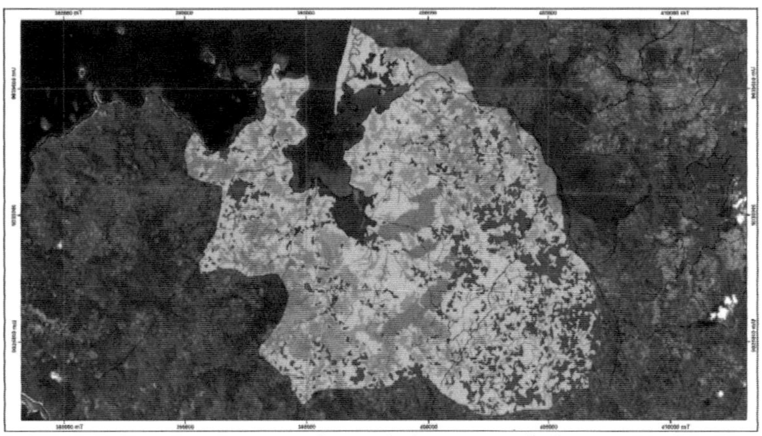

***Figure** 8. Crop factor of Jelateng watershed in 2015*

3.5. Support Practices (P)

Conservation factor can be obtained from land cover/land use map. Ground checking must be done before to perform the accuracy of the land cover/land use map and to validate the conservation and soil management factor. Interview method is required to determine the conservation technique that has been used in 1995 and 2015 are the same or equal.

***Figure 9**. Support practices factor of Jelateng watershed in 1995*

Based on this, it can generate three classes value of conservation and soil management factor that been used in 1995 and 2015. The most conservation factors that cause erosion is Condensed and Slippery which has a value of 1.3. Most of this value found at the upstream of Jelateng watershed itself. While conservation which widely used are pedestrian lane towards the top and bottom of the slopes that spread almost throughout Jelateng watershed. This can be seen in Figure 9 and Figure 10.

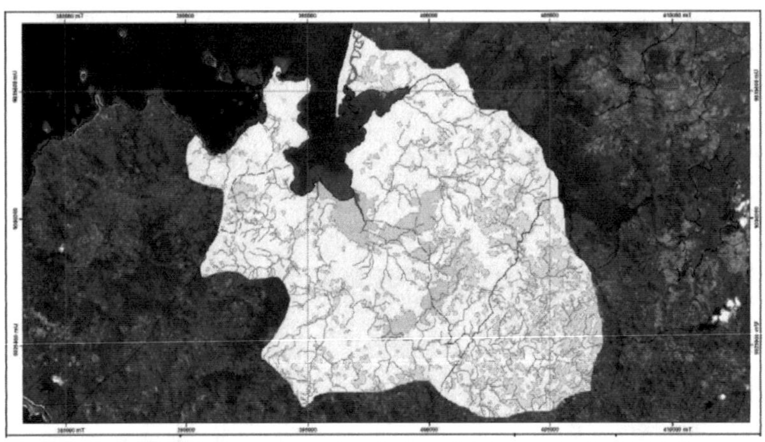

***Figure 10**. Support practices factor of Jelateng watershed in 2015*

3.6. Modelling of Erosion

Erosion modelling carried out after the parameters in the USLE equation predetermined their value, then the erosion rate in Jelateng watershed can be estimated by multiplying the factors of erosion through an empirical equation. There are two modeling is modeling erosion obtained for 1995 and 2015.

The first erosion model is modeling the erosion of 1995. It can be seen in Figure.11 that erosion many occur on the upstream Jelateng watershed. While on the upper and the middle Jelateng watershed are the locations that many established illegal mining (PETI). On the upper and the middle Jelateng watershed can be found their ground grinding process due to the low vegetation cover. The majority Jelateng watershed is a region with a low erosion rate. The low erosion rate less than 183.67 tons/hectare/year and high erosion rate between 408.34 to 633 tons/hectare/year.

The second erosion modelling is modelling the erosion of 2015 (Figure.12) which the erosion much occur on the upstream of Jelateng watershed. Meanwhile on the middle of the Jelateng watershed is the site of the former established illegal mining (PETI). Majority erosion is the erosion with lower classification. The low erosion rate of less than 432.2 tons/hectare/year and highest erosion rate from

615.64 to 1448.31 tons/hectare/year. The erosion rate with a lower classification caused by the activity of the local community to dredge the parent material for use as a building material and their mining activities.

Comparison between modeling erosion in 1995 and 2015 had a significant difference. In 1995 the regions with high erosion class is still quite a lot on the upstream side and middle side which a former illegal miners (PETI) locations. We can see on the figure that dark brown symbol which represent the medium class is still widely spread in many parts of Jelateng watershed. In contrast to modeling erosion 2015. The symbol can be seen predominantly light brown to dark brown. High erosion rate only spread in certain spots. In other words, 2015 is sufficient erosion control.

Differences in erosion between 1995 and 2015 are quite significant. This is because of differences in land cover/land use between these two years. In 1995 where dominanted by open land with moderate erosion rates. Land cover/land use of Jelateng watershed in 2015 detected as low density vegetation. This indicated progress in the conservation community from 1995 to 2015. It shows that a large enough public awareness to conserve land cover/land use where formerly open land, then began to be used to reduce the rate of soil erosion.

***Figure 11**. Soil erosion models in Jelateng watershed 1995*

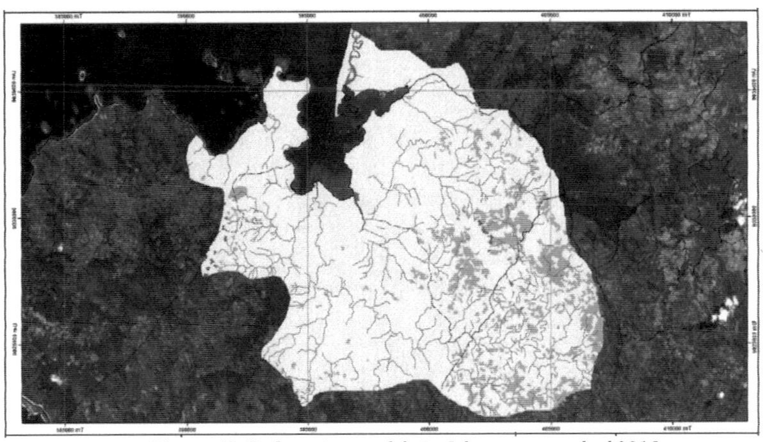

***Figure 12**. Soil erosion models in Jelateng watershed 2015*

4. Conclusion

Based on research that has been conducted and the results have been obtained, (1) LANDSAT-8 and LANDSAT-5 imagery were able to provide information related to modeling parameters erosion. The parameters can be extracted ie landforms, rocks, soil, and land cover/land use where the parameter is used to create land units maps. In modeling erosion in 1995 and 2015, information which has obtained from remote sensing imagery combined with field data and collaborated with laboratory analysis. (2) Results of this study shown that USLE model in IDRISI software were able to modeling soil erosion intensity at Jelateng watershed.

References

[1] Ardha N 2010 Memburu Emas di Sekotong Lombok Barat http://www.esdm.go.id/berita/artikel/56-artikel/4012--memburu-emas-di-sekotong-lombok-barat.html.
[2] Arsyad S 2000 Konservasi Tanah dan Air (Bogor : IPB Press)
[3] Asdak C 1995 Hidrologi dan Pengelolaan Daerah Aliran Sungai (Yogyakarta : Gadjah Mada University Press)
[4] Departemen Kehutanan 1998 Pedoman Penyusunan Rencana Teknik Rehabilitasi Teknik Lapangan dan Konservasi Tanah Daerah Aliran Sungai (Jakarta : Departemen Kehutanan)
[5] Hardiyatmo 2006 Penanganan Tanah Longsor dan Erosi (Yogyakarta : Gadjah Mada University Press)
[6] Suripin 2004 Pengembangan Sistem Drainase Yang Berkelanjutan (Yogyakarta : Andi Offset)

A comparison of lineament and fracture trace extraction from LANDSAT ETM+ panchromatic band and panchromatic aerial photograph in Gunungsewu karst area, Java-Indonesia

E Haryono[1], B S Widartono[1], H Lukito[2], and S B Kusumayuda[2]

[1]Faculty of Geography, The University of Gadjah Mada, Yogykarta, Indonesia
[2]Faculty of Mineral Technology, UPN Veteran University, Yogykarta, Indonesia

Email: e.haryono@geo.ugm.ac.id

Abstract. This paper aims at exploring interpretability of the panchromatic band of Landsat Enhanced Thematic Mapper Plus (ETM+) compared to the panchromatic aerial photograph in lineament and fracture trace extraction. Special interest is addressed to karst terrain where lineaments and fracture traces are expressed by aligned valleys and closed depressions. The study area is an single aerial photographic coverage in the Gunungsewu Karst, Java-Indonesia which is characterized by a lineament-controlled cone karst and labyrinth-cone karst. The result shows that the recording time of the Landsat ETM+ with respect to the shadow resulting from the sun illumination angle is the key factor in the performance of lineament and fracture traces extraction. Directional filtering and slicing techniques significantly enhance the lineament interpretability of the panchromatic band of Landsat ETM+. The two methods result in more lineaments and fracture traces which T-test affirm in 0.001 and 0.004 significant levels. Length based lineament analysis attains a better result compared to frequency based analysis.

1. Introduction

Aplication of remote sensing on lineament mapping has been the subject of study since the early development of remote sensing techniques. Lineament owes its important from the fact that lineaments are associated with structural. These are mostly the trace of discontinuities, such as bedding planes, faults, joints fractures with the ground surface [33]. O'Leary *et al.*, [16] defined a lineament as a simple or composite linear feature of surface whose parts are aligned in a rectilinear or slightly curvilinear relationship and which differ from the pattern of adjacent features and presumably reflects some sub-surface phenomenon. Parizek [18], considered a lineament is a 1.5 km length or more linear feature and a fracture trace is a linear features which is less than 1.5 km long.

Remote sensing techniques have more interest in lineament mapping due to their ability to provide more factual earth surface information, as compared to those of topographical maps. Lineament and fracture traces are readily interpreted and analyzed through pattern recognition in today broad coverage of remote sensing images. Moreover, satellite images with the availability of multispectral information and image enhancement techniques provide the opportunity to analyze relatively more reliable lineament and fracture traces. The wide ground coverage and better resolution of satellite images also enable regional as well as local lineament analysis [1, 14, 15, 19, 29, 30, 32]. Therefore interpretation of lineament and fracture traces from remote sensing images is much easier than that from topographical maps or even from field investigation.

Field investigation prepared by conventional field mapping techniques cannot identify the entire extent of lineaments existing in an area. On the contrary, remote sensing images enable to detect large lineaments which extent far beyond the field surveyor observation. The potential application related to lineament and fracture traces are mostly for groundwater exploration [4, 5, 13, 24].

Analyzing lineaments and fracture traces in karst regions has also been an interesting subject of study. The use of remote sensing in lineament analysis of karst areas provides meaningful information since karstification is often controlled by preexisting structures, such as joints, fractures, and faults. Those structures often control the doline or dry valley development leading to an alignement of dolines or dry valleys [2, 7, 13]. Lineaments and fracture traces are associated with underground river passages [4, 8, 9, 13, 18]. Lineaments and fracture traces have also meaningful information for cave development [10].

Lineament and fracture traces identification from remote sensing data can be achieved either using manual [2, 4, 7] or automated techniques [1, 15, 19, 31]. These two techniques have their own advantages and disadvantages. Lineament and fracture trace interpretation from aerial photograph or non-enhanced satellite images produce inconsistent results. Variations in lineament patterns can occur between interpretations of the same scene by one observer on multiple occasions as well as between interpretations of the scene by several observers. There is also the problem of assessing the similarity or the differences between interpretations from two different scenes having similar geological patterns [11]. Automated interpretation on the other hand provides more objective lineament features but is not able to differentiate geological lineaments from non-geological lineaments such as roads, land use borders or electricity lines.

This paper is aimed at exploring lineament interpretability of the panchromatic band of Landsat Enhanced Thematic Mapper Plus (ETM+) compared to the panchromatic aerial photographs.. Geologically, the study area is dominated by Miocene coral limestone of Wonosari Formation (Figure 1). The study area is a single aerial photographic coverage in the Gunungsewu Karst, Java-Indonesia which is characterized by a lineament-controlled cone karst and labyrinth-cone karst (Figure 2). The Wonosari Formation was uplifted during the late Pliocene and/or early Pleistocene and dips gently southwards at about 2° gradient. Dominant land covers of the area are dry fields and bare lands. Dry fields occupy the bottom of aligned dry valleys whereas bare lands with scattered perennial crops dominate the karst hill land covers. No land cover/land use changes have been reported taking place for the last four decades.

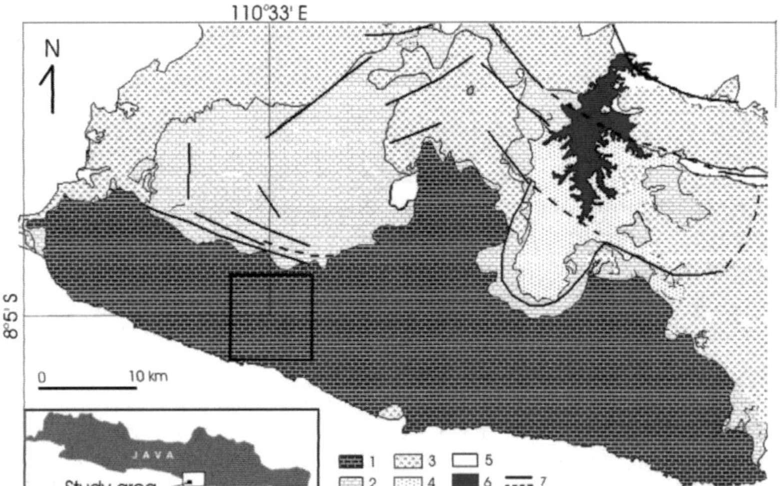

Figure 1. The study area and its geological setting : 1) Gunungsewu Karst 2) Non-karstified carbonate rocks 3) Tertiary volcanic clastic rocks, 4) Pleistocene alluvium, 5) Quaternary pyroclastics and alluvium, 6) Reservoir, 7) Fault and flexures [Compiled from 21, 32]

Figure 2. Photograph depicting part of Gunungsewu Karst which is characterized by conical hills and aligned dry valleys

2. Materials and Methods

The images used in this paper are panchromatic aerial photographs (1:50 000, September 1993) and Landsat ETM+ (row 066 path 119, June 2000). The image processing in the present paper was conducted using ILWIS Version 3.1, whereas lineament and statistical analysis were conducted using Rockworks99 and SPSS 10 respectively Before being processed, the aerial photograph was scanned in 200 dpi resolution. Such resolution is quite reasonable in this research since the original aerial photograph is 1:50,000 scale and the scanned image was interpreted on-screen at 1:25,000 scale. The scanned version was then geometrically corrected with polynomial methods. Seven Ground Control Points (GCPs) were taken in geometric correction resulting in 4.475 RMS errors. The geometric correction was also applied over the Landsat ETM+ raw images with 5 control points resulting in RMS error of 0.591. The selected Landsat ETM+ panchromatic band (band 8) was then stretched using histogram equalization techniques.

Two procedures were employed during interpretation. Firstly, interpretation was conducted by a group of 16 fourth year students from the Physical Geography Department. The students were divided into two groups where the first group interpreted the aerial photograph and the second one was asked to interpret the panchromatic band of Landsat ETM+. Time allocation for interpretation was 15 minutes. This procedure is intended to get a first impression of the aerial photograph and the original panchromatic band of Landsat ETM+ performances in depicting lineaments and fracture traces. Lineament interpretation was carried out on the printed aerial photograph and Landsat image on which the geometric correction and histogram equalization stretch technique have been applied. The printer used in this research is a 1200 dpi resolution printer. Both images were printed in the scale of 1: 50 000.

Secondly, interpretation was conducted by the first author through on screen digitizing over scanned aerial photographs and the pre-enhanced panchromatic band of Landsat ETM+. In this procedure the panchromatic band of Landsat ETM+ was first filtered to enhance the lineament features by applying Sobel kernels convolution in four principal directions (Table 1). These four spatial domain filters have been found very helpful in enhancing lineament features [29]. The filtered images were then visualized in black and white presentation through slicing procedure in order to depict extreme contrast visualization between shaded aligned valleys and illuminated aligned karst hills.

Table 1. Sobel kernel in four principle directions

N-S	NE-SW	E-W	NW-SE
-1 0 1	-2 -1 0	-1 -2 -1	0 1 2
-2 0 2	-1 0 1	0 0 0	-1 0 1
-1 0 1	0 1 2	1 2 1	-2 -1 0

In order to get more objective and consistent interpretation the author took into account some rules as follows:

a) the images being interpreted were zoomed in to 1 : 25 000 scale,
b) lineament delineation in the aerial photograph was performed along the bottom of an aligned valley (see Figure 3a),
c) the features considered as lineaments in the sliced-filtered images are continuous aligned black pixels without any interrupted white pixels (see Figure 3b),
d) lineaments which were less than 100 meters long were not included.

To avoid redundant lineaments, on screen interpretation of the four sliced-filtered images was performed consecutively by putting the image as background one by one starting from Sobel-NWSE and continued by Sobel-NS, Sobel-EW, and Sobel-SENW. If lineaments had been delineated from a certain filtered image, the same lineaments were not delineated again from the other filtered images where delineation only focuses on additional lineaments which were not depicted in the previously interpreted image. Thus, the lineament map derived from four filtered images was overlaid into a single lineament map.

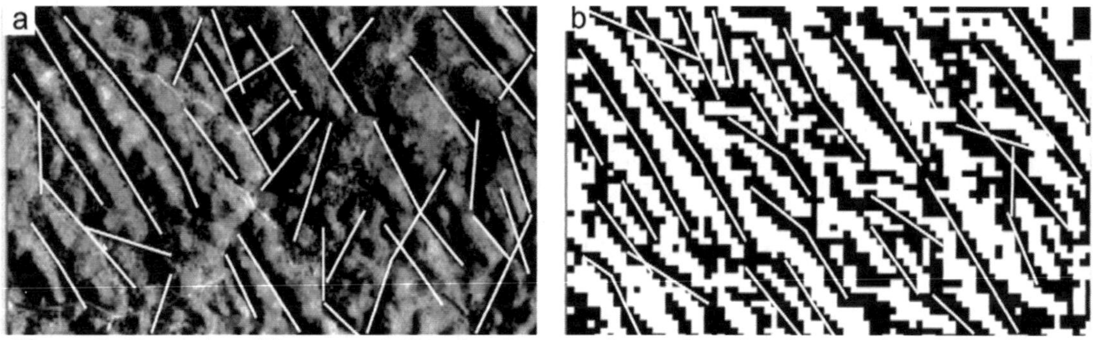

Figure 3. Examples of lineament delineation from panchromatic aerial photograph and slice-filtered image

3. Results and Discussion

3.1. Result

Sample areas derived from the image processing are presented in Figure 4. With a quick look at the images, it is immediately clear that the panchromatic aerial photograph depicts better features of lineaments and fracture traces compared to the band-8 of Landsat ETM+ (Figure 4a and 4b). Less detailed morphological features are the reason for less distinct lineaments in the stretched band 8 image compared to the aerial photograph. This is quite reasonable since the band 8 of Landsat ETM+ has coarser resolution. The less striking shadow must also be accounted for the explanation of the less distinctive lineament features of the band 8 in the Landsat ETM+. Observing Figure 4, the lineaments in the SW-NE azimuth depicted in the band 8 are not as clear as that in the aerial photograph.

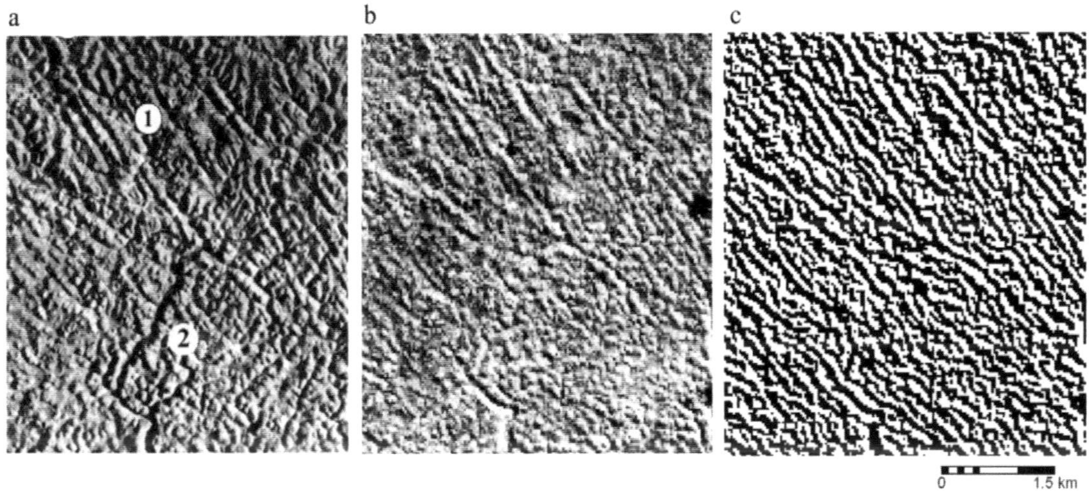

Figure 4. Lineaments depicted in (a) panchromatic aerial photograph, (b) panchromatic band of Landsat ETM+ and (c). Sliced-filtered band 8. Lineaments with SE-NW azimuth are presented clearly in both images (around no 1). Whereas, lineaments with NE-SW azimuth are not depicted well in the panchromatic band of Landsat ETM+. However they are still easily recognized in the aerial photograph (left of symbol 2 in 3a).

A better expression of lineaments and fracture traces in the panchromatic aerial photograph compared to the Landsat ETM+ band 8 is also confirmed by lineaments found by the students (Table 2). The group of seven students who were responsible for the aerial photograph interpretation produced more lineaments both in terms of the number and the length of lineaments and fracture traces. One tail T-test analysis indeed confirms the better result of aerial photographic interpretation in 0.002 significant level. However, the two groups have the same results with respect to dominant lineament azimuth. Variability in the delineation between the two groups as represented by the standard deviation does not indicate much difference. Such figures reveal that inconsistency among individuals is encountered during lineament delineation, in agreement with the findings suggested by Hutington and Raiche [11]. However, deviation in this paper is quite reasonable since the students were provided only approximately 15 minutes to do the interpretation. This short time is provided to make sure that results of interpretation are only derived from their first expression about the lineaments and fracture traces depicted in the images. Additional time and more experienced interpreters will obviously cause less variability.

Table 2. Lineament and fracture trace summary statistics acquired from the panchromatic aerial photograph and the band 8 of Landsat ETM+ (conducted by 16 students)

Type of Image	Numb interpreter)	Number of Lineaments		Length of Lineaments		Dominant Azimuth
		Average	Standard Deviation	Average (m)	Standard deviation	
Panchromatic Aerial Photograph	7	169.4	34	113 203.02	34 583.88	320-330
Band 8 of Landsat ETM+	9	132.6	49.2	98 084.04	21 338.46	320-330

Filtering and slicing techniques, however, result in better lineament and fracture trace appearance of Landsat ETM+ band 8 (Figure 4c). Contrast feature of binary presentation have made possible the lineaments and fracture traces be identified easier compared to that in the original Landsat ETM+ band 8 image as well as in the aerial photograph. This result confirms the previous work by Haryono *et al.*, [9] and Kresic [13], suggesting that slicing is successful in enhancing lineaments of karst terrain. Further and more careful interpretation conducted by the author over the slice-filtered band 8, nevertheless confirmed the above mention finding. Directional filtering and slicing enhancement techniques do increase significantly lineament interpretability of the Landsat ETM+ band 8 image. On screen presentation make the lineaments and fracture traces more obvious. Software facility of on screen zooming is very helpful especially within areas with dense lineaments.

Table 3 and Figure 5 show that more lineaments and fracture traces are extracted from the slice-filtered band 8 than from the panchromatic aerial photograph. This result is also affirmed by T-test statistical analysis upon the lineament length in one-tile significant level of 0.004. This paper shows that the slice-filtered band 8 is very helpful in the karst area where topographic expression of lineaments is not obvious, such as in the upright corner of the area (Figure 5). Another explanation which is accounted for the straightforward delineation from the sliced-filtered image is the contrast expression between aligned valleys (shade areas) and the opposite positive relief.

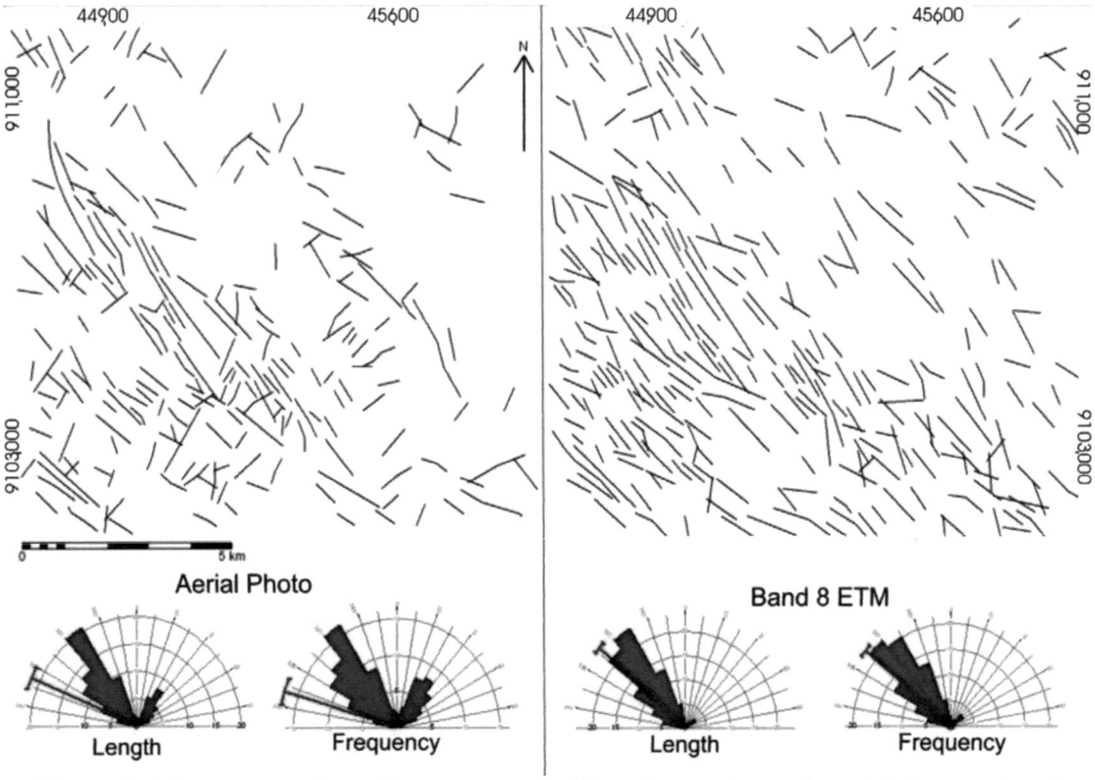

Figure 5. Lineaments and rose diagrams extracted from the panchromatic aerial photograph and Landsat ETM+ band 8 after being subjected to slicing and directional filtering

Table 3. Lineament and fracture trace summary statistics acquired from the panchromatic aerial photograph and the slice-filtered band 8 of Landsat ETM+ (conducted by the first author). Percentage in brackets

Type of Image	By Length (m)	Vector Statistics NS	NE-SW	EW	NW-SE	Mean	Dominant vector
Panchromatic Aerial Photograph	163 757.2 (100)	17 857.1 (10.9)	35 921.8 (21.9)	11 985.1 (7.3)	97 993.3 (59.8)	294	320-330

Band 8 of Landsat ETM+	232 075.6 (100)	20 196.8 (8.7)	16 430.4 (7.1)	22 484.1 (9.7)	174 122.6 (75)	131.7	320-330
	By Number						
Panchromatic Aerial Photograph	276 (100)	37 (13.4)	69 (25)	22 (7.9)	148 (53.6)	287	320-330
Band 8 of Landsat ETM+	346 (100)	32 (9.2)	30 (8,7)	32 (9.2)	252 (72.8)	310	310-320

Scrutiny over vector distribution, lineaments and fracture traces derived from aerial photograph and the slice-filtered image unveils a slightly different result which can be observed from the rose diagrams and from Table 3. The difference which is readily observed is higher frequency of the NE-SW vector in the aerial photograph than that in the slice-filtered band 8. Such result does not necessarily indicate that enhancing Landsat ETM+ band 8 through filtering and enhancing techniques is worse compared to the aerial photograph. The reason for this must be attributed to less shadows in the NE-SW direction in band 8 of Landsat ETM+. As mentioned previously, the original Landsat ETM+ band has intrinsically little shadow in the NE-SW direction. Hence Sobel directional filter applied in the present paper does not help much. Comparison between the two images within the NW-SE vector shows that result from the sliced-filtered image unveils more lineaments and fracture traces.

3.2. Discussion

The interpretability of both Landsat ETM+ band 8 and panchromatic aerial photograph in lineament and fracture trace extractions is governed by the way in which shade is expressed in the images. This is quite reasonable since interpretation of lineaments and fracture traces in Gunungsewu Karst is much dependent upon morphological expression of aligned karst hills cut by aligned dry valleys. As previously reported [20], shade is an important image characteristic in expressing relief. In that regard, recording time of the images that have never been discussed in previous works [1, 4, 5, 9, 13, 14, 17, 29, 32] with respect to the shadow resulting from the sun illumination angle is the key factor in the different interpretability of the images for lineament and fracture trace extractions. This reason is attributed to the lack of lineaments and fracture traces in the NE-SW vector resulting from the Landsat ETM+ Band 8 used in this research. The Landsat ETM+ band 8 was recorded in June where sun illumination was parallel to the aligned karst hills in the NE-SW direction. As a result, less shades are depicted in the Landsat ETM+ band 8. Whereas, the aerial photograph used was taken in September when sun illumination made angles to general lineament and fracture trace orientations of the studied area (Figure 6).

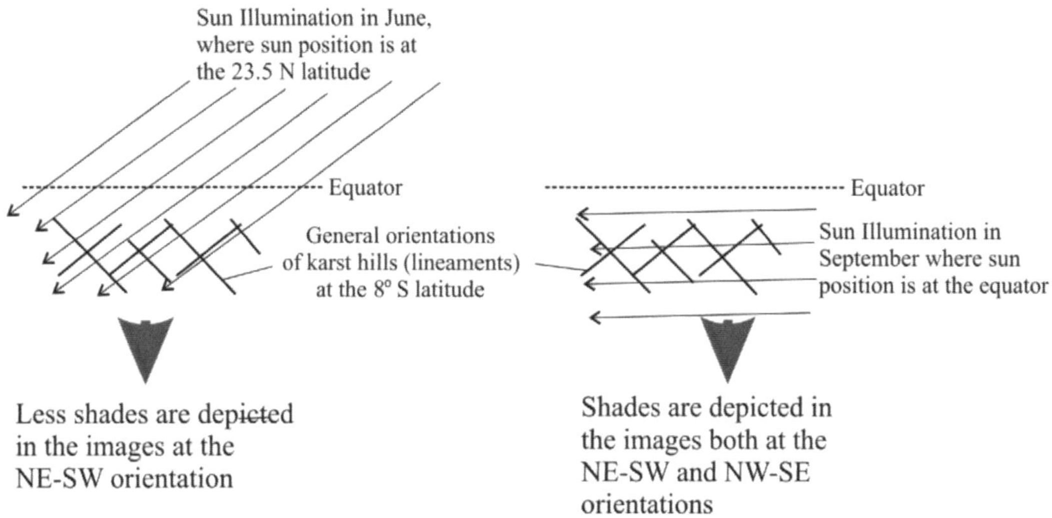

Figure 6. Illustration of sun illumination angles with respect to the shade expressed in the images

Undoubtedly, lineaments and fracture traces are better depicted in panchromatic aerial photographs compared to original Landsat ETM+ panchromatic band scenes, as shown by student results. Better resolution of aerial photographs results in detail morphological features including aligned karst hills and aligned dry valleys. This finding is in agreement with Gunawan [6], Kamel & Elsirafe [12], and Kresic [13]. However, Landsat ETM+ panchromatic band can be applied very favorably in lineament fracture trace extraction after being subjected to filtering as shown by the first author interpretation results. Filtering accompanied black and white presentation is found to be a powerful enhancement technique in lineament and fracture trace extractions. Directional filtering enhancing lineaments and fracture traces in this research echoes the results of Suzen & Toprak [29]. Slicing the filtered image to black and white presentation has made the identification of lineaments much easier compared to that from a stretched image as well as to that from aerial photographs. This result confirms the previous work by Haryono & Day [7], and Kresic [13], suggesting successful lineament enhancement of karst terrain from sliced images.

Fragmented lineaments as suggested by Suzen & Toprak [29], due to filtering are not always the case in the present paper. As long as the imposed filter direction is in line with the dominant lineament azimuth, fragmented lineaments are not likely to occur. Morphological expression of lineaments and fracture traces by shades of aligned karst hills in the study area is presumably being the other explanation of the unfragmented lineaments. However, the present research has shown that fragmented lineaments come into sight when the band 8 is convolved by Sobel kernel with NE-SW, EW and NS principal direction. Though a quantitative comparison is not carried out in this paper, the fragmented lineaments seem to be encountered when the direction of the domain spatial filter is not parallel to the dominant lineament azimuth.

Related to vector distribution analysis, care must be taken when deciding between length and frequency used in the analysis. Results of this paper demonstrate that presentation by length is more reliable than by frequency as can be seen from the dominant vector direction. Since filtering will modify the lineaments and fracture traces dimension, length-based analysis will be more reliable, especially when directional filter application causes lineaments fragmentation as it was encountered by Suzen and Toprak [29]. Let alone when the fragmentation does not arise in all direction of spatial domain filters. If this is the case, analysis based on frequency will lead to misinterpretation over the nature of the lineaments and fracture traces.

Frequency distribution of lineaments and fracture traces derived from aerial photograph confirms previous results [4, 7] suggesting dominant vectors of lineaments and fracture trace in the NW-SE and NE-SW directions. These lineaments and fracture traces are associated with joint systems, the so called Meratus System. These two joint sets are shear fracture resulted from north-south compression associated with tectonic plate convergence of the Hindian-Australian plate and the Eurasian Plate [3, 23, 25, 28, 32). They are said to be the reactivation of Oligo-Miocene deformation involving the underlying rocks. Faults are likely to occur in the area, however, their field evidence is difficult to be examined [22, 25]. These major structures are represented by dry valley systems of the area [4, 22].

4. Conclusions

The interpretability of both Landsat ETM+ band 8 and panchromatic aerial photograph in lineament and fracture trace extractions is governed by the way in which shade is expressed in the images. In that regard, recording time of the images with respect to the sun illumination angle is the key factor in the interpretability of the images for lineament and fracture trace extractions in structurally controlled conical karst terrain. Lineaments and fracture traces are better depicted in panchromatic aerial photograph compared to original Lansat ETM+ panchromatic band. However, Landsat ETM+ panchromatic band can be very favorable in lineament fracture trace extraction after being digitally enhanched. Filtering accompanied black and white presentation is found to be a powerful enhancement technique in lineament and fracture trace extractions. Presentation by length is more reliable than by frequency in the dominant lineaments and fracture trace direction analyses of conical karst terrain.

Acknowledgments

Thanks are due to the Higher Education Directorate of the Indonesian Government, who provided funding through Hibah Bersaing Scheme. We acknowledge the constructive comments on earlier drafts Dr. Paul M. van Dijk.

References

[1] Arellano-Baeza AA, Zverev AT, Malinnikov VA 2006 Study of changes in the lineament structure caused by earthquakes in South America by applying the lineament analysis to the Aster (Terra) satellite data *Advances in Space Research* **37** 690–697

[2] Bakose GYB 1994 Stereo photo interpretation of karst terrain in the Al Hatra area of *Iraq ITC Journal* **2** 139-144

[3] Balazs D 1968 Karst regions in Indonesia Karszt Barlangkutatas Volume V

[4] Brahmantyo B, Puradimaja D J, Bandono SSIA 1998 Interpretasi kelurusan dari citra SPOT dan hubungannya dengan pola pengaliran bawah tanah pada perbukitan Karst Gunungsewu Jawa Tengah bagian selatan *Buletin Geologi* **28** (1) 37-49

[5] Galanos I, Rokos D 2006 A statistical approach in investigating the hydro geological significance of remotely sensed lineaments in the crystalline mountainous terrain of the island of Naxos, Greece *Hydrogeology Journal* **14** 1569–1581

[6] Gunawan T 1997 The contribution of aerial photographs in evaluation of Bribin Catchment area, GunungKidul, Yogyakarta, Indonesia. *The Indonesian Journal of Geography* **29** 159-176

[7] Haryono E, Day M 2004 Landform Differentiation within the Gunungkidul Kegelkarst, Java-Indonesia *Journal of Cave and Karst Studies* **66** (2) 62-68

[8] Haryono E 2000 Some properties of epikarst drainage system in Gunung Kidul Regency Yogyakarta Special Province Indonesia *The Indonesian Journal of Geography* **32**(79-80) 75-86

[9] Haryono E, Nurcahyo AD, Gunawan T, Purwanto TH 2005 Underground river network modeling from lineaments and fracture traces by means of remote sensing and GIS *Proc of International Conference on Water Resources and Environmental Problems in Karst Belgrade and Kotor*

[10] Hung LQ, Dinh NQ, Batelaan O, Tam VT, Lagrou D 2002 Remote sensing and GIS-based analysis of cave development in the Suoimuoi Catchment (Son La - NW Vietnam) *Journal of Cave and Karst Studies* **64**(1), 23-33

[11] Hutington JF, Raiche, AP 1978 Multi-attribute method for comparing geological lineament

interpretations *Remote Sensing of Environment* **7**(2) 145-161

[12] Kamel AF, Elsirafe AM 1994 Delineation and analysis of the surface and subsurface structural lineament patterns in the North Lake Nasser area and its surroundings, Aawan, upper Egypt *International Journal of Remote Sensing* **15** (7) 1471-1493

[13] Kresic N 1995 Remote sensing of tectonic fabric controlling groundwater flow in Dinaric Karst *Remote Sensing of Environment* **53** 85-89

[14] Leech DP, Treloar PJ, Lucas NS, Grocott J 2003 Landsat TM analysis of fracture patterns a case study from the Coastal Cordillera of northern Chili *International Journal of Remote Sensing* **24** (19) 3709-3726

[15] Nyborg M, Berglund J, Triumf CA 2007 Detection of lineaments using airborne laser scanning technology Laxemar-Simpevarp Sweden *Hydrogeology Journal* **15** 29–32

[16] O'Leary DW, Friedman JD, Pohn HA 1976 Lineament, linear lineation some proposed new standards for old terms. *Geological Society of America Bulletin* **87** 1463–1469

[17] Parcharidis I, Psomiadis E, Stamtis G 1998 Using Landsat TM images to study the karstic phenomenon *ITC Journal* **2** 118-123

[18] Parizek RR 1976 On the nature and significance of fracture traces and lineaments in carbonate and other terranes *Proceeding Karst Hydrology and Water Resources Volume I Dubrovnic*

[19] Qari MHT 2010 Lineament extraction from multi-resolution satellite imagery: a pilot study on WadiBani Malik, Jeddah, Kingdom of Saudi Arabia, Arabian *Journal of Geoscience* 2010 DOI 10 1007/s12517-009-0116-3

[20] Raghavan V, Wadatsumi K, Masumoto S 1993 Automatic extraction of lineament information from satellite images using digital elevation data *Earth and Environmental Science* **2**(2) 148-155

[21] Rahardjo W, Sukandarrumidi, Rosidi HMD 1995 Geological Map of The Yogyakarta Sheet Java Geological Research and Development Centre: Bandung

[22] Samodra H 1997 Geological Map Surakarta Quadrangle Geological Directorate Bandung

[23] Samodra H 2006 Pembentukan lembah kering purba kawasan karst Contoh kasus Gunungsewu In Maryanto I Noerdjito M Ubaidillah R Eds Managemen Bioregional Masalah dan Pemecahannya Dilengkapi Kasus Jabotabek Puslit-Biologi LIPI Bogor

[24] Solomon S, Quiel F 2006 Groundwater study using remote sensing and geographic information systems (GIS) in the central highlands of Eritrea *Hydrogeology Journal* **14** 729–741

[25] Surono BT, Sudarno I, Wiryosujono S 1992 Geology of the Surakarta-Giritontro Quadrangles Java Geological Research and Development Centre. Bandung

[26] Šušteršič F 2002 Collapse dolines and deflector faults as indicators of karst flow corridors. *International Journal of Speleology* **31**(1/4) 115-127

[27] Šušteršič F 2006 Relationships between deflector faults collapse doline sand collector channel formation International *Journal of Speleology* **35**(1) 1-12

[28] Sutoyo 1994 Sikuen stratigrafi karbonat Gunungsewu *Proceedings Pertemuan Ilmiah Tahunan IAGI Ke 23 Jakarta*

[29] Suzen ML, Toprak V 1997 Filtering of satellite images in geological lineament analyses an application to a fault zone in central Turkey *International Journal of Remote Sensing* **19** (6) 1101-1114

[30] Tam VT, De Smedt F, Batelaan O, Dassargues A 2004 Study on the relationship between lineaments and borehole specific capacity in a fractured and karstified limestone area in Vietnam *Hydrogeology Journal* **12** 662–673

[31] Tripathi NK, Gokhale KVG 2000 Directional morphological image transforms for lineament extraction from remotely sensed images *International Journal of Remote Sensing* **21** (17) 3281-3292

[32] Van Bemmelen RW 1970 The Geology of Indonesia Volume 1A General Geology Martinus Nijhoff The Hague

[33] Zakir FA, Qary MHT, Mostafa ME 1999 A new optimizing techniques for preparing lineament density maps *International Journal of Remote Sensing* **20** (6) 1073-1085

Determination of The Water Catchment Area in Semarang City Using a Combination of Object Based Image Analysis (OBIA) Classification, InSAR and Geographic Information System (GIS) Methods Based On a High-Resolution SPOT 6 Image and Radar Imagery

Yudo Prasetyo[1] , Setyo Ardi Gunawan[2] and Zia Ul Maksum[3]

[1,2,3] Geodesy Department, Engineering Faculty, Diponegoro University, Jl. Prof. Soedharto, Tembalang, Semarang, 50277, Indonesia

[1] yudo.prasetyo@ft.undip.ac.id;[2] gunawan.sa@outlook.com;[3] ziaulmaksum@gmail.com

Abstract. Semarang is the biggest city in central Java-Indonesia which has a rapid and massive infrastructure development nowadays. In order to control water resources and flood, the local goverment has been built east and west flood canal in Kaligarang and West Semarang River. One of main problem in Semarang city is the lack of fresh water in dry season because ground water is not rechargeable well. Rechargeable groundwater ability depends on underground water recharge rate and catchment area condition. The objective of the study is to determine condition and classification of water catchment area in Semarang city. The catchment area conditions will be determine by five parameters as follows soil type, land use, slope, ground water potential and rainfall intensity. In this study, we use three methods approach to solve the problem which is segmentation classification to acquire land use classification from high resolution imagery using nearest neighborhood algorithm, Interferometric Synthetic Aperture Radar (SAR) to derive DTM from SAR Imagery and multi criteria weighting and spatial analysis using GIS method. There are three types optical image (ALOS PRISM, SPOT-6 and ALOS PALSAR) to calculate water catchment area condition in Semarang city. For final result, this research will divide the water catchment into six criteria as follows good, naturally normal, early critical, a little bit critical, critical and very critical condition. The result shows that water catchment area condition is in an early critical condition around 2607,523 Ha (33,17 %), naturally normal condition around 1507,674 Ha (19,18 %), a little bit critical condition around 1452,931 Ha (18,48 %), good with 1157,04 Ha (14,72 %), critical with 1058,639 Ha (13,47 %) and very critical with 75,0387 Ha (0,95 %). The distribution of water catchment area conditions in West and East Flood Canal have an irreguler pattern. In northern area of watershed consists of begin to critical, naturally normal and good condition. Meanwhile in southern area of watershed consists of a little bit critical, critical and very critical condition.

Keywords : Flood; GIS; InSAR; Segmentation Classification; Water Catchment Area.

1. Introduction

Water is an essential requirement for human survival. The overall amount of water on Earth at least 1.3 million km3 and is considered inexhaustible and can last up to millions of years in the future [14].

Apart from that, the presence of water is determined by the carrying capacity of nature is less to meet human needs. This is due to population growth continues to increase, while the amount of water tend to

IOP Conf. Series: Earth and Environmental Science **47** (2016) 012027 doi:10.1088/1755-1315/47/1/012027

be dynamic so as to increase the consumption of water will cause the water to become rarely. Lack of water affects all aspects, such as environmental, social, economic and politic.With the facts that occurred as above required a move to reduce the impact of risk on a less optimal watershed management. One problem solving that can be used is to determine the appropriate water catchment areas and to know distribution pattern. Methods of remote sensing and geographic information systems can be used as an approach step to determine the water catchment areas Ana distribution patterns. In remote sensing image processing performed by unsupervised classification methods and Object Based Image Analysis (OBIA) to obtain land cover maps of watershed areas (DAS) and extraction of DTM (Digital Terrain Model) of the image to get a digital elevation model and lop map in the watershed areas. While in data processing-based geographic information system using the overlay method and the weighting of the input data to determine the determination of water catchment areas and the spreading pattern.

2. Study area, data and methods

2.1. Description of the study area

Research area lied in the city of Semarang-Central Java Province with the geographical location between 6 ° 50 '- 7 ° 10' south latitude and 109 ° 35 '- 110 ° 50 east longitude with an area of 373.70 km^2 with a boundary to the north is the Java Sea, south boundary is Semarang regency, east boundary is Demak, and west boundary is Kendal.

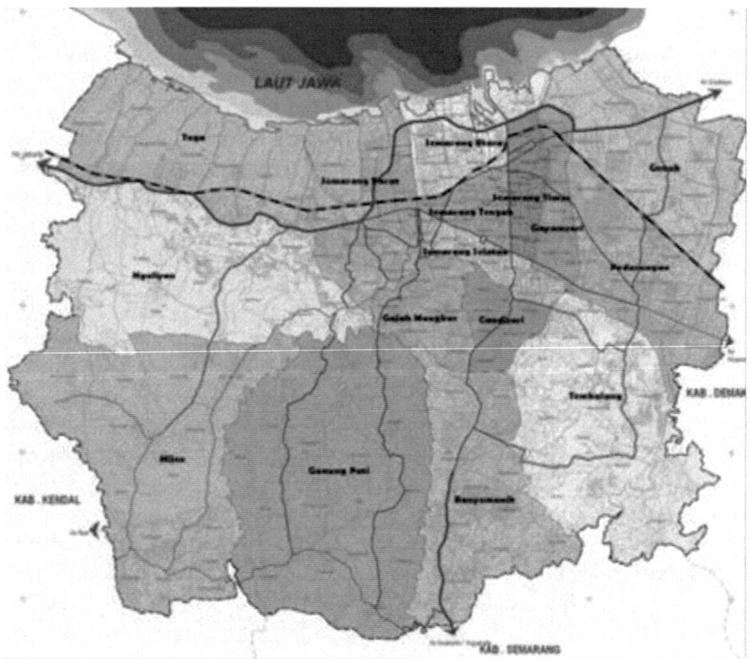

Figure 2.1. Research Area [4]

2.2. Research Data

In this study used several research data more clearly be seen in the Table 2.1.

Table 2.1.Research Data

No.	Data	Data Source	Year
1	ALOS PALSAR Image	LAPAN	2014
2	SPOT 6 Image	LAPAN	2012

3	GPS Survey and Observation Data	Field Surveying	2016
4	Land Use Map	SPOT Image OBIA Classification	2012
5	Slope and topographic Map	DTM extraction (ALOS PALSAR)	2014
6	Soil Type Map	BPDAS Pemali Jratun	2016
7	Underground Water Potential Data	Dinas ESDM	2016
8	Rainfall Intensity Data	Dinas PSDA	2016

2.3. *Object Based Image Analysis (OBIA) Classification*

Understanding of object-oriented classification technique which is essentially classify images based segments of the object segmentation into classes of land cover in accordance with the characteristics of the object [17].

More specific definition of the object based classification (Object-based Classification) or any other name Object Based Image Analysis (OBIA) [12] is the process of determining the object into a classroom where every object is regarded as an individual unit. By comparing objects with one another makes it possible to combine groups of similar objects into classes of concern for users. These classes form a region on the image after image classification that can be identified by color or symbol as seen in Fig 2.2.

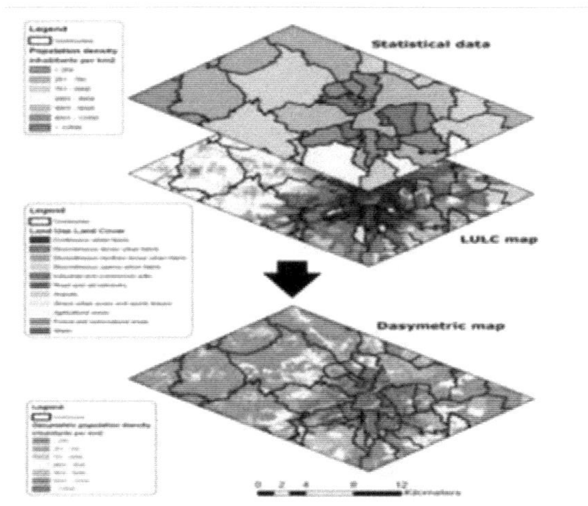

Figure 2.2. Object Based Image Analysis (OBIA) Classification [22]

Image segmentation is an initial step in image classification with object-based method. Image segmentation is used to classify the pixels that have the same structure, with the aim of making each individual structure into regions or individual regions [15]. Segmentation methods are generally categorized into two groups: technical edge-based and region-based. Multiresolution segmentation [3] is one of the region based segmentation algorithm most formidable that has been implemented in commercial software eCognition. In the software eCognition, segmentation is a semi-automated process in which the user can specify certain parameters that affect the size and shape of the segments generated imagery [10] as seen in Fig 2.3.

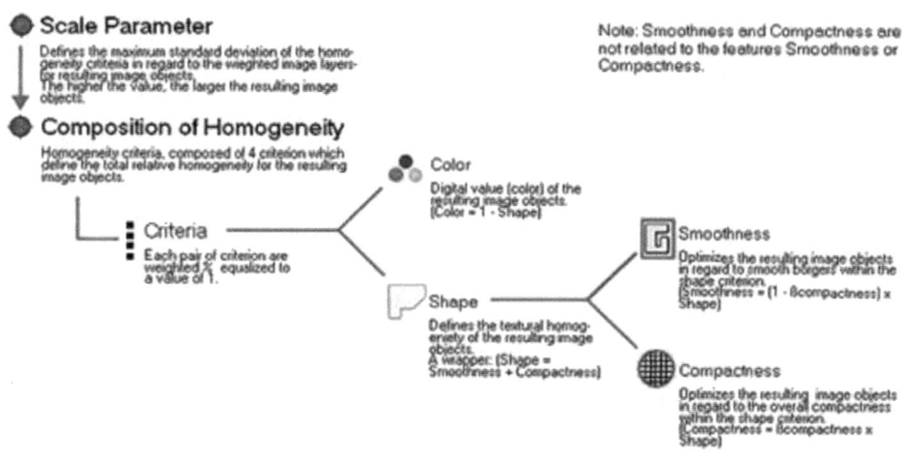

Figure 2.3. Segmentation Parameter Flowchart [6]

2.4. Water Catchment Area Modelling

Value weighting parameter based on the absorption of water in the Ministry of Forestry Regulation Republic of Indonesia No. 32 of 2009 on Procedures for Planning of Forest and Land Rehabilitation Engineering Watershed (RTkRLH-DAS).

The parameters include soil type, land use, slope and rainfall. Value weighting parameter water infiltration and water infiltration conditions classification criteria can be seen in Table 2.1-II.2.

Table 2.2. Water Catchment Value **Weighting**

No.	Parameter	Value Weighting
1	Soil Type	5
2	Land Use	4
3	Slope	3
4	Underground Water Potential	2
5	Rainfall Potential	1

Table 2.3. Water Cactchment Condition Criteria

No.	Total Score	Condition Criteria
1	>50	Good
2	44-50	Naturally Normal
3	38-44	Start Critical
4	33-38	Rather Critical
5	27-33	Critical
6	<27	Very Critical

For the implementation of the model of the catchment is done in accordance with the flow chart of research. In general, this study is divided into three main methods, namely the formation process map INSAR slope method, the model building process within the catchment using methods OBIA and GIS and field validation process. For field validation process is done by taking a sample of 40 points that represent each land cover classes.

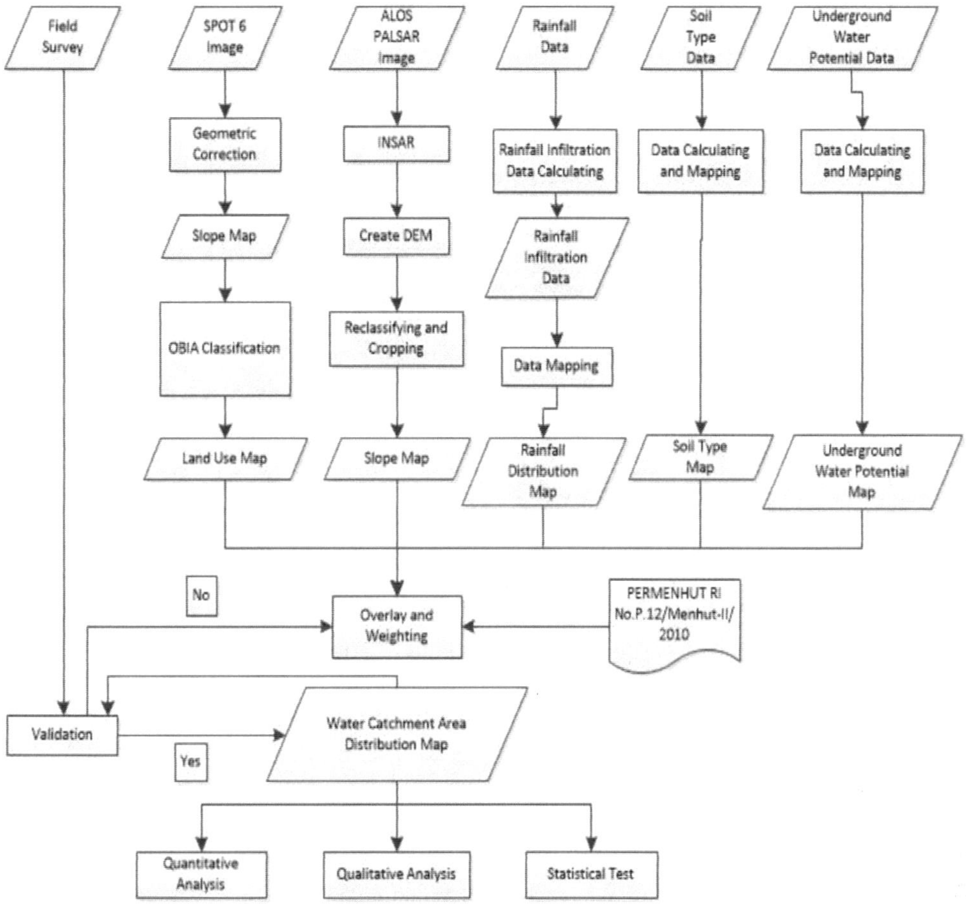

Figure 2.2. Research Methodology Flowchart

3. Results

3.1. Results and Land Use Classification Based OBIA Method

The results of the SPOT image segmentation using a multiresolution segmentation method produces an 1223 object segments for level 3, 5486 object segments for level 2, and 43170 object segments for level 1.

Fig 3.1 shows the segments are objects on SPOT imagery at level 1 is used for classification the image of the object based on the city of Semarang. Land use in the area of Semarang City neighborhood dominated by vegetation such as forests, parks and estates with an area of 17718.587 ha (45.47%), undeveloped land area as residential and industrial area of 14229.481 ha (36.52%), burden land area of 5128.4187 ha (13.16%) and water bodies such as rivers and reservoirs covering an area of 1888.5926 ha (4.85%). It can see in Fig 3.2.

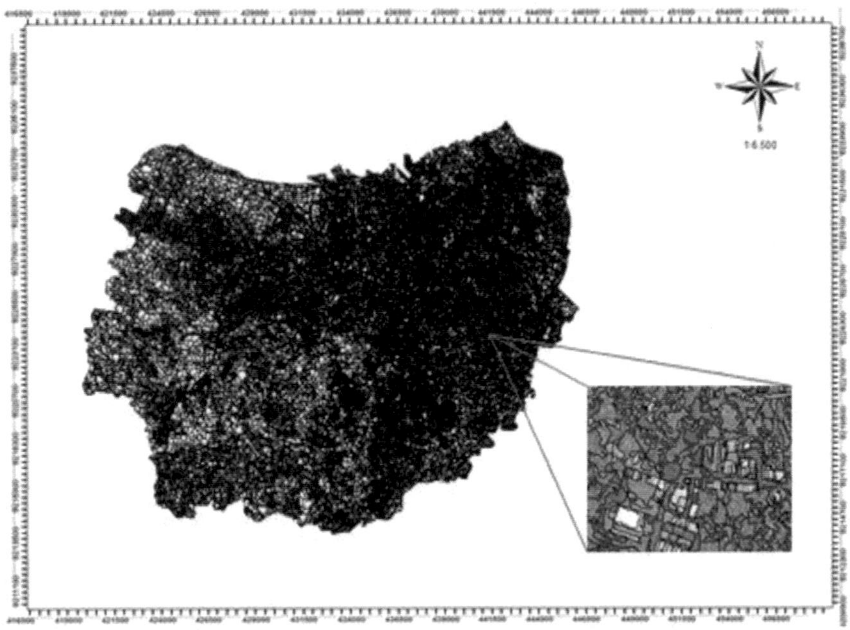

Figure 3.1. SPOT 6 Multiresolution OBIA Segmentation

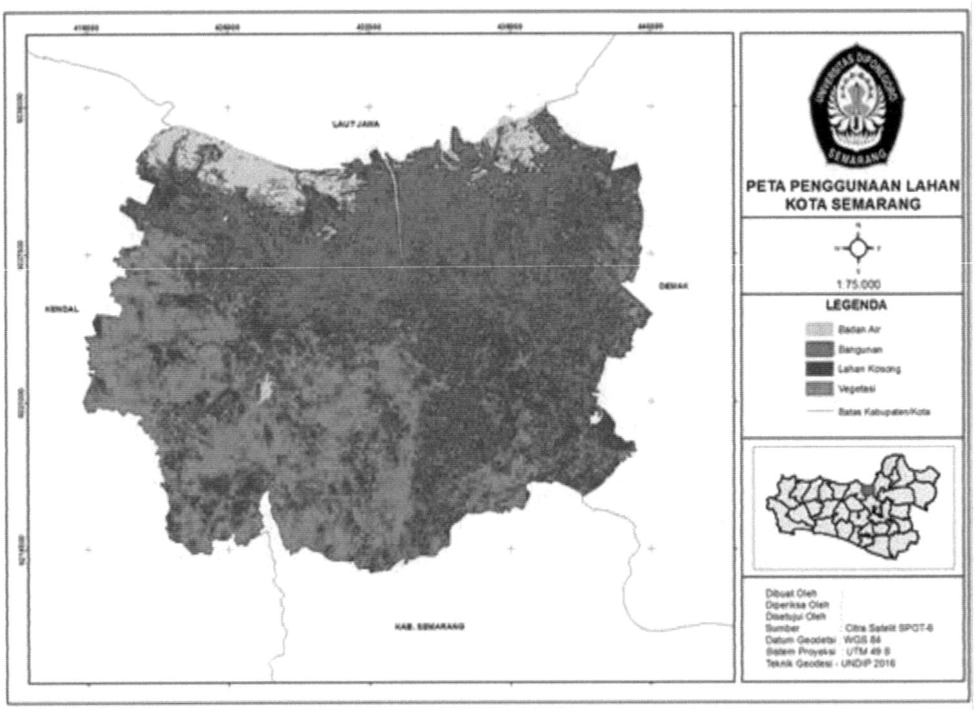

Figure 3.2. Land Use Classification Using OBIA Method in Semarang City

3.2. Soil Type Results and Analysis

Semarang city has largely alluvial soil types covering 13,992.54 ha (35.91%), alluvial soils are young soils derived from precipitation.

It is generally fertile because it contains minerals. Spreading in river valleys and coastal plains that are sensitive to erosion. The next city in the domination of land with an area of 10,880.63 ha mediteran (27.92%), and latosol land with an area of 9647.076 ha (24.75%). Mediteran soil has the same physiographic soil physiographic latosol is to indicate that the volcanic landscape is influenced by volcanic activity. The second composition of this land has undergone further weathering of rocks composed of volcanic breccia, lavas, tuffs which also affect how quickly the water infiltration rate. The las type is an area of land regosol 2619.186 ha (6.72%) and grumusol land with an area of 1825.647 ha (4.68%).

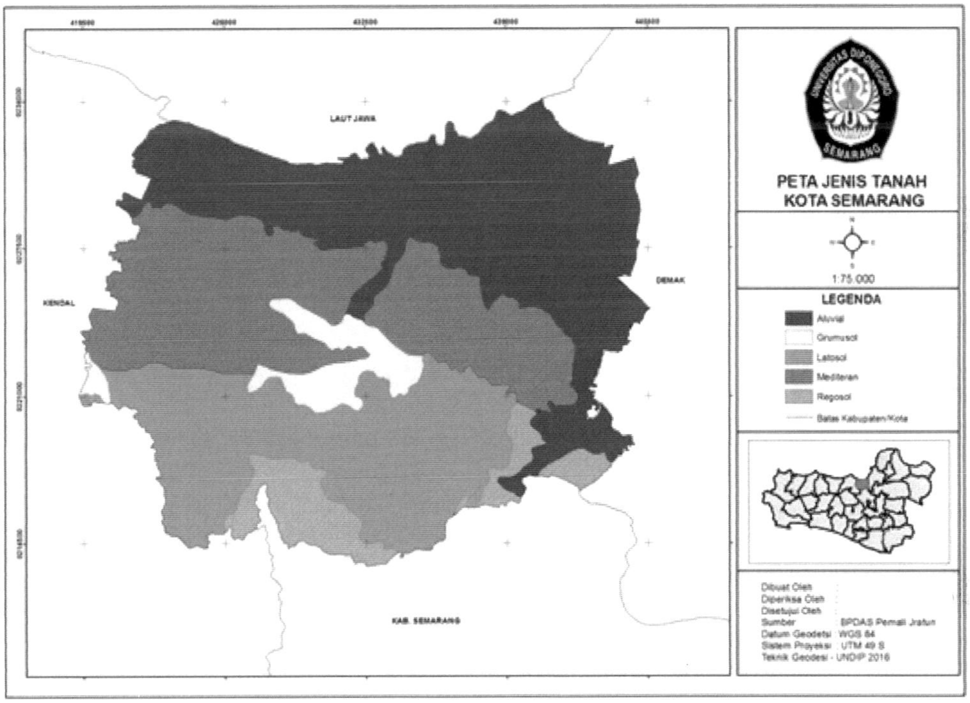

Figure 3.3. Soil Type Map in Semarang City

3.3. Slope Results and Analysis

Topographical conditions in Semarang mostly flat with an area of 28,296.38 ha (72.62%) of the total area of 38,965.079 ha.

The rest is filled by a sloping topography with an area of 6877.422 ha (17.65%), bumpy area of 2811.489 ha (7.21%), rather steep area of 790.243 ha (2.02%) and steep area of 189.546 ha (0.48%). Semarang city northern part has a flat topography. While the central and southern parts in addition to having a flat and hilly topography, as well undulating to steep. Rather steep undulating topography and can be found in the area Candisari, Banyumanik, and Tembalang, while the steep topography are found in Sub Gunungpati and Banyumanik southern part which is a mountainous area. Size slope of each class can be seen in Fig 3.4.

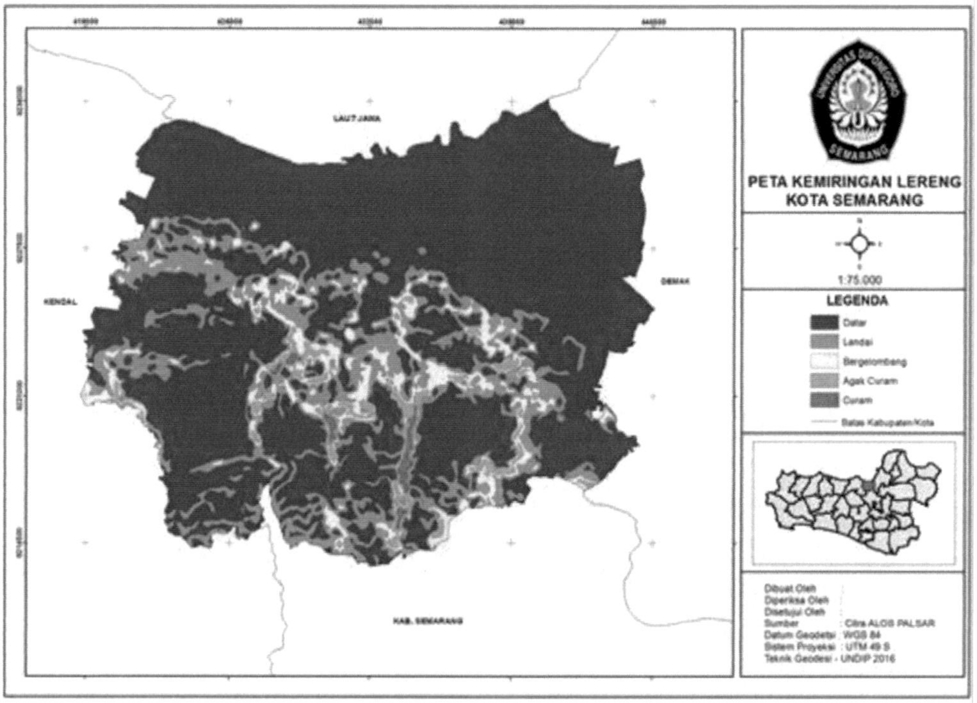

Figure 3.4. Slope Map in Semarang City

3.4. Underground Water Potential Results and Analysis

Whether or not a water catchment area depends on several factors such as soil type, land use, land slope (topography), the potential for ground water and rainfall.

The fifth parameter can not be separated from one another in analyzing a particular area of water infiltration. Based on the fifth parameter calculations results obtained catchment area condition, as shown in Table 3.1 and Figure 3.5, while in each of the districts in the city of Semarang in Table 3.8 and Figure 3.6.

Table 3.1. Water Infiltration Condition Percentage and Area Width

Water Infiltration Status	Width (ha)	Percentage (%)
Good	5633,603	14,46
Naturally Normal	8563,482	21,98
Start Critical	14367,776	36,88
Rather Critical	6852,834	17,59
Critical	3038,975	7,80
Very Critical	499,415	1,28
Total	38956,08	100

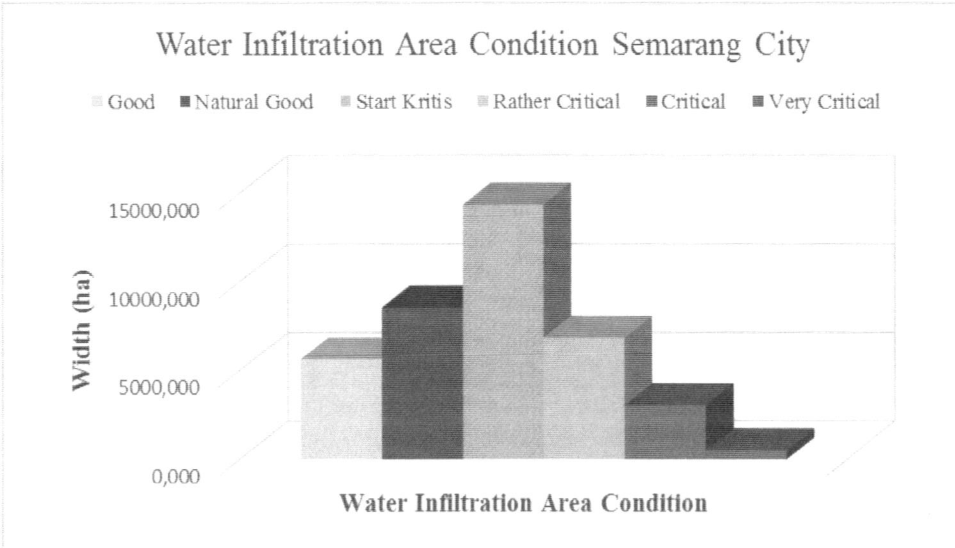

Figure 3.5. Water Infiltration Area Condition in Semarang City

Table 3.2. Sebaran Kondisi Kawasan Resapan Air Kota Semarang

No.	District	Condition (ha)						Total
		Good	Naturally Normal	Start Critical	Rather Critical	Critical	Very Critical	
1	Banyumanik	222,360	1237,971	715,684	719,079	146,312	13,594	3055,001
2	Candisari	0,232	55,090	208,919	175,880	245,479	16,658	702,258
3	Gajahmungkur	2,882	116,239	338,954	252,945	243,052	8,352	962,425
4	Gayamsari	129,157	5,657	456,522	24,623	0,000	0,000	615,959
5	Genuk	920,629	209,319	1316,282	339,716	0,000	0,000	2785,946
6	Gunungpati	1371,391	2234,334	943,256	932,668	426,207	169,091	6076,946
7	Kaliwungu	268,891	2918,874	1768,958	594,648	242,023	89,166	5882,561
8	Ngaliyan	13,212	278,022	1976,121	920,477	1045,902	169,044	4402,778
9	Pedurungan	722,726	163,595	1428,753	3,181	2,128	0,000	2320,383
10	Semarang Barat	230,843	93,318	1201,837	511,552	190,259	0,092	2227,902
11	Semarang Selatan	46,202	52,186	226,259	290,172	4,553	0,000	619,372
12	Semarang Tengah	49,184	3,238	475,189	10,840	0,104	0,000	538,554
13	Semarang Timur	72,377	0,846	463,597	24,190	0,000	0,000	561,010
14	Semarang Utara	85,636	19,665	954,746	160,539	0,000	0,000	1220,586
15	Tembalang	865,574	930,524	1050,743	793,769	313,319	28,985	3982,913
16	Tugu	632,306	244,604	841,957	1098,553	179,636	4,434	3001,490
Total		5633,603	8563,482	14367,776	6852,834	3038,975	499,415	38956,08

Based on the data above, the condition of water catchment area in Semarang mostly in critical condition with extensive start 14367.776 ha (36.88%), a natural normal condition with extensive 8563.482 ha (21.98%), good condition an area of 5633.603 ha (14.46%), the condition rather critical area of 6852.834 ha (17.59%), critical conditions measuring 3038.975 ha (7.8%) and the condition is very critical area of 499.415 ha (1, 28%).

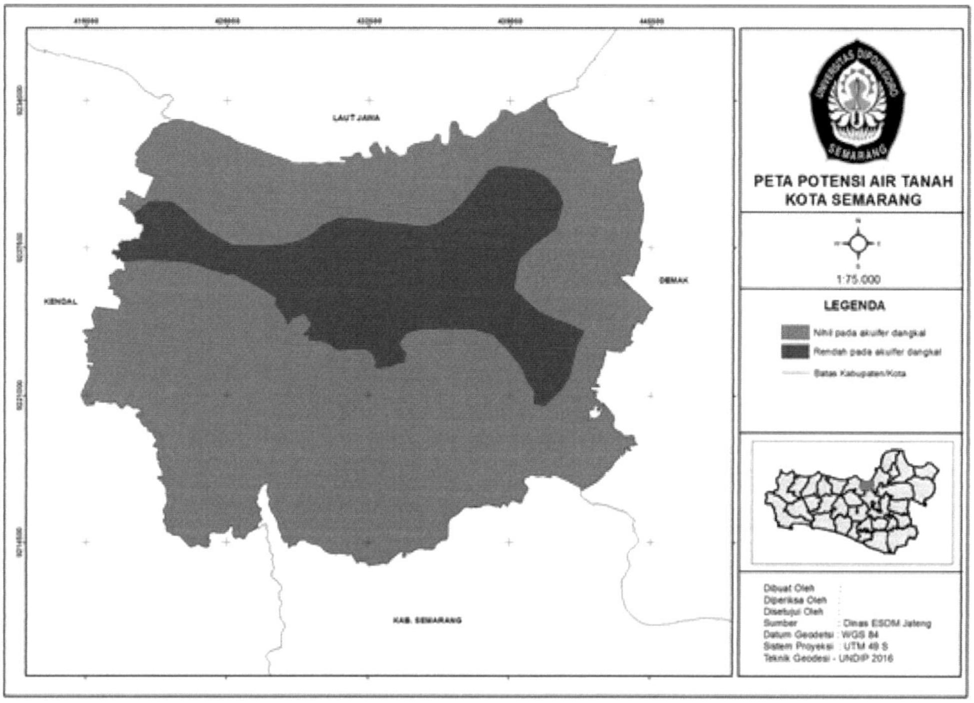

Figure 3.6. Underground Water Potential in Semarang City

4. Discussion and summary

In this study, the condition of water cacthment area in Semarang unfavorable. This is because the vast amount of water absorption good condition and normal to experience only 14197.085 ha (36%) of the total area of Semarang city.

While the condition critical to very critical start reaching 24759 ha (64%). Areas that could potentially be a water catchment area of Semarang is Gunungpati and Kaliwungu subdistrict. Gunungpati District has a good conditions water catchment area around 1371.391 ha and natural normal area of 2234.334 hectares, while the Kaliwungu District has good water catchment area around 268.891 ha and natural normal area of 2918.874 ha. From the total amount of water absorption good and natural normal condition in Semarang, a total of 6793.490 ha (47.9%) were in two districts. Gunungpati area and Kaliwungu have the same soil water potential, which has good quality water to a depth of 1-20 mbpt shallow aquifer. The depth of the groundwater table 1-15 mbpt, transmissivity coefficient of water 5 to 13.5 m^2/ day, the discharge of water from 0.03 to 0.13 liters / sec / m, and the optimum discharge from 0.2 to 1 ltr / sec. Based on this study, the estimated volume of ground water with good water infiltration condition and a value of transmissivity (T) of at least 5 m2 / day covering an area around 5633.603 ha was 102.8 trillion ltr / yr. The value of transmissivity (T) a maximum of 13.5 m^2 / day was 277.5 trillion liters / year. While the condition of water catchment areas with natural normal criteria is a minimum area of 8563.482 ha 156.3 trillion liters/year and a maximum of 421.9 trillion liters/year. The total volume of groundwater minimum and maximum can be seen in Figure 4.1.

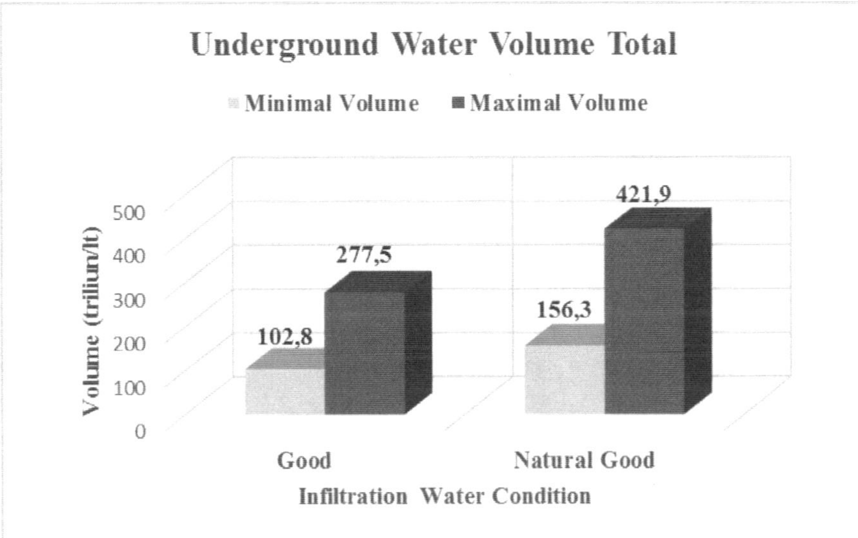

Figure 4.1 Total volume of Groundwater on Water Infiltration Condition Good and Normal Natural

Acknowledgement
This study is supported by grant from Diponegoro University-Engineering Faculty (DIPA 2016 schemes), Remote Sensing and Photogrammetry Laboratory and BAPPEDA Semarang City. We thank the reviewers for their comments and suggestions on the earlier version of the manuscript.

References

[1] Adibah N 2013 *Aplikasi Penginderaan Jauh Dan Sistem Informasi Geografis Untuk Analisis Resapan Air Skripsi* (Semarang: Universitas Diponegoro)
[2] Affandi A B 2013 *Pre-Processing Image (Composite, Cropping and Mosaic Image)* Departemen Geofisika dan Meteorologi Fakultas Matematika dan Ilmu Pengetahuan Alam. Institut Pertanian Bogor: Bogor
[3] Baatz M dan Schäpe A 1999 Object-Oriented and Multi-Scale Image Analysis in Semantic Networks *The 2nd International Symposium: Operationalization of Remote Sensing* New Methodologies ITC NL
[4] Bappeda Kota Semarang 2012 Wilayah dan Kondisi Geografis Kota Semarang Bappeda: Semarang
[5] Bhinneka 2016 Peralatan Komputer dan Elektronik. http://www.bhinneka.com/products/sku07016965/dell_precision_t3620__xeon_e3 1245v5___merchant_.aspx. Download July 20th 2016
[6] Definiens 2007 Definiens Developer 7 User Guide: Definiens: Munchen.
[7] Dirjen Reboisasi dan Rehabilitasi Lahan 1998 *Keputusan Direktorat Jenderal Reboisasi dan Rehabilitasi Lahan* 041/Kpts/V/1998 (Jakarta: Departemen Kehutanan)
[8] Environmental National Geographic 2015 A Freshwater Story. http://environment.nationalgeographic.com/environment/freshwater/freshwater-101-interactive/ accesed 20 Juni 2015
[9] Hastono F D 2012 *Identifikasi Daerah Resapan Air Menggunakan Teknologi SIG* Skripsi . Semarang: Universitas Diponegoro
[10] Herold M dan Scepan J 2002 Object-Oriented Mapping and Analysis of Urban Land Use/Cover Using IKONOS Data *Proceedings of 22nd EARSEL Symposium Geoinformation for European-wide integration*: Prague
[11] Lamo 2013 Estimasi Populasi Penduduk Berdasarkan Metode Unit Lingkungan Terkecil Menggunakan Metode Object Based Image Analysis (OBIA) Teknik Geodesi dan Geomatika.

Institut Teknologi Bandung: Bandung

[12] Modi M, Kumar R, Shankar, G Ravi, Martha, T R 2014 Land Cover Change Detection Using Object-Based Classification Technique: A Case Study Along The Kosi River, *Bihar ISPRS – International Archives of the Photogrammetry, Remote Sensing and Spatial Information Sciences* XL-**8** 2014: 839-843

[13] Nugroho SP 2002 *Kajian Terhadap Kebijakan Sumur Resapan Untuk Mengatasi Krisis Air Tanah di DKI Jakarta Tesis* (Bandung: Institut Teknologi Bandung)

[14] Prabowo A S E 2001 *Perencanaan Pengelolaan Kawasan Resapan Air Menggunakan Teknologi SIG (Studi Kasus DAS Garang Semarang) Tesis* (Depok: Universitas Indonesia).

[15] Syahbana *2012 Identifikasi Perubahan Tutupan Lahan Dengan Metode Object Based Image Analysis* (Bandung : Teknik Geodesi dan Geomatika Institut Teknologi Bandung)

[16] Suara Merdeka 2012 *Semarang Terancam Krisis Air* http://www.suaramerdeka.com/v1/index.php/read/cetak/2012/09/24/199827/Semarang-Terancam-Krisis-Air Download 20 Juni 2015

[17] Sutanto A, Trisakti B dan Arimurthy A M 2014 *Perbandingan Klasifikasi Berbasis Obyek dan Klasifikasi Berbasis Piksel pada Data Citra Satelit Synthetic Aperture Radar untuk Pemetaan Lahan* (Depok : Fakultas Ilmu Komputer Universitas Indonesia, Jakarta : Lembaga Penerbangan dan Antariksa Nasional)

[18] Tarantino 2004 *Land Cover Classification of Quickbird Multispectral Data with an Object-Oriented Approach Management Information Systems* (Southampton : WIT Press)

[19] Wasil 2012 *Identifikasi dan Perhitungan Luas Sawah dengan Citra Satelit Resolusi Tinggi Menggunakan Metode Object Based Image Analysis (OBIA)* (Bandung: Teknik Geodesi dan Geomatika Institut Teknologi Bandung)

[20] Wibowo S T and Suharyadi R 2012. Aplikasi Object-Based Image Analysis (OBIA) untuk Deteksi Perubahan Penggunaan Lahan Menggunakan Citra ALOS AVNIR-2 *Jurnal Bumi Indonesia* **1** 3 2012

[21] Wiwoho B S 2008 Model Penentuan Kawasan Resapan Air *Jurnal Hidrologi* 1-7 Woods Hole Oceanographic Institution 2015 A Freshwater Story. http://environment.nationalgeographic.com/environment/freshwater/freshwater-101-interactive/ Download 20 Juni 2015

[22] Xiaoxi S , Jixian Z. dan Zhengjun L 2004 *A Comparison of Object-Oriented and Pixel-Based Classification Approachs Using Quickbird Imagery* (Beijing : Chinese Academy of Surveiing and Mapping)

The Aplication of Landsat 8 OLI for Total Suspended Solid (TSS) Mapping in Gajahmungkur Reservoir Wonogiri Regency 2016

Apriwida Yanti[1], Bowo Susilo[2], and Pramaditya Wicaksono[3]

[1]Departement of Remote Sensing and GIS, Vocational School, Universtas GadjahMada, Yogyakarta 55281, Indonesia
[2]Departement of Geograhic Information Science, Faculty of Geographic, Universitas Gadjah Mada, Yogyakarta 55281, Indonesia
[3]Cartography and Remote Sensing, Departement of Geographic Information Science, Faculty of Geography, Universitas Gadjah Mada, Yogyakarta 55281, Indonesia

Email: apriwidayanti10@gmail.com, bowosusilo@ugm.ac.id, prama.wicaksono@geo.ugm.ac.id

Abstract. Gajahmungkur reservoir is administratively located in Wonogiri Regency, Central Java, with the main function as a flood control in the upstream of Bengawan Solo River. Other functions of the reservoir are as hydroelectric power plant (PLTA), water supply, irrigation, fisheries and tourism. Economic utilization of the reservoir is estimated until 100 years, but it is begun to be threatened by the silting of the reservoir. Eroded materials entering water body will be suspended and accumulated. Suspended Material or TSS (Total Suspended Solid) will increase the turbidity of water, which can affect the quality of water and silting the reservoir. Remote sensing technology can be used to determine the spatial distribution of TSS. The purposes of this study were to 1) utilize and compare the accuracy of single band Landsat 8 OLI for mapping the spatial distribution of TSS and 2) estimate the TSS on Gajahmungkur reservoir surface waters up to the depth of 30 cm. The method used for modelling the TSS spatial distribution is the empirical modelling that integrates image pixel values and field data using correlation analysis and regression analysis. The data used in the empirical modelling are single band of visible, NIR, and SWIR of Landsat 8 OLI, which was acquired on 8 May 2016, and field-measured TSS values based on the field data collection conducted on 12 April 2016. The results revealed that mapping the distribution and the estimated value of TSS in Reservoir Gajahmungkur can be performed more accurately using band 4 (red band). The determinant coefficient between TSS field and TSS value of image using band 4 is 0.5431. The Standard Error (SE) of the predicted TSS value is 16.16 mg/L. The results also showed that the estimated total TSS of May 2016 according to band 4 is 1.087,56 tons. The average estimation of TSS value in up to the depth of 30 cm is 61.61 mg/L. The highest TSS distribution is in the northern parts, which was dominated by eroded materials from Keduang River.
Keyword: TSS, Estimation, Mapping, Landsat 8 OLI

1. Introduction

One of many problems in the maintenance of reservoir is the silting of the reservoir. Eroded materials entering the water will be suspended and accumulated. Suspended Material or TSS (Total Suspended

Solid) will increase the turbidity of water, which can affect the quality of water and silting the reservoir. Monitoring the silting process is difficult, because the reservoir area is large and the process of materials entering the reservoir cannot be seen clearly. The problem will be obvious when floods occur due to decreased capacity of the reservoir.

Currently, technology of remote sensing has been constantly evolving and can be used to sense water body. Landsat 8 OLI have a unique sensor, which can be used in sensing the condition of a body of water. Temporal resolution of Landsat 8 OLI is 16 days, freely available, and is effective to be used to monitor water conditions in reservoir continually.

Utilization of Landsat 8 OLI to monitor the sedimentation of reservoir is via the estimation of TSS in water. If the number of TSS is high, the accumulation of sediment at the base of the reservoir is also high. Besides this, the recording of Landsat 8 OLI is approximately 185 km x 185 km, it will be identified the areas with high potential occurs silting, or the area that become the potential location of entrance of sedimentary material in reservoir.

Landsat 8 has many bands used for the interpretation of various objects on the Earth's surface. Eleven single band consisting of visible bands, infrared bands, panchromatic, cirrus and thermal band, have unique function in the interpretation of object. Based on the ability of penetration and sensitivity of each single band to interpreting the suspended materials, visible and infrared bands are the most effective. To perform the mapping and estimation of TSS in reservoir, we need to know which band has the best performance.

Gajahmungkur Reservoir was selected as the study area (Figure 1) because the Gajahmungkur Reservoir has silting reservoir problem. Based on JICA study, the average annual sediment in period of 1993-2004 amounted to 3.18 million m^3[1]. Based on studies conducted by the Research Directorate of Water Problems (DPMA) 1982 [2] the erotion rate is estimated at 8.58 mm/year. In addition, Sub Center for Rehabilitation and Conversion Land of Solo (Sub Balai RLKT Solo, 1985[2]) stated that the erotion rate is estimated at 26.00 mm/year. With the erosion rate of that value, Faculty of Geography-SBRLKT Solo 1996 [2] estimated that the economic life of Gajahmungkur Reservoir is reduced from the planned use of 100 years to just 27 years.

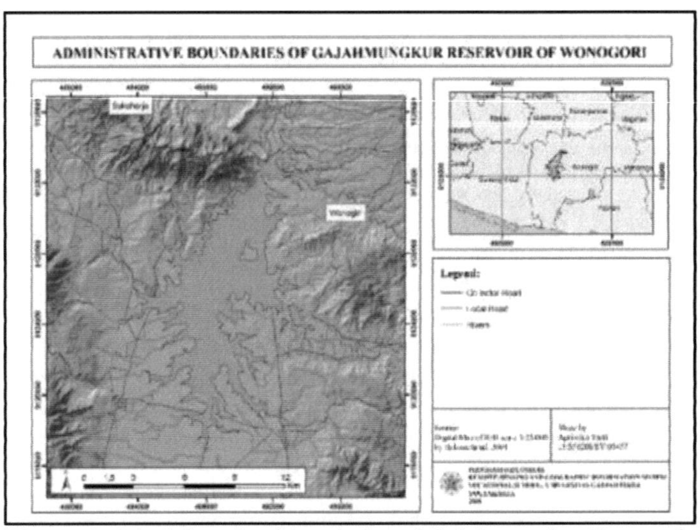

Figure 1. Administrative boundaries of Gajahmungkur Reservoir

2. Objectives

The objectives of this study are 1) evaluate which spectral band have better accuracy in the estimation of TSS. We integrated image pixel values and TSS field data from Gajahmungkur Reservoir water of the same coordinate to perform empirical modeling. From individual field data, we performed accuracy assessment, and 2) map and estimate of distribution of TSS using spectral band with the

IOP Conf. Series: Earth and Environmental Science **47** (2016) 012028 doi:10.1088/1755-1315/47/1/012028

highest accuracy. Only the TSS in the water column up to the depth of 30 cm from the surface was estimated.

3. Methods

3.1. Image Data

Two images of Landsat 8 OLI used in this study were acquired in February 18th, 2016 and May 8th, 2016. The Landsat 8 OLI (path 119 row 65) covers the whole Gajahmungkur Reservoir. Landsat 8 OLI acquired on February 18th, 2016 was used to determine the distribution of sample location before field survey, while Landsat 8 OLI acquired on May 8th, 2016 was to perform the modeling and mapping.

3.2. Field Data

Field survey was conducted on April 12nd, 2016 using speedboat and GPS handheld to get to each sample point based on field survey plan map. The date of field survey and image acquisition has difference of two months. It was because only in this date of acquisition the image has the smallest cloud cover before field survey conducted. The difference had little influence, because this image was only used to see the pattern of TSS prior to field survey.

Water was sampled up to the depth of 30 cm from the surface. Total samples obtained are 49 out of 54 planned samples, which consist of 25 sample points and 24 of validation points. The sampled water were measured on April 14th, 2016 in the Laboratory of Hydrology and Water Quality of the Faculty of Geography, Universitas Gadjah Mada by using SNI-06-6989.3-2004 Gravimetry method.

3.3. Radiometric Correction

Radiometric image correction is required to improve the quality of the image and improve the pixel values that do not correspond to the value of the spectral reflectance of the actual object [3]. Radiometric corrections conducted in this study are a correction Top of Atmosphere (ToA) and Dark Subtraction (DOS). Correction to ToA changed the pixel values of the image into the reflectance values by normalizing the angle and intensity of solar energy, while the DOS correction intended to remove atmospheric path radiance. DOS correction assumes that there is a pixel containing 0% reflectance [4].

3.4. Classification

Classification was done to get the water turbidity classes by Landsat 8 OLI acquired in February 18th, 2016. Classification was done by density slice method based on spectral reflectance histogram. That was because the peak of the histogram represent one type of water turbidity. Density slice is one method of classification that can be done by selecting the brightness level of the entire value of existing pixels in the image. Danoedoro [3] has the assumption that every object in the image of a particular channel has a brightness value of a certain range, the range of brightness values of 0-255 can be sliced into several intervals that describe the appearance of the object in general. The classification process resulted in nine different classes with clear water on the first class and the most turbid water is in the ninth (Figure 2). The range of water turbidity level are as follow: The Ist has a density range of -0.0244 to -0.0032, IInd has a density range of -0.0032 to -0.0008, IIIrd has a density range of -0.0008 to 0.0020, IVth has a density range of 0.0020 to 0.0075, Vth has a density range of 0.0075 to0.0106, VIth has a density range of 0.0106 to 0.0145, VIIth has a density range of 0.0145 to 0.0259, VIIIth has a density range of 0.0259 to 0.0393, IXth has a density range of 0.0393 to 0.0758**.** From these nine classes of turbidity, we determined the distribution of modeling and validation samples.

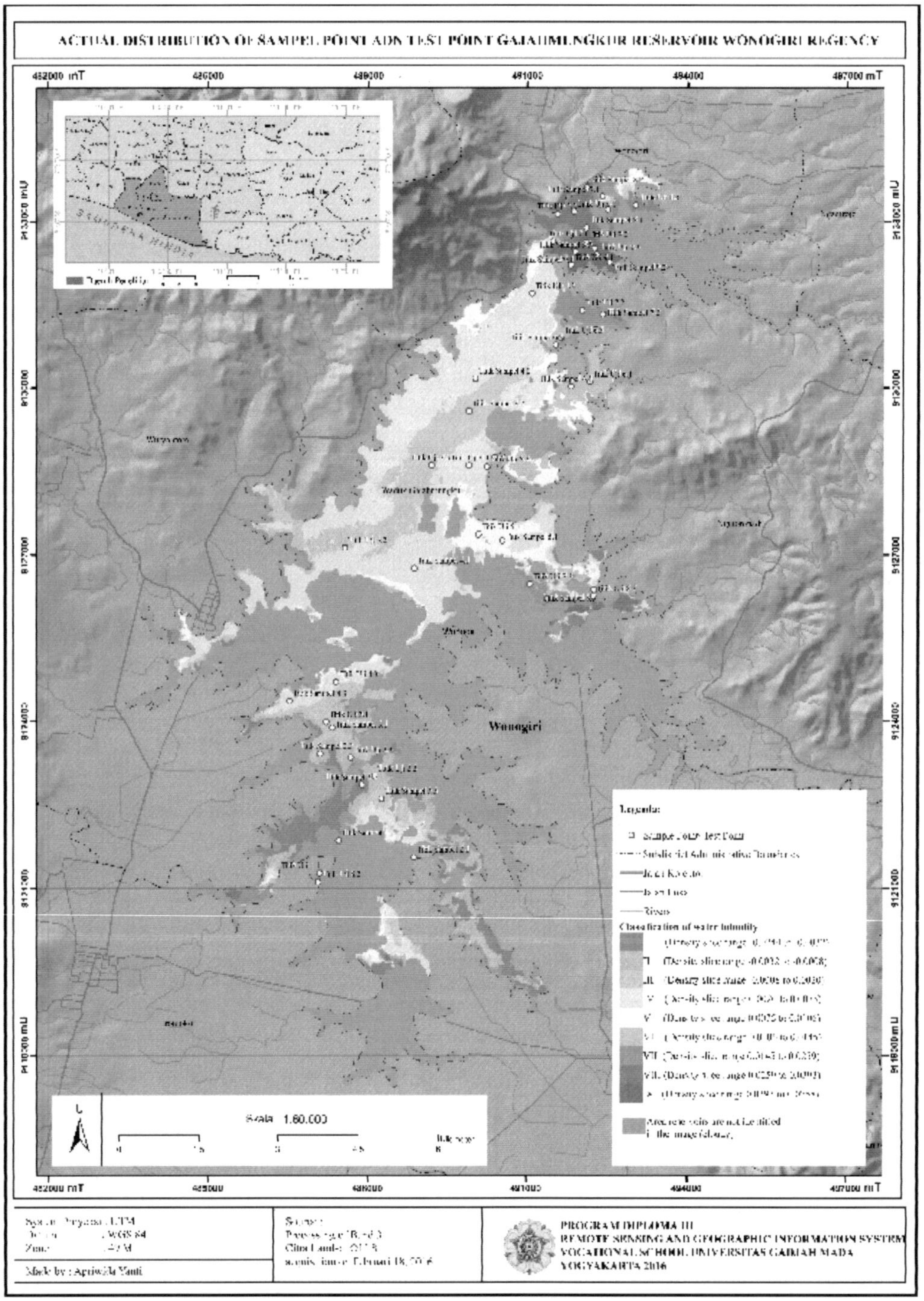

Figure 2. Actual distribution of Sample Point and Test Point in the field survey

3.5. Correlation Analysis

Pixel values of Landsat 8 OLI acquisition May 8th, 2016 was extraction based on actual coordinate points of field survey. Correlation analysis was done between pixel value of image and TSS field value of laboratory measurement result.

For 25 samples, the significant threshold at 95%CL is 0.396. Table 1 shows the correlation analysis of 25 data between TSS field data and the pixel value. The *r* of band 1, band 5, band 6 and band 7 was <0.396 and not the relationship is not significant, which can be caused by several factors. Field survey conducted on April 12, 2016 has a difference of ± 1 month from the date of acquisition of the image on May 8, 2016. This affected the distribution of TSS pattern in the water reservoir. The intensity of the rainfall that occurred during the one month is also quite high, so the spatial distribution of TSS can change. The field TSS values which not correspond to the image condition need to be repaired via contextual editing.

Table 1. Correlation analysis between TSS values in the field with the pixel values of band 1 - band 7 before contextual editing. The *r* value with mark (*) is significant.

Number of samples (n) = 25

Band	Modeling samples		Accuracy assessment samples	
	Corelation Coefficient (*r*)	Coefficient of Determination (R^2)	Corelation Coefficient (*r*)	Coefficient of Determination (R^2)
Band 1 – Ultra-blue	0.361	0.130	0.278	0.077
Band 2 – Blue	0.413*	0.170	0.297	0.088
Band 3 – Green	0.449*	0.201	0.289	0.083
Band 4 – Red	0.459*	0.211	0.283	0.080
Band 5 – NIR	0.365	0.133	0.340	0.115
Band 6 – SWIR 1	0.349	0.122	0.307	0.094
Band 7 – SWIR2	0.357	0.127	0.325	0.105

3.6. *Regression Analysis*

Regression analysis was used to obtain resultant regression function to model the TSS (Figure 3 – Figure 9). The Independent Variables are pixel value and the Dependent Variables are TSS field values. Regression analysis used was linear equation model. The basic formula for the resultant regression function is as follow:

$$Y = A + BX \quad (1)$$

Explanation:

Y = The dependent variable (field TSS)
A = The constants
B = The slope of regression
X = The independent variables (pixel values)

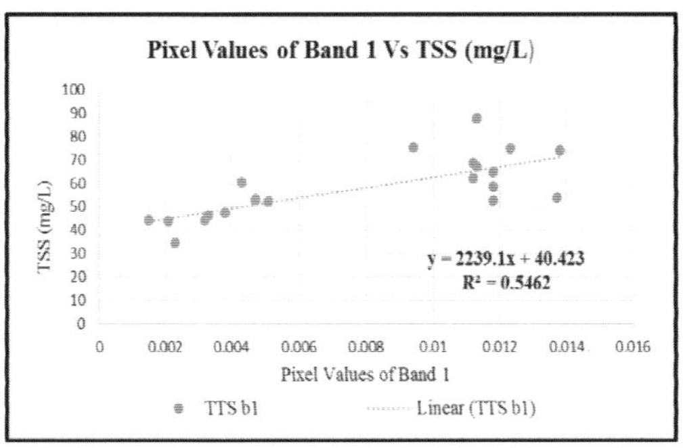

Figure 3. Scatter plot of band 1 vs field TSS (mg/l)

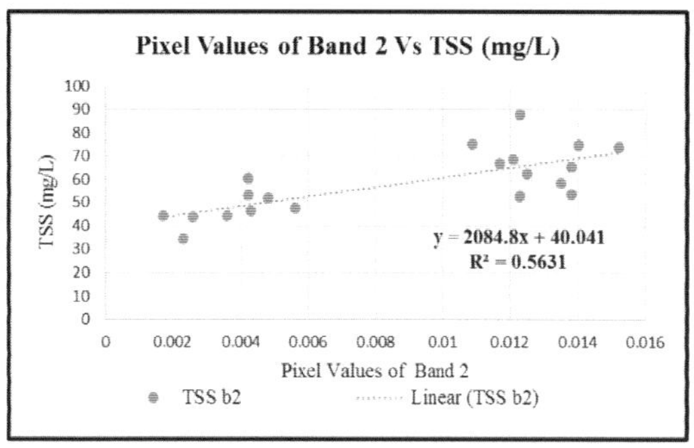

Figure 4. Scatter plot of band 2 vs field TSS (mg/l)

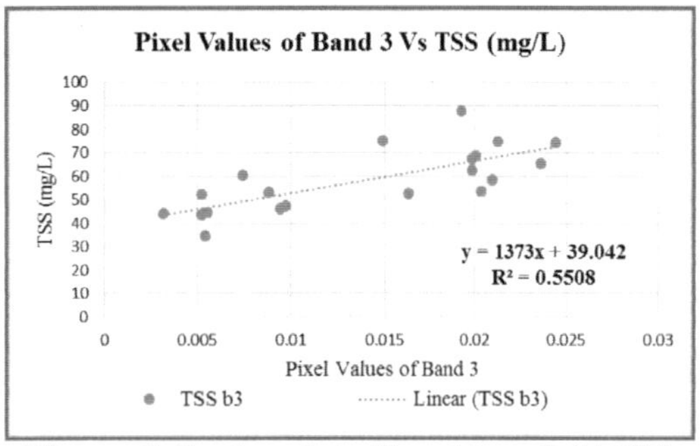

Figure 5. Scatter plot of band 3 vs field TSS (mg/l)

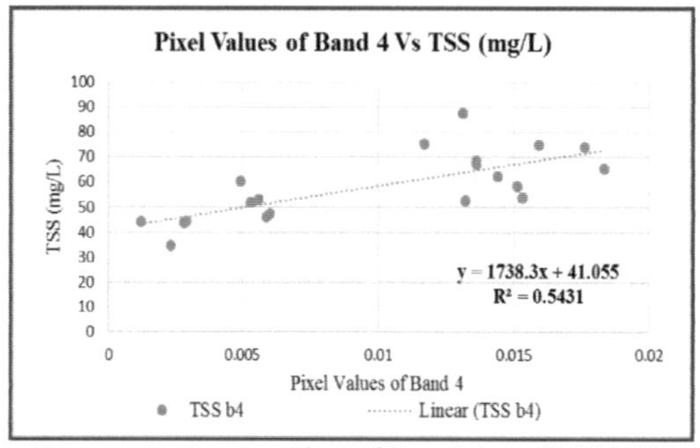

Figure 6. Scatter plot of band 4 vs field TSS (mg/l)

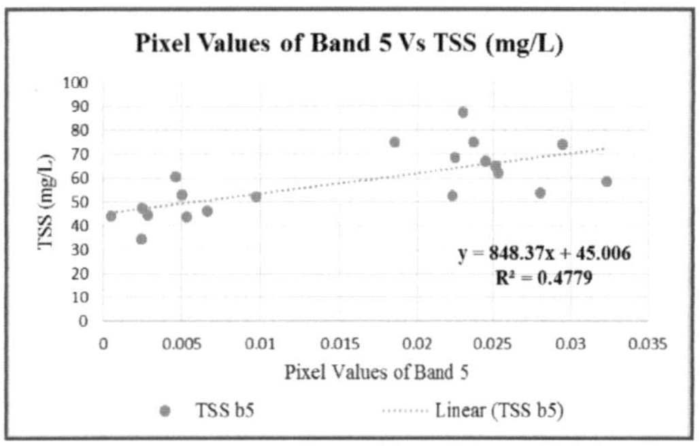

Figure 7. Scatter plot of band 5 vs field TSS (mg/l)

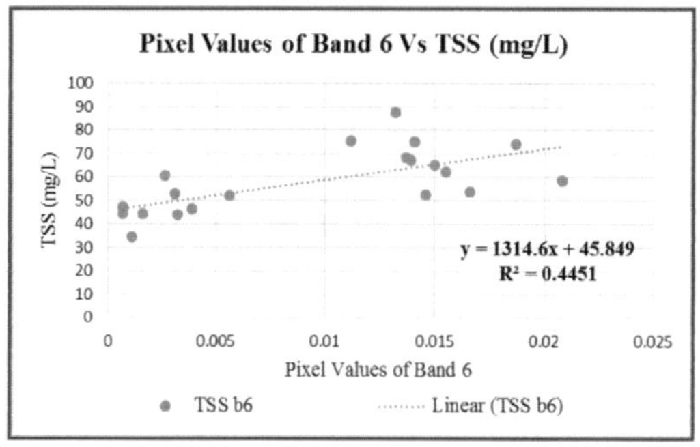

Figure 8. Scatter plot of band 6 vs field TSS (mg/l)

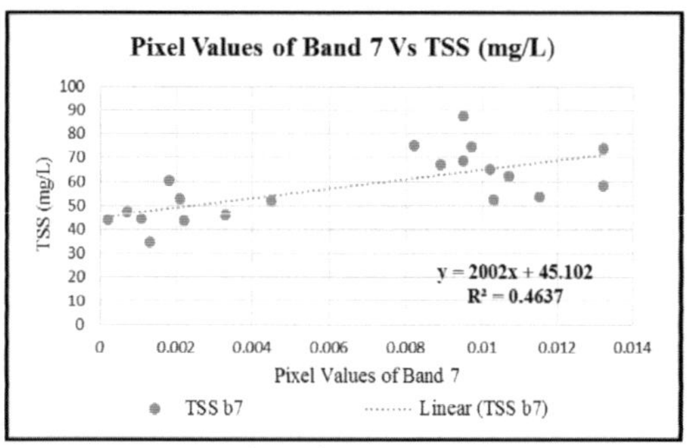

Figure 9. Scatter plot of band 7 vs field TSS (mg/l)

3.7. Contextual editing

Contextual editing is a process to eliminate, modify data or information that was less relevan. Contextual editing applied in this study is to eliminate the TSS samples that are not relevant with the pixel values. Irrelevant data adversely affected the correlation analysis. The process of elimination was done based on the visualisation of pixel value and the corresponding TSS value in the field. The assumption used is if the pixel is bright, then the value of TSS is also higher. This is because the pixel values of the water will be more reflective if the content of TSS is high.

There are five samples removed for modelling samples and four samples removed from validation samples. As seen on Table 2, contextual editing significantly improved the correlation coefficients.

Table 2. Correlation analysis between TSS values in the field and the pixel values of band 1 - band 7 at sample locations after contextual editing

Number of samples (n) = 20				
	Modeling samples		Accuracy assessment samples	
Band	Corelation Coefficient (*r*)	Coefficient of Determination (R^2)	Corelation Coefficient (*r*)	Coefficient of Determination (R^2)
Band 1 – Ultra-blue	0.739	0.546	0.536	0.287
Band 2 – Blue	0.750	0.563	0.581	0.337
Band 3 – Green	0.742	0.550	0.593	0.352
Band 4 – Red	0.736	0.543	0.601	0.361
Band 5 – NIR	0.691	0.477	0.625	0.390
Band 6 – SWIR 1	0.667	0.445	0.630	0.397
Band 7 – SWIR2	0.680	0.463	0.647	0.419

3.8. Accuracy Assessment

Accuracy assessment was done to understand the error of prediction between modeled TSS with field TSS as a reference. Accuracy test was done on each band. The equation for the calculation of error value is:

$$SE = \sqrt{\frac{(\varepsilon error)}{n-2}} \quad (2)$$

Explanation :
SE = Standard error of estimate (mg/L)
∑ error = Total of error of predicition on each band
error = The calculation of error at one point (TSS in the field – TSS of image)^2
n = The number of sample

SE value shows the deviation between TSS in the field and modeled TSS in mg/L. High SE can be interpreted that the TSS of image are highly overestimated or underestimated from the actual TSS in the field.

3.9. Estimation of Total Sediment Content

The calculation of total sediment has been done by calculate the TSS of image from applying the regression equation every single band. TSS value being modelled has a unit of mg/L. The unit needs to be converted to g/m^3. The conversion was done by multiplying each pixel value of TSS with the number of occurrences of pixels and also the depth of measurement. The estimation was only performed on a model with the lowest SE. It is considered that the total estimate of TSS of the most accurate model represents the actual TSS value in the field.

3.10. Research flowchart
The research flowchart is shown in Figure 10.

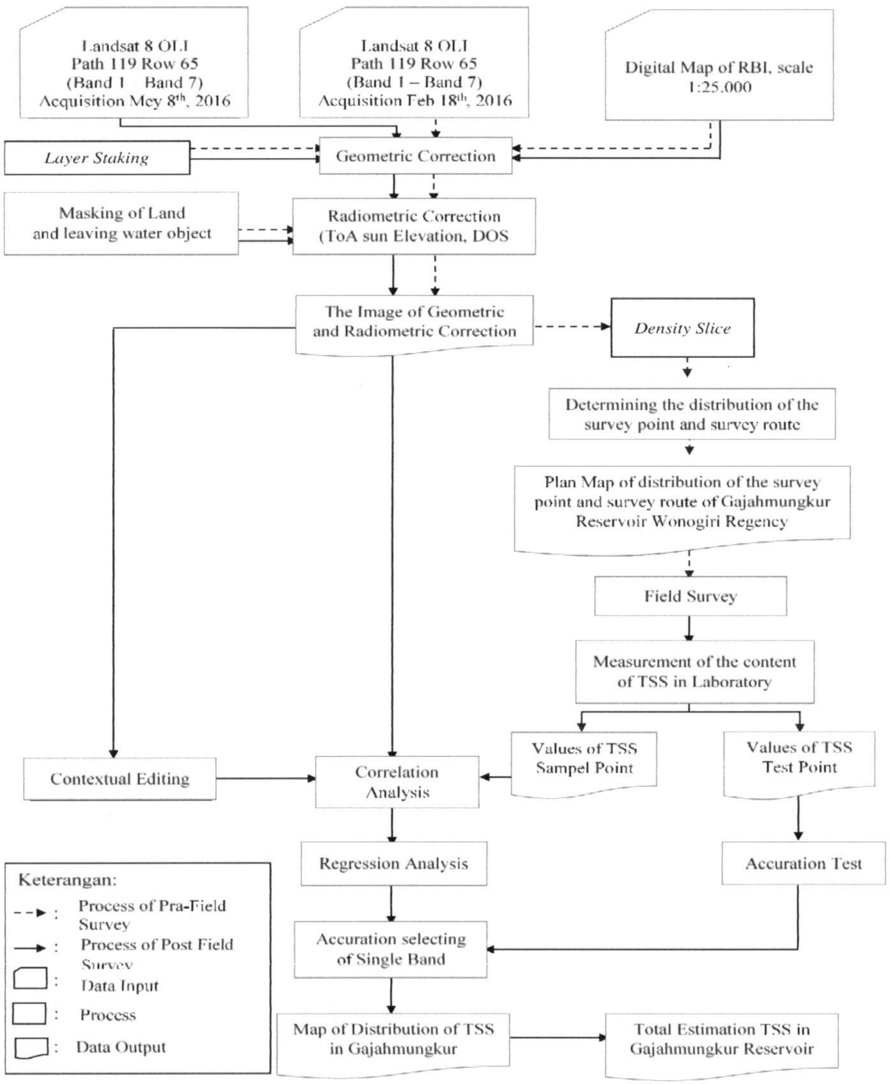

Figure 10. Research flowchart

4. Discussion

4.1. Distribution of TSS

The spatial distribution of TSS from each band is generally almost similar, but the estimated value of TSS is different. It is due to the difference spectral reflectance properties of each band on water. The water reflectance values can illustrate the condition and quality of the water. TSS turbidity level became one of the factors of the spectral properties of water. Turbid water will have higher reflectance values than clear water [6]. Band 1 and band 2 showed that the northern part of the reservoir has > 70 mg/l, which higher than the other bands. That was because band 1 (0.43-0.45 μm) and band 2 (0.45 – 0.51μm) have high the capability to penetrate the water. According to Butler et al. [5], blue (0.45 μm - 0.52 μm) and green energy (0.52 μm - 0.60 μm) has the best water column penetration ability. Water penetration capability will decrease with the increasing turbidity of water column. Band 1 and band 2 can reflect the materials of TSS on the deeper water column. Thus, the detection of TSS in the water column by band 1 and band 2 is more detail than other bands. Just note that band 1 has lower ability than other visible bands to reflecting the material of TSS in the water, but more sensitive to clear water. This is evident in the minimum value in the estimation of the band 1 or band 2, which produced lower minimum value of the estimated TSS compared to the other bands. In this case, it is because this energy can penetrate deeper into the water column but unable to strongly return the signal to the sensor in large quantities, because the lack of sensitivity to the material of TSS in the water.

The general appearance of the visualization TSS from band 3 (0.51 – 0.59μm) is quite different from other bands. The northern part of the reservoir was dominated by a TSS class of 60-70 mg/l. It was just that the estimated maximum value of TSS from band 3 is quite high compared to the band 1 and band 2, but the estimated minimum value of the TSS in the image is the lowest of all. That was because band 3 has a lower water penetration capability but is better capable of reflecting TSS in the water. The estimated minimum value of band 3 is the lowest, it is because the ability of reflecting the water with less material of TSS is lower then other band.

The appearance of the band, 5, 6 and 7 is generally similar and not much different from the band 1 and 2. Those bands have maximum estimated TSS value quite high compared to bands 1, 2 and 3. In general, the ability to penetrate water of band 4, 5, 6 and 7 are lower than bands 1, 2 and 3. Thus, they only detect TSS based on the information on water surface. It is just the sensitivity of these bands to TSS is higher than bands 1, 2 and 3. This made the estimated minimum value in band 4, 5, 6 and 7 is higher than bands 1, 2 and 3.

Table 3. The result of TSS estimation (mg/l) from each band

Band	Maximum	Minimum	Mean
Band 1	120.94	37.18	56.29
Band 2	140.87	36.24	56.47
Band 3	170.41	33.17	57.03
Band 4	270.38	40.50	61.60
Band 5	289.66	44.80	56.30
Band 6	269.41	45.47	56.05
Band 7	205.79	44.94	55.84

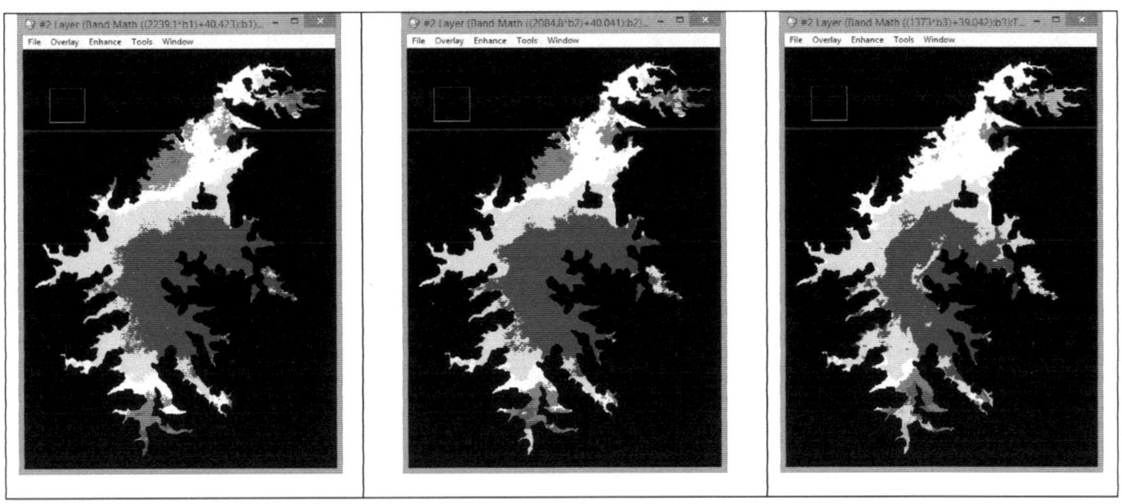

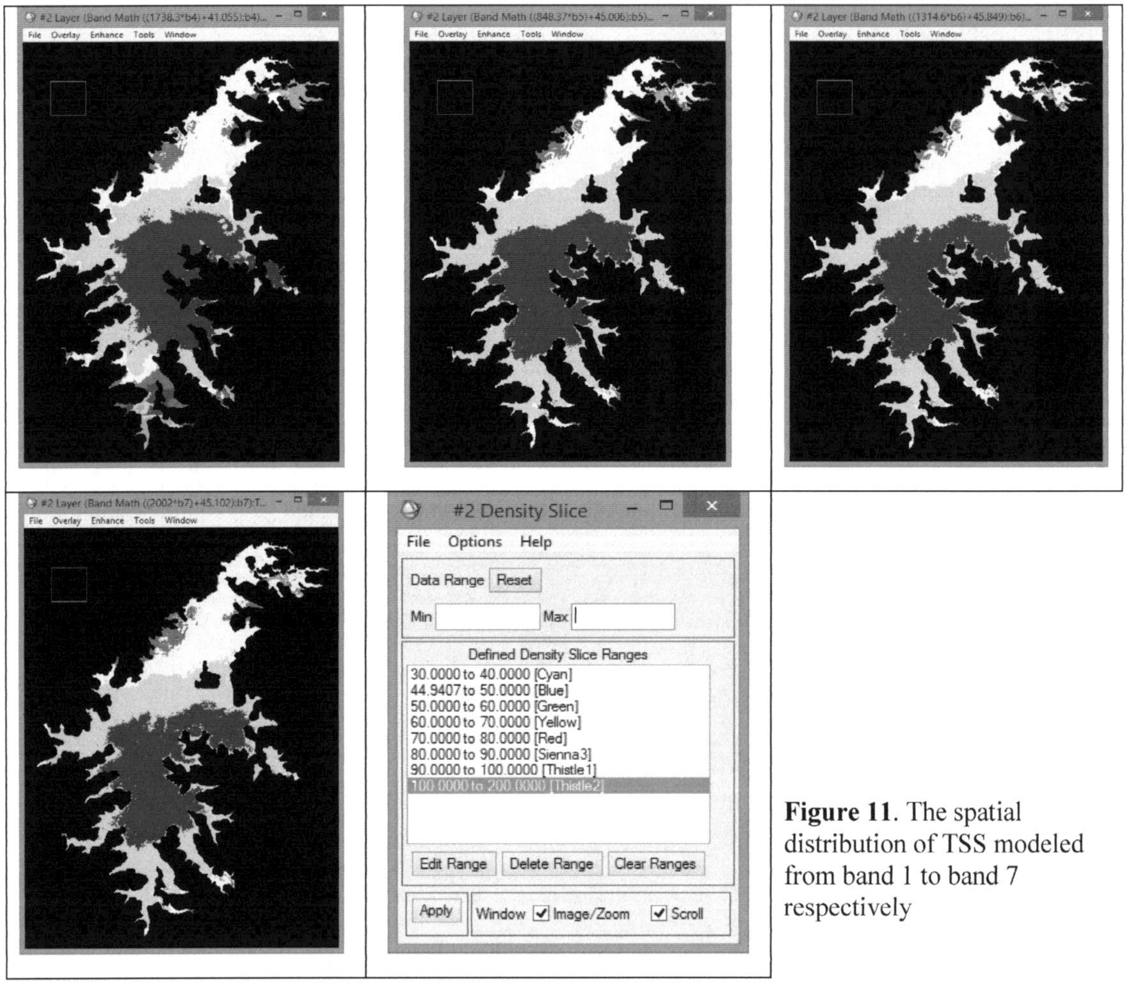

Figure 11. The spatial distribution of TSS modeled from band 1 to band 7 respectively

4.2. Accuracy Test

The accuracy assessment indicates that band 4 has the smallest deviation in the estimation of TSS image with ±16.16 mg/l. Meanwhile, band with the highest deviation in estimating TSS is band 6 with SE of ±17.21 mg/l. This can be caused by the penetration ability and susceptibility to materials of TSS or object of water.

Table 4. The result of accuracy assessment

Band	Total Error	SE (mg/l)
Band 1 – Ultra-blue	5289.89	17.14
Band 2 – Blue	5202.18	17.00
Band 3 – Green	5093.80	16.82
Band 4 – Red	4702.81	16.16
Band 5 – NIR	5262.05	17.10
Band 6 – SWIR 1	5334.28	17.21
Band 7 – SWIR2	5151.07	16.92

4.3. *Mapping*

Information of TSS spatial distribution were presented using continuous symbol at 1: 60.000 map scale (Figure 12). Based on that map, Gajahmungkur Reservoir was dominated by waters with TSS value of 40.50 mg/l - 60.50 mg/l and the there are many variations in the water with TSS value of >120.50 mg/l. TSS value >90.503 mg/l concentrated in the northern and southern ends of the reservoir. The northern part is the estuary area of Keduang river, which contributes the highest sediment of all seven sub-watershed to Gajahmungkur Reservoir. The southern part of the reservoir is a community fish cage area and the river estuary of Temon sub-watershed and Alang sub-watershed. TSS value <90.503 mg/l to >60.503 mg/l is uniformly dispersed in most of the northern parts of the reservoir, especially in the northwestern parts of the pier and community fish cages activities. TSS value <60.50 mg/l to 40.50 mg/l are distributed in middle of the reservoir towards the eastern part of the reservoir. Based on field observations, the eastern parts of the reservoir have more vegetation than the western parts.

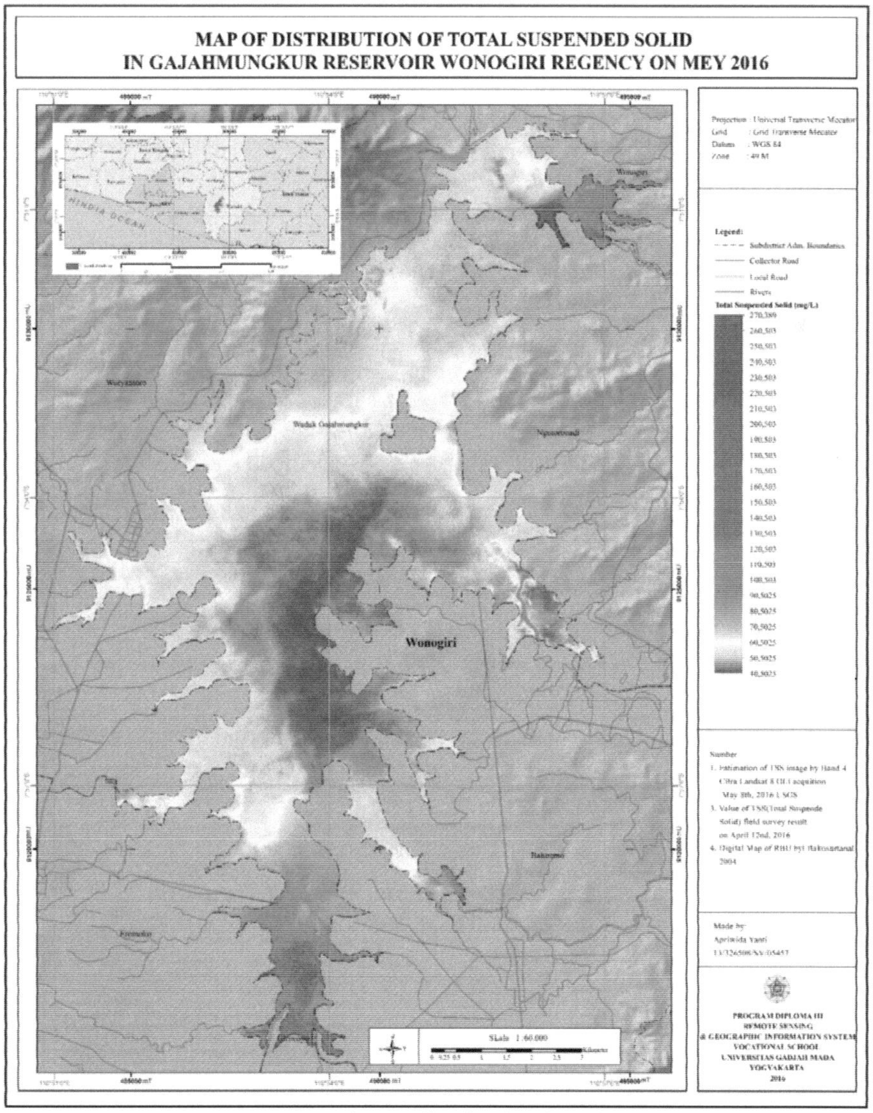

Figure 12. Map of TSS distribution based on Band 4

4.4. Estimation
According to the extraction of statistical data of modelled TSS based on band 4, the lowest TSS value is 40.5 mg/l and the highest value is 270.39 mg/l. The mean values of TSS modelled by band 4 is 61.61 mg/l. The total value of TSS in the reservoir is 1087.56 tons. This value means that there are 1087.56 tons of suspended solids in Gajahmungkur Reservoir up to the depth of 30 cm on May 2016.

5. Conclusion
This research shows that Landsat 8 OLI band 4 has the best accuracy with SE of ±16.16 mg/l. The distribution of TSS in Gajahmungkur Reservoir on May, 2016 based on band 4 of Landsat 8 OLI has an estimated value of TSS as follow: 1) the lowest is 40.50 mg/l, 2) the highest is 270.39 mg/l, and 3) the average value is 61.61 mg/l. The TSS is high in the northern and southern parts of the reservoir, especially in the estuary of the river. The estimated TSS in the Gajahmungkur Reservoir up to the depth of 30 cm modelled from band 4 of Landsat 8 OLI is 1087.56 tons.

References
[1] Rahman M M, Harisuseno D & Sisinggih D 2008 *Studi Penanganan Konservasi Lahan di Sub DAS Keduang DAS Bengawan Solo Kabupaten Wonogiri* Malang Master Program in Water Engineering Universitas Barwijaya

[2] Wiryanto 2014 Model Pengelolaan Perairan Waduk Berdasarkan Tingkat Kesuburan dan Pencemaran Air (Kasus di Waduk Gajahmungkur Wonogiri Jawa Tengah) Dissertation Universitas Gadjah Mada

[3] Danoedoro P 2012 *Pengantar Penginderaan Jauh Digital* Yogyakarta ANDI Yogyakarta

[4] Pat s Chaves Jr 1968 An Improved Dark-Object Subtraction Technique for Atmospheric Scattering Corection of Multispectral Data. Remote Sensing of Environment (Online) (https://www.researchgate.net/publication/22379584 3,accessed on April 8th,2016).

[5] Butler M J A, Mouchot M C, Barole V & Le Blanc C 1988 *The Apllications of Remote Sensing Technology to Marine Fisheries: An Introductory Manual.* Roma FAO Fisheries Technical Paper No 295 165

[6] Hasyim B 1997 Optimasi Penggunaan Data Inderaja dan Sistem Informasi Geografi untuk Pengawasan Kualitas Lingkungan Pantai akibat Limbah Industri (Jakarta The National Research Council Office of the State Minister for Research and Technology)

Total Suspended Solid (TSS) Mapping of Wadaslintang Reservoir Using Landsat 8 OLI

Muhammad Fauzi [1], Pramaditya Wicaksono [2]

[1] Experimental Station For Hydrology And Water Management, Research Center For Water Resources, Ministry Of Public Work And Housing, Indonesia
[2] Cartography and Remote Sensing, Department of Geographic Information Science, Faculty of Geography, Universitas Gadjah Mada, Indonesia

fauzie.em@gmail.com[1], prama.wicaksono@geo.ugm.ac.id[2]

Abstract. Lakes, water reservoirs, and ponds are sources of fresh water that support the life of all living beings as well as human's social and economic activities. A common problem on reservoirs in Indonesia is the siltation due to the discharge of river materials into the reservoir. This process of sedimentation declines the function of the reservoir. Total Suspended Solid (TSS) can be used as the indicator of sediment in the reservoir, which usually consists of silt, fine sand and microorganisms. The high concentration of TSS in the water column can be used as the indication that the process of sedimentation in the reservoir is also high. This study aims to 1) map the spatial distribution of TSS and 2) estimate the TSS contained in Wadaslintang Reservoir using Landsat 8 OLI. The algorithm of TSS for Wadaslintang Reservoir was created based on the empirical relationship between field TSS data and reflectance values. Based on this research, it was found that the ratio of red and green band produced the best correlation with the TSS ($r = 0.533$). The resultant regression function of the regression analysis between ratio of red and green band and field TSS data was used to model the spatial distribution of the TSS on Wadaslintang Reservoir with standard error of 10.8 mg/l. The average TSS in the reservoir is estimated to be 22.61 mg/l.

1. Introduction

The availability of water resources is fundamental to support regional economic development. The limitation of water resources can impede and restrict the activities of development and economic prosperity of the people. Lake/reservoir water is used for various purposes, including the utilization of raw drinking water, irrigation water, power generators, flushing out, fisheries, and so on. The aforementioned functions of the lake water show just how important fresh water from reservoirs/Lakes to human life.

Wadaslintang is one of the reservoirs located in the province of Central Java, precisely located in Kecamatan Wadaslintang, Wonosobo Regency. This reservoir is of multifunctional use such as for raw water supply, power generation, irrigation and also tourism. With reservoirs have reached the age of 28 years, sedimentation is certainly have happened to this reservoir. One of the factors that indicate the occurrence of sediment in the reservoir is the presence of Total Suspended Solid (TSS), which is usually made up of silt and fine sand as well as microorganisms. The higher content of TSS in the water reservoir can indicate that the process of sedimentation in the reservoir is also high.Remote Sensing Technology is growing with the introduction of a wide range of satellite systems with a variety of technologies, resolutions, and mission. Satellite remote sensing has been able to provide data and information on the natural resources of land and water regularly and periodically. This study aims to utilize remote sensing technology, in particular Landsat 8 OLI, to determine the spatial distribution of TSS and also to determine the amount of TSS contained in Wadaslintang Reservoir.

2. Methodology

This study is divided into three stages, namely the pre field survey, field survey and post field survey. Pre field survey including data collection, the correction of the image that will be used, and image processing to determine the sample locations. Field survey activity is the sampling of water in the reservoir Wadaslintang, for TSS levels calculation. Finally, the post-field stage is the laboratory analysis of TSS and analysis of TSS by using satellite imagery.

3. Analysis

3.1 Image Correction

Geometric correction was not performed in this study because the image used has been geometrically correct for the given scale, and thus no geometric correction is performed. Radiometric correction is an important process in digital satellite image processing and interpretation . In this research, Landsat 8 OLI DN was converted in Top-of-Atmospheric (TOA) reflectance using the following formula:

$$\rho\lambda' = M\rho Qcal + A\rho \qquad (1)$$

$\rho_{\lambda'}$ = TOA reflectance of band λ
Q_{cal} = The value of the pixels (DN), as the band used
M_P = Constant rescalling (REFLECTANCE_MULT_BAND_x, where x is the band used)
A_P = Constant power (REFLECTANCE_ADD_BAND_x, where x is the band used)
The value of M_p and A_p are obtained from the image metadata

3.2 Preparation Of The Sample Point Field

To determine the location of TSS sample in Wadaslintang Reservoir, the water pixels of the reservoirs were classified based on their brightness levels using dencity slice technique, which is one of the simplest methods in categorizing or classifying objects, particularly for a single band analysis. The histogram of the Red band was used as the basis for the density slice, by identifying the major normal distribution curve of the histogram. The pixel value interval for the each resulting smaller normal distribution curve was used classify the image accordingly.

The interval of pixel value for each class is irregular, depending on the characteristic and variation of water clarity distribution across the scene. The of irregular intervals because of spectral reflection of material objects of the body of water on the interval for each class is as follows:
I : 0,053 - 0,055 (Red)
II : 0,055 - 0,057 (Green)
III : 0,057 - 0,059 (Blue)
IV : 0,059 - 0,061 (Yellow)
V : 0,061 - 0,064 (Cyan)
VI : 0,064 - 0,069 (Magenta)
VII : 0,069 - 0,075 (Maroon)
VIII : 0,075 - 0,087 (Sea Green)

The result of the dencity slice was used as the basis to determine location of the sample. Based on the eight density slice class, 64 sample locations were determined, where there are eight samples for each class. This eight samples per class was further divided into four samples for empirical modeling and four samples for accuracy assessment.

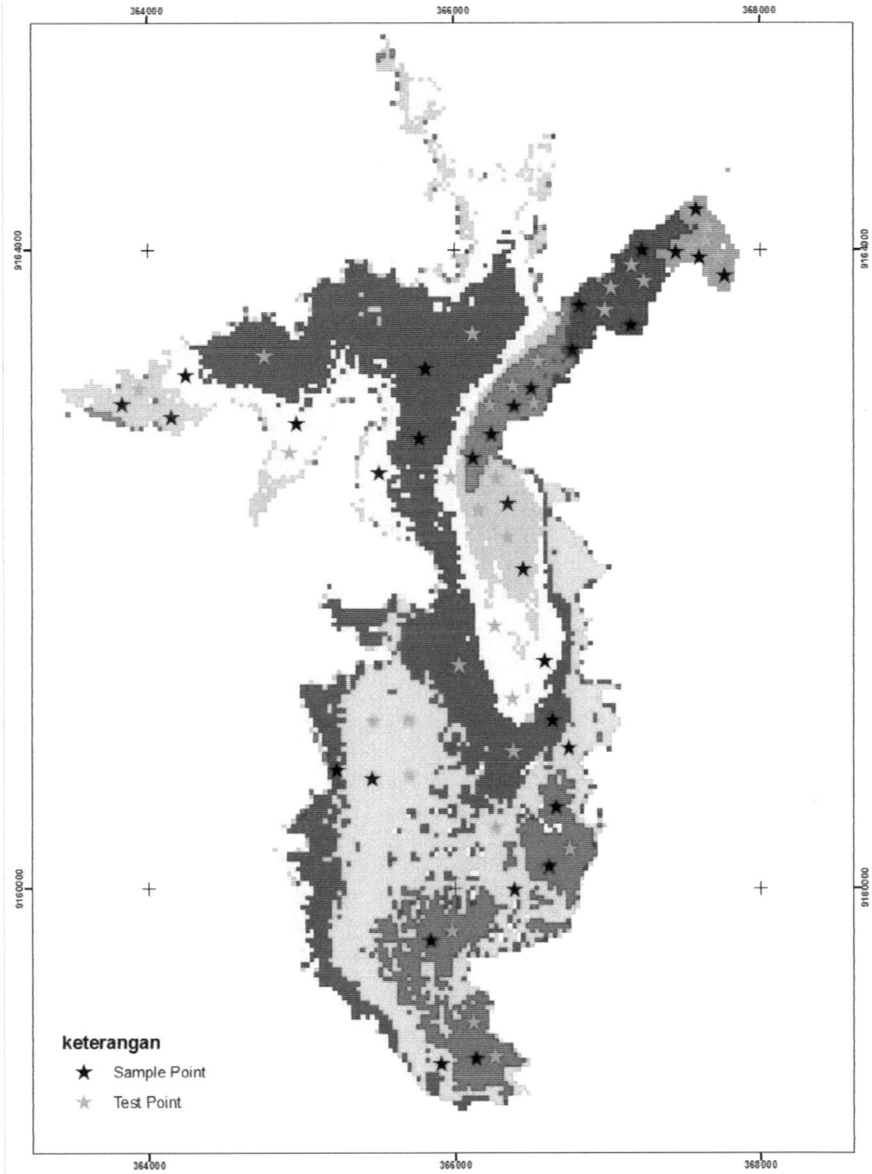

Figure 1. Location of samples for model and accuracy assessment

3.3 Field Survey

Field survey was performed on 6 August 2015. Motor boat was used as to visit each sample location. The actual location of the sample location was recorded using handheld GPS. The amount of water sample collected per site is 600 ml up to the depth of 30 cm. In total. 58 number of samples were collected during field survey.

3.4 TSS laboratory measurement

The TSS of the water samples was measured at the laboratory of Hydrology and Water Quality of the Faculty of Geography Universitas Gadjah Mada. Gravimetric method was used to measure the TSS (SNI-06-6989-3-2004).

Table 1. The result of laboratory measurement of TSS. Sample points were used to create an empirical model and test points were used to calculate the standard error of the estimated TSS.

No	Sample Code	TSS (mg/l)
1	Sample Point 1.1	13.8
2	Test Point 1.1	18.4
3	Sample Point 1.2	42.1
4	Test Point 1.2	33.6
5	Sample Point 1.3	12.3
6	Test Point 1.3	31.3
7	Sample Point 1.4	19.5
8	Sample Point 2.1	4.1
9	Test Point 2.1	37.5
10	Sample Point 2.2	37.7
11	Test Point 2.2	42.5
12	Sample Point 2.3	21.3
13	Test Point 2.3	15.1
14	Sample Point 2.4	38.8
15	Test Point 2.4	2.9
16	Sample Point 3.1	43.5
17	Test Point 3.1	20.8
18	Sample Point 3.2	11
19	Test Point 3.2	15.7
20	Sample Point 3.3	3.7
21	Test Point 3.3	7.4
22	Sample Point 3.4	7.2
23	Test Point 3.4	12.5
24	Sample Point 4.1	2.2
25	Test Point 4.1	4.6
26	Sample Point 4.2	2.7
27	Test Point 4.2	9.9
28	Sample Point 4.3	21.6
29	Test Point 4.3	8.3
30	Sample Point 4.4	10.3
31	Test Point 4.4	30.4
32	Sample Point 5.1	14.9
33	Test Point 5.1	15.4
34	Sample Point 5.2	11.7
35	Test Point 5.2	9.5
36	Sample Point 5.3	10.7
37	Test Point 5.3	11.7
38	Sample Point 5.4	10.5
39	Test Point 5.4	12.2
40	Sample Point 6.1	5.7
41	Test Point 6.1	23.8
42	Sample Point 6.2	29.9
43	Test Point 6.2	12.2
44	Sample Point 6.3	28.1
45	Test Point 6.3	1.6
46	Sample Point 6.4	17.5
47	Test Point 6.4	15.1
48	Sample Point 7.1	28.1
49	Test Point 7.1	32.8
50	Sample Point 7.2	11.7
51	Test Point 7.2	18.1
52	Sample Point 7.3	52.7
53	Test Point 7.3	42.8
54	Sample Point 7.4	23.9
55	Test Point 7.4	25.7
56	Sample Point 8.1	14.6
57	Test Point 8.1	94.1
58	Test point 8.3	15.3

3.5 TSS Data Processing Result

To obtain the empirical model of TSS using remote sensing image, on TSS, the pixel value of each TSS sample location was extracted. The results of this extraction was compared with the value of the field TSS data. Afterward, the Pearson correlation analysis was performed between field TSS and TOA reflectance value of Landsat 8 OLI bands i.e. 1 (Ultra blue), Band 2 (blue), (green) Band 3, Band 4 (red), Band 5 (Near infrared), Band 6 (SWIR), 7 Band (SWIR), Band 9 (Cirrus).

Table 2. The results of pixel values extraction from Landsat 8 OLI

No	X	Y	Ultra Blue	Blue	Green	Red	NIR	SWIR	SWIR	Cyrus
1.1	366015	9158935	0.0216	0.0223	0.0225	0.0165	0.0076	0.0047	0.0025	0.0014
1.2	366015	9158935	0.0216	0.0223	0.0225	0.0165	0.0076	0.0047	0.0025	0.0014
1.3	366585	9160195	0.0210	0.0214	0.0231	0.0158	0.0035	0.0042	0.0025	0.0011
1.4	366555	9160585	0.0211	0.0215	0.0232	0.0162	0.0013	0.0038	0.0025	0.0009
2.1	366375	9160015	0.0228	0.0232	0.0246	0.0175	0.0041	0.0042	0.0021	0.0015
2.2	366735	9160915	0.0221	0.0225	0.0245	0.0175	0.0048	0.0053	0.0033	0.0013
2.3	365475	9160765	0.0228	0.0233	0.025	0.0177	0.0021	0.0031	0.0019	0.0011
2.4	365925	9158935	0.0229	0.0229	0.0236	0.0172	0.0089	0.0056	0.0028	0.0012
3.1	366675	9161155	0.0223	0.0227	0.0261	0.0191	0.0035	0.0045	0.0029	0.0011
3.2	365745	9163225	0.0237	0.0247	0.0279	0.0190	0.0025	0.0033	0.0019	0.0014
3.3	365715	9162835	0.0236	0.0240	0.0282	0.0196	0.0026	0.0035	0.0023	0.0011
3.4	365325	9160765	0.0229	0.0232	0.0251	0.0179	0.0037	0.0034	0.0021	0.0012
4.1	366585	9161485	0.0233	0.0245	0.0281	0.0207	0.0037	0.0047	0.0031	0.0013
4.2	364275	9163225	0.0237	0.0245	0.0288	0.0202	0.0087	0.0056	0.0027	0.0014
4.3	364935	9162895	0.0238	0.0251	0.0282	0.0202	0.0077	0.0055	0.0029	0.0014
4.4	365535	9162625	0.0241	0.0246	0.0279	0.0204	0.005	0.0047	0.0026	0.0012
5.1	366435	9161995	0.0246	0.0260	0.0306	0.0222	0.0038	0.0043	0.0027	0.0012
5.2	366285	9162475	0.0241	0.0258	0.0303	0.0222	0.0036	0.0039	0.0023	0.0014
5.3	363825	9163045	0.0262	0.0264	0.0310	0.0223	0.0133	0.0071	0.0036	0.0016
5.4	364125	9162925	0.0260	0.0270	0.0317	0.0231	0.0145	0.0081	0.0039	0.0011
6.1	366135	9162745	0.0258	0.0272	0.0341	0.0254	0.0039	0.0042	0.0021	0.0014
6.2	366225	9162895	0.0262	0.0279	0.0355	0.0264	0.0067	0.0051	0.003	0.0014
6.3	366375	9163075	0.0256	0.0277	0.0351	0.0267	0.0065	0.0051	0.0029	0.0010
6.4	366465	9163165	0.0259	0.0281	0.0354	0.0267	0.0073	0.0055	0.0032	0.0012
7.1	366735	9163405	0.0265	0.0283	0.0364	0.0279	0.0086	0.0055	0.0033	0.0015
7.2	367125	9163585	0.0259	0.0285	0.0376	0.0293	0.0114	0.0062	0.0036	0.0011
7.3	367215	9163975	0.0277	0.0294	0.0394	0.0312	0.0128	0.0068	0.0039	0.0015
7.4	366825	9163645	0.0265	0.0286	0.0373	0.029	0.0088	0.0053	0.0029	0.0012
8.1	367515	9164005	0.0278	0.0304	0.0426	0.036	0.0130	0.0059	0.0033	0.0014
8.2	367575	9163945	0.0282	0.0317	0.0453	0.0398	0.0147	0.0063	0.0035	0.0011

These pixel values were used to buildan empirical model of TSS using remote sensing data.

3.6 The Pearson Correlation Analysis

The correlation analysis was conducted to look at the relationship as well as the direction of the relationship of two or more variables. This was done in order to find spectral bands with significant relationship with the field TSS, which is necessary to perform regression analysis. Only spectral bands with significant correlation coefficient (r) that can be used as input in the regression analysis. The input pixel values for the empirical model are the value of a single band, band ratio, and band multiplication.

3.7 Regression Analysis

Regression analysis is used to predict the value of TSS using image pixel values. The scatter plot of the model with the highest R^2 is provided in Figure 2.

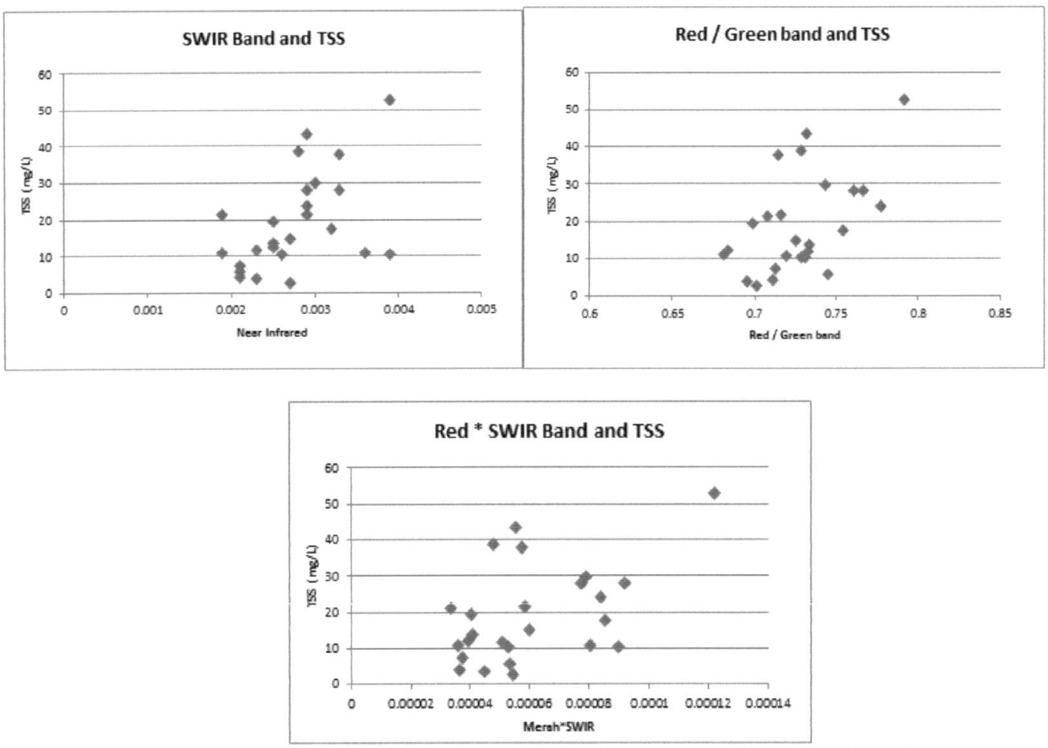

Figure 2. scatter plot showing the regression analysis of band 7, red/green band, and red*SWIR band with field TSS (informasi axis pakai bahasa inggris, lalu, garisnya dikeluarin Zi)

From the analysis, the ratio of red and green bands band produced the highest R^2. The resultant regression function is TSS = (255.78*(ratio red/green) - 166.89).

3.8 Accuracy assessment

The resultant regression function of the best model was applied to Landsat 8 OLI image to obtain the spatial distribution of TSS in the Wadaslintang reservoir. To calculate the accuracy of the modeled TSS, the SE (Standard error of estimate) technique was used as follows:

$$SE = \sqrt{\frac{\sum(\text{TSS Field} - \text{TSS model})^2}{n-2}}$$

Description:
SE (mg/L)
n = number of data

Table 3. The accuracy assessment of the TSS modeled from Red/Green band

No	X	Y	Reference TSS (mg/L)	Modeled TSS (mg/L)	(Reference TSS – Modeled TSS)^2
1	366105	9159205	18.4	16.457	3.776
2	365895	9159775	33.6	12.724	435.791
3	366645	9160345	31.3	8.888	502.283
4	365505	9161065	15.1	13.661	2.072
5	365685	9160675	2.9	18.121	231.674
6	366435	9160705	20.8	13.481	53.571
7	366105	9163525	15.7	10.476	27.293
8	364785	9163375	7.4	9.764	5.588
9	366045	9161395	12.5	18.362	34.368
10	366345	9161185	4.6	15.554	120.001
11	364935	9162745	9.9	13.038	9.847
12	365985	9162535	8.3	16.047	60.008
13	366225	9161665	30.4	18.572	139.890
14	366285	9162265	15.4	19.717	18.635
15	366255	9162565	9.5	24.532	225.975
16	363945	9163135	11.7	16.531	23.340
17	366225	9162385	12.2	20.026	61.249
18	366225	9163045	23.8	23.823	0.001
19	366495	9163075	12.2	28.793	275.317
20	366525	9163315	15.1	24.034	79.818
21	367005	9163615	32.8	37.078	18.303
22	367215	9163825	18.1	37.870	390.855
23	367155	9163945	42.8	37.217	31.167
24	367035	9163795	25.7	32.775	50.056
				Σ (Reference TSS – Modeled TSS)^2	2800.877
				SE	10.803 mg/L

The results of the accuracy assessment produced standard error of 10.8 mg/l. That value is the representation of the average deviation of the modeled TSS. The median the median TSS modelling results is 20.3 mg/l. The value of SE is smaller than the average value of the modeled TSS.

4. Results & Discussion

Total TSS in Wadaslintang Reservoir was estimated from the TSS values of the modeling results. In calculating the total number of TSS, the number of pixel values and the content of TSS were used. The content of TSS is in mg / l, as measured by the TSS laboratory results. From the statistical results, the average TSS in the Wadaslintang reservoir on 2015 was 22.61 mg / l.

TSS mapping is intended to see the spatial distribution of TSS. The spatial distribution of TSS is the result of empirical modeling that was carried out before. Through spatial distribution map of TSS, the distribution of the magnitude of TSS in Wadaslintang Reservoir can be identified. The result also shows that the area that contains the highest TSS on Wadaslintang reservoir is at the headwaters of the reservoir, which is located at the inlet of the main Wadaslintang reservoirs on the river Bedegolan. The TSS map also indicates that the content of TSS in that area is gradually increasing. The high TSS is observed in the field, where the area is very muddy compared to water in other areas of the reservoir.

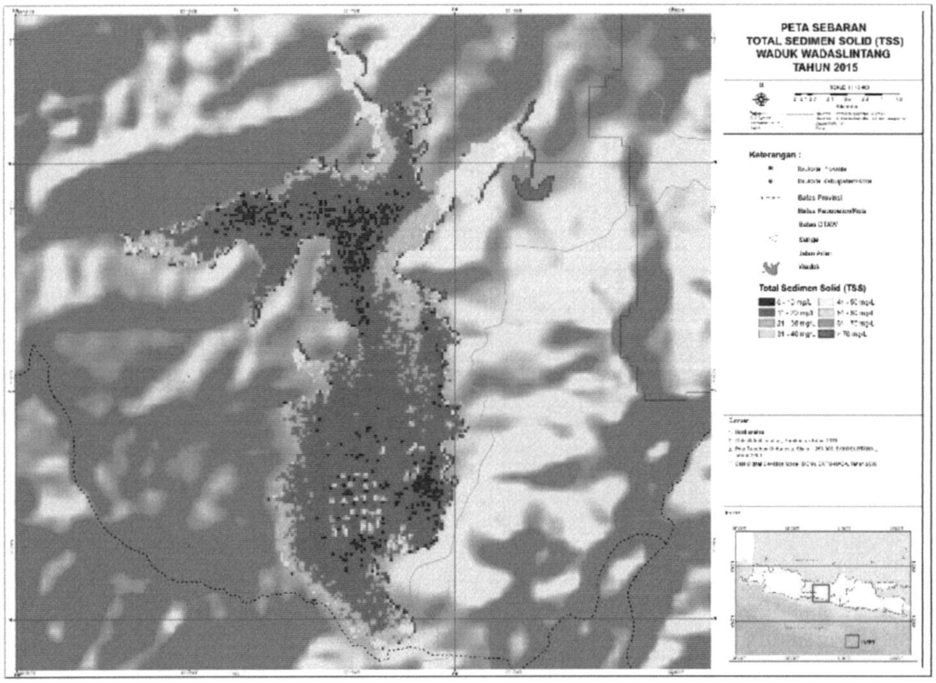

Figure 3. Total Suspended Solid Map in Wadaslintang Reservoir

Conclusion

The results of TSS mapping in the reservoir Wadaslintang using remote sensing image shows that the area with the highest content of TSS is in the region upstream of the dam, that is located at the inlet of the main Wadaslintang reservoirs on the river Bedegolan. TSS spatial distribution maps showed that the content of TSS in that area are gradually increasing, and accumulated in that area. The accuracy assessment on the accuracy of TSS results with independent field test showed that the value of the Standard Error is 10.8 mg / l. This value shows the average deviation of modeled TSS from the actual TSS value in the field. The average of modeled TSS is 20.3 mg / l.

References

[1] Asdak C 2007 Hidrologi dan Pengelolaan Daerah Aliran Sungai (Yogyakarta : Gadjah Mada Univ Press)
[2] Danoedoro Projo 2012 Pengantar Penginderaan Jauh Digital (Yogyakarta : Penerbit Andi)
[3] Lillesand dan Kiefer 1990 Penginderaan Jauh dan Interpretasi Citra (Terjemahan) Gadjah Mada University Press: Yogyakarta
[4] Sembiring R K Analisis Regresi Edisi Kedua (Bandung : Penerbit ITB)
[5] Sugiyono 2010 Metode penelitian kuantitatif kualitatif &RND (Bandung :Alfabeta)
[6] Suripin 2004 Pelestarian Sumber Daya Tanah dan Air (Yogyakarta: Andi)
[7] Sutanto 1986 Penginderaan Jauh Jilid 1 Gadjah (Yogyakarta: Mada University Press)
[8] Sutanto 1987 Penginderaan Jauh Jilid 2 (Yogyakarta : Gadjah Mada University Press)

IOP Conf. Series: Earth and Environmental Science **47** (2016) 012030 doi:10.1088/1755-1315/47/1/012030

The effect of choosing three different C factor formulae derived from NDVI on a fully raster-based erosion modelling

Bambang Sulistyo[1)]

[1)]Department of Soil Science, Faculty of Agriculture, University of Bengkulu
Jalan WR Supratman, Kandang Limun, Bengkulu 38371, Indonesia
E-mail: bambangsulistyounib@gmail.com

Abstract. The research was aimed at studying the efect of choosing three different C factor formulae derived from NDVI on a fully raster-based erosion modelling of The USLE using remote sensing data and GIS technique. Methods applied was by analysing all factors affecting erosion such that all data were in the form of raster. Those data were R , K, LS, C and P factors. Monthly R factor was evaluated based on formula developed by Abdurachman. K factor was determined using modified formula used by Ministry of Forestry based on soil samples taken in the field. LS factor was derived from Digital Elevation Model. Three C factors used were all derived from NDVI and developed by Suriyaprasit (non-linear) and by Sulistyo (linear and non-linear). P factor was derived from the combination between slope data and landcover classification interpreted from Landsat 7 ETM$^+$. Another analysis was the creation of map of Bulk Density used to convert erosion unit. To know the model accuracy, model validation was done by applying statistical analysis and by comparing E_{model} with E_{actual}. A threshold value of ≥ 0.80 or $\geq 80\%$ was chosen to justify. The research result showed that all E_{model} using three formulae of C factors have coeeficient of correlation value of > 0.8. The results of analysis of variance showed that there was significantly difference between E_{model} and E_{actual} when using C factor formula developed by Suriyaprasit and Sulistyo (non-linear). Among the three formulae, only E_{model} using C factor formula developed by Sulistyo (linear) reached the accuracy of 81.13% while the other only 56.02% as developed by Sulistyo (non-linear) and 4.70% as developed by Suriyaprasit, respectively.

1. Introduction

Degraded land is one of the global environmental problems that must be overcome [1] [2]. Number of degraded land is increasing from year to year , while the rate of improvement can not cope with the pace of the damage. Destruction of Natural Resources in Indonesia occured because all commodities of natural resources being exploited without regard to the carrying capacity [3]. Planning to conserve degraded land requires good and accurate data, one of them is the availability of erosion map.

Generally, erosion data is predicted using a model because to gain actual erosion requires much resources (timely, costly and labour intensive). USLE is one of the existing erosion model applied worldwide, including Indonesia [4]. Nevertheless, so far erosion analysis conducted is based on analysis using vector-based maps. This method involves simplification, either algorithms or procedures, and subject to subjectivity, so the result has high uncertainty [5].

With the the technological advance in Remote Sensing (RS) and Geographical Information System (GIS) these uncertainties can be minimized, that is by applying a fully raster-based erosion modelling [6]. This is in a line with Fistikoglu and Harmancioglu who state that erosion modelling which is estimated using USLE will be more *reliable* when the analysis is conducted using small raster-based data because initially USLE is developed at small areas [7]. Raster-based erosion modelling can be conducted objectively, using established algorithms and mathematical formulae, and no simplification is needed [8]. Slope data can be analyzed more accurately and more faster by utilizing *Digital Elevation Model* (DEM) in a GIS environment, while C factor can be derived through the analysis of vegetation index of remote sensing data [4].

USLE is applied worldwide because this model is easily managed, relatively simple and the number of required parameters is relatively less as compared to other more complex erosion modelling [4]. In Indonesia, its usage has been started since 1972 by Soil Research Agency in Bogor, meanwhile Ministry of Forestry also applies USLE to assess degraded land and has been adopted national wide [10]. Morgan and Nearing has proven that USLE has higher accuracy compared to RUSLE (*Revised USLE*) and the more complex model of WEPP (*Water Erosion Prediction Project*) [9]. USLE erosion model is predicted using equation as follows [11]:

$$A = R\ K\ L\ S\ C\ P \tag{1}$$

where

A = mean annual soil erosion rate (ton/hectare/year)
R = rainfall erosivity factor (R factor) (MJ mm/ha/h/year)
K = soil erodibility factor (K factor) (ton hectare/MJ/mm)
LS = slope length and steepness factor (LS factor) (dimensionless)
C = cover and management factor (C factor) (dimensionless)
P = support practice factor (P factor) (dimensionless)

A fully raster-based erosion modelling is an erosion modelling using data input that are all in raster format, not in raster format as a result of *Vector to Raster Conversion* algorithm [4]. From 5 parameters, LS, C and P are factors that can be directly as data input using raster format, while R and K factors can have raster format through spatial interpolation available in almost all GIS software. Spatial interpolation is the process of using points with known values to estimate values at other points [12].

Of the 5 factors in the formula of USLE, the C factor is the factor most difficult to determine at a very large region [13]. This expresses influence of vegetation upon soil erosion, ranging from 1 to 0, where higher values indicate no cover effect hence more erosion while lower C means a very strong cover effect resulting in no erosion. In practice, a C factor is determined by using existing C factor which is the result of research at different location from the location where the erosion will be calculated [14]. C factor information obtained is then assigned to the landuse map which was previously prepared. This method produces a constant of C factor on a relatively wide area, and do not reflect the variation of vegetation [15]. This will certainly have an impact on the calculation of erosion. By using remotely sensed digital data, C factor can be derived through the analysis of vegetation index called Normalized Different Vegetation Index (NDVI).

The research was aimed to study the efect of choosing three different C factor formulae derived from NDVI on a fully raster-based erosion modelling in Merawu watershed, Banjarnegara, Central Java.

2. Materials and Methods

Data required for fully raster-based erosion modelling were: topographical map, landform map, monthly data/report on sediment yield in watershed outlet during 24 months (June 2004 to May 2006), remotely-sensed data of Landsat 7 ETM^+ recorded on 21 May 2003 and on 20 June 2006, rainfall data during 24 months (June 2004 to May 2006) recorded in Merawu watershed and surroundings, other data and reports which support the activity. To analyze and handle these data various GIS software were used: ILWIS (*Integrated Land and Water Information System*) version 3.4 and Arc/View version 3.5. Meanwhile, some hard wares were also required consisting of drafting tablet, equipment used for field work such as: binoculars, compass, hagameter, soil munsell color, tape, ring sample, auger, *Global Positioning System* (GPS), and digital camera.

Research area was located in Merawu watershed lies between 109°41'24" – 109°50'24" E and 7°10'12" – 7°22'12" S and administratively located in Banjarnegara district, Central Java Province (Figure 1). Merawu watershed covers ± 22.734 hectares with 3 main rivers flowing through the area from north to south that are: Merawu, Urang and Penaraban rivers. Among the watersheds in the area, Merawu watershed resulted the most of sediment yield to Sudirman Reservoir (11 mm yr-1) [16].

Methods were applied by analysing factors affecting erosion in GIS environment using fully raster-based format. The pixel size for the study was 30 m by 30 m to account for the spatial resolution of Landsat 7 ETM$^+$ which was 30 m by 30 m.

Diagrametrically, a fully raster-based erosion modelling is presented in Figure 2.

Monthly rainfall data which recorded between June 2004 and May 2006 (from 8 rainfall stations located within and surrounding of Merawu watershed) was computed to get R factor based on formula developed by Abdurachman [17] as follow:

$$R_m = (Q^{2.263} * Pm^{0.678})/(40.056*D^{0.349}) \quad (2)$$

where:

- R_m = monthly average of rain erosivity index (EI_{30})
- Q = monthly average of rainfall (cm/month)
- Pm = maximum daily rainfall average (cm)
- D = monthly average of the number of rainfall days

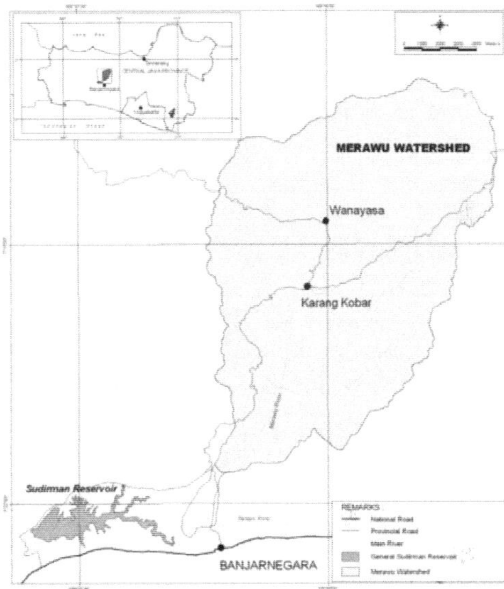

Figure 1. Location of the study.

The result of R factor then was plotted on a map for each station according to its position, digitised, transformed and spatially interpolated using *Moving Average* technique to gain map of R factor of the study area.

Soil erodibility factor (K factor) was determined using formula as follow:

$$K = \{2.17 \times 10^{-4} \times (12\text{-}OM) \times M^{1.14} + 4.20 \times (s\text{-}2) + 3.23 \times (p\text{-}3)\}/100 \quad (3)$$

where K is soil erodibility (ton.hectare.hour/ (hectare.MJ.mm)), OM is percentage of organic matter, **s** is soil structure class, **p** is soil permeability class and M is {(% silt + % very fine sand) x (100 - % clay)}.

Thirty soil samples, distributed evenly according to the landform, were taken in the field. The result of K computation for each sample then was plotted according to its position, digitised, transformed and spatially interpolated using *Kriging* technique to gain map of K factor of the study area. To apply spatial interpolation using *Kriging* technique it needs information about *sill*, *nugget* and *range* values that can be obtained by executing *spatial correlation* analysis.

Slope is derived directly from DEM from which also can be derived L factor by taking *flow direction* from each pixel into consideration. Slope distance for each pixel is equal to 30 meters long for *flow direction* directed to the South, West, North and East, and is equal to 42.43 meters long for *flow direction* directed to Southeast, Southwest, Northwest and Northeast.

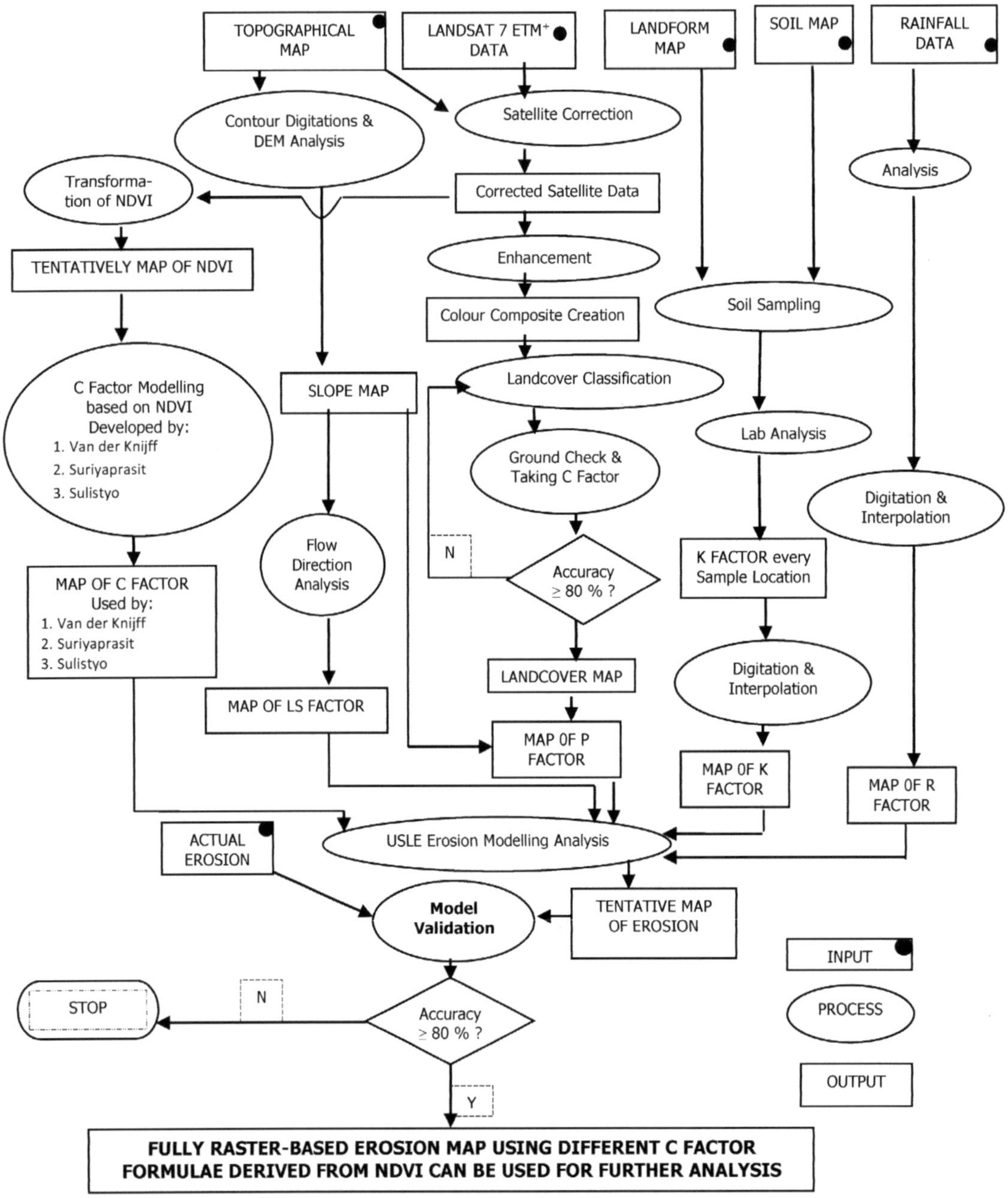

Figure 2. Diagram flow of the study

Slope length and steepness factor (LS factor) for slope < 20% is computed using the formula of Schwab *et al.* [17]:

$$LS = \sqrt{\{(L_a) \times (1.38 + 0.965\ s + 0.138\ s^2)/100\}} \quad (4)$$

while for slope > 20% LS factor is computed using the formula of Goldman *et.al.* [17]:

$$LS = [(65,41 \times s^2) / (s^2 + 10.000) + (4,56 \times s) / (s^2 + 10.000)^{0,5} + 0,065]\ [(L_a/2,21)^m] \quad (5)$$

where L_a is actual slope length (in meters), s is slope (in %) and m is a constant value which is dependt on the slope, those are m = 0.1 if $s \leq 1\%$; m = 0.3 if s > 1% and if $s \leq 3\%$; m = 0.4 if s > 3% and if s < 5%; and m = 0.5 if $s \geq 5\%$.

Cover and management factor (C factor) was derived from NDVI and formulated as follow:
Developed by Sulistyo [4]:

(a) Linear equation

$$C = 0.6 - 0.77 \text{ NDVI} \tag{6}$$

(b) Non-Linear equation

$$C = 5.77 * \exp\{-5.62 * \text{NDVI}\} \tag{7}$$

Developed by Suriyaprasit [19]:

$$C = 0.227 * \exp\{-7.337 * \text{NDVI}\} \tag{8}$$

NDVI is formulated as:

$$\text{NDVI} = (\text{NIR} - \text{R}) / (\text{NIR} + \text{R}) \qquad [9]$$

where NIR and R indicate channel or band of Landsat 7 ETM^+ which are near infrared and visible red respectively.

Landcover change analysis was done using two satellite Landsat 7 ETM^+ (recorded on 20 June 2006 and 21 May 2003) to know the rate of changes to justify the correction factor used for computing monthly C factor for 24 months in a line with the months of rain erosivity.

Generally, for the shake of ease and practicality, support practice factor (P factor) is assigned value as 1 for the whole area under studied. In this research P factor is derived from the combination between slope data from Digital Elevation Model and landcover classification interpreted from Landsat 7 ETM^+ using criteria developed by Abdurachman *et al.* [20] as shown in Table 1.

Table 1. P factor value based on Abdurachman

No.	Landcover	P factor
1	Agricultural Area with Slope ≤ 8 %	0.50
2	Agricultural Area with Slope between ≥ 8 % and 20 %	0.75
3	Agricultural Area with Slope ≥ 20 %	0.90
4	Shrub, Secondary Forest and Forested Area	1.00

Another analysis supporting the research activity was the creation of bulk density map which was generated through plotting bulk density data according to its position, digitised, transformed and spatially interpolated using *Kriging* technique. Map of bulk density was used to convert erosion unit as from a Mg ha-1 $month^{-1}$ to mm $month^{-1}$ [21].

After whole data were analysed, then erosion can be calculated. The result of erosion using USLE is assumed only to gain *sheet* and *rill erosion*. In order to get total erosion in a watershed (*gross erosion*), other erosion such as *gully erosion* and *channel erosion* are determined according to the result of previous research done by Piest *et al.* and Seyhan [18] stated that *gully erosion* was one-fifth (1/5) of the total sediment occured, while *channel erosion* was about 10% of *sheet* and *rill erosion*.

Technically, estimated total soil loss (A) of Merawu watershed is the result of multiplication among USLE parameters previously described. Pixel value of map of erosion from USLE (A) is soil loss as a result of *rill erosion* and *sheet erosion* for the area of 30 m x 30 m, in Mg ha^{-1} $month^{-1}$. By multiplying (and then summing them up for the whole Merawu watershed) pixel value of map of erosion from USLE (A) with pixel area (900 m^2 = 0.09 hectare) and divide it by bulk density, watershed area (22,734 ha = 227,340,000 m^2) and constant number of 10 will result real soil loss in a watershed ($A_{watershed}$) in mm $month^{-1}$.

$$A_{watershed} = (A \times 0.09 / \text{bulk density} / 227{,}340{,}000 / 10)_{1-n} \tag{10}$$

However, USLE is assumed only to gain *sheet* and *rill erosion*. In order to get total erosion in a watershed (*gross erosion*), other erosion such as *gully erosion* and *channel erosion* are determined according to the result of previous research done by Piest *et al.* and Seyhan [18] who stated that *gully erosion* was one-fifth (1/5) of the total sediment occured, while *channel erosion* was about 10% of *sheet* and *rill erosion*.

$$E = (A + G + C) \qquad [11]$$

where E is *gross erosion*, A is *sheet* and *rill erosion* resulted from USLE, G is *gully erosion* and C is *channel erosion*.

A fully raster-based erosion modelling is a new model in which it needs model validation. Comparison between erosion as a result of modelling (E_{model}) with actual erosion (E_{actual}) can be done using statistical analysis (ANOVA or Correlation Analysis) or direct comparison (substracted E_{model} from E_{actual}). A threshold value of > 0.8 or > 80% is chosen to determine whether or not a model is accepted or refused. Actual erosion data for this study was suplied by PT Indonesia Power which regularly monitor sediment yield in outlet of Merawu watershed.

3. Results and Discussion

3.1. Map of R Factor

Eight rainfall stations located within and surrounding of Merawu watershed were used for the study to compute R factor. Example of the pattern of some R factor as a result of spatial interpolation using *Moving Average* technique is presented in Figure 3.

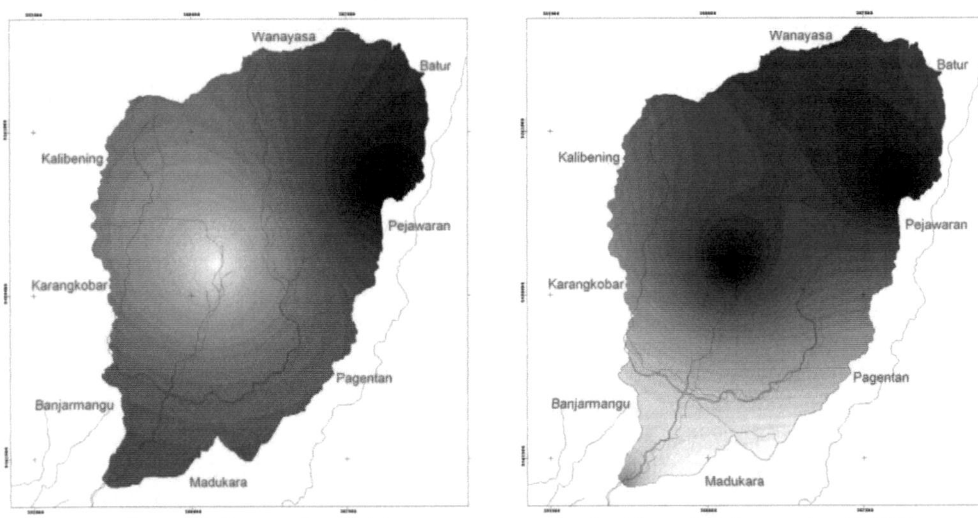

a. Map of R factor, Oct 2004 b. Map of R factor, April 2005

Figure 3. The pattern of some R factors of Merawu watershed

3.2. Map of K Factor and Map of Bulk Density

To apply spatial interpolation using *Kriging* technique, to get map of K factor and map of Bulk Density, it needs information about *sill*, *nugget* and *range* values that can be obtained by executing *spatial correlation* analysis. After some *trial and error*, finally to map K factor the values of 0.000; 0.013; and 8.000 were chosen as *sill*, *nugget* and *range*. While to map the distribution of bulk density the values of 0.025; 0.150 and 8.500 were chosen as *sill*, *nugget* and *range*. The result of map of K factor map of bulk density is shown in Figure 4. Merawu watershed has soil erodibility in average of 0.29 (minimum: 0.08 and maximum: 0.54), while their bulk density average is 1.60 (minimum 1.03 and maximum 2.16).

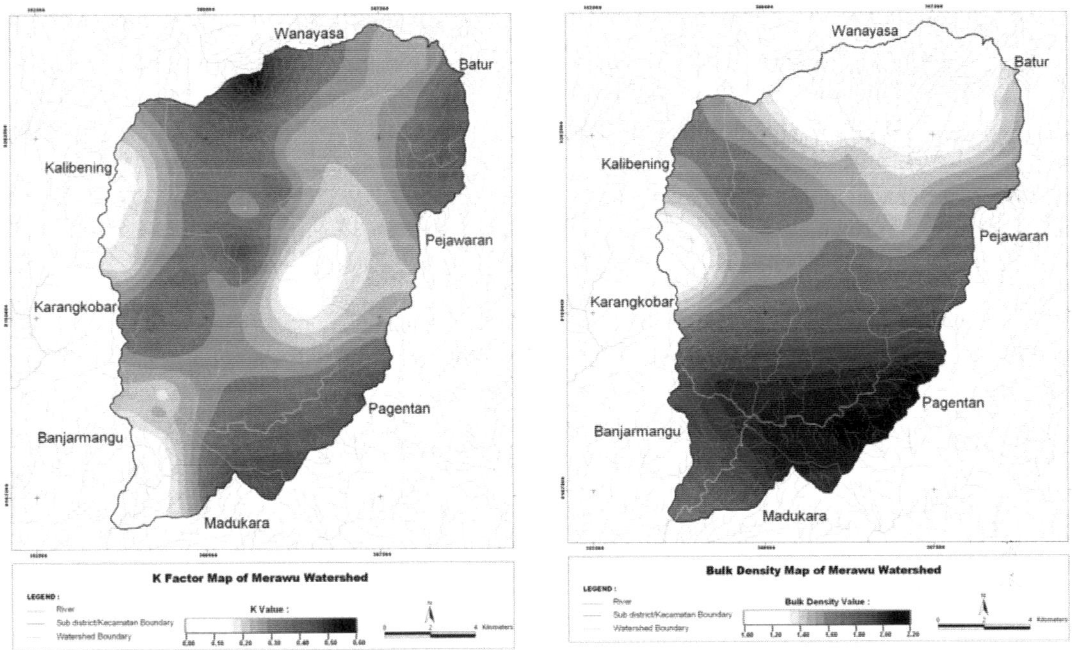

Figure 4. Map of K factor (left) and Map of bulk density (right) of Merawu watershed

3.3. Map of LS Factor

The area of LS factor is presented in Table 2, while its distribution is shown in Figure 5. From Table 3 it can be inferred that Merawu watershed was dominated by LS factor < 20 with average covering area of 16,465 ha, while the rest area had LS factor > 20.

Table 2. Area of Merawu watershed according to its LS factor

No	LS factor	Area
1	$0 < LS < 20$	16,468
2	$20 \leq LS < 40$	3,437
3	$40 \leq LS < 60$	1,680
4	$60 \leq LS \leq 80$	1,150
	Total	22,734

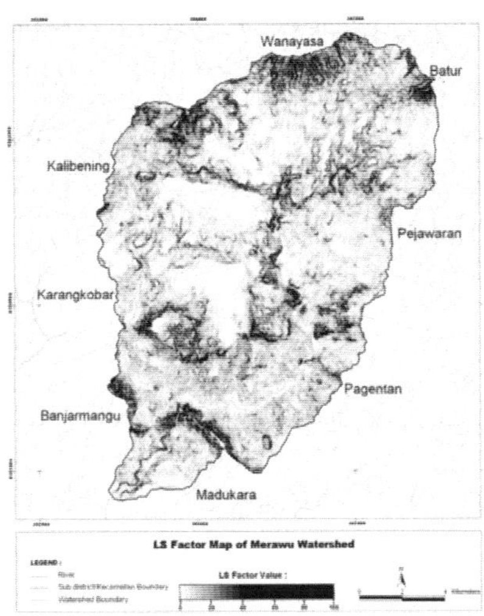

Figure 5. Map of LS factor of Merawu watershed

3.4. Map of C Factor

The area of C factor developed by Suriyaprasit and Sulistyo (linear and non-linear) is shown in Table 3, while its distribution, especially which is developed by Sulistyo, is presented in Figure 6.

To interpolate C factor every month in accordance with the months used in computing R factor, analysis of landcover change was done. This analysis was used by overlaying NDVI recorded on 20 June 2006 on NDVI recorded on 21 May 2003. The result shows that the total number of unchange pixel is 218,947 (86.69%). This means that the pixel changed is 13.31% during 36 months, or it can be concluded that the rate of change is 0.3698% mo^{-1}. This value is used for interpolating monthly C factor between May 2006 and June 2004.

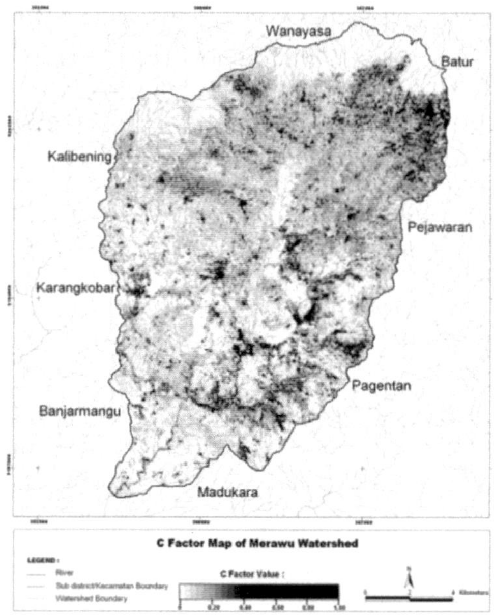

Figure 6. Map of C factor derived from NDVI as developed by Sulistyo

3.5. Map of P Factor

The result of P factor is presented in Table 6, while its distribution is shown in Figure 7.

Table 3. The area of C factor in Merawu watershed (hectare)

No.	Interval class of C factor	Developed by:		
		Sulistyo (non-linear)	Suriyaprasit	Sulistyo (linear)
1.	0 - 0.1	15,109	22,672	12,988
2.	0.1 - 0.2	4523	55	5,804
3.	0.2 - 0.3	1265	6	2,291
4.	0.3 - 0.4	627	0	1,116
5.	0.4 - 1.0	1,206	0	535
	Total	22,734	22,734	22,734

Table 4. The area of P factor in Merawu watershed

No	Landcover	Area (Ha)	Area (%)
1.	Agricultural Areas with P factor = 0.50	2,760	12.14
2.	Agricultural Areas with P factor = 0.75	5,604	24.65
3.	Agricultural Areas with P factor = 0.90	6,824	30.01
4.	Non Agricultural Areas with P factor = 1	7,548	33.2
	T o t a l	22,734	100,00

From table 4 it can be inferred that Merawu watershed is dominated by agricultural area with P factor < 1 covering 15,186 ha (66.8%), while the rest is non agricultural area with P factor = 1 covering 7,548 ha (33.2%).

3.6. Fully Raster-Based Erosion Modelling by Using three different C Factor formulae derived from NDVI

The gross erosion estimated using USLE generated from three different C factor formulae derived from NDVI (E_{model}), actual erosion (E_{actual}) and the result of validation is presented in Table 5.

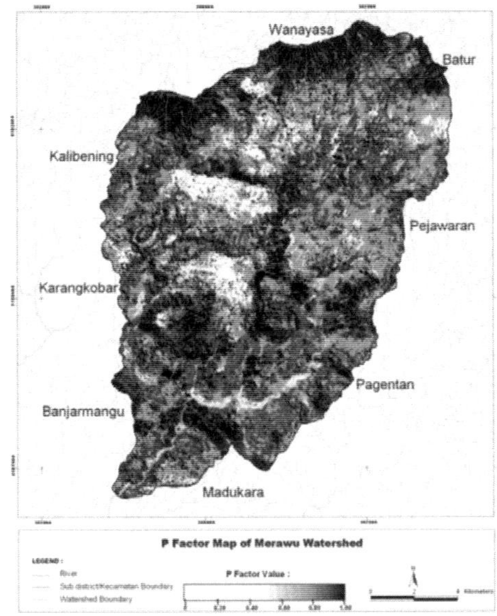

Figure 7. Map of P factor of Merawu watershed

From Table 5 it can be inferred that all E_{model} using three different C Factor formulae derived from NDVI have high coefficient of correlation with E_{actual} (r = 0.87). The result of analysis of variance (F test) shows that there was significantly difference between E_{model} and E_{actual} when using C factor formula developed by Suriyaprasit and Sulistyo (non-linear), indicated by the $F_{computation}$ value (26.16 and 4.24 respectively) which was greater then F_{table} (4.06) using degree of freedom 1 and 46 at α 5%. Nevertheless, among the 3 formulae, only E_{model} developed by Sulistyo reached the accuracy of 81.13% while the other only 56.02% as developed by Sulistyo (non-linear) and 4.70% as developed by Suriyaprasit, respectively. Its mean that model developed by Sulistyo (linear) can be used for further analysis (such as for planning purposes, research, or other analysis).

Table 5. Summary of Estimated USLE erosion, Actual Erosion and the result of validation

No	Month	E_{actual} (mm/month)	E_{model} (mm/month) Sulistyo (Non-Linear)	Suriyaprasit	Sulistyo (Linear)
A.	Average	1.027	2.63	0.05	1.22
B.	Coefficient of Correlation		0.87	0.87	0.87
C.	ANAVA test:	$F_{computation}$	4.24	26.16	0.24
		F_{table}	4.06	4.06	4.06
D.	Absolute Accuracy (%)		56.02	4.70	81.13

However, it must be realized that a model will be very suitable to be developed in accordance with the conditions of the location where the model is developed. Necessary modifications is needed if the model is applied to another location. The model developed by Suriyaprasit conducted in Thailand. Actually, the model developed by Suriyaprasit similar to the model developed by Sulistyo, namely C factor that was derived from the formula after applying regression analysis between NDVI of remote sensing data and C factor measured directly on the field. The difference is that Suriyaprasit only using non-linear model, while Sulistyo using both linear and non-linear model. Linear model apparently better than those of non-linear model.

4. Conclusions

All E_{model} using three formulae of C factors have coeeficient of correlation value of > 0.8. The results of analysis of variance showed that there was significantly difference between E_{model} and E_{actual} when using C factor formula developed by Suriyaprasit and Sulistyo (non-linear). Among the three formulae, only E_{model} using C factor formula developed by Sulistyo (linear) reached the accuracy of 81.13% while the other only 56.02% as developed by Sulistyo (non-linear) and 4.70% as developed by Suriyaprasit, respectively.

References

[1] Bohre P and Chaubey OP 2014 Restoration of degraded lands through plantation forests *Globa l Journal of Science Frontier Research* **14** 18-27

[2] Sulistyo B 2015 Kajian Perubahan Tingkat Kekritisan Lahan Sebagai Akibat Proses Eliminasi Unit Lahan *Proceeding of National Seminar* Masyarakat Biodiversitas Indonesia, Yogyakarta **4** pp 828-33

[3] Kartodihardjo H 2008 Pengelolaan sumberdaya alam dan lingkungan hidup Seminar Lingkungan Hidup dan Sumberdaya Alam Universitas Bengkulu Bengkulu

[4] Sulistyo B 2011 *Pemodelan spasial lahan kritis berbasis raster melalui integrasi Landsat 7 ETM+ dan sistem informasi geografis dissertation* Gadjah Mada University

[5] Eweg HPA, Lammeren LV, Deurloo H and Woldai Z 1998 Analysing degradation and rehabilitation for sustainable land management in the Highlands of Ethiopia *Land Degradation Development* **9** 529-42

[6] Sulistyo B 2009 Toward a fully and absolutely raster-based erosion modelling by using RS and GIS *Indonesian Journal of Geography* **41** 149-70

[7] Fistikoglu O and Harmancioglu NB 2002 Integration of GIS with USLE in assessment of soil erosion *Water Resource Management* **16** 447-67

[8] Hadmoko DS 2007 Toward GIS-based integrated landslide hazard assessment: A critical overview *Indonesian Journal of Geography* **39** 55-77

[9] Wainwright J and Mulligan M 2002 *Environmental Modelling Finding Simplicity I Complexity* (London: John Wiley & Sons Ltd)

[10] Indonesian Ministry of Forestry 2009 *Peraturan Menteri Kehutanan RI Nomor: P 32/Menhut-II/2009 tentang Tata Cara Penyusunan Rencana Teknik Rehabilitasi Hutan dan Lahan Daerah Aliran Sungai* (Jakarta: Ministry of Forestry)

[11] Wischmeier WH and DD Smith 1978 Predicting Rainfall Erosion Losses: A Guide to Conservation Planning USDA *Agriculture Handbook* No 37

[12] Chang KT 2008 *Introduction to Geographic Information Systems* (New York: McGraw-Hill International Edition)

[13] Alejandro MA and Amasa K 2007 Estimation of vegetation parameter for soil erosion using linear Spectral Mixture Analysis of Landsat ETM data *Photogrammetry and Remote Sensing Journal* **62** 309-24

[14] Folly A, Bronsveld MC and Clavaux M 1996, A knowledge-based approach for C-factor mapping in Spain using TM and GIS *International Journal of Remote Sensing* **12** 2401-15

[15] Wang G, Wente S, Gertner GZ, and Anderson A 2002 Improvement in mapping vegetation cover factor for the universal soil loss equation by geostatistical methods with Landsat Thematic Mapper images *International Journal of Remote Sensing* **23** 3649-67

[16] PT Indonesia Power 2009 *Laporan Penyelidikan Sedimentasi Waduk PLTA PB Soedirman. Unit Bisnis Pembangkitan Mrica* (Banjarnegara: PT Indonesia Power 2009)

[17] Abdurachman A 2008 Teknologi dan strategi konservasi tanah dalam kerangka revitalisasi pertanian *Pengemb Inovasi Pert* **2** 105-24

[17] Asdak C 2007 *Hidrologi dan Pengelolaan Daerah Aliran Sungai* (Yogyakarta: Gadjah Mada University Press)

[18] Santoso MR 2005 *Aplikasi teknik interpretasi citra landsat TM dan SIG untuk pemetaan erosi menggunakan metode USLE pada DAS Oya DIY* Thesis Gadjah Mada University, Yogyakarta

[19] Suriyaprasit M 2008 *Digital terrain analysis and image processing for assessing erosion prone areas* Thesis,International Institute for Geo-Information Science and Earth Observation (ITC), Enschede The Netherlands

[20] Abdurachman A, Barus A and Kurnia U 1985 Pengelolaan tanah dan tanaman untuk usaha konservasi tanah. *Pemb Penel Tanah Pupuk* **3** 7-12

[21] Arsyad S 2000 *Konservasi Tanah Dan Air* (Bogor: IPB Press)

IOP Conf. Series: Earth and Environmental Science **47** (2016) 012031 doi:10.1088/1755-1315/47/1/012031

Spatial Temporal Land Use Change Detection Using Google Earth Data

Adi Wibowo[1,2], Khairulmaini Osman Salleh[2], F.Th.R Sitanala Frans[1], Jarot Mulyo Semedi[1]

[1]Department of Geography, Faculty of Mathematics and Natural Sciences, University of Indonesia, 16424 Depok, Indonesia
[2]Department of Geography, Faculty of Arts and Social Sciences, University of Malaya, 50603 Kuala Lumpur, Malaysia

adi.w@sci.ui.ac.id, jarot.mulyo@ui.ac.id

Abstract. Land use as representation of human activities had different type. Human activity needs land for home, food, school, work, and leisure. Land use changed depends on human activity in the world within spatial and temporal term. This study aims to identify land use change using Google Earth data spatially and temporally. To answer the aim of this research, Google Earth data within five-year used for the analysis. This technique use for detection and mapping the land use change. The result saw the spatial-temporal land use change each year. This result addressed very importance of Google Earth Data as spatial temporal land use detection for land use mapping.

Keywords: Land Use, Google Earth Data, Change Detection, Spatial-Temporal

1. Introduction

Urbanization including residential, commercial, and industrial developments initiated one of the most dramatic human-induced changed of a natural ecosystem: a natural landscape, often containing transpiring vegetation and a pervious surface, is converted to a built, largely impervious landscape made up of rigid, sharp-edged roughness elements [1]. With rapid urbanization, there has been a tremendous growth in population and buildings in cities. In Singapore, rapid population influx has led to demands for converting natural areas to public housing [2]. The Urban conurbation within urbanized in Selangor has a remarkable related area in 1988 compared to the latest urban land cover in 1999 that urban land cover can be associated with buildings, road pavement, highways, green parks and also bare soil due to earthwork activities [3]. Land use type used climate information [4] as land use cover had ability absorb and reradiate sun radiation to generated urban heat ([2], [5], [6]).

Those phenomenon land use change as natural as consequences of human dimension within urban land uses cover types ([7], [8], [9], [10], [11]). Even though its natural phenomena in urban area, it could be threat in tropical cities ([9], [12], [13], [14]). Sun heat stored and reradiated related with land surface and urbanized area related with high temperature ([15], [16]), the other hand vegetated surfaces with moist soil related with low temperature [16] and without vegetated [15] of the city-core was higher temp than its surrounds covered by vegetation. Analysis spatial temporal ([1], [5], [17],

[18]) used to analysis spatial temporal of the highest vegetation cover on land use cover type ([5], [17], [18]).

The Google Earth (GE) shows the area at fine spatial resolution, revealing clearly that it were dominated by a few institutional structures, which a little investigation identifies as a medical school and close-up of the anomalous tract, showing its current mix of park and institutional land use [19]. The GE imagery were selected for a case study in Wuhan City to perform an object-based land use/cover classification with classification accuracy was assessed by using 570 validation points, the results showed that GE has an overall classification accuracy of 78.07% and strongly proved the potentials of GE in land use/cover mapping [20]. This research aim to identifying land use change using Google Earth Data. For detection and collecting data land cover can used Google Earth data. Google Earth has entered the realm of everyday life with powerful engaging (with) the world and an integral part of the processes of how we come to know spaces, places and sites [21].

Land use change is a phenomenon caused by human activities as a spatial temporal impact in line with human dynamic growth.

2. Material and method

This research study area is at Tangerang City on Banten Province, Depok City on West Java Province (the fast growing city near Jakarta) and University of Indonesia and University of Malaya (University of Inodonesia near Jakarta, Indonesia and University of Malaya near Kuala Lumpur, Malaysia). The focus of this research is on what are the Spatial and Temporal Land Use Change Detecting Using Google Earth Data (figure 1).

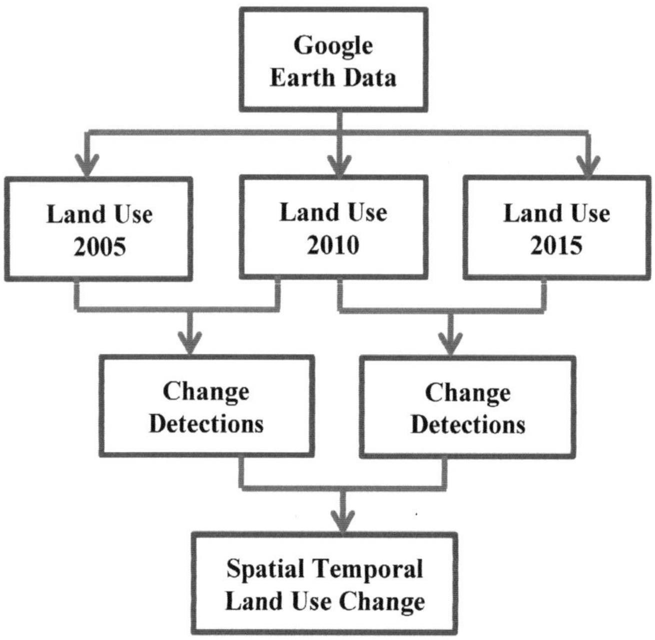

Figure 1. Diagram of Research Methodology

Data collection and extraction start from downloaded satellite imagery data of Google Earth as resource data. "*Google Earth software's success at tangibly impacting what is happening on the ground that's way Geographers, political theorists, and media scholars have begun to consider used Google Earth Data*" [22]. The satellite image data from Google Earth is mosaicked using ArcGIS software. The result of the mosaicked images does not have geographic references. A georeferenced tool is used to the mosaicked images to get a georeferenced image based on Universal

Transform Mercator (UTM) projection. Land use type from the Google Image was digitized based on 10 m x 10 m square grid ([6], [14], [16]) and the attribute on every land use types also being collected (figure 2). After finish created land use cover data from Google Earth then used survey validated the land use cover. Thus data land use cover types save in geodatabase storage and easy next processed summary of total area each land use cover type, setting within standard symbol and color, and the las processed is layout used Arc GIS with cartographic standard processes until ready presentation as land cover map ([6], [14], [16]).

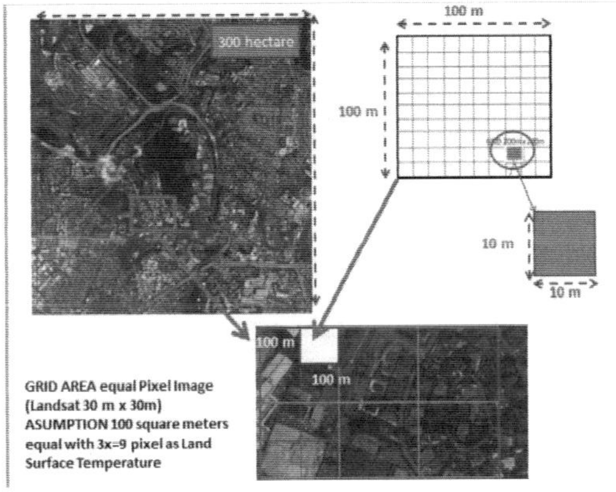

Figure 2. Digitize Land Use using grid 10 x 10

3. Result and discussion

3.1. Google Earth Data

Geographic Information Systems (GIS) Data is spatial data that is georeferenced to real-world locations on the Earth and there are two types of GIS data, **vector data** (vector datasets consist of points, lines, or polygons representing objects on the map) data and **raster** data (raster datasets are regular grids of data, which can represent images such as satellite or aerial photography, continuous surfaces such as elevation models, or thematic classes such as land cover or habitat maps) (https://www.google.com/earth/outreach/ tutorials/importgis.html). The accelerated rate of urban growth in tropical cities highlights the critical necessity of creating more outdoor spaces for leisure and recreation activities of citizens [23]. The Google Earth had data temporal depend on data archive already used with previous study in Pinang Sub-Districts, Tangerang City [24], Kukusan Depok City [25] and also studied at University of Indonesia ([14], [16]).

This research used data sample at area near Situ Cipondoh in Tangerang City used temporal archived Google Earth data between 2004, 2010 and 2015. Those data used five years to get different land use type, but the archive Google data did not have on 2005 then it changed into Google Earth data on 2004. Based on figure 3, area near Situ Cipondoh had land use on North Area within home, government official, road, river, and another land use type in urban area. This land use information representation on 2015 Google Earth Data on archive (7/8/2015).

Figure 3. Google Earth Data 2015

Based on figure 4, area near Situ Cipondoh had land use on North Area within home, government official, road, river, and another land use type in urban area. This land use information representation as 2010 Google Earth Data archive (2/15/2010). In general, both Google Earth Data is very similar, but in detail within 10 x 10 square meters, Google Earth Data on data 2015 and Google Earth Data on 2005 had different land use type.

Figure 4. Google Earth Data 2010

Based on figure 5, area near Situ Cipondoh had Land Use on North Area within home, government official, road, river, and another land use type in urban area. This land use information representation on 2004 Google Earth Data archive (3/24/2004) with different information based on compare data with Google Earth Data on 2010 and Google Earth Data on 2015.

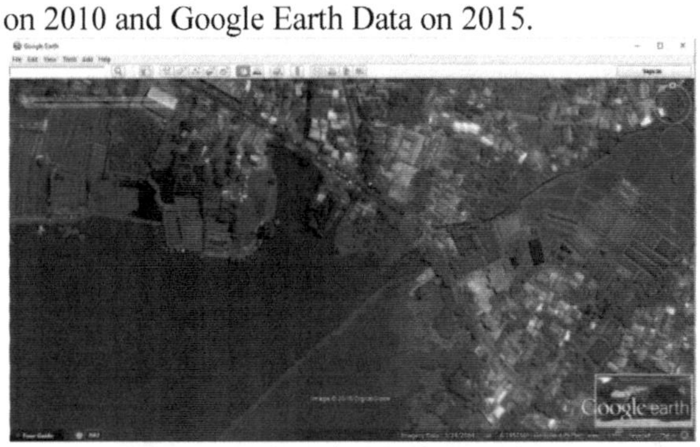

Figure 5. Google Earth Data 2004

3.2. Change Detection Using Google Earth Data

Google Earth Data archived could be used to identify land use change on specific area. For example, land use changed on area near Situ Cipondoh in the past time (temporal analysis) could identify used Google Earth Data and easy to get the information of land use type existing in the past time. Based on figure 6 saw on circle red color were land use changed from land use type farming used Google Earth Data on 2004 compare to land use type home and office used Google Earth Data on 2010.

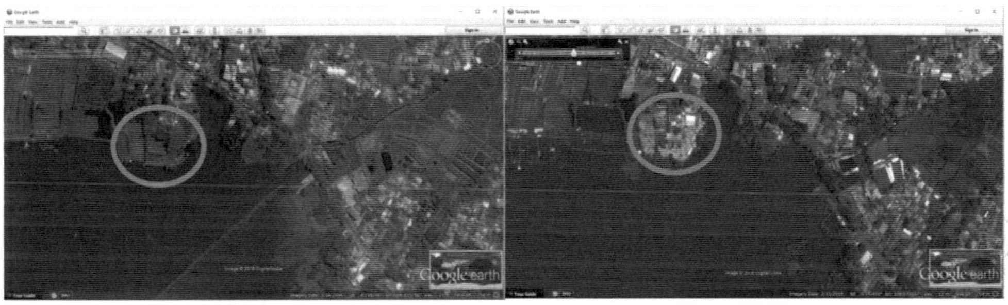

Data 2004 Data 2010

Figure 6. Change Detection Using Google Earth Data 2004 and 2010

Data 2015 Data 2010

Figure 7. Change Detection Using Google Earth Data 2010 and 2015

The previous study [11] in Tangerang City saw on figure 8 and table 1. Based on figure 8 saw on rectangle with red color were land use changed used Google Earth Data on 2001 compare to land use used Google Earth Data on 2012. These change detection using Google Earth Data archived very useful to get information of spatial-temporal land use change on the research study.

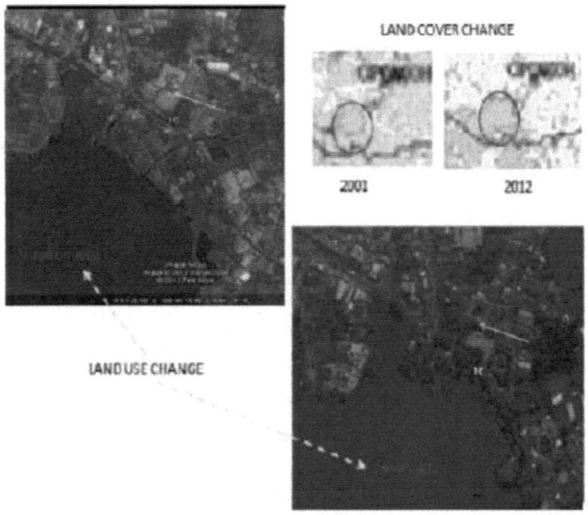

Figure 8. Change Detection Using Google Earth Data 2001 and 2012 at Situ Cipondoh, Tangerang City, Banten Province

Based on data land use change on table 1, the total area land cover change can detection, for example open space change negative 2,607.3 hectares and water bodies plus 143.2 hectares

Table 1. Land Use Changed 2001 and 2012 in Tangerang City, Banten Province

No.	General Land Cover Type	2001 (hectare)	2012 (hectare)	Changed (hectare)
3.	Street	1.945.58	1,945.58	0.0
4.	Open Space	2.572.10	35.17	-2,607.3
5.	Build up area	7,065.41	11,252.58	+4.187.2
6.	Water body	28.31	171.55	+143.2
7.	Vegetation cover	6,428.63	5,159.49	-1.269.1

The other research used Google Earth Data ([14], [16]) saw on figure 9. This research used Google Earth Data detection land use changed in university campus. The red and yellow rectangle gave the information new building at University of Indonesia and the purple rectangle as land use changed from park become sport area. The green rectangle land use changed from part become building and the black rectangle as land use change from parking area become building, both land use changed at University of Malaya.

IOP Conf. Series: Earth and Environmental Science **47** (2016) 012031 doi:10.1088/1755-1315/47/1/012031

2010 2013 2015

2010 2013 2015

Figure 9. Change Detection Using Google Earth Data 2010, 2013 and 2015 at Univ. of Indonesia and Univ. of Malaya

The other research used Google Earth Data [23] saw on figure 10 and table 2. This research used Google Earth Data detection land use changed near university campus. The black rectangle (figure 10) gave the information building near toll road in 2015 (Google Earth Data 9/29/2015) more than building in 2012 (Google Earth Data 3/27/2012). Those are another Google Earth Data 5/17/2004 on figure 9 saw the information of land use before the new exit toll development in black rectangle area.

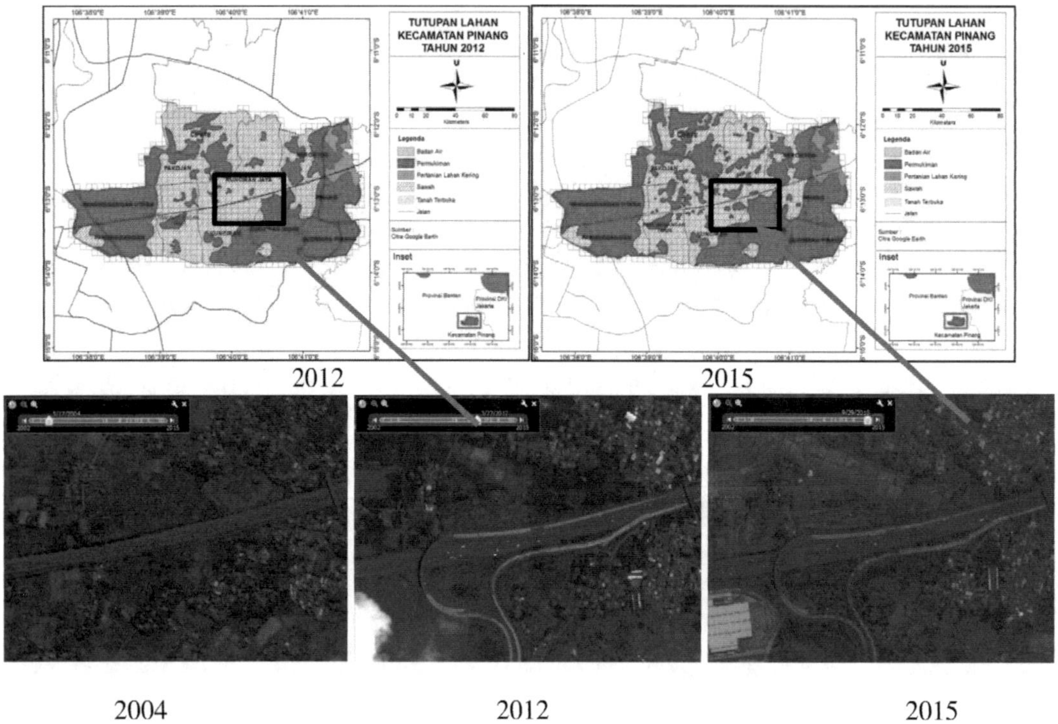

Figure 10. Change Detection Using Google Earth Data 2004, 2012 and 2015 at Pinang District, Tangerang City

The research saw on table 2 total building density changed between 2012 and 2015 at Penang District. This research used Google Earth Data detection land use changed within building changed saw with black rectangle (figure 10) gave the information of land use change within road and building added at Penang District.

Table 2. Building Density Changed 2012 and 2015 in Penang District, Tangerang City

No.	Building Density	2012 (hectare)	2015 (hectare)	Changed (hectare)
1.	Low	948.96	482.31	-466.65
2.	Moderate	1,104.10	1,214.10	+110.00
3.	High	88.02	456.93	+368.01

The other research used Google Earth Data [24] saw on figure 10 and table 3. This research used Google Earth Data detection land use changed near university campus. The red circle gave the information building near university campus (500 meters) in 2015(Google Earth Data 7/18/2015) more than building in 2005 (Google Earth Data 4/6/2005).

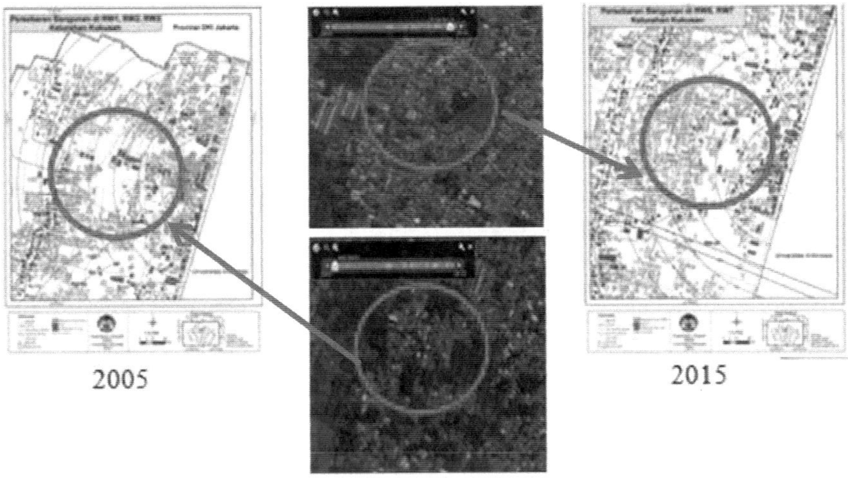

Figure 11. Change Detection Using Google Earth Data 2005 and 2015 at Kukusan Village, Depok City

The research saw on table 3 with total building changed between 2005 and 2015 within radius 100 meters from University of Indonesia gate at Kukusan Village. This research used Google Earth Data detection land use changed near university campus. The red circle (figure 11) gave the information building near university campus (500 meters) in 2015 more than building in 2005. The land use change as building added at Kukusan Village used Google Earth Data had detection that highest building change or added with radius 100-300 meters near UI campus.

Table 3. Building Changed between 2005 and 2015 in Kukusan Village, Depok City

No.	Radius (meters)	Total 2005	building 2015	Changed (hectare)
1.	<100	57	116	+ 59
2.	100-200	211	293	+ 82
3.	200-300	350	435	+ 85
4.	300-400	461	530	+ 69
5.	400-500	518	558	+ 40

Those all research at Tangerang, Depok and UM and UI saw that Google Earth Data very importance used to detection of land use changes.

4. Conclusion

The conclusion this research on the spatial-temporal land use change detection, using the Google Earth Data, is very importance and very useful as spatial temporal of land use detection for land use mapping.

Acknowledgements

The authors are deeply grateful to Department of Geography, Faculty of Mathematics and Natural Sciences, University of Indonesia for support of the research.

References

[1] Tran Hung, Daisuke Uchihama, Shiro Ochi and Yoshifumi Yasuoka 2006 Assessment with satellite data of the urban heat island effects in Asian mega cities *International Journal of Applied Earth Observation and Geo-information* **8** Issue 1 p34–48

[2] Wong N H and Yu C 2005 Study of Green Areas and Urban Heat Island in a Tropical City *Journal of Habitat International* **29** 547-558.

[3] Ahmad S and Noorazuan Md Hashim 2007 Effects of Soil Moisture on Urban Heat Island Occurrences: Case of Selangor Malaysia *Humanity & Social Sciences Journal* **2** p132-

[4] Chao, Rena, Tejo, Spit, Sanda, Lenzholzerc, Hung, Lam, Steve, Yimd, Bert, Heusinkvelde and Lutz Katzschnerf 2012 Urban climate map system for Dutch spatial planning *International Journal of Applied Earth Observation and Geoinformation* **18** p207–221

[5] Manat Srivanit and Kazunori Hokao 2013 Evaluating the cooling effects of greening for improving the outdoor thermal environment at an institutional campus in the summer *Building and Environment* **66** p158-172

[6] Adi Wibowo, Agung Raditya, Djoko Harmantyo, and Jarot Mulyo Semedi 2015 *Land Surface Temperature as Urban Hazard in Education Area (A Case Study: University of Indonesia) Proceeding the First International Conference of Indonesian Society for Remote Sensing Surabaya Indonesia*

[7] Parham A Mirzae and Haghighat F 2010 Approach to study urban heat island: abilities and limitation. *Journal of Building and Environment* **45** p2192-2201

[8] Rizwan A Memon Dennis Y C Leung and Chun-Ho Liu 2009 An Investigation of Urban Heat Island Intensity (UHII) as an indicator of urban heating *Journal of Atmospheric Research,* **94** Issue 3 p491-500.

[9] Ichinose Toshiaki Matsumoto Futoshi Kataoka and Kumi 2008 Urban thermal environment and its mitigation through urban planning process *Geographical Reports of Tokyo Metropolitan University* **43** p33-40.

[10] Kim Y H and J J Baik 2005 Spatial and Temporal Structure of the Urban Heat Island in Seoul *Journal American Meteorological Society* **44** p591-605

[11] Adi Wibowo, Andry Rustanto and Iqbal Putut Ash Shidiq 2013 Spatial-Temporal Analysis of Urban Heat Signature in Tagerang City *Journal Indonesian Geography* **45** 2 p101-112

[12] Laras Tursilowati, Josaphat Tetuko, Sri Sumantyo, Hiroaki Kuze and Erna Sri Adiningsih 2012 Relationship between urban heat island phenomenon and land use/land cover changes in Jakarta, Indonesia *Journal of Emerging Trends in Engineering and Applied Sciences* **3** 4 p645-653

[13] Laras Tursilowati 2008 Urban Heat Island and their contribution on climate change and relationship with land use change *Proceeding National Seminar on Global Warming and Global Change: Fact, Mitigation and Adaptation.* ISBN 978-979-17490-0-8

[14] Adi Wibowo and Khairulmaini Osman Salleh 2016 *Land Use Cover and Its Effect on Urban Heat Signature: A Spatial Temporal Analysis Proceeding 13th International Asian Urbanization Conference Yogyakarta Indonesia*

[15] Taha Haider 1997 Urban Climates and Heat Islands: Albedo, Evapotranspiration, and Anthropogenic Heat *Energy and Buildings* **25** 2 p99-103

[16] Adi Wibowo and Jarot Mulyo Semedi 2016 *Spatial Temporal Analysis of Air Surface Temperature Behavior on Small City (University) Proceeding the 13th International Asian Urbanization Conference Yogyakarta Indonesia*

[17] Wong N H Jusuf S K Aung L W Htun Kyaw Thu To Syatia and Wu Xuchou 2007 Environmental Study of the Impact of Greenery in an Institutional Campus in the Tropics *Journal of Building and Environment* **42** p2449-2970

[18] Wong N H and Jusuf S K 2008 GIS-based greenery evaluation on campus master plan *Landscape and Urban Planning* **84** p166-182

[19] M F Goodchild 2008 The use cases of digital earth *International Journal of Digital Earth* **1**

Issue **1** p31-42
[20] Shim, David 2014 Remote sensing place: Satellite images as visual spatial imaginaries. *Geoforum* **51** p152–160.
[21] Hu Q, Wu W, Xia T, Yu Q, Yang P, Li Z, and Song Q 2013 Exploring the use of Google Earth imagery and object-based methods in land use/cover mapping. *Remote Sensing* **5** 11 p6026-6042.
[22] Park, L 2009 Digging into Google Earth: An analysis of ''Crisis in Darfur'' *Geoforum* **40** p535–545
[23] Nastaran Makaremi, Elias Salleh, Mohammad Zaky, Jaafar Amirhosein, Ghaffarian Hoseini 2012 Thermal comfort conditions of shaded outdoor spaces in hot and humid climate of Malaysia *Building and Environment* **48** p7-14
[24] Anom Cahyo Galih Pranoto Adi Wibowo Djoko Harmantyo 2016 *Analysis Spatial Temporal Urban Heat Island in 2012 and 2015 at Pinang Sub-Districts, Tangerang City Department of Geography Faculty Mathematics and Natural Sciences University of Indonesia Depok West Java Indonesia*
[25] Dwi Fanny Wulandari Adi Wibowo Dewi Susiloningtyas 2016 *Spasial Pattern Studentification in Kukusan Depok West Java* Department of Geography Faculty Mathematics and Natural Sciences University of Indonesia Depok West Java Indonesia

Coupling of Markov chains and cellular automata spatial models to predict land cover changes (case study: upper Ci Leungsi catchment area)

K. Marko, F. Zulkarnain, and E. Kusratmoko

Department of Geography, Faculty of Mathematics and Natural Science, Universitas Indonesia, Kampus UI Depok, Depok 16424, Indonesia

Email : kuswantoro@ui.ac.id; kusrif@yahoo.com

Abstract. Land cover changes particular in urban catchment area has been rapidly occur. Land cover changes occur as a result of increasing demand for built-up area. Various kinds of environmental and hydrological problems e.g. floods and urban heat island can happen if the changes are uncontrolled. This study aims to predict land cover changes using coupling of Markov chains and cellular automata. One of the most rapid land cover changes is occurs at upper Ci Leungsi catchment area that located near Bekasi City and Jakarta Metropolitan Area. Markov chains has a good ability to predict the probability of change statistically while cellular automata believed as a powerful method in reading the spatial patterns of change. Temporal land cover data was obtained by remote sensing satellite imageries. In addition, this study also used multi-criteria analysis to determine which driving factor that could stimulate the changes such as proximity, elevation, and slope. Coupling of these two methods could give better prediction model rather than just using it separately. The prediction model was validated using existing 2015 land cover data and shown a satisfactory kappa coefficient. The most significant increasing land cover is built-up area from 24% to 53%.

1. Introduction

Land cover changes is a complicated, dynamic, and spatially explicit phenomena. Land cover changes affected by many factors related spatial-temporal phenomena, for example: proximity, physical conditions, and also socio-economics factors [1,2,3]. In order to detect land cover changes, geographic information system (GIS) and remote sensing are powerful tools to generate land cover changes effectively and efficiently [4,5,6]. Furthermore, advanced GIS technique can even predict land cover changes [7].

Over last decades, some studies trying to find out the best method for predicting land cover changes. In general, there are three types of land cover changes modelling i.e. empirical/statistical model, dynamic model, and hybrid model. A hybrid model known as a better model than empirical/statistical model and dynamic model because it combined two advantages both empirical/statistical model and dynamic model [8]. One of hybrid model that commonly used to predict land cover changes is geographic information system (GIS) based Markov chains-cellular automata model. This research is coupling two different methods: Markov chains, which is empirical/statistical model, and cellular automata, which is dynamic model that incorporate using GIS platform.

The purpose of this research is to predict future land cover changes in upper Ci Leungsi catchment area based on Markov chains-cellular automata spatial model, and to review the effectiveness of MC-CA spatial model for predicting land cover changes in catchment area scale. This research is important to recognize the future condition of upper Ci Leungsi catchment area with assumption by business as usual approach (BAU). In addition, the result of this research can be used as a reference to formularize best management practices of integrated water resources management in the future.

2. Area of Study and Data

Upper Ci Leungsi catchment area is located in Bogor Regency, about 15 kilometers from Bekasi City and 25 kilometers from Jakarta, the capital city of Indonesia. This catchment area is a subsystem of Bekasi watershed. The total catchment area is approximately 231 km^2 that cover five districts namely Gunung Putri, Citeureup, Klapanunggal, Babakan Madang, and Sukamamakmur.

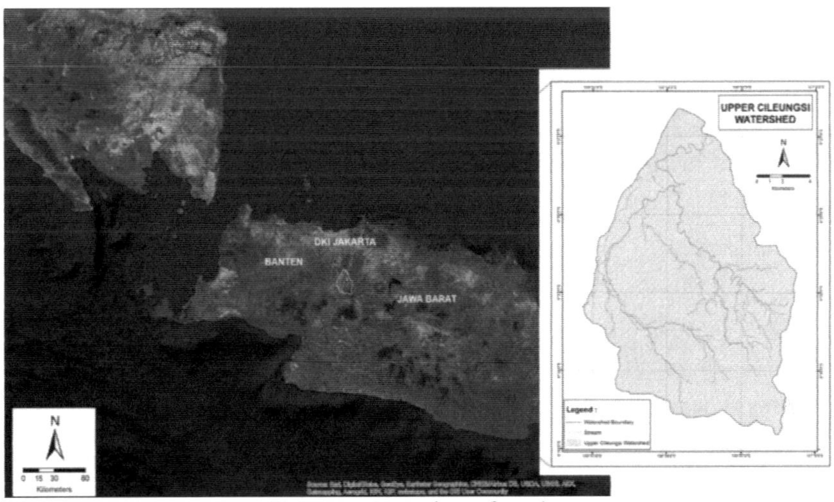

Figure 1. Location of study area

This research used multi-temporal land cover data to process the Markov chains - cellular automata spatial models. Multi-temporal land cover obtained by digital image processing from Landsat 5 and Landsat 8 that downloaded from http://earthexplorer.usgs.gov. This research also used physical characteristics maps such as: slope, elevation, and infrastructure proximity maps such as road network, central business district locations, and highway gate locations. Physical characteristics and infrastructure maps is using as driving factors to develop cellular automata while multi-temporal land cover is using to produce Markov transitional probabilistic map. Here is a list of data that was used in this research:

Table 1. List of datasets

Datasets	Source	Date or Year	Resolution
Landsat imageries • LT51220642005183BKT01 • LT51220652005183BKT01 • LT51220642010213BKT01 • LT51220652010213BKT01 • LC81220642014256LGN00 • LC81220652014256LGN00	United States Geological Survey	Acquisitions : • 02/07/2005 • 02/07/2005 • 01/08/2010 • 01/08/2010 • 13/09/2014 • 13/09/2014	Raster (30 meters)
Digital elevation model	ASTER Imagery	2009	Raster (30 meters)
Road network map	Badan Informasi Geospasial	2009	Vector

3. Methodology

Markov chains is a dynamic processes based on Markovian random process that calculate the probability of changes from particular object (for example: forest) into another objects (for example: built-up area). Markov chains describes as transitional probability matrix [1]:

$$p = (P_{ij}) = \begin{matrix} p_{11} & p_{12} \cdots & P_{1n} \\ P_{21} & P_{22} \cdots & P_{2n} \\ P_{n1} & P_{n2} & P_{nn} \end{matrix}$$

The probability of changes from particular object (ith) into another object (jth) is described as a transformation probability (P_{ij}); n is the land cover types of the catchment area. P_{ij} must meet the following condition below [1]:

$$0 \leq P_{ij} \leq 1 (i, j = 1, 2, 3, \ldots, n)$$
$$\sum_{i=1}^{n} P_{ij} = 1 \ (i, j = 1, 2, 3, \ldots, n)$$

Based on non-after effect of Markovian random processes and probability of Bayes condition, Markov chains model is obtained by [1]:

$$P_{(n)} = P_{(n-1)} P_{ij}$$

Where:

$P_{(n)}$ is state probability of any times
$P_{(n-1)}$ is preliminary state probability

Cellular automata developed by Von Neumann for self-reproducible systems [8]. Cellular automata characteristics are spatially explicit and dynamics that can simulation the evolution of two-dimensional objects [9, 10]. GIS raster data is known have resemblance with cellular automata concept hence many studies generate future land cover using cellular automata [9]. In order to generate future land cover map, there are many variables that drive the change of objects, called driving force [11]. Driving force consists of many factor for example: proximity, elevation, slope, etc. [1, 2, 3, 11].

Markov chains is just calculating the probability of changes but cannot represent the changes in a spatial-explicit term [12, 13]. Combining Markov chains and cellular automata is effort to minimalize the weakness of Markov chains [13]. Cellular automata represent the spatial-explicit term and location of changes while Markov chains predict changes quantitatively. Combining two methods would give better result than perform a model separately [14]. The procedure of Markov chains-cellular automata model is presented in figure below:

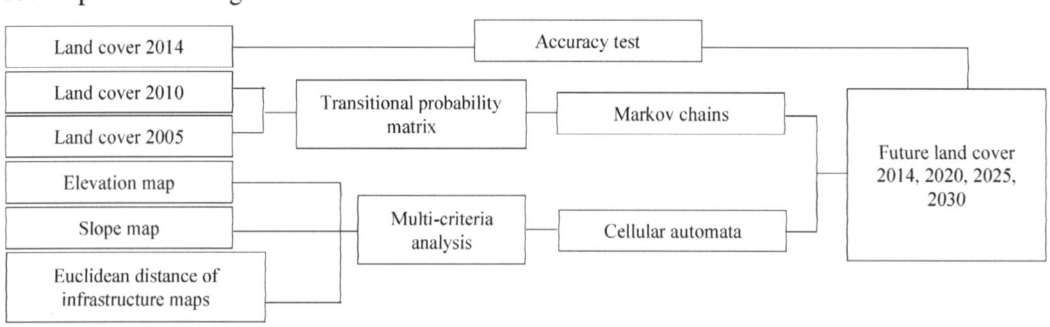

Figure 2. The procedure of Markov chains – cellular automata model

Accuracy test was conducted in order to know how the reliability, capability, and limitation of Markov chains is – cellular automata spatial models to predict land cover changes. Accuracy test of land cover model is an important part of the model that become the standard evaluation model whether the result of the model can be used for policy maker or not [21, 22]. The accuracy of the predicted land cover model was assessed using the kappa statistics. Using the same methods and steps, Markov chains – cellular automata spatial model was applied to create the projection of 2014 land cover based on 2005 land cover and 2010 land cover then the 2014 projected model is being evaluated with kappa statistics based on 400 ground truth derive from high resolution satellite imagery, Quickbird images. The kappa statistics calculated by equation below:

$$K = \frac{N \sum_{i=1}^{r} X_{ii} - \sum_{i=1}^{r} (X_{i+}.X_{+i})}{N^2 - \sum_{i=1}^{r} (X_{i+}.X_{+i})}$$

4. Result and Discussions

4.1. Land Cover Changes Analysis

4.1.1. Spatial Pattern of land cover changes

The classification of land cover divided into 7 types i.e. built-up area (BA), forest (FO), shrubs (SH), bare land/mining (BM), cultivated area (CA), paddy field (PF), and water bodies (WB). Over 10 years, built-up area is growing rapidly in Upper Ci Leungsi catchment area. Generally, the spreading of built-up areas concentrated in the central and northern part of the catchment. The spatial pattern of built-up area also indicated the linearity following the road network. Land cover changes can be seen in figure below:

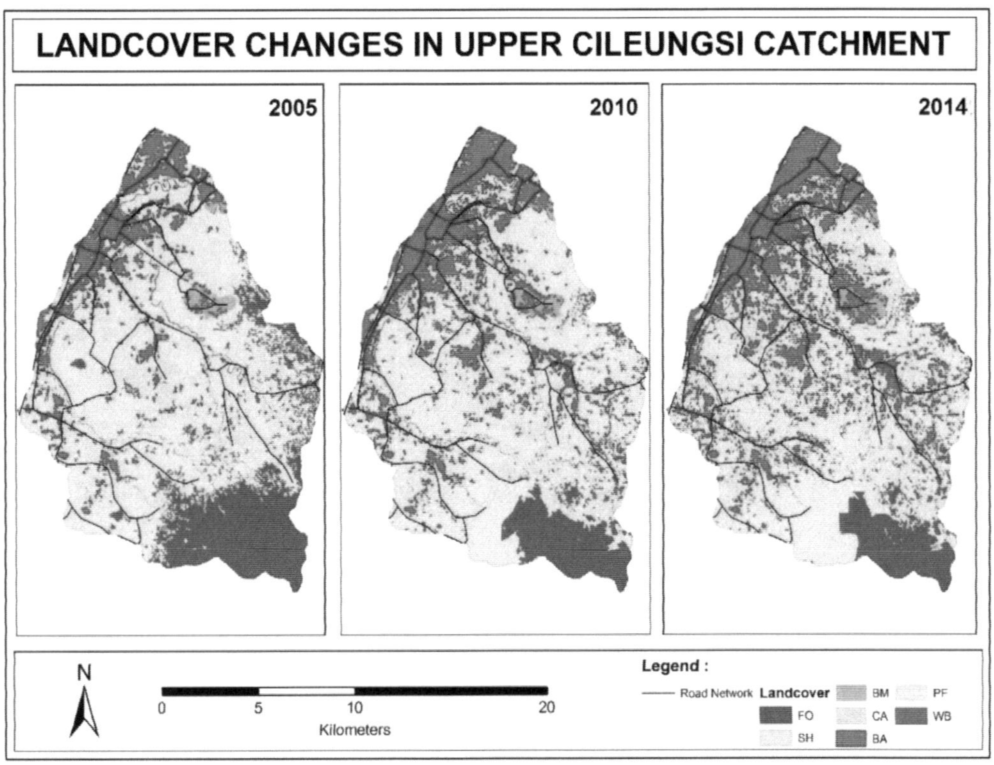

Figure 3. Land cover changes between 2005 to 2010 and 2010 to 2014

The figure above shows that built-up area spreads significantly from 14% to 23% in period time 2005 to 2010 and 23% to 28% in period time 2010 to 2014. Over a decade, built-up area grew double. On the other hand, other types of land cover such as shrubs and forest is declining distinctly. Forest was declining from 18% to 9% while shrubs was also declining from 58% to 52% for over ten years. It is an evidence that there were much area of shrubs and forest that converted into built-up area between 2005 to 2014. Other types of land cover i.e. bare land/mining, cultivated area, paddy field, and water bodies was not significantly changing that just declining or increasing only 1% between 2005 to 2014.

4.1.2. Transitional probability matrix of land cover changes

Transitional probability matrix is matrix that calculate the probability changing of each types of pixel of land cover into another types of land cover based on Markovian random process [15, 16]. The probability describes in a range of value between 0-1 which zero value means that the event of changing is impossible happen and 1 means is certainly happening. The transitional probability matrix of is shown below:

Table 2. Transitional probability matrix of land cover

			To						
	Period	Land cover	FO	SH	BM	CA	BA	PF	WB
From	**05-10**	FO	0.54	0.29	0.00	0.16	0.01	0.01	0.00
	10-14		0.85	0.15	0.00	0.00	0.00	0.00	0.00
	05-10	SH	0.01	0.83	0.00	0.03	0.10	0.03	0.00
	10-14		0.00	0.75	0.01	0.05	0.12	0.07	0.00
	05-10	BM	0.00	0.03	0.69	0.00	0.29	0.00	0.00
	10-14		0.00	0.01	0.88	0.00	0.11	0.00	0.00
	05-10	CA	0.00	0.68	0.00	0.19	0.04	0.01	0.00
	10-14		0.01	0.62	0.01	0.30	0.04	0.02	0.00
	05-10	BA	0.00	0.00	0.00	0.00	1.00	0.00	0.00
	10-14		0.00	0.00	0.00	0.00	1.00	0.00	0.00
	05-10	PF	0.00	0.00	0.00	0.00	0.24	0.12	0.00
	10-14		0.08	0.00	0.00	0.06	0.27	0.07	0.00
	05-10	WB	0.00	0.00	0.00	0.00	0.00	0.00	1.00
	10-14		0.00	0.01	0.00	0.00	0.00	0.00	1.00

Based on transitional probability matrix of land cover above, five types of changing have higher probability (higher than 0.2) than the other i.e. forest becomes shrubs, bare land/mining becomes built-up area, paddy field becomes built-up area. The other transitional probability is relatively low therefore the probability of the changes is not significant.

4.2. Land cover Prediction

4.2.1. Driving Factor

Driving factors are used to determine the location that may changes from each pixel of land cover into another type [17]. Driving factors divided into three types, slope, elevation, and infrastructure's proximity. Infrastructure's proximity then itemized into specific types that could stimulates the changes such as proximity from road, proximity from highway gates, and proximity from central business district.

Slope is one of important factor that driving the land cover changes [13]. A steep slope could become limiting factor for built-up area. Most of built-up area is found in relatively flat area because related to the easiness of structure and building construction. Furthermore, steep slope is vulnerable for mass movement that could stimulate landslide so it become another reason why slope is a limiting factor for built-up area [18].

Proximity also becomes important factor particular for built-up area invasion. Built-up area will occupy the land that close to vital infrastructure i.e. road networks, central business districts, and highway gates. Determining proximity of highway gates that become one of driving factor is unique because highway is the most important infrastructure for residents to access the capital city of Indonesia, Jakarta where the majority of Ci Leungsi catchment's residents work there. Ci Leungsi catchment located in peri-urban of Jakarta that many residents of this catchment depends on Jakarta and use highway as a major road to access Jakarta [19, 20]. All of driving factor then will be used for cellular automata based on multi-criteria analysis.

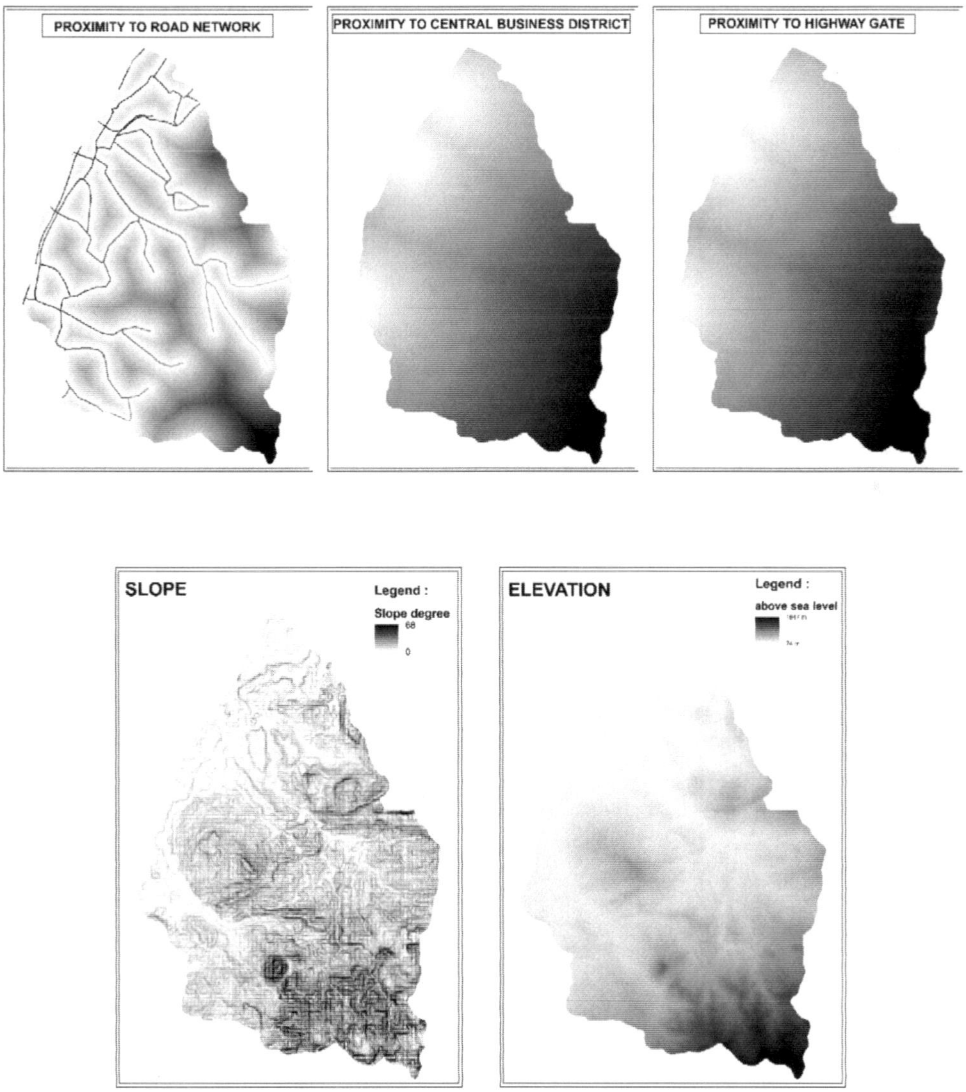

Figure 4. Driving factor maps

4.2.2. Land cover Change Prediction

Land cover change prediction in 2020, 2025, 2030 is shown in figure 7. The prediction shows expansion of built-up area in the central of catchment area. The expansion might be happening because the central of catchment area is suitable for the development of built-up area since it is physical characteristics (low to moderate elevation and flat slope) and its proximity to the infrastructure. It also shows the disappearing of paddy field and shrubs that might be converted into built-up area in the future. The forest predicted still exist because pixels of forest located in limiting factor, which has characteristics: steep slope, high elevation, far from road network, central business districts, and highway gate. This limiting factor makes forest might still exist and not decreasing significantly in the future. Based on figure 8 below, built-up area is predicted increasing from 28% to 39%, 46%, and 53% in 2020-2030. Almost all of types except built-up area and water bodies is predicted decreasing. The most decreasing area is shrubs, disappears about 24% from 2014 to 2030.

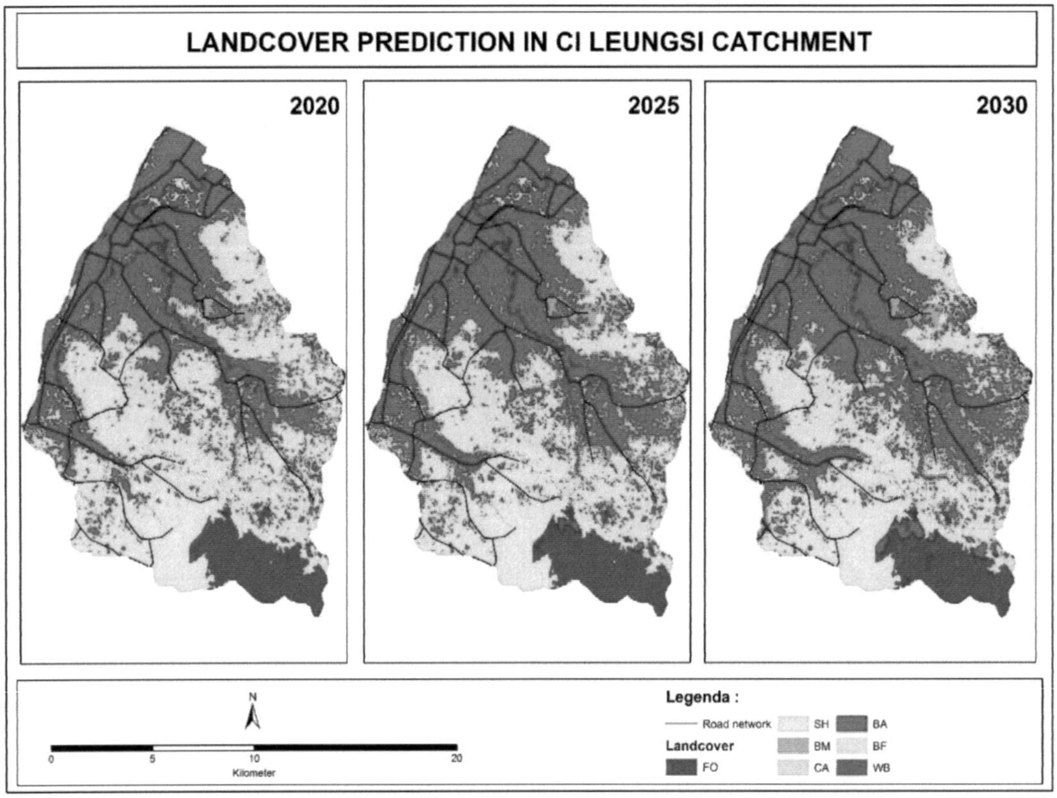

Figure 5. Land cover prediction (left: 2020, center: 2025, right: 2030)

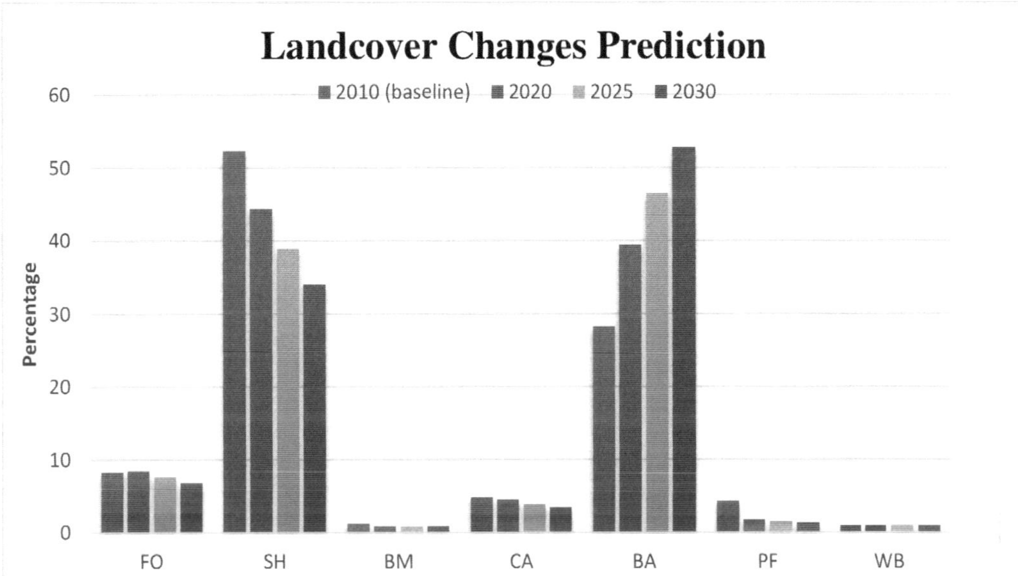

Figure 6. Land cover changes prediction charts

4.2.3. Accuracy Test of Prediction Model

The overall kappa coefficient test table is shown that the prediction model is 82.58%, tested from 400 ground truth data that acquired from Quickbird image. The best user's accuracy is built-up area, forest, and water bodies which above 90%. Other types of land cover i.e. shrubs and bare land/mining has adequately accuracy which above 80% but cultivated area and paddy field has lower accuracy than the other land cover, just about below 80%.

Table 3. Kappa coefficient test table

land cover	ground truth							ground Truth	user's accuracy (%)
	ref. 1 (FO)	ref.2 (SH)	ref.3 (BM)	ref. 4 (CA)	ref. 5 (BA)	ref. 6 (PF)	ref. 7 (WB)		
FO	37	3	-	-	-	-	-	40	92.50
SH	-	107	-	5	8	-	-	120	89.17
BM	-	5	24	1	-	-	-	30	80.00
CA	-	12	-	18	-	-	-	30	60.00
BA	-	9	-	-	111	-	-	120	92.50
PF	-	-	-	4	7	19	-	30	63.33
WB	-	-	-	-	-	-	30	30	100.00
Ground Truth	37	136	24	28	126	19	30	400	82.50
Kappa Coefficient									82.58

5. Conclusion and Recommendation

A business as usual land cover projection model in Ci Leungsi catchment was successfully executed using coupling of Markov chains and cellular automata spatial model with adequately kappa coefficient. The result of projection models showed that built-up area will increase almost doubled from 24 to 53%. Almost of land cover types predicted accurately except cultivated area and paddy field. For next research, developing some land cover change scenarios based on land use policy and best land use management probably can improve the accuracy of projection model and also finding the most influential driving factor that changing the land cover. There are some opportunities to apply the projection model for improving best management practices in Ci Leungsi Catchment to support integrated water resources management program. This research is also very useful as a reference for land use planning in watershed scale for government and stakeholders.

Acknowledgments
The authors would like to thank the Department of Geography, Faculty of Mathematics and Natural Science, University of Indonesia which has facilitated the research activities through excellent research grants.

References
[1] Guan D, Li H, Inohae T, Su W, Nagaie T, Hokao K 2011 Modeling urban land use change by the integration of cellular automaton and markov model Ecological Modelling **222** 3761-3772
[2] Al-Sharif A A A and Pradhan B 2014 Monitoring and predicting land use change in Tripoli Metropolitan City using an integrated Markov chain and cellular automata model in GIS Arabian Journal of Geosciences **7** 4291-4301
[3] Fuglsang M, Münier B, Sten Hansen H 2013 Modelling land-use effects of future urbanization using cellular automata: An Eastern Danish case Environmental Modelling & Software **50** 1-11
[4] Leao S, Bishop I, Evans D 2004 Simulation urban growth in a delevoping nation's region using a cellular automata-based model Journal of Urban Planning and Development **130** 145-158
[5] Batty M and Xie Y 1997 Possible urban automata Environmental and Planning B: Planning & Design **24** 175-192
[6] Dadras M, Shafri H Z M, Ahmad N, Pradhan B, Safarpour S 2015 Spatio-temporal analysis of urban growth from remote sensing data in Bandar Abbas city, Iran The Egyptian Journal of Remote Sensing and Space Sciences **18** 35-52
[7] Batty M, Xie Y, Sun Z 1999 Modeling urban dynamics through GIS-based cellular automata Computers, Environment and Urban Systems **23** 205-233
[8] Von Neumann J and Burks A W 1966 *Theory of Self-reproducing* Automata (Urbana: University of Illinois Press)
[9] Ye B and Bai Z 2008 Simulating land use/cover changes of Nenjiang county based on CA-Markov model Computer and Computing Technologies in Agriculture **1** 321-329
[10] Long Y, Jin X, Yang X, Zhou Y 2014 Reconstruction of historical arable land use patterns using constrained cellular automata: A case study of Jiangsu, China Applied Geography **52** 64-77
[11] Lauf S, Haase D, Hostert P, Lakes T, Kleinschimt B 2012 Uncovering land-use dynamics driven by human decision-making – A combined model approach using cellular automata and system dynamics Environmental Modelling & Software **27-28** 71-82
[12] Lopez E, Bocco G, Mendoza M, Duhau E 2001 Predicting land-cover and land-use change in the urban fringe A case in Morelia city, Mexico Landscape and Urban Planning **55** 271-285
[13] Shafizadeh Moghadam H and Helbich M 2013 Spatiotemporal urbanization processes in the megacity of Mumbai, India: a Markov chains-cellular automata urban growth model Applied Geography **40** 140-149
[14] Arsanjani J J, Helbich M, Kainz W, Boloorani A D 2011 Integration of logistic regression, Markov chains and cellular automata models to simulate urban expansion International Journal of Applied Earth Observation and Geoinformation **21** 264-272
[15] Eastman J R 2009 IDRISI 16: The Taiga Edition (Worcester: Clark University)
[16] Brown D G, Pijanowski B C, Duh J D 2000 Modeling the relationship between land use and land cover on private lands in the Upper Midwest, USA Journal of Environmental

Management **50** 247-263

[17] Halmy W W A, Gessler P E, Hicke J A, Salem B H 2015 Land use/land cover change detection and prediction in the north-western coastal desert of Egypt using Markov-CA Applied Geography **63** 101-112

[18] Vanacker V, Vanderschaeghe M, Govers G, Willems E, Poesen J, Deckers J, De Bievre D 2003 Linking hydrological, infinite slope stability and land-use change models through GIS for assesing the impact of deforestation on slope stability in high Andean watersheds Geomorphology **52** 299-315

[19] Winarso H, Hudalah D, Firman T 2015 Peri-urban transfomation in the Jakarta metropolitan area Habitat International **49** 221-229

[20] Firman T 2009 The continuity and change in mega-urbanization in Indonesia: A survey of Jakarta-Bandung Region (JBR) development Habitat International **33** 327-339

Effectiveness of Vegetation Index Transformation for Land Use Identifying and Mapping in the Area of Oil palm Plantation based on SPOT-6 Imagery (Case Study: PT.Tunggal Perkasa Plantations, Air Molek, Indragiri Hulu)

H A Setyowati[1], S H Murti B S[2], and Sukentyas E S[3]

[1]Graduate student from Departement of Cartography and Remote Sensing, Faculty of Geography, Universitas Gadjah Mada, Sekip Utara, Sleman, Yogyakarta 55281, Indonesia
[2]Lecturer from Departement of Cartography and Remote Sensing, Faculty of Geography, Universitas Gadjah Mada, Sekip Utara, Sleman, Yogyakarta 55281, Indonesia
[3]Researcher from Technology And Data Center, Agencies and National Space Flight (LAPAN), Pekayon, Pasar Rebo, Jakarta 13710, Indonesia

heratania.a.s@mail.ugm.ac.id

Abstract. The reflection of land surface, atmosphere and vegetation conditions affect the reflectance value of the object is recorded on remote sensing image so that it can affect the outcome of information extraction from remote sensing imagery one multispectral classification. This study aims to assess the ability of the transformation of generic vegetation index (Wide Dynamic Range Vegetation Index), the vegetation index transformation that is capable reducing the influence of the atmosphere (Atmospherically Resistant Vegetation Index), and the transformation of vegetation index that is capable of reducing the influence of the background soil (Second Modified Soil Adjusted Vegetation Index) for the identification and mapping of land use in the oil palm plantation area based on SPOT-6 archived on June 13, 2013 from LAPAN. The study area selected oil palm plantations PT. Tunggal Perkasa Plantations, Air Molek, Indragiri Hulu, Riau Province. The method is using the transformation of the vegetation index ARVI, MSAVI2, and WDRVI. Sample selection method used was stratified random sampling. The test method used mapping accuracy of the confusion matrix. The results showed that the best transformation of the vegetation index for the identification and mapping of land use in the plantation area is ARVI transformation with a total of accuracy is 96%. Accuracy of mapping land use settlements 100%, replanting 82.35%, 81.25% young oil palm, old oil palm 99.46%, 100% bush, body of water 100%, and 100% bare-soil.

Keywords: Vegetation Index Transformation, Multispectral Classification, Land Use, Oil Palm Plant, SPOT-6 Imagery

1. Introduction

Central data and Agriculture Information System [2] states that Perkebunan Besar Swasta (PBS) dominate the large of Oil palm area, followed by Perkebunan Rakyat (PR) and Perkebunan Besar

Negara (PBN). The total large of Oil palm area in 2011 is approximately 8,91 million hectare with the large of PBS area of 4,65 million hectare (52,22%), PR 3,62 million hectare (40,64%) and PBN 0,64 million hectare (7,15%). Indragiri Hulu district is a region in province of Riau which has Oil palm as its main commodity [4] with number one contribution of total oil palm production in Indonesia [9]. The spread of Oil palm plantation in Indonesia started from Sumatra East Coast and Aceh [7] and continued to most land in Indonesia [15]. Oil palm *(Elaeis guineensis jac)* is plant from Nigeria and Brazil which has specific characteristics [5]. Composed in Fresh Stem Fruit [1] these plants planted in specific pattern and space therefore the image is easily detected by satellite imagery. Activity of plantation management needs a detail mapping for a large area which impacts to a big budget while high resolution image can only cover narrow area. Remote sensing method, one of them is maximum likelihood multispectral classification, is considered as adequate to classify object accurately in large area. Unfortunately, remote sensing system has weakness to atmosphere impact. Weather condition, climate, rain, cloud, fog, soil background and the side effect of Oil palm industry activity such as smoke, could affect image condition therefore correction to overcome some disruptions is needed. The use of varies transformation of vegetation index like generic, atmosphere effect control, and soil background effect control based on SPOT-6 image is considered to be able to fix obtained spectral value. This study aims to assess the ability of generic vegetation index transformation (Wide Dynamic Range Vegetation Index), the vegetation index transformation which is capable of reducing the influence of the atmosphere (Atmospherically Resistant Vegetation Index), and the transformation of vegetation index that is capable of reducing the influence of the background soil (Second Modified Soil Adjusted Vegetation Index) for the land use identification and mapping in oil palm plantation area based on SPOT-6 archived on June 13, 2013 from LAPAN.

2. Research Area

This research covers part of Indragiri Hulu district, province of Riau at Oil palm plantation of PT. Tunggal Perkasa Plantation as seen in **Picture 1.**

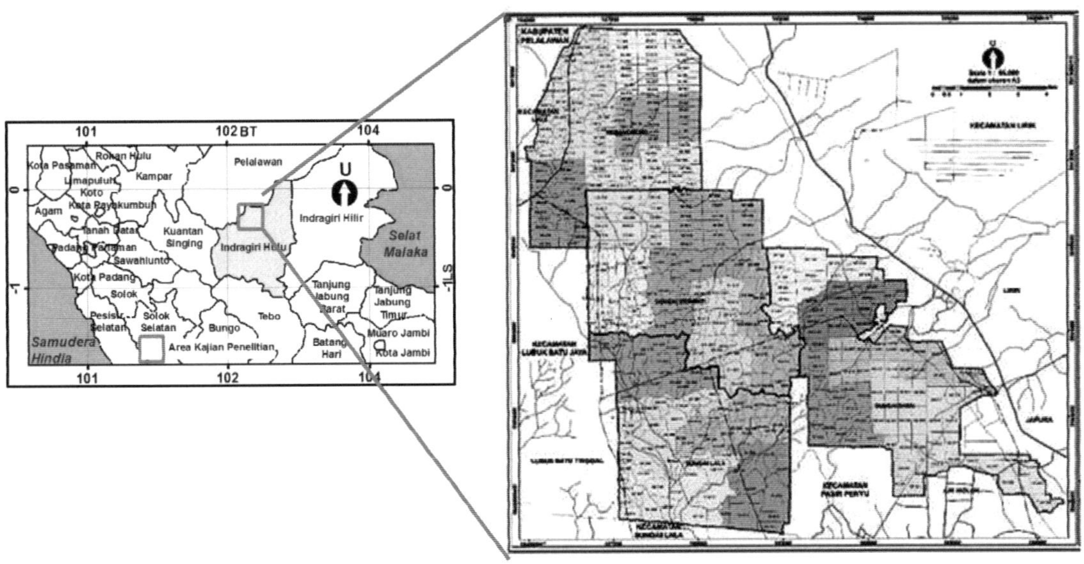

Picture 1. Research Location
Source : Basic Research Material ,2015

3. Materials and Methods

3.1. Data

This research used multispectral SPOT-6 Imagery with 6 meters spacial resolution archived on June 13, 2013 from LAPAN. Other used data is Indonesia Topographical Map Scale 1 : 50.000 Page 0915-41 Simpang Kelayang, Indonesia Topographical Map Scale 1 : 50.000 Page 0915-42 Air Molek, Indonesia Topographical Map Scale 1 : 50.000 Page 0915-43 Ukui Satu, Indonesia Topographical Map Scale 1 : 50.000 Page 0915-44 Pangkalan Kapau and Plantation Block Map which are used to extract area boundaries and basic informations.

3.2. Methods

3.2.1 ToA radiometric correction

Technology and Remote Sensing Data Center [13] in Agencies and National Space Flight release that radiometric correction information for satellite imagery SPOT-6 can be done with radiometric correction Top Of Atmosphere (TOA). TOA correction erase obscurity therefore produced imagery data is more focus and visualy clear. TOA correction needs zenithal angle input (cos θ) (90- elevation angle), elevation angle, inciden angle, canal gain (Gi), canal bias (Bi), canal solar irradiance (solar ESUNi), which are in metadata file attaching in each scene of satellite recording result and earth-sun space (d^2) which can be calculated from those parameters. All input is entered in formula applying in each canal (DNi) with the formula below:

$$\boldsymbol{Reflectance\ (Top\ Of\ Atmosfer)}: \frac{\pi(\frac{DNi}{Gi}+Bi)d^2}{ESUNi.\cos\theta} \quad (1)$$

3.2.2 Transformation of Vegetation Index

Transformation of vegetation index used in this research is Transformation of Generic Vegetation Index and Transformation of Vegetation Index which Press the Atmosphere Effect with formula as below [6] with additional combination Hatfield et.al [8] in Susetyo [16] :

a. Generic Transformation of Vegetation Index
WDRVI (*Wide Dynamic Range Vegetation Index*)

WDRVI is transformation of vegetation index using deviation value of close infrared by adding constants formula as below

$$\boldsymbol{WDRVI}: \frac{(0.1Near\ Infra\ Red-Red)}{(0.1Near\ Infra\ Red+Red)} \quad (2)$$

b. Transformation of Vegetation Index Pressing Soil Background Effect
MSAVI2 (*Second Modified Soil Adjusted Vegetation Index*)
MSAVI is a transformation of vegetation index developed to minimize soil reflection effect as in SAVI and MSAVI formula with formula as below:

$$\boldsymbol{MSAVI2}: \frac{(2(NIR)+1)-\left(\sqrt{(2(NIR)+1)^2}\right)-(8(NIR-Red))}{2} \quad (3)$$

c. Transformation of Vegetation Index Pressing Atmosphere Effect
ARVI (*Atmospherically Resistant Vegetation Index*) is transformation of vegetation index developed to decrease sensitivity of atmosphere effect by applying normalization to blue radiance, red radiance and infrared close to formula below: Gamma values 1

$$\boldsymbol{ARVI}: \frac{(Near\ Infra\ Red-rb)}{(Near\ Infra\ Red+rb)} \quad (4)$$

$$\boldsymbol{rb}: \boldsymbol{Red-(Gamma(Red-Blue)} \quad (5)$$

3.2.3 Oil palm Identification

This identification is used to help multispectral classification which will be done if there was similar appearance but factually different. For example the appearence of Oil palm and non-Oil palm which has similarity. This interpretation is done by help from same imagery data or other imagery data processed with certain composite therefore the land cover is seen. This process is done by showing the image parallel. Oil palm plantation has different pattern and texture from other objects surround the area. In comparison with other objects surround, Oil palm plantation has lined cluster pattern therefore it has regular and neat pattern that make multispectral classification easier.

3.2.4 Multispectral Classification

Classification method used is supervised classification with maximum likelihood algorithm which algoritm homogen object statistically shows normal distribution of histogram (Bayesian) [6]. Number of samples as training area help differentiate objects. Training area creation can be done with ROI (Region of Interest) with ENVI 5.1 software. Therefore 7 classes of land covers are found, which are residential use, replanting, young oil palm, old oil palm, grass, body of water and bare soil.

3.2.5 Majority Filter

After multispectral classification process, there will be anomali pixels in certain class. For example in Oil palm classification class, there are some pixels of body water thus the classification is not as expected, therefore Majority Filter operation is needed to make classification result more compact. Multiscpectral classification process and majority filter is chosen since both are general classification methods and suitable for wide are coverage. There are also some weaknesses if there were different land covers but had the same spectral numbers, they went to the same class, while in majority filter, if there were an object like a long and narrow garden path, it was generalized with other classes.

3.2.6 Confusion Matrix

Field accuracy test method used is by confusion matrix. This method uses independent data collection therefore it is logically acceptable. This accuracy test involve land cover result classification test.

4. Result, Analysis, and Discussion

4.1 Identification of Oil Palm

Oil palm identification process is done by appearance observation on image to choose ROI or region of interest as Oil palm identification key based on multispectral classification step. To ease Oil palm identification process, interpretation keys are used where old Oil palm has dark green color (in composite 321 true color) or dark red color(in composite 432 false color) while young Oil palm has light green color (in composite 321 true color) or light red color (in composite 432 false color). The appearance distinguish can be seen in Picture 2.

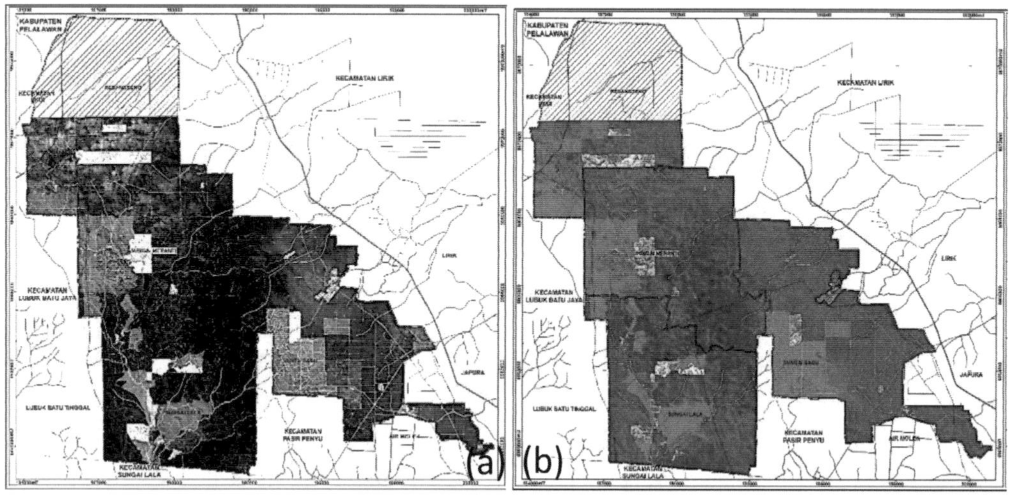

Picture 2. Visual Comparison Imagery SPOT-6 (a) in composite condition 321 True Color and composite condition 432 False Color
(Source: Imagery Processing Result SPOT-6, 2015)

PT. Tunggal Perkasa Plantation is plantation owned by privat company with inti plantation and plasma plantation management (cooperation with local community). Before now, this Oil palm plantation was ruber plantation so there are some ruber trees owned by local people in the borderline of company Oil palm plantation. Beside that, this plantation is also share borderline with protected forest used as arboretum land as seen in **Picture 3.**

Picture 3. Plant Appearance (a) Antigonon leptopus (b) Turnera subulata (yellow) (c) Turnera subulata (white) (d) Arachis pintoi
(Source: Field Survey Documentation, 2015)

Field condition shows that the main plants are oil palms but to keep soil ecology function and environment as well as to keep soil humidity for oil palm media so some other plants are planted in same field such as mucuna beans, ferns *Nephrolepis bisserrata*, flower like *Turnera subulata*, *Euphorbia heterophylla*, *Cassia tora*, and *Antigonon leptopus*, and also natural fertilization in ilfunction inter-row with prunned palm frond, empty stem and etc as in **Picture 4**. The plants appearance will possibly disturb land use classification since it located near Oil palm and probably has same color.

Picture 4. Plant Appearance (a) Ferns (Nephrolepis biserrata) (b) Empty Stem Fertilizer Application (c) Ilfunction Inter-row (d) Beans Muccona sp
(Source: Field Survey Documentation,2015)

Oil palm has regular and neat pattern because of good plantation management. Plantation operational standard requires specific plant space therefore it results in specific pattern such as plantation blocks. Oil palm location or site has close connection with block shape where each block is separated with plantation road. Different from other plantation plants, by looking at the plantation road, it could be guaranteed as Oil palm plantation. The pattern of local community Oil palm will be different in appearance because of the lack of plantation management system.Old Oil palm texture is rougher than young Oil palm texture. This is because old Oil palm has tight canopy where young Oil palm has thin canopy, especially for a very young Oil palm. ROI selection for Oil palm identification determine how land cover classification produced. Selection of pixels of samples which are more homogenous and overspread will result in better classification then heterogenous pixels. Heterogenous pixels will result in worse classification because of mixing pixels in training area which do not represent field appearance. Oil palm identification and other land cover are easier with compisite image 432 than composite image 321.

4.2. Multispectral Classification and Majority/Minority Filter

Classification used is multispectral supervised classification with maximum likelihood method and continued with final majority/minority filter classification which generalizetiny objects in bigger class. This method is the most common method in classifying activity. This classification method could classify object based on object spectral value with pixels. Image composite used in this multispectral classification is composite 432 as seen in **Picture 2**. Which shows red color for vegetation and cyan color for buiding. Land use determination is based on SNI 7645-1:2014 [11] for small scale mapping and middle scale mapping. Multispectral classification process is done with visual interpretation help which has done for Oil palm identification and land cover before. Purwadhi and partners [11] said location for Oil palm is suitable in some land covers like field, bushes, bare soil, convertion production forest and settlement of other function areal status; as well as the Oil palm plantation itself both mix plantation or not. This is appropiate with land use determination in research area where after the appearance identification both Oil palm and other land covers are gained seven land covers class in research area, which are body water, settlements, bush, bare-soil, replanting, old Oil palm and young Oil palm.Each transformation of vegetation index has different spectral value range for every object recorded. **Table 1**. Shows comparison spectral value range formed in image which in general the value can devide into two main object, which are Oil palm vegetation and non-vegetation. Oil palm object of vegetation is devided into two, which are young Oil palm and old Oil palm.

Table 1. Range of Spectral Value in Vegetation Object and Non Vegetation
(Source: SPOT-6 Imagery Process Result with Some Transformation of Index , 2015)

Object/ Index Transformation of Vegetation	MSAVI2	ARVI	WDRVI
Non Vegetation	-0.013 s/d -0.884	-0.045 s/d 0.425	-0.833 s/d -0.625
Young Oil palm	0.881 s/d -0.594	0.596 s/d 0.689	-0.148 s/d -0.016
Old Oil palm	-0.884 s/d -0.669	0.425 s/d 0.596	-0.625 s/d -0.148

PT. Tunggal Perkasa Plantation as reseach object area has clear borderline with homogenous plants which is Oil palm therefore the classification process is easier. Oil palm differentiate only by age because the appearance of Oil palm is difficult to see even with 6 metres resolution imagery. Old Oil palm is assumed as productive plant and young Oil palm is assumed as non-productive plant. Oil palm differentiationespecially young Oil palm is not enough with only by visual interpretation and spectral value because young productive plant has same spectral with young non-productive plant. Filter majority process result in **Picture 5.** shows same objects classification therefore the result is better and closer to the real appearance then only applying multispectral classification. While in **Picture 5.** shows classification result and majority filter in transformation of vegetation index. The closest classification result from the real appearance is formed by ARVI transformation.

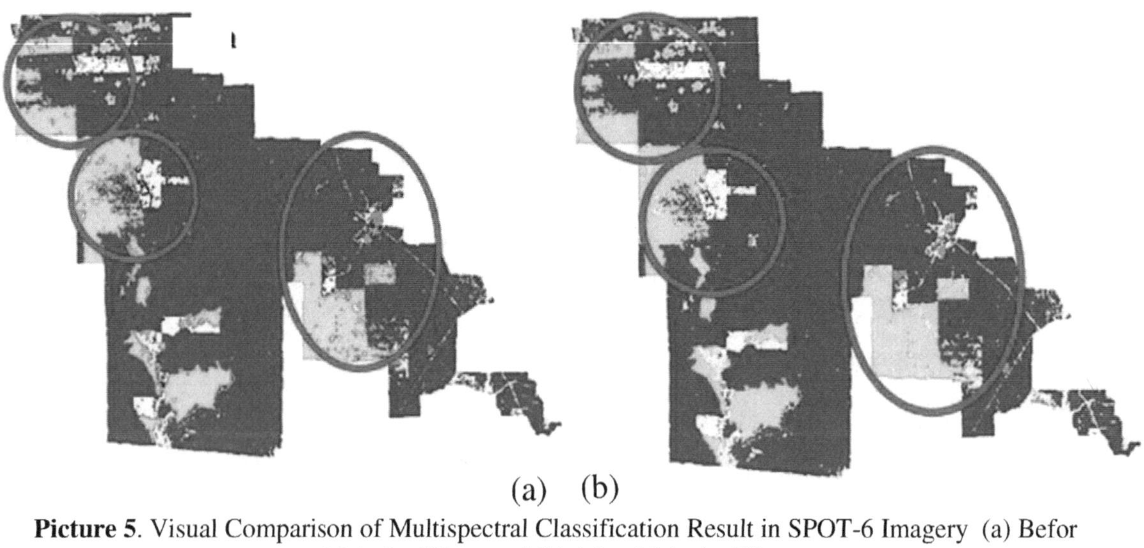

(a) (b)

Picture 5. Visual Comparison of Multispectral Classification Result in SPOT-6 Imagery (a) Befor Majority Filter, and (b) After Majority Filter
(Source: SPOT-6 Imagery Processing Result, 2015)

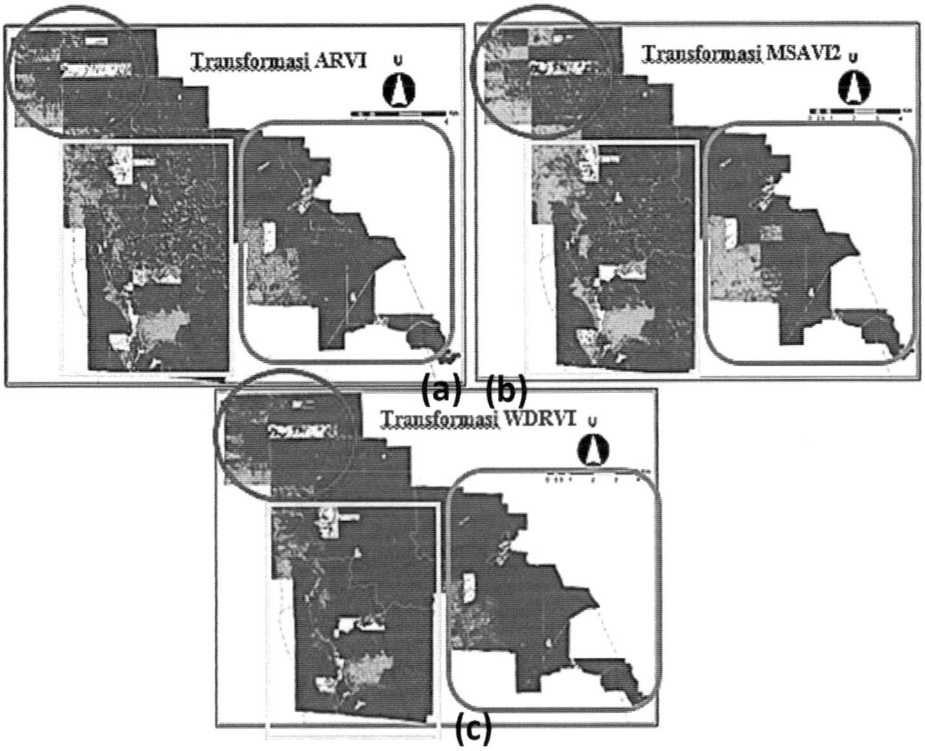

Picture 6. Visual Comparison of Multispectral Classification Result and Majority Filter in SPOT-6 Imagery (a) With ARVI Transformation of Vegetation Index, (b) With MSAVI2 Transformation of Vegetation Index, and (c) With WDRVI Transformation of Vegetation Index (Source: SPOT-6 Imagery Processing Result, 2015)

4.3. Mapping Accuracy Test with Confusion Matrix

The accuracy of multispectral classification result is counted with confusion matrix table. This accuracy test table compares appearance of classification result object with the real object in field. Based on classification and that majority filter then image result from ARVI transformation is the closest to real image then other transformation, therefore the land cover wide is attain, they are old Oil palm for 10.111,77 hectare, young Oil palm for 2.132,32 hectare, bare-soil for 125,65 hectare, replanting for 420, 84 hectare, settlements for 66,98 hectare, bush for 25,23 hectare, manmade-lake for 5,58 hectare and smog for 136,92 hectare. Accuracy test result in **Table 2.** shows general result for old Oil palm or productive plant classification with felicitous result with accuracy to 99.46%.

Table 2. Table of Land Cover Multispectral Classification Accuracy Test
(Source : Calculation Result of Interpretation Accuracy, 2015)

		FIELD									
		Settlements	**Replanting**	**Young Oil palm**	**Old Oil palm**	**Bush**	**Body Water**	**Bare Soil**	**Total**	**User Accuracy**	**Error Comission**
CLASSIFICATION	Settlements	64	0	0	0	0	0	0	64	100.00	0.00
	Replanting	6	28	0	0	0	0	0	34	82.35	17.65
	Young Oil palm	0	0	26	6	0	0	0	32	81.25	18.75
	Old Oil palm	1	0	0	185	0	0	0	186	99.46	0.54
	Bush	0	0	0	0	3	0	0	3	100.00	0.00
	Bush	0	0	0	0	0	3	0	3	100.00	0.00
	Bare Soil	0	0	0	0	0	0	5	5	100.00	0.00
	Total	71	28	26	191	3	3	5	327	Overall	
	Product Accuracy	90.14	100.00	100.00	96.86	100.00	100.00	100.00		Accuracy :	0.960
	Error Omission	9.86	0.00	0.00	3.14	0.00	0.00	0.00		Kappa :	0.935

The accuracy is affected by sampling selection and numbers of selected sampling. For example in replanting classification with accuracy 82.35% where from 34 samples, there are 6 samples of settlements. This is because those 6 samples have almost similar rooftop color with replanting land. A block of settlements in plantation around the hill made them do not look different with replanting land. Then in old Oil palm classification with 99.46% accuracy, where from 186 samples, there are 1 sample goes to settlements.

Moreover in young Oil palm classification with 81.25% accuracy with total 32 samples, there are 6 samples of old Oil palm or productive plant. This is because new-old Oil palm looks like young Oil palm. In fact, visual appearance and spectral from young Oil palm could mean some more apearance like old Oil palm but youngOil palm (for example TM 5, then young Oil palm (for example TM 1-3)or land cover plants like beans and nephrolepis.

Land covers like settlements, bare-soil, bush, and body water have high accuracy to 100%. This means no single classification mistake or it can also means sample selection for classification accuracy test is felicitous even with random. Interpretation of the object is easier because the appearance is clearer and has specific interpretation key. **Picture 7, Picture 8,** and **Picture 9** show visual appearance of an object in image which possibly have different field information therefore the interpretation result and classification as well as filtering could make accuracy mistakes.

Picture 7. Visual appearance result of settlements (a) Multispectral classification spektral in SPOT-6 imagery with composite 432 with (b) real appearance of water of body in field (c) real appearance of Oil palm industrial complext in field, dan (d) staff settlements complex with living facillities (Source : Field Survey Result of Multispectral Classification Check [14]

Picture 8. Old Oil palm Visual Appearance (a) Marihat Variety in SPOT-6 Imagery Composite 432, (b) Socfind Variety in SPOT-6 Imagery Composite 432, (c) Real Appearance of Marihat Variety in land, (d) Real Appearance of Socfind Variety in land (Source: Field Survey Result of Multispectral Classification [14]

Picture 9. Visual Comparison of Multispectral Classification Result of SPOT-6 Imagery (a) Before Majority Filter, and (b) After Majority Filter
Source: Field Survey Result of Multispectral Classification Test [14]

Accuracy test result of multispectral classification with real condition and overall accuracy have approximately 96% while classification accuracy for each land cover has different accuracy. Kappa index shows in 0.935. This means in this multispectral classification proces, the writer could avoid interpretasi mistake of 93.5% therefore the probability of mistake is only about 6.5%. According to [6] agronomy researchers use accuracy limit as much as 90% for plant classes that also could be said that overall land cover classification results have high accuracy.

5. Conclusion

Research result shows SPOT-6 imagery is capable to indentify and map land use in Oil palm plantation with high accuracy as high as 96% with best multispectral classification from ARVI transformation. Mapping accuracy in land use for settlements is 100%, replanting is 82.35%, young Oil palm is 81.25%, old Oil palm is 99.46%, bush is 100%, body water is 100% and bare-soil is 100% therefore land wide for old Oil palm is 10.111,77 hectare, young alm oil is 2.132,32 hectare, bare-soil is 125,65 hectare, replanting 420,84 hectare, settlements 66,98 hectare, bush 25,23 hectare and body water 5,58 hectare.

6. Acknowledgments

The writers would like to say thank you to Head of LAPAN Data Processing, Mr. Drs. Kustiyo, M.Si who gave access to SPOT-6 imagery for this research and to all staff in Technology and Data Processing LAPAN. I would also say thank you to PT. Tunggal Perkasa Plantation administration, Mr. Sumarno who gave research permitand to all staff who helped during research and Head of Afdeling and Supervisor who helped field activity during research and last but not least, Mr. Hadi Purnomo Aulia as field supervisor.

References

[1] Astra Agro Niaga 1995 *Pedoman Brevet Dasar-1 Tanaman Pabrik Teknik.* (Jakarta : PT. Astra Agro Lestari, Tbk)

[2] Badan Pusat Statistik 2013 *Proyeksi Penduduk Indonesia 2010-2035* (Jakarta : Badan Pusat Statistik Jakarta-Indonesia)

[3] Pusat Statistik 2014 *Berita Resmi Statistik Kabupaten Riau : Perkembangan Ekspor dan Impor Riau Agustus 2014* Last access at http://riau.bps.go.id/ October 12, 2014

[4] Badan Standardisasi Nasional 2014 *Klasifikasi Penutup Lahan- Bagian 1 : Skala Kecil dan Menengah. SNI (Standar Nasional Indonesia) 7645-1:2014.2014* (Jakarta: Badan Standardisasi Nasional. ICS 07.040)

[5] Corley R H V and Tinker P B 2003 *The Oil palm (fourth edition)* (USA :Blackwell Science Ltd- Blackwell Publishing Company)

[6] Danoedoro P 2012 *Pengantar Penginderaan Jauh Digital* (Yogyakarta: ANDI Yogyakarta.

[7] Fauzi Y, Widyastuti Y, Setyawibawa I, dan Hartono R 2002 *Kelapa Sawit: Budidaya, Pemanfaatan Hasil dan Limbah , Analisis Usaha dan Pemasaran. Edisi revisi* (Jakarta: Penebar Swadaya)

[8] Hatfield J L, Gitelson A A, Schepers J S and Walthall C L 2008 Application of Spectral Remote Sensing for Agronomic Decisions *Agronomy Journal* S-117

[9] Maulidar and Indri 2014 *Kemendag : Eropa Tanggapi Positif CPO Indonesia* Last access at www.tempo.co October 7, 2014

[10] *Peran Industri dan Perkebunan Sawit Bagi Perekonomian Bangsa* Last access at www.investasikelapasawit.com October 7, 2014

[11] Purwadhi, Florentina S H, Haryani, Nanik S, Siwi, Sukentyas E 2010 *Informasi Spasial Sebaran dan Potensi Perkebunan Kelapa Sawit dari data Penginderaan Jauh di Propinsi Sumatera Selatan* Majalah Berita Inderaja Pekayon Jakarta Timur : Pusat Data Penginderaan Jauh LAPAN. Vol.IX No.16 Juli 2010 ISSN 1412-4564

[12] Pusat Data dan Sistem Informasi Pertanian 2013 *Informasi Ringkas Komoditas Perkebunan No.01/01/I, 7* Januari 2013 Last access at http://pusdatin.setjen.pertanian.go.id/ October 12, 2014

[13] Pusat Data dan Teknologi Penginderaan Jauh, PUSTEKDATA LAPAN 2014 *Koreksi Radiometrik Data Satelit Spot-6* Last access at http://pustekdata.lapan.go.id/ November 12, 2014

[14] Setyowati dan Heratania A 2015 *Aplikasi Citra SPOT-6 Berbasis Transformasi Indeks Vegetasi untuk Estimasi Produksi Kelapa sawit (Elaeis guineensis jacq) (Kasusu Perkebunan Kelapa Sawit PT. Tunggal Perkasa Plantations, Air Molek, Kaupaten Indragiri Hulu, Propinsi Riau, Sumatera) Skripsi* (Yogyakarta : Fakultas Geografi Universitas Gadjah Mada)

[15] Sunarko 2007 *Petunjuk Praktis Budidaya dan Pengolahan Kelapa Sawit* (Jakarta: Agromedia Pustaka)

[16] Susetyo dan Imam 2012 *Permodelan Estimasi Produksi Tanaman Karet (Hevea brasiliensis Muell.-Arg) Berbasis citra Satelit, Potensi genetik, dan Data Satuan Medan. (Stusi Kebun Getas, PTPN IX Salatiga, Kabupaten Semarang, Jawa Tengah) Thesis* (Yogyakarta Fakultas Geografi Universitas Gadjah Mada)

Linear Spectral Mixture Analysis of SPOT-7 for Tea Yield Estimation in Pagilaran Estate, Batang Central Java

F Fauziana[1], P Danoedoro[2], and S Heru Murti[3]
[1] Remote Sensing, Geography Faculty, Gadjah Mada University, Yogyakarta 55281, Indonesia
[2] PUSPICS Geography Faculty, Gadjah Mada University, Yogyakarta 55281, Indonesia
[3] Cartography and Remote Sensing, Geography Faculty, Gadjah Mada University, Yogyakarta 55281, Indonesia

[1] fatmawati.fauziana@gmail.com, [2] projo.danoedoro@geo.ugm.ac.id,
[3] sigit@geo.ugm.ac.id

Abstract. Remote sensing has been utilized especially for agriculture yield estimation. Tea yield is effected by biology characteristic including crown density. The challenge of tea yield estimation uses multispectral remote sensing data is the presence of object beside tea. This mixed pixel problem can disturb spectrally to recognize tea tree, so it is necessary to use pixel approach. The aims of this research are (1) to determine fraction of tea and non-tea; (2) to estimate crown density percentage based on tea Normalized Difference Vegetation Index (NDVI); (3) to estimate tea yield based on crown density. SPOT-7 was utilized for this application. Linear Spectral Mixture Analysis (LSMA) has applied to determination fraction percentage each pixel. Each pure endmember was read the NDVI value. NDVI of tea tree has sensitivity with crown density. Counting tea NDVI was applied for NDVI mixed pixel. Linear regression analysis has applied for estimating crown density and tea yield. The results of this research are SPOT-7 which can recognize tea, tree shade, impervious and soil each pixel with accuracy 99,84%. Although it produced high accuracy, it has overestimate at certain tea estate because of the attendance of impervious. Regression analysis of crown density and NDVI showed coeffisien determination 52%. This model result 4-100% crown density percentage, where crown density 4–55% were located beside tea tree or pruned-tea block. Regression analysis of crown density and tea yield relation showed coeffisien determination 45%. This model produced 161,34–1296,8 kg/ha. Each this model resulted Root Mean Square Error (RMSE) 14,27% and 551,52 kg/ha.

1. Introduction

Remote sensing imagery both passive and active sensor have been used to obtain various informations on earth surface. Remote sensing imagery has potential to estimate tea yield. In Indonesia, tea (*Camellia sinensis* (L). O. *Kuntze*) is one of strategic commodity and support economic country. Traditionally, estimation agriculture yield by collecting census data and field survey. Thus will be expensive and spend a lot of time if applied at wide estate [9].

The estimation agriculture yield is necessary for making the decision like price and export and import. Tea yield is determined by wide tea estate and productivity. There is a nuisance spectrally when determining wide tea estate. The objects which exist at tea estate not only tea tree but also shade tree, asphal, stoned-street, soil, and tea factory. Based on estate management, tea trees need shade tree to keep stability of shoot tea growing. Fit spatial resolution SPOT-7, there are other objects beside tea tree at one pixel, hence it is necessary be resolved by separation mixed-pixel. Linear Spectral Mixture Analysis (LSMA) has been built to resolve mixed-pixel beside fuzzy classification [10]. LSMA assumes dominant objects (i.e. endmember) in every pixel at tea estate is linear. LSMA has successed be applied for vegetation study [2]. This method also can be applied like [3] said that tree shadow between tea canopy causes mixing interclass and nuisance remote sensing imagery efectiveness.

Tea productivity is interaction between environment condition like altitude and pruning-year. Pruning-year denotes vegetative growth including tree crown since tea trees have been pruned at certain block. Near-infrared has sensitivity to a number of leaves reflection. It is assumed that by increasing a number of leaves will increase digital number of infrared. Increasing tea leaves are interaction between branch growth and prining year. Crown density increases along with increasing pruning year and assumed tea yield will increase. Vegetation indices (VI) are widely employed remote sensing technique for measurement crown density. The most VI which employed is Normalized Difference Vegetation Index (NDVI) [4]. The VI of value every pixel does not always show the pure tea tree. Hence, this research demonstrate to combine LSMA and VI to produce tea crown density.

The objectives of this study are (1) to determine the tea fraction and non-tea fraction, (2) to count the tea crown density based on NDVI, and (3) to count the tea yield based on tea crown density.

2. Study area

This study area, located at Pagilaran Estate exactly Pagilaran unit production. It is located in Batang, Central Java, Indonesia (Figure 1). The estate area covers approximately 1.132,25 ha which devided three afdeling thus are Pagilaran, Kayulandak and Andongsili. Beside tea, Pagilaran also cultivates clove and quinine. Pagilaran unit production located in Kemulan mount slope at an altitude 740-1600 m above sea level. The soil type is dominated andosol that has high characteristics to binding water, loose, and structured crumb. Rainfall between 3500-6000 mm/year.

3. Data

SPOT-7 image acquired on 15th of March 2015. There is no cloud covers this area, therefore it helps to make easy interpretation the objects. SPOT-7 imagery has dynamic range acquisition 12 bit per pixel. There are 2 detectors within SPOT-7, multispectral and pancromatic, each images has spatial resolution 6m and 1,5m. SPOT-7 satelite deliver images over 4 bands spectral thus are blue (0,450–0,520 μm), green (0,530–0,590 μm), red (0,625–0,695) and near-infrared (0,760–0,890 μm). Pancromatic mode works on 0,450–0,745 μm. SPOT-7 satelite constellation has a revisit time every 26 days. This data has orthorectified correction. The digital numbers data were convertd to radiance data using metadata. Radiance data four images were converted to reflectance data using Top of Atmosphere (ToA). Google Earth used to assess accuracy fraction percentage.

4. Methods

4.1 Image processing

LSMA was performed on SPOT-7 image after radiometric correction using ENVI 5.2 classic (Figure 2). The important assumption SMA is spectral signature endmember each pixel is linear [1][7]. LSMA model require two principal points. First, endmember proportion each pixel must be equal 1,00 which mathematically formulated as follows:

$$\sum_{i=1}^{N} F_i = F_1 + F_2 + \cdots + F_N = 1 \qquad (1)$$

Second, each spectral bands, brightness value (BVλ) each pixel is amount of BV pure pixel each endmember multiple by weight fraction endmember plus one.

$$BV_\lambda = F_1 BV_{\lambda,1} + F_2 BV_{\lambda,2} + \cdots + F_N BV_{\lambda,N} + E_\lambda \qquad (2)$$

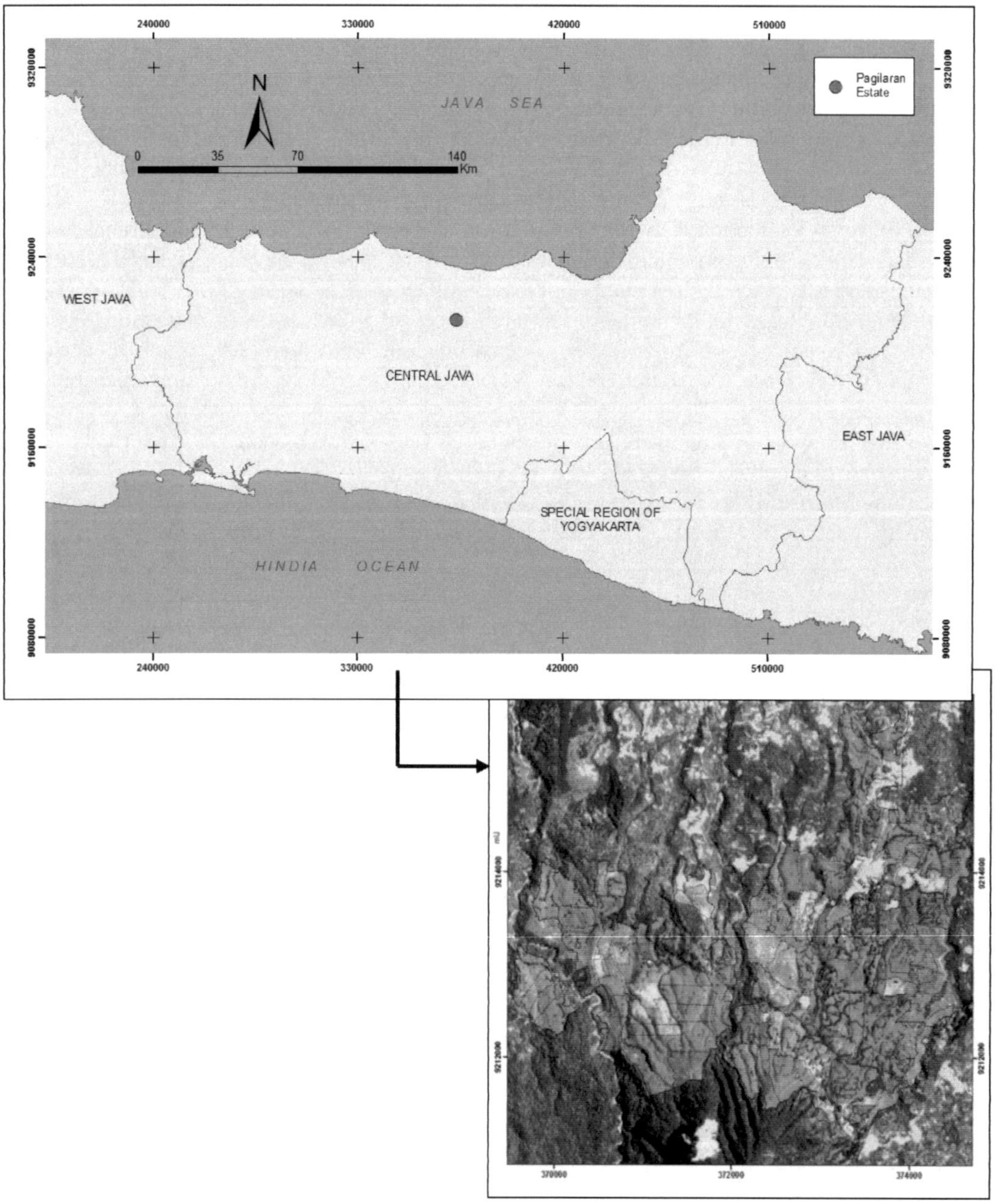

Figure 1. Location map of Pagilaran Estate showed some tea block were pruned and begin growing after pruning. SPOT-7 is displayed in false colours.

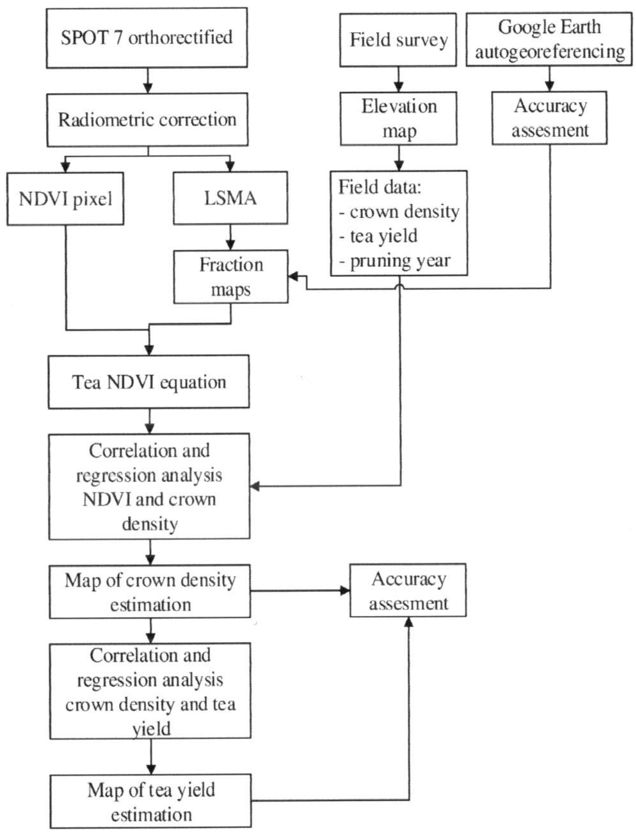

Figure 2. Flow chart shows methodology this research

The stages of LSMA as followed:

4.1.1. Minimum Noise Fraction (MNF)

MNF was known to remove data redundancy and present good quality imagery. Disturbance identification was shown by eigenvalue. Research area covers 2 tiles, hence these were applied separately. Statistic calculated from both tiles shows MNF 1 has the highest eigenvalue and MNF 4 has the lowest eigenvalue (Table 1). It means MNF 1 has lower noise than MNF 4. Only 1 tile continued to the next step, due to the eigenvalue over 1 only 1 MNF transformation whereas expected endmembers were 4

Table 1. Statistic data of MNF transformation

Tile	Band	Minimum value	Maximum value	Mean	St.Dev	Eigenvalue
R2C2	MNF 1	-10.509	55.023960	-2.59	3.227136	24,987
	MNF 2	-20.456	13.006959	-2.06	3.728131	6,186
	MNF 3	-7.157	26.850830	0,53	1.459634	3,380
	MNF 4	-13.435	24.751259	-0,65	1.510371	1,605
			Total			36,157
R3C2	MNF 1	-10.509	55.023	-2.60	3.229	23,974
	MNF 2	-20.456	13.007	-2.07	3.732	0,488
	MNF 3	-7.157	26.851	0.53	1.462	0,182
	MNF 4	-13.435	24.751	-0,66	1.512	0,072
			Total			24,716

4.1.2. Pixel Purity Index (PPI)

There two ways to select endmember, these are by spectral library and PPI process. PPI was applied to obtain pure pixel each endmember. PPI worked by the iteration to detect pure pixel. The result of this process is the pure pixel value that brightness rate denote how many pixel is recorded extremely. The advantages of MNF and PPI processes separate pure pixel and mixed pixel, decrease analized pixel and recognize endmember easily [1].

4.1.3. Endmember selection

Endmember selection is the important step to obtain meaningful classification of spectral mixture analysis. Endmember output depends on input bands that are used. Selected endmember including tea tree, shade tree, impervious and soil. Four objects are assumed to represent object at Pagilaran, note that the clove and quinine are ignored. PPI leads to choose endmember based on the majority iteration each endmember. Tea tree was derived at tea block, shade tree was derived at (*Albizia saman (Jacq.) Merr*), impervious was derived at tea factory and soil was derived at pruned-tea block. Although at the pruned-tea block was not found majority soil due to residu pruning tea at local block.

4.1.4. LSMA implementation

The result of LSMA process is the fraction percentage each endmember and error imagery shown by root mean square error (RMSE). Based on this map fraction percentage, it can be recognized pure pixel and mixed pixel. Constraint unmixing was applied for this process that the total fraction of every pixel is 1.

4.1.5. Normalized Difference Vegetation Index (NDVI) Transformation

NDVI has sensitivity to vegetation density. This index is applied to near infrared and red. Near-infrared portion shows higher reflectance in multiple leaves than single leaf. This is due to addictive reflectance from second and next layer of leaves (Howard, 1991). This condition is found on tea tree, the old leaves participate reflectance portion, whereas only shoot tea which is plucked, it can add NDVI and likely does not indicate high plucking. NDVI equation can be expressed as,

$$NDVI = \frac{NIR-red}{NIR+red} \quad (3)$$

where NIR is near-infrared reflectance shown at band 4 in SPOT-7 and Red is red reflectance shown at band 3 in SPOT-7.

The result of this equation is NDVI pixel that indicate mixed-pixel, hence to compute tea NDVI should be expressed by formula which adapted from Brown, 2001,

$$VI_{pixel} = f_{tea}*VI_{tea}+f_{shade\ tree}*VI_{shade\ tree}+ f_{impervious}*VI_{impervious}+f_{soil}*VI_{soil}+ \epsilon \quad (4)$$

Where VI denotes NDVI each endmember fs denotes fraction each endmember and ϵ is RMSE. Based on Eqs. (4) can be expressed the equation as,

$$VI_{tea}= VI_{pixel}- f_{shade\ tree}*VI_{shade\ tree}- f_{impervious}*VI_{impervious}-f_{soil}*VI_{soil} - \epsilon \quad (5)$$

Eqs. (5) was obtained from the spectral mixture analysis which each pixel is linear. NDVI pure pixel of endmember were displayed in table 2. The highest value is tea and the lowest is impervious.

Table 2. NDVI each pure endmember

Endmember	NDVI pixel
Tea	0.635938
Shade tree	0.367396
Impervious	-0.001394
Soil	0.323242

4.1. Crown density and tea yield computation

Empirical relation between the NDVI and crown density were investigated, whereas relation between the crown density and tea yield was tried to be investigated. Model was applied by correlation and regression anlisys. Crown density was obtained from field measurement, which ignore soil, tree shadow and branch. Field measurement was executed by plotting 12x12 meters and samples were determinated by altitude. Sampling method was proportional stratified sampling. The informations obtained from each samples beside tea crown density and tea yield are pruning-year and clon.

5. Results

5.1 Fraction accuracy assessment

The result of LSMA process shows there are minus value and value more than 1. Ideally, each pixel from constraint unmixing is 1 or 100%. It is due to the objects at the Pagilaran estate not only four objects based on endmember and they have characteristics spectrally. Figure 4 displays the endmember percentage.

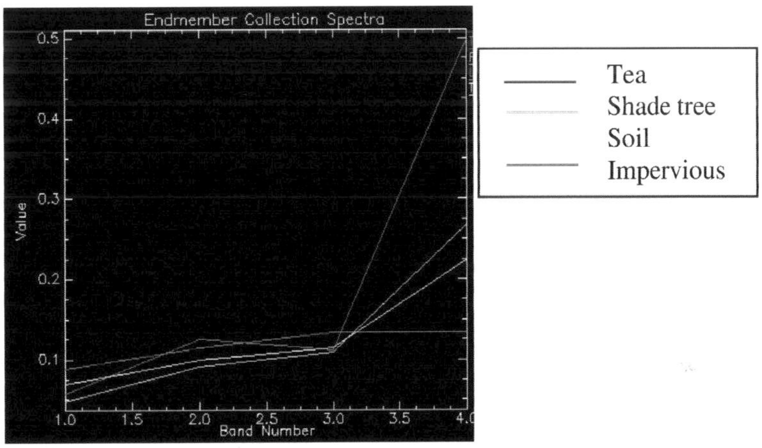

Figure 4. Spectral reflectance each endmember

Tea tree has the highest spectral reflectance in green and near-infrared region, it means healthy tea on good condition and tea leaves reflectance in some layers and also was controlled by tea internal structure. Tea curve reflectance is higher than shade tree because tea crown is denser than shade tree crown. Impervious reflectance increase in green region and constant in red and near-infrared. Soil reflectance increase along with the increasing of band region especially until near-infrared.

Accuracy assessment to exam fraction estimation is mean absolute error (MAE). Validation process was based on Google Earth image with higher spatial resolution. Accuracy was applied in permanent object. MAE for tea was 0,28, shade tree 0,16, impervious 0,10 and soil 0,11. Overall accuracy was 99,84. Although overall accuracy was high but there are pixels had impervious overestimate due to brightness pixel indicate tea or impervious.

5.2 Crown density computing based on NDVI

Tea crown density on field was input to produce crown density map based on NDVI. Statistic analysis show crown density was influenced 52% by tea NDVI. NDVI is a simple vegetation index to computing vegetation density, almost majority of researches used it. Figure 5 (left) tends to show interaction between pruning-year and crown density that can be divided into two classess. First class has characteristics, majority from seed and pruning-year is 1 which on going to vegetative growth. Pruning year 1 denote a few tea branch and has gap in tea tree. Second class was dominated seed and pruning-year 2 until 4. This is the optimum condition of tea growth. Any amount whatever of NDVI, if at tea block with optimum growth will produce same crown density. Pruning-year 2 experienced growing branch and addictive dense leaves, until reach constant growth in pruning-year 4. Crown density was measured by ignoring plant disease and weather condition. The similarity seed for first and second class is tens old moreover hundreds old. Any amount whatever of tea age will produce good vegetative growth if in a good estate management.

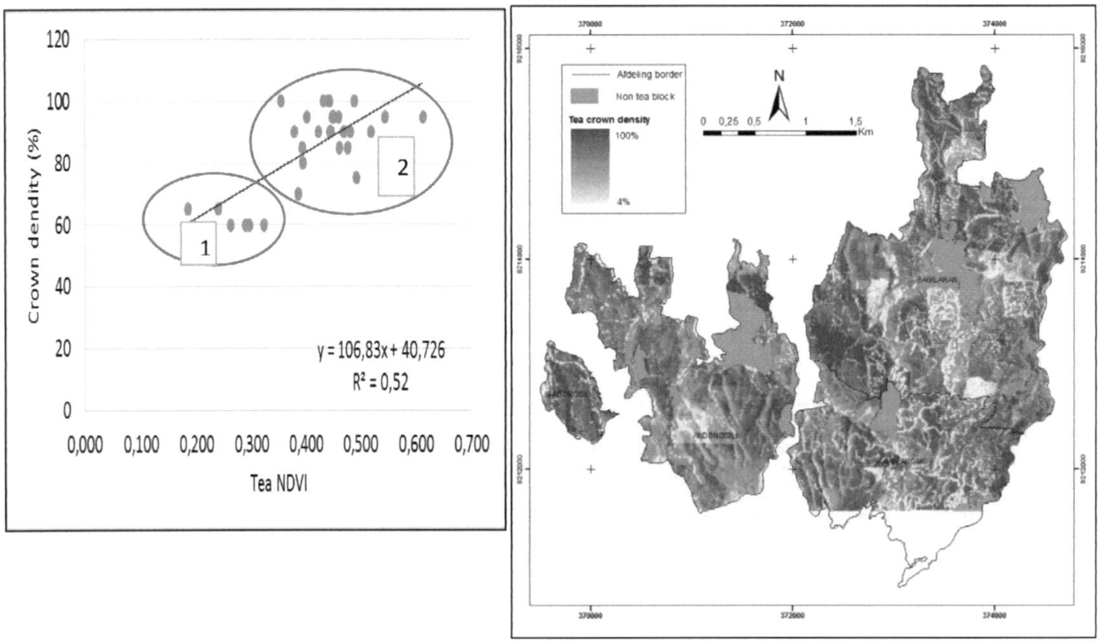

Figure 5. Regression graphic between tea NDVI and tea crown density (left) and map of distribution of the tea crown density estimation (right)

Figure 6. The average condition of first class tea crown density (left) and second class tea crown density (right)

The lowest estimated tea crown density was 4% and the highest tea crown density was 100%. Crown density 4% until 55% were found at shade tree and pruned-tea block (Figure 5 (right)). RMSE was 14,27% and accuracy assesment 86,37%. The model was accepted to be input tea yield estimation model, although there are overestimate and underestimate crown density percentage. Overestimate denote different pruning-year tea block that was before pruning, while field measurement on pruning-year 1.

5.3 Tea yield computing based on crown density

Different tea yield although at the same pruning-year is non-uniformity of the shoot tea growth. Although tea block has dense leaves but the distribution shoot tea growth is not uniform, will produce different tea yield. It is due to unfit plucking technique by shear. Sometimes tea pluckers do not pluck young shoot because limits of scope or they leave drop young shoot in gaps of tea tree. Tea yield was influenced 45% by crown density, it means not only crown density influence tea yield. Figure 7 (left) displayed 2 classes tea yield. First class was dominated by pruning-year 1 tea trees and over but vegetative growth do not produce enough dense shoot tea. Second class was dominated by pruning-year 2 tea trees and over.

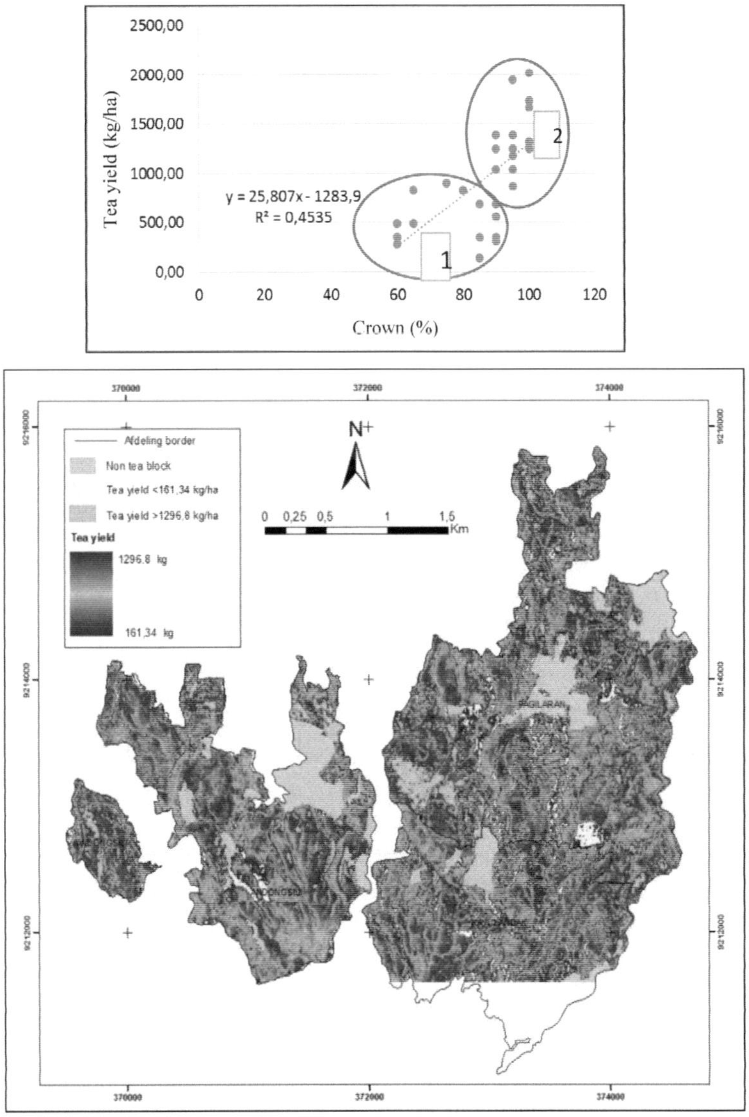

Figure 7. Regression graphic between tea crown density and tea yield (left) and map of distribution of the tea yield estimation (right)

Figure 8. The average condition tea of first class tea yield (left) and second class tea yield (right)

The overall model result caused pixels beside tea fraction has participation to tea yield. Hence, masking was applied in 56%-100% crown density. It assumed that minimum 56% crown density is enough to produce tea yield. The effect of masking is remove tea yield at slope tea block. Actually, this tea block has leaves density but LSMA of SPOT-7 recognize this place as shadow. Minimal tea yield was 161,34 kg/ha produced by 56% crown density and maximum tea yield was 1296,8 kg/ha produced by 100% crown density (Figure 7 (right)). RMSE was 551,52 kg/ha and accuracy assessment 35,65%. Difference tea block between acquisition image and field survey produce high standard deviation. The examples, Pulosari II block was not pruned on March 2015 yet but on December 2015 was pruning-year 1, it produce low tea yield. Pekandangan IA 1 block was like pruned on the image whereas based on field data this block did not produce tea yield from last March until April.

6. Discussion

Digital number of SPOT-7 is mixed-pixel of the objects in the tea estate. The separation fraction of each pixel help to remove the error determination of tea width and vegetation index transformation. This model show high accuracy of LSMA even though there are overestimate impervious pixel in tea block. It is caused high albedo fraction including stoned-road between tea blocks, non-uniformity topography and endmember selection which do not include all objects in tea planation. [10] remove soil from high albedo fraction and distinguish impervious with water and high dense vegetation.

The lack LSMA of this study is unavailable spectral library to recognize spectral particular endmember, it trade on how many iteration show pure pixel whereas there is tea block which considered uniform. [5] summarize hyperspectral data can distinguish tea growing in sunlit and shade condition in green and near-infrared region. Blue region distinguish tea growth and healthy condition. Red and near-infrared regions can discriminate pruned and unpruned-tea.

Tea block with high tree population and high dense leaves contribute higher reflectance energy than shade tree in near-infrared region. Actually, high reflectance tea estate do not related with tea yield directly, due to only plucking tea shoot might not reach 20% from leaves population every tea tree. Crown density is an approach that assume many tea leaves will increase tea yield. Crown density has stronger correlation with NDVI than TVI (Transformed Vegetation Index) and RVI (Ratio Vegetation Index) [7].

Tea yield produces from young leaves which increase from pruning-year 1 until 3 and constant in pruning year 4. This condition does not prevail in all tea block. Tea shoot can not be plucked in unhealthy condition, long dry season and error of plucking. The periodicity of tea plucking dependon altitude. High altitude causes long plucking cycle, it is due to lack of sunlit and slow photosynthesis.

7. Conclusion

LSMA can seperate objects in tea estate using SPOT-7 image including tea tree, shade tree, impervious and soil and produce accuracy 99,84%. Crown density was computed based on NDVI, it produce 4%

until 100% crown density and RMSE 14,27% and accuracy assessment 86,37%. Tea yield estimation based on crown density produce 161,34 kg/ha until 1296,8 kg/ha and RMSE was 551,52 kg/ha and accuracy assessment 35,65%. Low accuracy is caused plucking and human interference including pruning.

8. Acknowledgement

The authors would like to thank LAPAN with SPOT-7 Pagilaran estate, all staffs Pagilaran estate, Dina Ardihani and Sugeng Riyadi who assisted during field data collection.

References

[1] Asis A M De and Omasa K 2007 Estimation of vegetation parameter for modeling soil erosion using linear Spectral Mixture Analysis of Landsat ETM data *Photogrammetry and Remote Sensing* **62** 309–324
[2] Brown D G 2001 *A Spectral Unmixing Approach To Leaf Area Index (LAI) Estimation At The Alpine Treeline Ecotonein GIS and Remote Sensing Applications in Biogeography and Ecology A. C. Millington, S. J. Walsh, and atrick E. Osborne* (New York: Springer Science and Bussiness Media)
[3] Dutta R 2011 *A Spatio - Temporal Analysis of Tea Productivity and Quality in North East India Dissertation* University of Twente The Netherlands
[4] Jensen J R 2014 *Remote Sensing of the Environment An Earth Resource Perspective* Second Edition (New York: Pearson)
[5] Kumar A, Manjunath K R, Meenakshi, Bala R, Sud R K, Singh R D and Panigrahy S 2013 Field hyperspectral data analysis for discriminating spectral behavior of tea plantations under various management practices International *Journal of Applied Earth Observations and Geoinformation* **23** 352–359
[6] Muningsih R, Indradewa D and Sulistyaningsih E 2014 *Karakter Fisiologis dan Hasil Pucuk Teh pada Beberapa Umur Pangkas Produksi dan Tinggi Tempat. Ilmu Pertanian.***17(1)** 25–36
[7] Rohman A S 2002 *Analisis Digital Data Citra Landsat TM dan SIG untuk Estimasi Produksi Teh di Sebagian Wilayah Bogor, Cianjur dan Sukabumi Propinsi Jawa Barat* Skripsi Gadjah Mada University
[8] Somers B, Asner, Gregory P, Tits L, and Coppin P 2011 Endmember Variability in Spectral Mixture Analysis: A review *Remote Sensing of Environment* **115(7)** 1603–1616
[9] Tripathi N K, Rajapakse S S, and Honda K 2004 Tea Yield Modeling Based on Satellite Derived LAI *Geocarto International* **19(3)** 51–54
[10] Weng Q 2010 *Remote Sensing and GIS Integration: Theories, Methods, and Application* (New York: McGraw-Hill)

Applications of Remote Sensing and Geographic Information System to Identify Rice Planting Season During El Nino Years: Case Study in the Pringsewu District, Province of Lampung

S D Ratri, E Kusratmoko, and F S Hardiyanti
Magister Program of Geography, Indonesia University, Build H, FMIPA UI, New Collage, Depok, 16421, Indonesia

sariedwika@gmail.com

Abstract. Spatial Information about rice planting season (RPS) in a wide areas, particularly during periods of El Nino, is important to support an information about the availability of rice continously. Application of remote sensing and geographic information system (GIS) technology can support it's information continuously and accurate. In this study, we attempted to identify the rice planting season during El Nino years of 1997, 2006 and 2015 in the Pringsewu district, Lampung and we compared with meteorological drought index. Spatial information of the RPS obtained through Interpretation of multitemporal Landsat data aquired in 1997, 2006 and 2015 using normalized difference vegetation index (NDVI) and the humidity index. While standardized precipitation index (SPI) is used as a indicator of meteorological drought. This study has shown that the application of remote sensing and GIS could accurately monitor the rice planting season during the periods of El Nino in 1997, 2006 and 2015. The fallow land dominated during the El Nino years and there were no significant difference between years. While drought information based on SPI values showed different results between years of El Nino events. In this paper we also discussed the relationship between distribution of fallow land and meteorological drought in a spatial perspective.

Keywords: Rice planting season, Remote Sensing Application, Meteorological drougt, Pringsewu, Lampung.

1. Introduction

Global warming is a problem faced today as the impact of long-term accumulation of atmospheric climate change pollution. Studies related to the impact of climate change on agriculture by the National Academy of Science / NAS (2007), showed that agriculture in Indonesia has been influenced significantly by the variations of the annual rainy and between years by Australia-Asia Monsoon and El Nino-Southern oscilation (ENSO). The strong influence of ENSO it can be proved by looking at the incidence of droughts and drought in various regions of Indonesia, which coincided with El Nino events [1].

The agricultural sector is the sector most vulnerable to the impacts of climate change because agriculture is very dependent on the climate in its activities [2]. Meteorological drought in this study to determine the level of rainfall over a season in cultivated agricultural land, especially in paddy fields. Meteorological drought is the first indication of a drought [4].

Pringsewu is one district that has the potential of food crops, especially rice which is a priority in improving agricultural production in Lampung [3]. In connection with the remote sensing and Geographic Information System (GIS) can help in knowing the cropping patterns in years of El Nino, so the results are expected to help provide accurate information for the sustainability of agriculture and capable of supporting District Pringsewu in meeting the target of rice self-sufficiency. In knowing the indicators of drought when the El Nino years can be classified by the SPI (standardized precipitation index). Thus, in this study aims to analyze and classify the cropping pattern of rice paddy during a drought in El Nino years with the help of SPI through Landsat TM with the matriculation of NDVI and humidity index.

2. Method

The study was conducted in the District Pringsewu using Landsat time series when the occurrence of El Nino in 1997, 2006 and 2015 with the spatial approach. The variables used in this study is a meteorological drought and cropping of paddy. The data used is secondary data that is processed by several methods, namely:

2.1 SPI (Standardized Precipitation Index)

According to Hayes et al (1999) that the determination of drought criteria do with rainfall data processing for 30 years (1986-2015). The data is then calculated by the method of Standardized Precipitation Index (SPI) to determine the meteorological drought index is happening with SPI drought index. SPI uses software processing methods scopic with a choice of time of 1 month to determine the meteorological drought in 7 classification SPI. The drought classification based on SPI values are presented in table 1.

Table 1. Classification scale SPI value

Score	**Classification**
≥2,00	Really wet
1,50 ~ 1,99	Wet
1,00 ~ 1,49	Dampy
0,99 ~ -0,99	Normal
-1,00 ~ -1,49	Bit dry
50 ~ -1,99	Dry
≤ -2,00	Very dry

Source: Mc Kee, 1993 (treated by BMKG)

a. NDVI (Normalized Difference Vegetation Index)

Processing to obtain NDVI using ENVI 4.2 software by the equation:

$$\text{NDVI} = \frac{\text{Band 4} - \text{Band 3}}{\text{Band 4} + \text{Band 3}}$$

Based on the research results Widagdo (2000), the vegetation index during the growth of rice plants from early planting to be harvested parabolic shape. NDVI value of the current rice plants aged 3-4 MST to 16 show the shape of the curve with a peak at the age when rice (phase) optimum vegetative - rice bunting (aged about 70-80 days after planting, or about 10-11 MST).

b. Image enhancement for moisture

Obtained from the processing of Landsat 5 and Landsat 8 with rasioning band 3 and 5 on the band math using ENVI 4.2. Bright image processing results indicate that the image has a high wetness. It can be used for paddy area experiencing drought during the dry season or below normal rainfall [5].

c. Overlay analysis using ArcGIS 10.3

The results of the processing of Landsat NDVI and humidity intersect using ArcGIS 10.3.2. to determine the activity of planting paddy rice by the following matrix.

Table 2. Matrix Determination Period Planting Rice

No	NDVI	Indeks Kelembaban				
		Sangat Kering	Kering	Lembab	Sangat Lembab	Tergenang
1	Tidak ada vegetasi	Lahan Bera	Lahan Bera			Penggenangan Lahan
2	Kehijauan sangat rendah	Lahan Bera	Lahan Bera	Fase Vegetatif	Fase Vegetatif	Penggenangan Lahan
3	Kehijauan rendah	Lahan Bera	Lahan Bera	Fase Vegetatif	Fase Vegetatif	Fase Vegetatif
4	Kehijauan sedang	Fase Pematangan	Fase Pematangan	Fase Reproduktif	Fase Reproduktif	
5	Kehijauan tinggi	Fase Pematangan	Fase Pematangan	Fase Reproduktif	Fase Reproduktif	

Source: Rizatus and Dwi, 2007 (Treated)

3. Results and Discussion

3.1. Pringsewu Climate

Meteorological drought that occurred in Pringsewu cause drought index different. Based on calculation of SPI, the meteorological drought occurred in the Kebupaten Pringsewu are presented in the following table.

Table 3. Drought meteorological District Pringsewu In El Nino years

No	Rain Station	SPI (*Standardized Precipitation Index*)																							
		July				August				September				October				November				December			
		1994	1997	2006	2015	1994	1997	2006	2015	1994	1997	2006	2015	1994	1997	2006	2015	1994	1997	2006	2015	1994	1997	2006	2015
1	Banyuwangi	K	K	K	AK	AK	AK	AK	N	K	K	K	N	AK	K	K	AK	N	N	AK	AK	B	N	AK	N
2	Fajaresuk	K	K	N	AK	N	N	N	N	N	AK	AK	N	N	AK	AK	AK	K	AK	AK	N	N	N	N	N
3	Gading Rejo	K	AK	N	AK	N	N	N	N	N	N	N	N	AK	AK	AK	AK	N	AK	K	N	N	K	N	N
4	Gisting	AK	N	N	N	K	N	N	N	N	AK	N	N	N	N	N	K	N	N	N	AK	N	N	N	N
5	Kunyir	K	K	N	N	N	AK	N	N	N	AK	N	AK	N	K	N	N	N	N	N	N	N	N	N	AK
6	Panutan	N	N	N	N	N	N	N	N	N	N	AK	N	N	N	K	N	N	N	N	N	N	N	AK	N
7	Podorejo	K	AK	N	N	AK	AK	AK	N	AK	AK	AK	N	N	K	AK	AK	N	SK	AK	N	N	N	N	N
8	Pringsewu	SK	N	N	N	K	AK	K	AK	AK	K	K	N	N	SK	K	SK	AK	SK	SK	N	N	N	AB	N
9	Way Gatel	K	AK	N	N	N	N	N	N	N	AK	AK	N	N	K	AK	AK	AK	K	K	N	N	N	N	N
10	Way Harong	N	N	N	N	AK	N	AK	N	N	N	N	N	N	AK	N	N	N	N	N	N	N	N	N	N
11	Wonokriyo	AK	AK	N	N	N	N	N	N	N	AK	N	N	AK	AK	AK	AK	AK	K	K	N	N	K	N	N

Keterangan:

SK : Very Dry N : Normal
K : Dry AB : Dampy
AK: Bit Dry B : Wet

Source: Data processing, 2016

The years of El Nino in July to December each year it appears that the District Pringsewu have meteorological drought levels from normal to very dry. In July of meteorological drought occurred in the classification of normal to very dry. In a very strong El Nino, which in 1997 showed the level of dryness in July in the index rather dry to dry. In strong El Nino in 2015, drought in July was dominated by the normal index. Meteorological drought classification rather dry to very dry more dominant place in November. This shows the peak drought during El Nino in the District Pringsewu occurred in November.

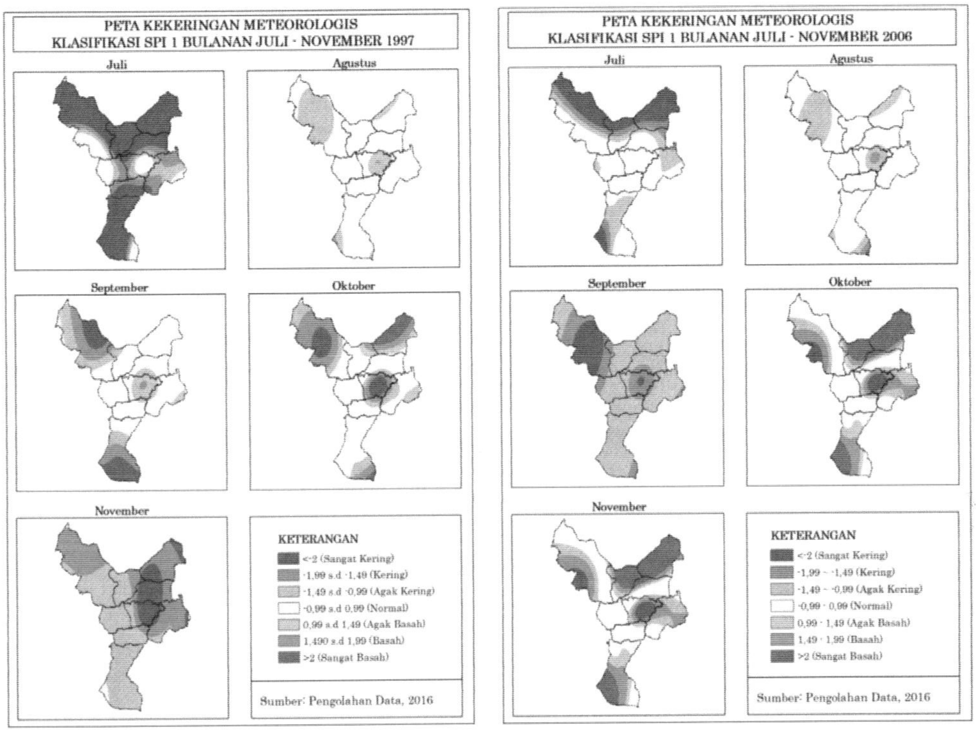

Figure 1. SPI Drought In July-November 1997 and 2006 in the District Pringsewu

Based on Figure 1, the El Nino of 1997 appears a strong influence in the district Pringsewu. El Nino was very strong in 1997 led to almost all areas affected by drought meteorological Pringsewu very dry in July. In the months that followed in Regency Pringsewu experiencing meteorological drought in the classification of normal to very dry to a peak in November. During the five-month drought continues, only in October just started to appear wet to very wet classification is in the region Adiluwih districts and parts of Ambarawa.

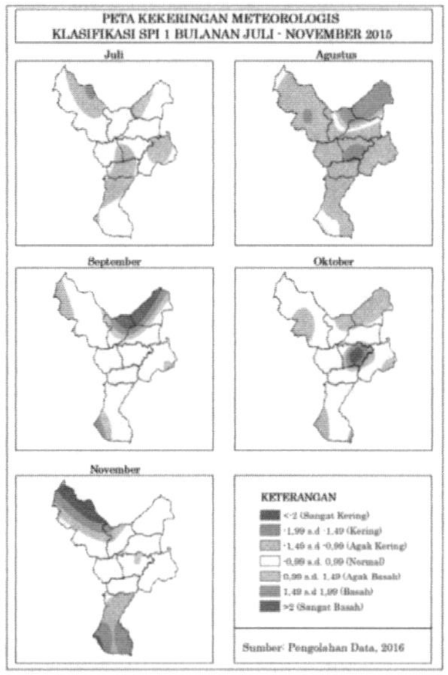

Figure 2. SPI Drought In July-November 2015 in the District Pringsewu

2015 El Nino going strong again, as shown in Figure 2. In 2015 drought meteorological in district Pringsewu very dry classification is seen only in a small area in the district of Pringsewu. Drought meteorological bit dry almost scattered throughout the region in August, just in Adiluwih visible in the wet classification. In 2015 the area suffered least of meteorological drought dry classification is Adiluwih and Ambarawa.

3.2. Rice Planting Pattern in District Pringsewu

The cropping pattern is influenced by rainfall, because rice production is dependent on water. Some farmers explained that the existing irrigation techniques is still inadequate in some areas. When the dry season paddy fields which are not traversed irrigation experience dryness. In addition to the presence of irrigation, irrigation for paddy can use the river water for rice fields, which are located not far from the river.

Based on interviews at the Department of Agriculture and Forestry District Pringsewu, explaining that the cropping pattern paddy rice in the District Pringsewu conducted three times in one year. The majority of wetland also contained two rice cropping pattern interspersed with crops. Wetland conditions in El Nino years compared with normal conditions can be demonstrated through the interpretation of Landsat imagery. The data processing of Landsat TM 5 and 8 using ENVI 4.2 software is made to calculate the value of vegetation index and moisture index.

In connection with the cropping pattern, imagery used to see in the area Pringsewu planting season is in March. As the wet months are used to compare the cropping patterns in a normal year with El Nino year. November is the beginning of the main cropping season paddy rice in Pringsewu. November is the month when the El Nino weather anomalies, which can be used to compare the current cropping patterns normal year with El Nino. In figure 3 is presented wetland conditions at planting time in the district Pringsewu in normal and El Nino year in March and November. It is apparent that the condition of El Nino (1997, 2006 and 2015) largely fallow rice field in a condition. The ratio of the condition of the fields during El Nino in 1997 and 2015 are presented in the following table.

Table 4. Size Wetland Condition In November At The El Nino of 1997 and 2015

November 1997		November 2006		November 2015	
Planting season	Large (Ha)	Planting season	Large (Ha)	Planting season	Large (Ha)
Fallow	17974	Fallow	13794,8	Fallow	13217
Reproductive phase	22	Maturation	38	-	-
Total	17996	Total	13742,8	Total	13217

Source: Data processing, 2016

November in normal climate is the main cropping paddy fields in the district Pringsewu. Shown with the planting season in November in a normal year (2014 and 1996), rice paddy in the vegetative to reproductive phase. Conditions fallow in November in El Nino years showed the effect of El Nino impact on changes in the growing season. The absence of rain resulted in water supply for agricultural productivity is inadequate in several districts Pringsewu.

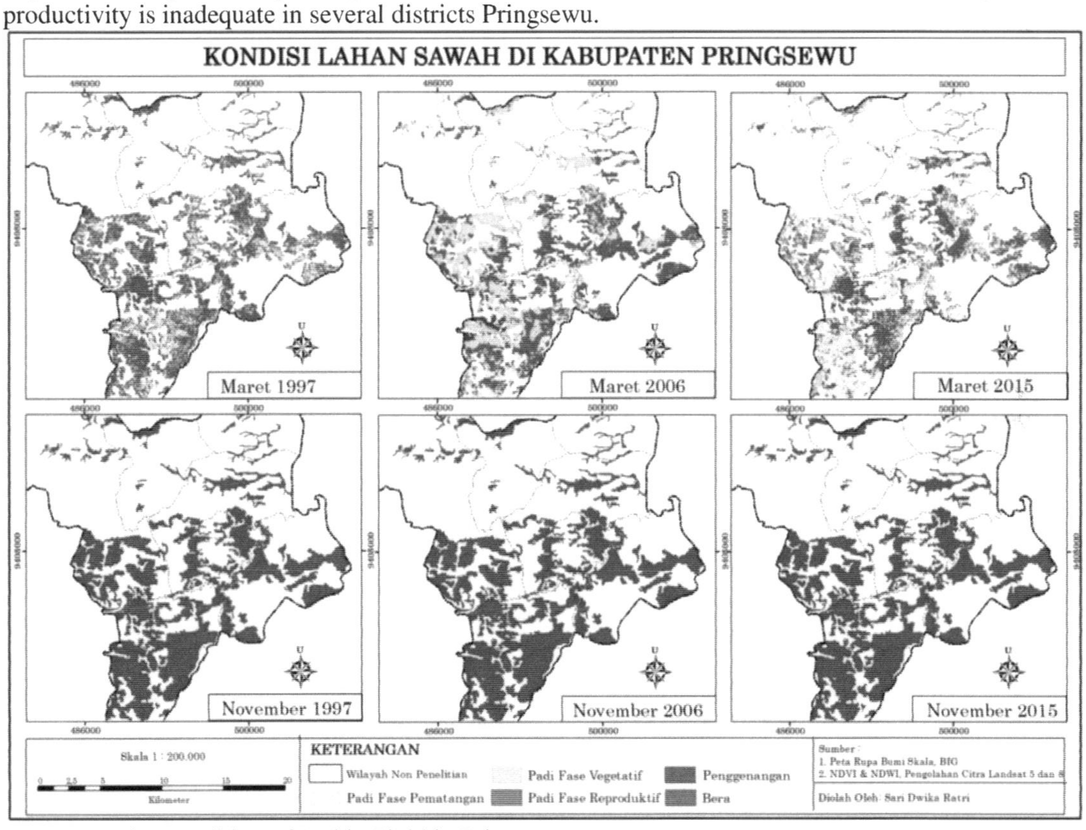

Figure 3. The Condition of Paddy Field in Pringsewu

The drought that hit Pringsewu when El Nino caused a shift in the main planting season, which in November which is a transition month dry season to the wet season. Based on the results of the processing of Landsat 8 recording dated December 19, 2015 main paddy rice planting time become a retreat for 4-5 dasarian, is presented in Figure 4.

Shifting cropping patterns that starting from 1 November 2015 occurred between rice fields that exist in each district is different. Activities rice fields occur in December, there is still in a state of fallow land. Shifting planting season occurs faster in Sukoharjo districts, Banyumas, North Exhibition, and Adiluwih.

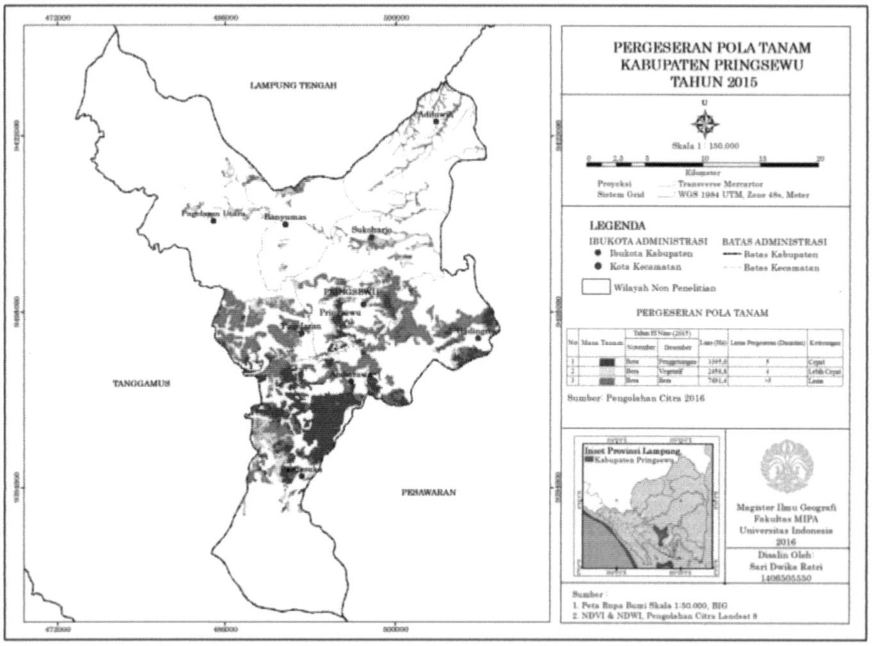

Figure 4. Map Shift Cropping Patterns in Pringsewu

4. Conclusion

Drought occurred in 1997, 2006 and 2015 in the district Pringsewu generally covers the whole region. The impact of El Nino in 2015 is relatively weaker than in 1997 and 2006.Drought in El Nino effect on lowland rice cropping pattern in the district Pringsewu. The results of processing using Landsat imagery and GIS show in an El Nino paddy fields in the district of Pringsewu dominated by fallow. The pattern of planting paddy rice in El Nino years of 1997, 2006 and 2015 are always changing as a result of drought conditions paddy fields. It affects the withdrawal of the main planting season in November. Shifting the growing season can reach> 4 dasarian, so that in December became the main planting season in El Nino.

SPI drought index is based on spatially seen from Landsat imagery processing and GIS shows the areas most often experience faster fallow season El Nino is the dominant rice field located in the wetland units in the form of river terrace and the plains.

References

[1] Abawi Y, I Yasin S, Dutta T, Harris M, Ma'shum D, Mc Clymont I, Amien and R Sayuti 2002. *Capturing the benefit of seasonal climate forecast in agricultural management: Subproject 2- Water and Crop Management in Indonesia Final Report to ACIAR QCCA-DNRM* Toowoomba Australia

[2] Boer R, Herniwaty D, Haryanti dan A Wulansari 2010 *Analisis Dampak Perubahan Iklim di Semarang* Yayasan Bintari dan Erca Semarang

[3] BPS Kabupaten Pringsewu 2013 *Pringsewu Dalam Angka* BPS Pringsewu

[4] Danu, Triatmoko, Armi, Susandi, Musa, Ali, Mustofa, Erwin dan E S Makmur 2012 Using Standardized Precipitation Index Method For Identification Meteorogical Drought In Pantura West Java Area *Jurnal Program Studi Meteorologi Institut Teknologi Bandung*

[5] Purwadhi S H F 2001 *Interpretasi Citra Digital* Grasindo Jakarta

[6] Rizatus Shofiati dan Dwi Kuncoro G P 2007 *Inderaja Untuk Mengkaji Kekeringan di Lahan Pertanian Balai Besar Penelitian dan Pengembangan Sumberdaya Lahan Pertanian.* Informatika Pertanian Volume 16 Number 1

Protection of Paddy Field and Recommendation of Regional Planning in Cianjur Regency, West Java, Indonesia

Khursatul Munibah[1], Rani Yudarwati[2], Enni Dwi Wahyunie[3] and Hermanu Wijaya[4]

[1, 2, 3, 4]: Department of Soil Science and Natural Resource, Faculty of Agriculture, IPB, Indonesia

khursamun@yahoo.co.id, rani.yudarwati@yahoo, edwiwahjunie@yahoo.com, hermanu_wijaya@yahoo.com

Abstract. Cianjur Regency is one of *"lumbung"* paddies in West Java Province that can contribute to rice sufficiency for West Java 13.5%. However conversion of paddy field into other land use still happen in Indonesia because of land rent of paddy field less than other land use and also the low commitment of Government to protect the paddy field to get food self-sufficiency. Objectives are analysis of paddy field protection and recommendation of regional planning. Paddy field protection was determined based on existing paddy field, land suitability, economic value and Regional Spatial Planning (RTRW). Recommendation of regional planning was determined based on priority level of paddy field protection, Regional Spatial Planning (RTRW) and rice sufficiency status. The results showed that land suitability, economic value and also allocating land for paddy field in the RTRW can support realization of paddy field protection. The paddy field that included into the first and third priority is 30.14% and 38.45%, respectively. The other priorities of paddy field protection are around 15%. This research is recommended that 87.5% of the paddy field existing can be protected to get rice sufficiency with surplus around 48.782 ton.

1. Introduction

In the period 2003-2013, the paddy field Indonesia increased with the rate of 23.553,8 ha/year due to the construction of new paddy field, but in West Java decreased with the rate of 909,8 ha/year (BPS of Cianjur Regency 2014). Likewise, the paddy field in Cianjur Regency also decreased 5, 438.4 ha for 2000-2015 because of conversion into built land and dryland (Chairunnisa 2016). On the other hand, Cianjur Regency is one of *"lumbung padi"* in West Java that produced 839.776 ton in 2011 and increased to 882.622 ton in 2012 (BPS of Cianjur Regency, 2013).

Many factors that influence on paddy field conversion but the general factors are (a) land rent of paddy field less than the other land use, (b) the low commitment of Government to protect the paddy field to get food self-sufficiency. The land rent of paddy field, mixed garden and settlement are Rp. 44,12- Rp. 1070, 44 per m2 per year, Rp. 51,33 - Rp. 1.493,5 6 per m2 per year and Rp. 208,33 - Rp. 35.069,44 per m2 per year respectively (Rumiris, 2008). A comparison between land rent of paddy field and oil palm are 1: 1,11 (Hamdan, 2012) and according to Asni (2005) showed that land rent of paddy field and oil palm are Rp 1.387.577/hectare and Rp 5.735.203/hectare, respectively. Based on Sugandi et al., 2013 showed that land rent of paddy field and oil palm on Bengkulu are Rp7.650.590/hectare/year and Rp 11.104.076/hectare/year.

The low commitment of Government to protect the paddy field showing by building construct permit (*IMB*) on paddy field for settlements, industry and public facilities such as hospital, mosque, church. Indonesia has approved an agreement between states on Word Trade Organization (WTO) in Bali on 2013. It means that began in 2015 Indonesia have to accept foodstuffs from other countries. This would greatly disrupt the Government's efforts to achieve food sovereignty.

Therefore, the study on paddy field protection is urgently needed. The regulation on the paddy field protection has been published as Law No.41 of 2009 that recommends for each regency to establish sustainable paddy field. Cianjur Regency had planned 21.000 ha to be sustainable paddy field and has written on the Regional Spatial Planning Map (RTRW), however it is not realized yet in 2015 Objectives of this research are analysis of paddy field protection and recommendation of regional planning.

2. Material and Method

2.1. *Study area*

The study area (figure 1) is Cianjur Regency, West Java, Indonesia and has coordinate of 6° 21'- 7° 25' S and 106° 42'-107° 25' E. The total areas cover 361.435 ha, comprising 32 districts. Capital of Cianjur Regency is located in Cianjur District.

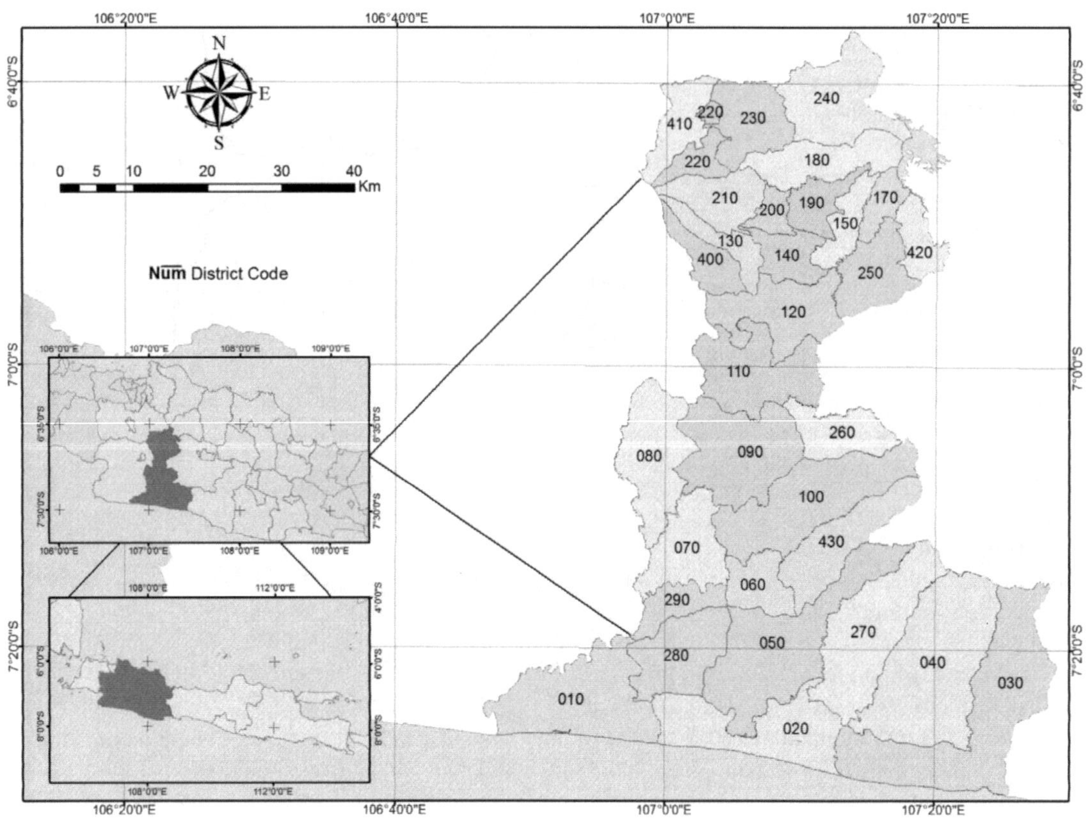

Figure 1. Location of study area

2.2. Distribution of paddy field on its land suitability and the Regional Spatial Planning Map

Distribution of paddy field on its land suitability and the regional planning of Cianjur Regency were obtained by overlapping between the Paddy Field Map, the Land Suitability Map and the Regional Spatial Planning Map (RTRW) and Administration Map. The secondary data includes Paddy Field Land Suitability Map with scale 1:50.000 (Munibah et al., 2015),

Regional Spatial Planning Map (RTRW) with scale 1:50.000 (Bappeda of Cianjur Regency 2015), Paddy Field Map with scale 1:10.000 (Department of Crop Farming and Horticulture, 2010), and Administration Map with scale 1:25000 (Agent of Geospatial Information).

2.3. Calculating an economic value of paddy field

Parameters for calculating an economic of paddy field were obtained by interviewing with farmers about (1) budged for buying paddy seed, fertilizer and pesticides; (2) labors budged for tillage, seedbed, fertilizing, spraying pesticides, grass weeding, post-harvest, (3) productivities of paddy field, (4) selling price of paddy (Rp/ton). The samples size for primary data collection is 94 farmers distributed in all district in Cianjur Regency. The interviews with farmers were guided by a questionnaire.

2.4. Determination of paddy field protection

Determination of paddy field protection was based on land suitability, regional planning and R/C value with certain rules (Table 1). Based on the ease (low conflict) to protect the paddy field, it is determined priority level, such as priority 1, 2, 3 and 4. The smaller number showed the greater priority to be protected.

Table 1. Determination rules of paddy field protection

Land use	**Regioanal Spatial Planning (RTRW)**	**R/C value**	**Suitable (S2)**	**Marginally suitable (S3)**	**Not suitable (N)**
Irrigated or rainfed	Conservation forest area	≥ 1	Priority_4	Priority_4	Priority_4
	Protected forest area		Priority_4	Priority_4	Priority_4
	Production forest area		Priority_3	Priority_3	Priority_4
	Industrial area		Priority_3	Priority_4	Priority_4
	Water bodies		Priority_4	Priority_4	Priority_4
	Plantation/annual crop area		Priority_3	Priority_4	Priority_4
	Rural settlement area		Priority_3	Priority_4	Priority_4
	Urban settlement area		Priority_3	Priority_4	Priority_4
	Wetland agriculture area		Priority_1	Priority_1	Priority_1
	Sempadan sungai		Priority_1	Priority_2	Priority_3
Irrigated	Upland agriculture area		Priority_1	Priority_2	Priority_3
Rainfed			Priority_2	Priority_3	Priority_4

For example: priority_1 means paddy field in the first priority to be protected.

2.5. Determination of rice sufficiency status

Determination of rice sufficiency status based on supply and demand of rice, with formula:

$$Sk_i = (A_{si} * Y_{pi} * IP_{pi} * C) - (P_i * S_j)$$ where:

Sk_i = Rice sufficiency status (surplus/minus)
A_{si} = Paddy field area (ha)
Y_{pi} = Average productivity (ton/ha)
IP_{pi} = Average cropping index (times/year)
P_i = Total population (people)
S_j = Average rice consumption per capita for Cianjur Regency (105. 85 kg/capita/year)

IOP Conf. Series: Earth and Environmental Science **47** (2016) 012036 doi:10.1088/1755-1315/47/1/012036

C = Conversion factor from paddy into rice = 0,6274

2.6. Determination of regional planning recommendation
Determination of paddy field protection was based on existing paddy field and the Regional Spatial Planning Map (RTRW) with certain rule (Table 2). There are three recommendations on actual paddy field: (1) paddy field protected, (2) reserve paddy field protected, and (3) paddy field that can be allocated according to the Regional Spatial Planning Map.

Table 2. Detemination rule of regional planning recommendation

Regional Spatial Planning Map (RTRW)	Paddy field at priority level:			
	Priority_1	Priority_2	Priority_3	Priority_4
Conservation forest area				(1) Reserve paddy field protected OR (2) Paddy field that can be allocated according to the Regional Spatial Planning Map (RTRW)
Protected forest area				(1) Reserve paddy field protected OR (2) Paddy field that can be allocated according to the Regional Spatial Planning Map (RTRW)
Production forest area			Reserve paddy field protected	Reserve paddy field protected
Industrial area			Paddy field that can be allocated according to the Regional Spatial Planning Map (RTRW)	Paddy field that can be allocated according to the Regional Spatial Planning Map (RTRW)
Water bodies				Paddy field that can be allocated according to the Regional Spatial Planning Map (RTRW)
Plantation/unnual crop area			Reserve paddy field protected	Reserve paddy field protected
Rural settlement area			Paddy field that can be allocated according to the Regional Spatial Planning Map (RTRW)	Paddy field that can be allocated according to the Regional Spatial Planning Map (RTRW)
Urban settlement area			Paddy field that can be allocated according to the Regional Spatial Planning Map (RTRW)	Paddy field that can be allocated according to the Regional Spatial Planning Map (RTRW)
Wetland agriculture area	Paddy field protected			
Sempadan sungai	Paddy field protected	Paddy field protected	Reserve paddy field protected	
Upland agriculture area	Paddy field protected	Paddy field protected	Reserve paddy field protected	Reserve paddy field protected

3. Result and Discussion

3.1 Distribution of paddy field on its land suitability
The paddy field in Cianjur Regency have been occupying the land with suitability class S2 (suitable) and class S3 (marginally suitable) with an area of 21.54% and 78.33% respectively. Only a little of paddy field were found on the land with suitability class N (not suitable) with an area 0.13%. The limitation factor of land suitability class S2 and S3 were drainage (r), C-organics, pH, (f) and N total (n).

Table 3 and figure 2 showed that paddy field was distributed in Northern Cianjur (47.5%), Southern Cianjur (30.0%) and Centre Cianjur (22.5%). Based on the land suitability analysis where paddy field in Cianjur Regency has high potential to be protected.

Table 3. Distribution of paddy field on its land suitability

Land suitability for paddy field	Northern Ciajur		Centre Cianjur		Southern Cianjur		Total	
	ha	%	ha	%	ha	%	ha	%
Suitable (S2)	6,017.1	9.21	4,957.7	7.59	3,102.4	4.75	14,077.1	21.54
Marginally Suitable (S3)	24,941.0	38.16	9,773.5	14.95	16,477.4	25.21	51,191.9	78.33
Not Suitable (N)	79.7	0.12	1.8	0.00	6.6	0.01	88.2	0.13
Total (ha)	**31,037.8**	**47.5**	**14,733.0**	**22.5**	**19,586.4**	**30.0**	**65,357.2**	**100.0**

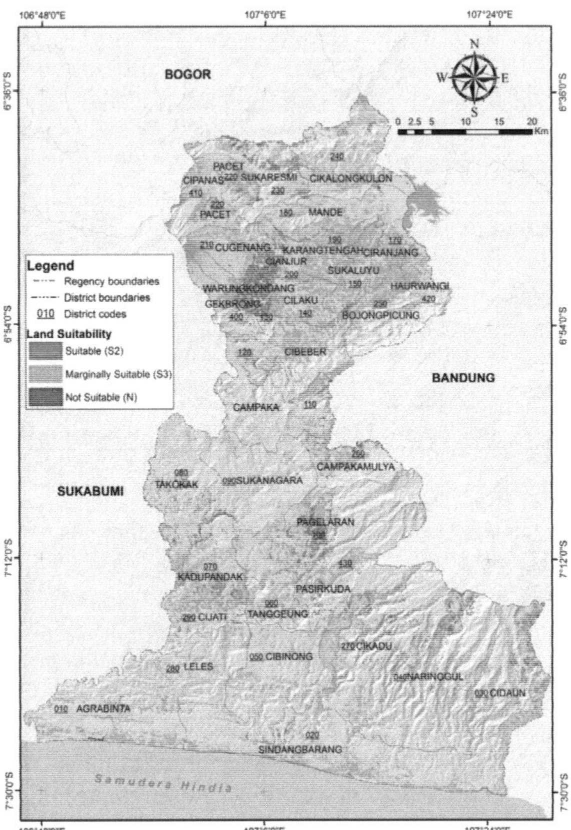

Figure 2. Spatial distribution of paddy field based on its land suitability

3.2, Distribution of paddy field on the Regional Spatial Planning Map (RTRW)

Table 4 and figure 3 showed that paddy field in Cianjur Regency have been occupying on land that accord with the Regional Spatial Planning Map (RTRW), such as wet agriculture area (28.34)% and dryland agriculture area (43.72)%. While, paddy field have been occupying on land that did not accord with the Regional Spatial Planning (RTRW), such as rural settlement, *sempadan sungai*, urban settlement and plantation/annual crop area in range (4.13–8.71)%.

Based on the Regional Spatial Planning Map where Cianjur Regency still has the large paddy field to be protected, especially paddy field in wet agriculture area (28.34%) and dryland agriculture area (43.72). The irrigated paddy field dominantly found in the Northern Cianjur and the other hand the high conversion of paddy field will be happened in the Northern Cianjur too. Therefore, the Government's commitment needs to be improved in term of paddy field protection.

Table 4. Distribution of paddy field based on the Regional Spatial Planning Map

Regioanal Spatial Planning Map (RTRW)	Northern Cianjur		Central Cianjur		Southern Cianjur		Total	
	ha	%	ha	%	ha	%	ha	%
Conservation forest area	120.2	0.2	0.1	0.0	318.9	0.5	439.2	0.67
Protected forest area	8.7	0.0	401.5	0.6	560.1	0.9	970.3	1.48
Production forest area	414.9	0.6	314.5	0.5	902.0	1.4	1,631.4	2.50
Industrial area	52.2	0.1	0.0	0.0	0.0	0.0	52.2	0.08
Water bodies	21.4	0.0	25.3	0.0	10.1	0.0	56.9	0.09
Plantation/annual crop area	1,985.0	3.0	1,552.8	2.4	2,154.3	3.3	5,692.1	8.71
Rural settlement area	1,295.6	2.0	1,140.7	1.7	263.6	0.4	2,699.9	4.13
Urban settlement area	3,965.1	6.1	0.0	0.0	0.0	0.0	3,965.1	6.07
Wetland agriculture area	17,001.2	26.0	1,055.4	1.6	462.5	0.7	18,519.1	28.34
Sempadan sungai	738.2	1.1	990.7	1.5	1,027.6	1.6	2,756.5	4.22
Dryland agriculture area	5,435.3	8.3	9,252.0	14.2	13,887.3	21.2	28,574.5	43.72
Total (ha)	**31,037.8**	**47.5**	**14,733.0**	**22.5**	**19,586.4**	**30.0**	**65357.2**	**100.00**

3.3. Economic value of paddy field

Economic of paddy field that was calculated based on R/C value (Avenue/Cost), if R/C value >1 means the land rent of paddy field is still profitable. Based on Table 5 showed that R/C value of the land rent of paddy field on all districts in Cianjur Regency >1, it means that paddy field is still profitable that can support the program of paddy field protection.

Table 5 showed that average of R/C value is 2.18 with the profit of Rp 11,203,006,-/ ha/planting season, while the R/C maximum and minimum are 2.73, with the profit of Rp 15,535,750,-/ha/planting season and 1.67 with the profit 7,086,100, respectively. Other studies have shown that R/C value of irrigated paddy field in Cihea village, Cianjur Regency is 2.13 with the profit Rp 7,342,200,- /ha/planting season (Kurniadiningsih 2012). However, the profits of the land rent of paddy field can't be sufficient the life needs of farmer family; it is supported by the data in which almost all farmers have other jobs besides farming, such as trade, *ojeg*.

The R/C value of paddy field in Northern Cianjur is slightly higher than in Central and Southern Cianjur because of productivity difference. Most of Irrigated paddy field was found in Northern Cianjur that has productivity slightly higher than Rainfed paddy field that found in Central and Southern Cianjur

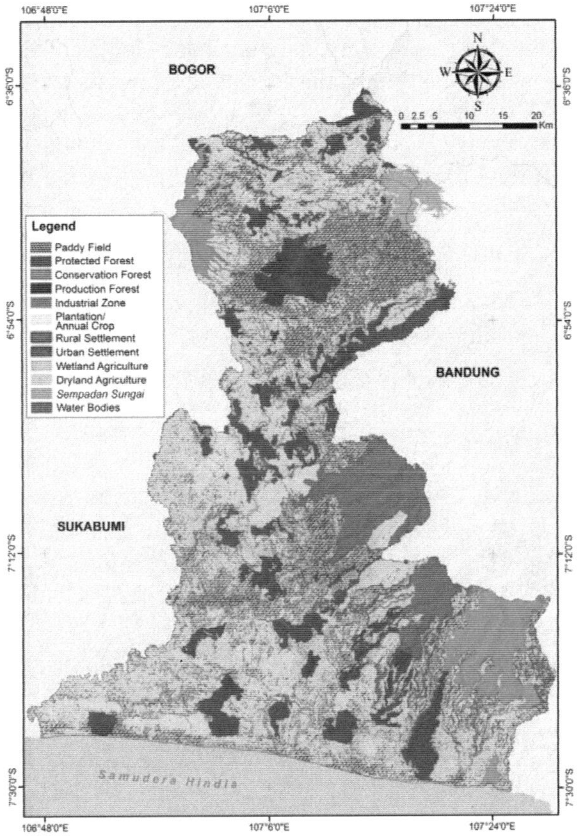

Figure 3. Spatial distribution of paddy field based on the Regional Spatial Planning Map

Table 5. The R/C value of the paddy field land rent

Code		District	R/C Value	Code		District	R/C Value
250		Bojongpicung	2.12	110		**Campaka**	2.06
200		Cianjur	2.10	260		Campakamulya	1.85
120		Cibeber	1.67	070		Kadupandak	1.73
240		Cikalongkulon	2.49	100	Central Cianjur	Pagelaran	1.86
140		Cilaku	2.53	430		Pasirkuda	2.68
410		Cipanas	1.91	080		Takokak	1.97
170		Ciranjang	2.26	060		Tanggeung	1.85
210	Northern Cianjur	Cugenang	2.26	090		Sukanagara	2.20
400		Gekbrong	2.41	010		**Agrabinta**	1.97
420		Haurwangi	2.07	050		Cibinong	1.87
190		Karangtengah	2.54	030		Cidaun	2.58
180		Mande	1.87	290	Southern Cianjur	Cijati	2.49
150		Sukaluyu	2.59	170		Cikadu	1.90
220		Pacet	2.30	040		Naringgul	2.60
230		Sukaresmi	1.69	280		Leles	2.23
130		Warungkondang	2.73	020		Sindangbarang	2.30

3.4. Paddy field protection and recommendation of regional planning

Based on the land suitability, the Regional Spatial Planning Map, R/C value and certain rules in table 1 were obtained the priority level of paddy field protection. Based on table 6 and figure 4 showed that area of the paddy field in Cianjur Regency included into the first priority for protected is 30.14% of which 27.4% found in Northern Cianjur. The paddy field that included into the third priority for protected is 38.45% of which dominant found in Central and Southern Cianjur. The other priorities of paddy field protection are around 15%.

Table 6. Priority level of paddy field protection

Priority level of paddy field protection	Northern Cianjur		Central Cianjur		Southern Cianjur		Total	
	ha	%	ha	%	ha	%	ha	%
The first priority (priority_1)	17,898.4	27.4	1,287.0	2.0	513.9	0.8	19,699.4	30.14
The second priority (priority_2)	2,793.6	4.3	4,026.9	6.2	3,485.0	5.3	10,305.4	15.77
The third priority (priority_3)	4,907.6	7.5	7,557.9	11.6	12,663.6	19.4	25,129.1	38.45
The fourth priority (priority_4)	5,438.3	8.3	1,861.1	2.8	2,923.9	4.5	10,223.4	15.64
Total	**31,037.8**	**47.5**	**14,733.0**	**22.5**	**19,586.4**	**30.0**	**65,357.2**	**100.0**

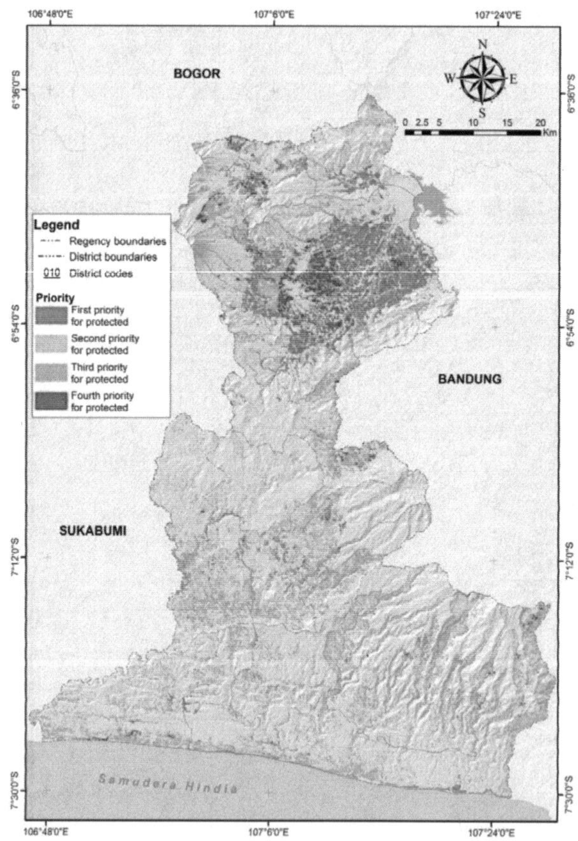

Figure 4. Priority level of paddy field protection

Based on the priority level of paddy field protection, the Regional Spatial Planning Map and certain rules in table 2 were obtained 4 recommendations of regional planning that showed by table 7.

Table 7. Recommendation of paddy field protection

No.	Recommendation of paddy field protection	Paddy field (ha)		Total	
		Irrigated	Rainfed	ha	%
1	Paddy field protected	22,283.9	7,720.9	30,004.8	45.91
2	Reserve paddy field protected	1,714.7	25,454.1	27,168.9	41.57
3	Reserve paddy field protected OR allocated based on the Regional Spatial Planning Map (RTRW)	3.3	1,406.3	1,409.6	2.16
4	Paddy field area that can be allocated based on the Regional Spatial Planning Map (RTRW)	5,220.2	1,553.8	6,774.0	10.36
	Total (existing paddy field)	**29,222.1**	**36,135.1**	**65,357.2**	**100.0**

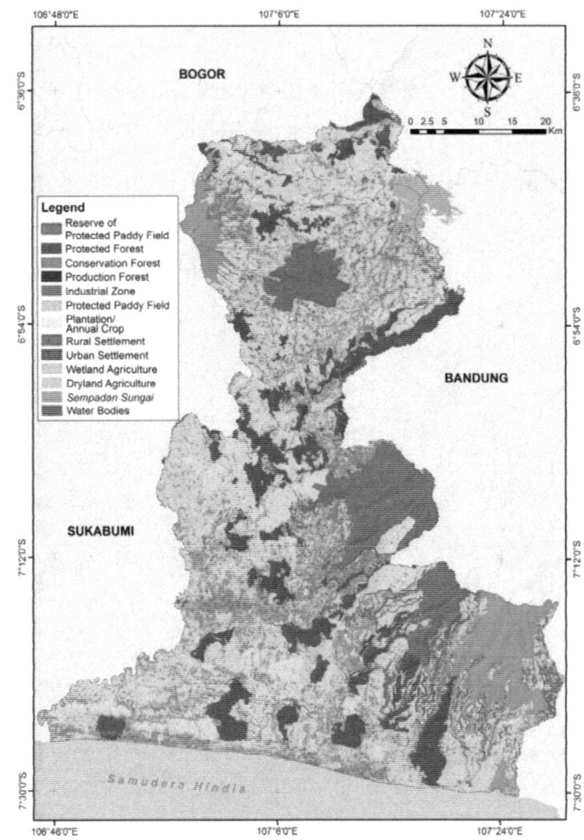

Figure 5. Recommendation of regional planning that protect paddy field

Status of rice sufficiency was calculated based on ratio between supply and demand with assumptions:
1) Average of paddy productivity is 5.6 ton/ha
2) Cropping index for irrigated and rainfed paddy field are 2 times/year and 1 times/year, respectively
3) Total population is 2,225,316 people
4) Average rice consumption per capita for Cianjur Regency is 105,85 kg/capita/year
5) Conversion factor from paddy into rice = 0,6274

Table 8 showed that surplus status of rice sufficiency will be achieved if (a) existing paddy field (scenario 1) and (b) paddy field protected and reserves paddy field protected (scenario 3) will not be converted into other landuse.

Table 8. Status of rice sufficiency based on supply and demand

Scenarios	Paddy field (ha)		Total Population (people)	Supply (ton)	Demand (ton)	Status	
	Irrigated	Rainfed					
1. Existing paddy field	29,220.6	36,135.1	2,225,316	331,031	235,550	95,481	Surplus
2. Paddy field protected	22,283.9	7,720.9	2,225,316	182,859	235,550	-52,691	Minus
3. Paddy field protected and reserves paddy field protected	23,998.6	33,175.0	2,225,316	284,332	235,550	48,782	Surplus

4. Conclusion

Conclusions of this research are:
1) Land suitability, economic value and also allocating land for paddy field in the RTRW can support realization of paddy field protection.
2) The paddy field that included into the first and third priority is 30.14% and 38.45%, respectively. The other priorities of paddy field protection are around 15%.
3) This research is recommended that 87.5% of the paddy field existing can be protected to get rice sufficiency with surplus around 48.782 ton.

References

[1] Asni 2005 *Analysis of production pendapatan and land convertion on Batu Labuan Regency West Java Province* [thesis] Medan [ID] North Sumatera University
[2] Bappeda [Regional Planning and Development Agency] of Cianjur Regendy 2015 *Local regulation No.12 Regional Spatial Planning Map (RTRW)* Cianjur Regency West Java
[3] BPS [Central Bureau of Statistic] of Cianjur Regency 2013 *District in numbers of Cianjur Regency* Cianjur [ID] West Java
[4] BPS [Central Bureau of Statistic] of Cianjur Regency 2014 *District in numbers of Cianjur Regency* Cianjur [ID] West Java
[5] Chairunnisa C 2015 *Recommendation of wetland protection priority in Cianjur Regency to support rice contribution of West Java* [thesis] Bogor [ID] Bogor Agricultural University
[6] Hamdan 2012 *Economic of paddy field convertion into oil palm on South Seluma District Seluma Regency Bengkulu Province* [Thesis] Bogor [ID] Bogor Agricultural University
[7] Kurniadiningsih 2012 *Economic evaluation of paddy field using Rice Intensification System (SRI) Method in Cihea Village Cianjur Regency West Java* [thesis] Bandung [ID] Faculty of Civil Engineering Bandung Institute of Technology
[8] Munibah K, Widiatmaka and Wijaya H 2015 *Carrying capacity of paddy field and independence domestic food* Institute of research and service community (LPPM) Bogor [ID] Bogor Agricultural University

[9] Rumiris 2008 *Analysis of land use change and land rent of agriculture and non agriculture on Dramaga District Bogor Regency* [Scripsi] Bogor [ID] Bogor Agricultural University
[10] Sugandi D, Ishak A and Hamdan 2013 F*actors effecting to paddy field conversion into oil palm and its control strategy on Bengkulu Province* Bengkulu [ID] Agency of Agriculture Technology Assessment

Role of Remote Sensing and Geographyc Information System Mapping for Protected Areas Land Rice Field Subak, Buffer Zones, and Area Conversion (Case Studies In Gianyar Regency, Bali Province)

Indayati Lanya[1] and N. Netera Subadiyasa [2]

[1] Lecturer of Study Program Agroecotechnology, Faculty of Agriculture, Udayana University, Indonesia

[2] Study Program Agroecotechnology, Faculty of Agicultural, Udayana University, Indonesia

Corresponding author Telp: 0361 421573, e-mail : *indahnet@yahoo.co.id*

Abstract. Conversion of rice fields in Bali 2579 ha/year, Law Number 41 of 2009 [1] and five of Government Regulation (GR), mandates the Local Government (LG) has a Regional Regulation (RR) or Rule Regent/Mayor, on the protection of agricultural land sustainable food (PALSF). Yet none provincial government of Bali has PALSF; although Subak as world cultural heritage. Similarly, Gianyar regency development strategy directed to integrate agriculture with tourism. Landsat 8 images, Word View Coverage 2015 Gianyar district and ArcGIS 10.3 software used for of rice field mapping and zoning of land protection Subak. Ten thematic maps (watersheds, land use, irrigation, relief/slope, rainfall, spatial planning, land suitability, productivity, the distance from downtown) as a variable parameter, weighted and balanced numerically. Numerical classification agricultura land using for the overlay menu and reselek. The total value of >125 as rice need to be protected, 100-125 value for buffer zone, and the value of <100 rice fields can be converted to 40 next year. For the 20 next year, the total value of> 100, 50-100 and <50 respectively to rice fields that need to be protected, wetland buffer, and rice fields can be converted. Region Subak sustainable of rice field protection, buffer and can be converted in a row for the next 20 years is 10973 ha, 3855 ha and 311 ha. For the next 40 years Subak conserved of rice field (8019 ha), buffer (5855 ha), and can be converted (3124 ha). Subak land pattern of spread can be converted to an supply of land for non-agricultural development of the region downstream to the access road Ida Bagus Matera (Jln. Province / national) in the coastal areas of Gianyar.

Keywords:: *Remote sensing, thematic map, protected areas, buffers, paddy field, land conversion, Subak*

1. Introduction

Gianyar regency is one area of the National Tourism Development Sarbagita (Denpasar, Badung, Gianyar and Tabanan) in Presidential Regulation No. 51 of 2014 [2]. Research 2014 [3] produce data and information, that: development of tourism in Bali negative impact on rice field function transfer 800 ha year^{-1}. Conflicts of interest between agriculture and tourism continues, consequently wetland continuously pressured by the interests of non-agricultural development, both for residential as well as for infrastructure objects and tourist destinations. Conversion of agricultural land have occurred because the supply of land in Bali Generally, and Gianyar especially is a rice field.

Development of tourism in Gianyar very rapidly, particularly in tourism centers, such as coastal areas and tourism infrastructure. Rice fields in Bali Subak incorporated in the territory. Therefore, Subak rice fields not only as agricultural resources, but contain the environmental balance function, social and agrarian culture that is very noble. Subak agreed as local wisdom of Bali, a nationally recognized, and as a world cultural heritage [4], which is able to implement the philosophy of Tri Hita Karana (THK).

Along with advances in the tourism sector, and the development of population in urban areas is a conflict of interest utilization of water resources. Originally springs as a source of irrigation water, has changed functions for household needs, hotel, restaurant and other tour industry which have an impact on the lack of irrigation water and rice field function transfer. The existence of rice fields as well as a basic component in food production, the loss of rice paddies have a negative impact on food security. For the existence rice paddies need to be protected by law such as the Regional Regulation (Perda), to be used as agricultural land resources in a sustainable manner.

The preparation of Regional Regulation has been mandated by the Act, the Government Regulation (PP) and the Regulation of the Minister of Agriculture (Permentan), relating to the Protection of Land Pertanaian sustainable Food (PLP2B). PLP2B legal products, needs to be supplemented with spatial data that is accurate and timely, and accessible to stakeholders through information technology, for ease in implementation. The use of high resolution satellite imagery that can record any objects on the Earth's surface, and software Geographical Information System (GIS) is able to answer the needs mentioned above.

Research the role of remote sensing and GIS for mapping of rice field zoning as LP2B (sustainable Subak rice field, buffer, and can be converted) very urgent to protect of Subak rice field. Paddy fields as a determinant component of food sovereignty, culture and nature conservation, normasilasai ecosystems, as well as maintain world heritage, and is able to apply the Balinese philosophy of THK namely: the balance between nature, man and God.

The research objective is: (1) make a zonation map paddy fields (zoning map) through classification of numerical-based remote sensing and Geographic Information Systems (GIS), (2) get food balance, and (3) criteria for Subak land management in each zone area (sustainable/ protection, buffer, and converted).

2. Material and Methods

The research location in the district of Gianyar, Is one of eight districts in the regency of Bali Province. Located between 08 °8'48"-08°38'58" south latitude and 115°13'29"-115°22'23" east longitude. Its borders by Denpasar and Badung regency in West side, Bangli Regency in the north side, Bangli and Klungkung Regency at eastward, then Badung Strait and Ocean Indonesia in the southern (Figure 1). Satellite Imagery WorldView Gianyar District 2015 is presented in Figure 2. Is located at altitude of 250- 950 meters above sea level. The area of Gianyar Regency 368 km^2 or 6.53% of the Bali Province. Land Area 2.540 hectares of rice fields and agricultural land 1039.00 hectares of non-rice fields. Rainfall in the area ranges from 1,900 -2,500 mm/th (Gianyar Central Bureau of Statistics, 2013) [5]. There is a tourism center on the south coast, the area of Ubud, Sukawati, and Tampak Siring, as well as natural and cultural attractions other. Gianyar Regency was divided into 7 districts, that is Sukawati District (12 villages, 111 villages/ banjar), Blahbatuh District (9 villages, 67 banjar), Gianyar District (17 villages, 96 banjar), Tampaksiring District (8 villages, 70 banja), Ubud District (8 villages, 80 banja), Tegallalang District (7 villages, 65 banja), and Payangan District (9 villages, 59 banjar).

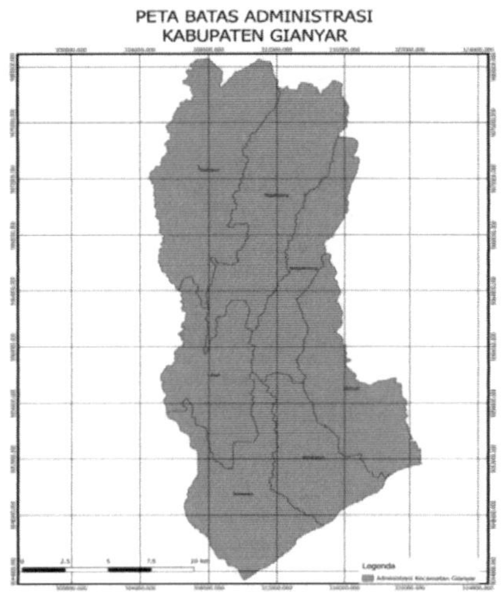

Figure 1. Map of the District Administration in Gianyar (Topographic maps analysis, 2015).

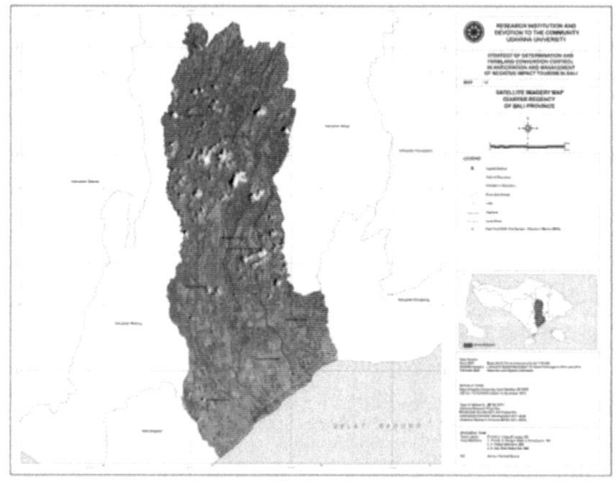

Figure 2. Satellite Imagery Worldview 2015 Gianyar Regency Research Area (analysis, 2015)

Materials used include: topographic maps (Bakosurtanal, 2002) [6] for the analysis: watershed, high places, relief, distance from the town center, and irrigation canals. Thematic maps (land suitability, rainfall map, map of rice production, the distance from the city center) and maps of Gianyar Regional Spatial Plan (RTRW) 2012-2032 (Local Regulation/Perda No. 16 of 2012) [7] . Satellite images World View Gianyar Coverage 2015. Remote sensing technology is used to create spatial maps of land use. This research tool in the form of specific set of computers with the ArcGIS 10.1 program, is used for the digitization and analysis of thematic maps.

Methods used include: (1) the interpretation of satellite imagery for mapping land use, (2) field survey, (3) digital mapping thematic maps based on GIS, (4) analisis statistik neraca pangan, (5) the classification of numerical paddy fields, conducted through weighting 1-10 and scores 1-3, each parameter to get a zoning map using ArcGIS 10.1 program using the overlay method, intersect and query analysis, (6) the preparation of paddy fields classification criteria (sustainable, buffer, and can be converted).

3. Result and Discussion

3.1. Thematic Map

Interpretation of satellite imagery and topographic maps by using software ArcGIS 10.1 produces 10 kinds of thematic maps, namely: 1) land use maps, (2) map suitability of land with a map RTRW (protected areas and agriculture), (3) map the morphology of Regions watershed, (4) the height of the map, (5) relief map, (6) the rainfall map, (7) irrigation map, (8) agroekosistem land suitability maps, (9) a map of rice production, and (10) map distance from the city center.

Land use maps obtained from the analysis of satellite imagery and field observations are presented in Figure 3. The level of accuracy of the identification rice fields reached 98%, due to the fallow rice fields. Figure 4 overlay between rice fields map (Figure 3)

IOP Conf. Series: Earth and Environmental Science **47** (2016) 012037 doi:10.1088/1755-1315/47/1/012037

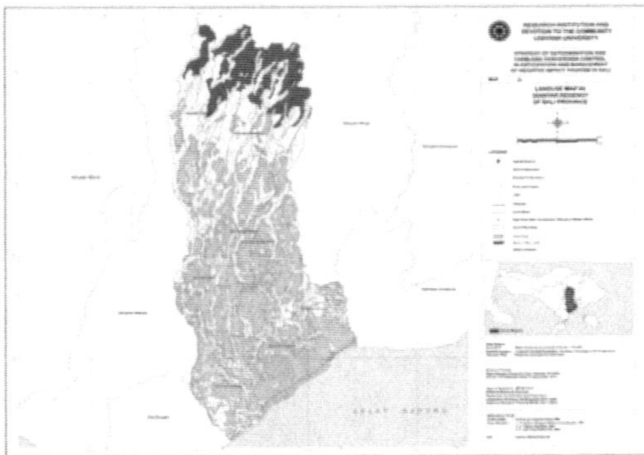

Figure 3. Land use map (analysis, 2015).

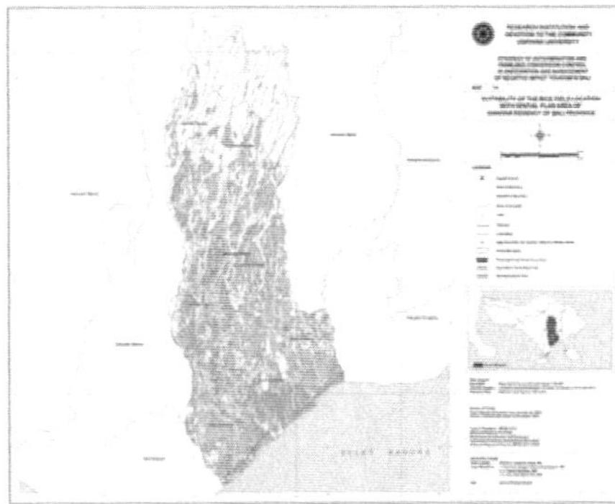

Figure 4. Land suitability map of rice field, with RTRW Map (analysis, 2015)

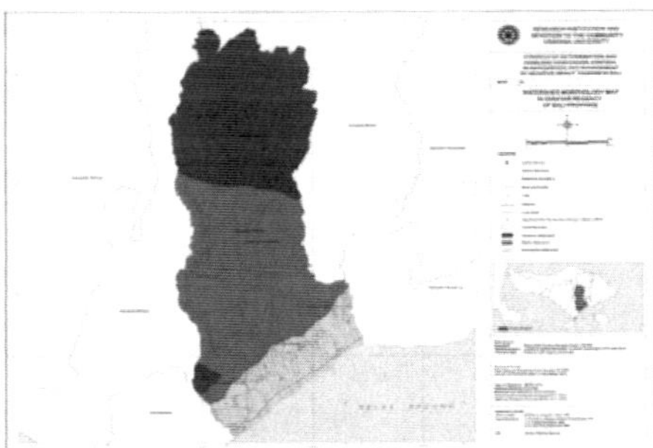

Figure 5. Morfologi Watershed Map (analysis, 2015)

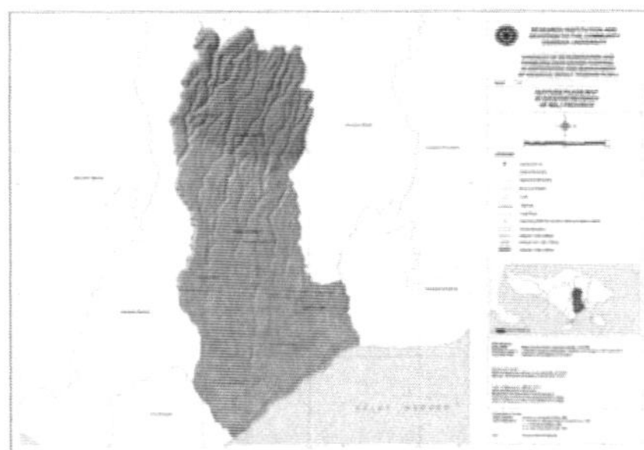

Figure 6. Elevation Map (analysis, 2015)

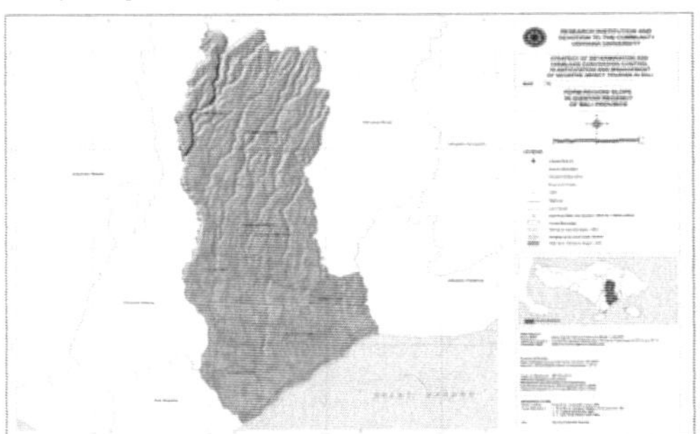

Figure 7. Peta reilief (analysis, 2015)

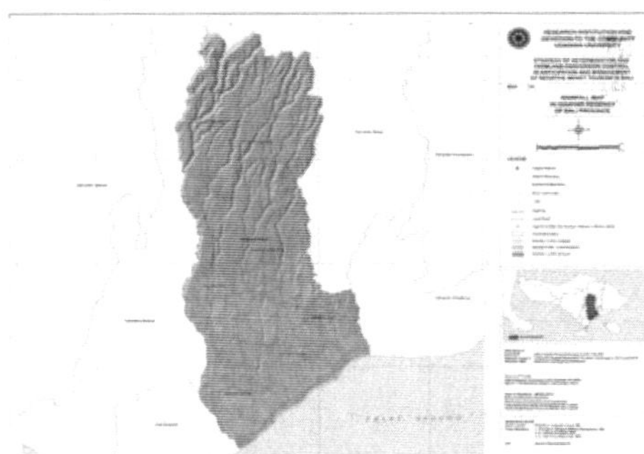

Figure 8. Peta Curah hujan tahunan (analysis, 2015)

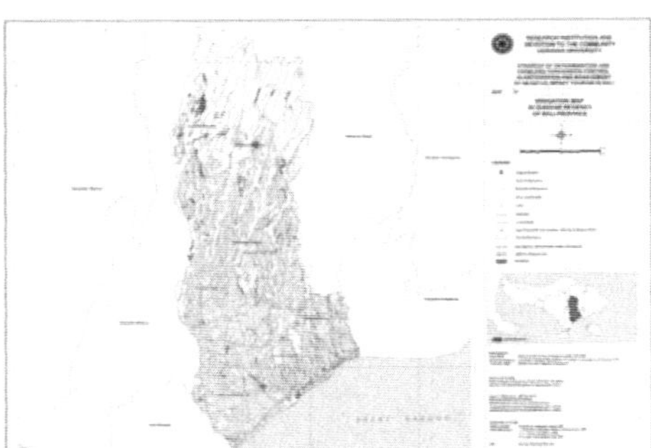

Figure 9. Map of the area irrigated (analysis, 2015)

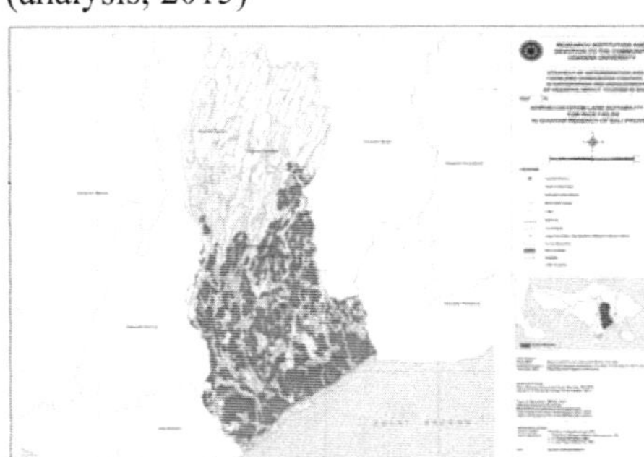

Figure 10. Agroekosistem land suitability maps (analysis, 2015)

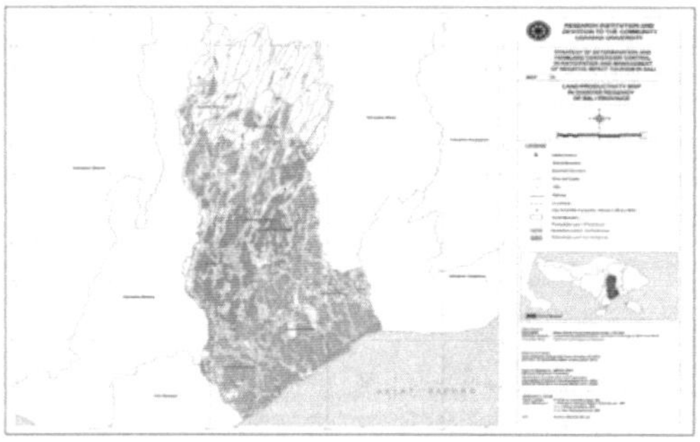

Figure 11. Map of the average production of paddy ton ha-1 (analysis 2015)

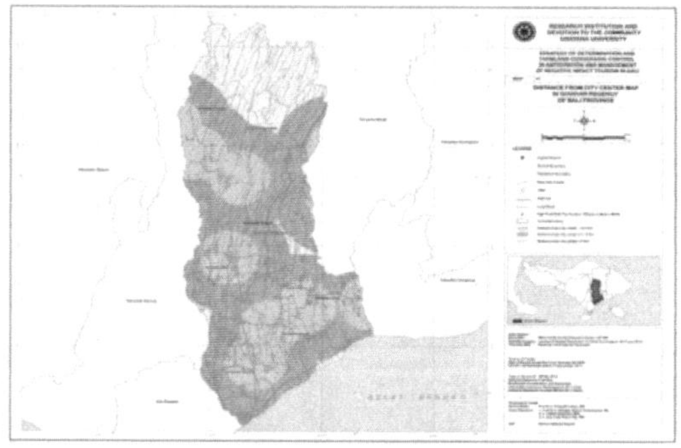

Figure 12. Distance Map from the city center (analysis, 2015)

with a map of the RTRW, generate suitability maps of rice field with green open space (RTH) and the farm area. Figure 5 maps the morphology of the Watershed is the result of the analysis of topographic maps with satellite imagery. Figure 6 (a map of high places) and 7 (relief map) obtained from the analysis of topographic maps; Figure 8 (precipitation map) the results of the analysis of rainfall data of the last 10 years; Figure 9 (a map of the region irrigated) the results of the analysis of of rice field with irrigation networks; Figure 10 (a map of the agro-ecosystem land suitability for paddy rice) the results of the field survey and soil analysis of land suitability classification of agro-ecosystem; Figure 11 (a map of rice production results of field survey and analysis of statistical data), and Figure 12 is a map of the distance from urban centers, the results of the analysis of topographic maps and satellite imagery.

Based on the maps above show that: the upstream area is dominated by the dry land, while the middle and downstream are dominated by rice fields. The use of of rice field consistent with the high point, the area of irrigation, and agro-ecosystem land suitability. The higher and the more upstream, as well as diminishing returns of rice field area. The lower the level of the land suitability result of factors limiting the temperature decreases. The upstream is dry land farming and mixed farms, generally found at an altitude of more than 500 m asl. Based on environmental sustainability factors, the upstream part is necessary to land conservation as rain water catchment and water supply downstream. Their irrigation channels are still good, then the use of rice field still exist, in all regions of the Gianyar Regency. Rice production data is quite high, averaging 6 ton ha^{-1} crop^{-1}. The challenge remains the development of housing and urban infrastructure that interfere with irrigation channels.

3.2. Supplies and Needs Food (Food Balance Sheets)

Inventory analysis and food needs based on projected population and of rice field, as well as the rate of land conversion. Figure 13 the population projection data, the need for food, rice field area, and food supplies for 2015, 2020, 2030, 2040, and 2050. Results of regression analysis needs and food supply (Graph food balance sheet) are presented in Figure 14. The figures show a spacious rice field and food supplies inversely proportional to population growth and food needs. The higher the food needs followed the lower the food supply. Rice field area decreased 34, 61 ha yr^{-1}, followed by a decline in food supply 1.730 tons of rice yr^{-1}. While the population increased by 12 ,560 people by the year $^{-1}$ and increased food needs tons of rice 1.6328 tons year $^{-1}$. This data is used as a basis in determining the zoning classification of of rice field.

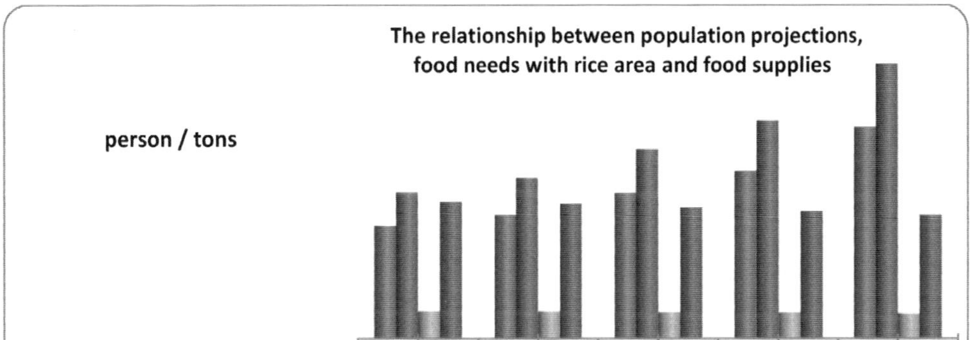

	2015	2020	2030	2040	2050
Projected Population x 1,000 people	520,763	571,749	673,721	775,693	979,637
Projected food needs (0.1 tons)	676,99	743,27	875,84	1008,4	1273,53
Projected rice field area x 100 ha	126,63	124,93	121,53	118,12	114,473
Food supplies (ton rice)	633,15	624,65	607,65	590,6	572,365

Figure 13. The relationship between population projections food needs with rice area and food supplies.

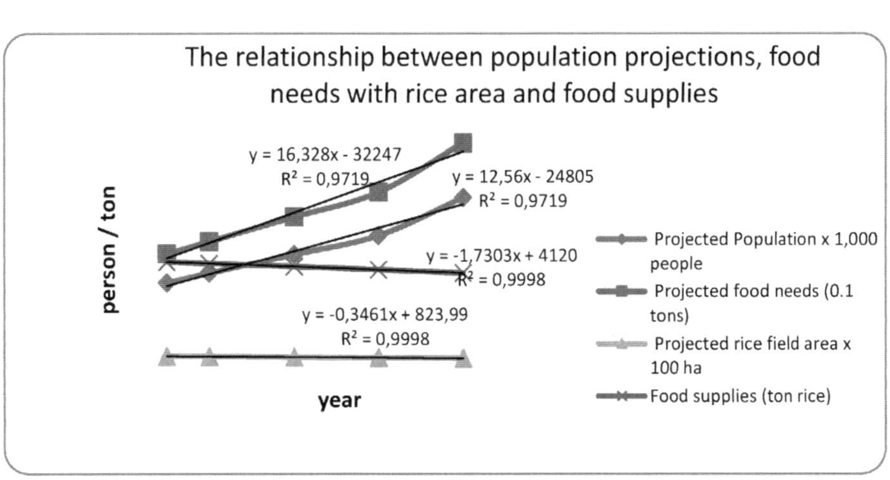

Figure 14. The relationship between population projections, food needs with rice area and food supplies.

3.3. Numerical classification (Zoning Map) PLP2B.

Numerical classification using the ArcGIS 10.1, the results of the overlay 10 thematic maps, which have been weighted and do the scoring as parameters in accordance with its role in the conservation of agricultural land resources based environment. The method uses intersect menu and query analysis, produced maps of rice field Subak zoning (zoning map) sustainable, buffer (limited conversion). Paddy field that can be converted to 20 years or 1x Long Term Development Plan (RPJP) forwards included in Figure 15 and for the next 40 years (2 x RPJP) in Figure 16.

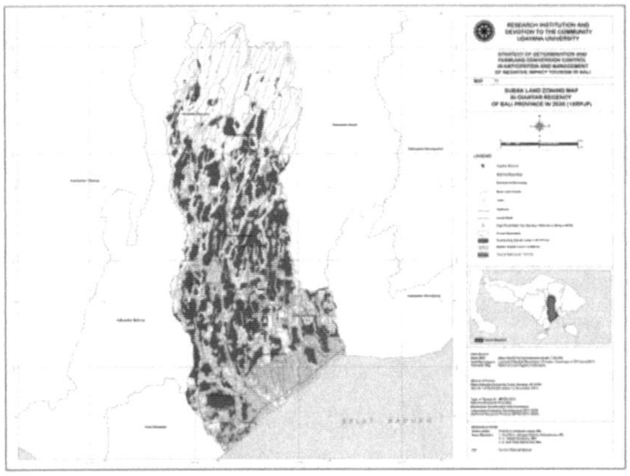

Figure 15. Land Zoning Map Subak: protected, protected, buffer and conversionfor 20 years the (analysis, 2015)

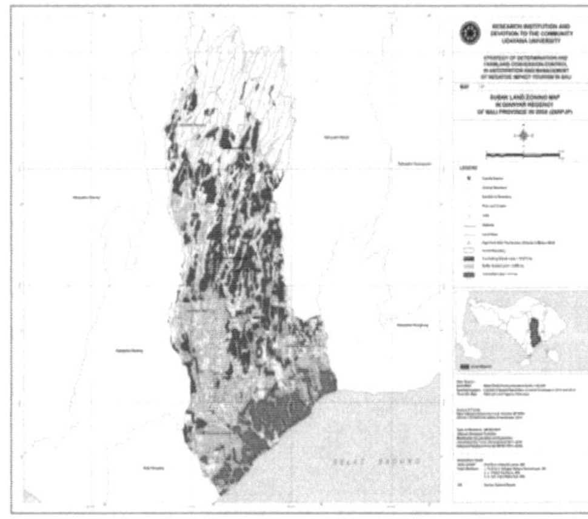

Figure 16. Land Zoning Map Subak: protection, buffer and conversion next 40 years (analysis, 2015)

Region Subak sustainable of rice field protection, buffer and can be converted in a row for the next 20 years is 10,973 ha, 3,855 ha and 311 ha. For the next 40 years Subak conserved of rice field (8019 ha), buffer (5,855 ha), and can dikonvers (3,124 ha). Subak land pattern of spread can be converted to an inventory of land for non-agricultural development of the region, downstream, with an access road Ida Bagus Matera (Jln. Provincial / national), the coastal areas and the tourist area of Ubud Gianyar.

Results of numerical classification using the ArcGIS 10.1 overlay 10 thematic maps which have been weighted and scored as parameters in accordance with its role in conservation of agricultural land resources based environmen, using menu intersect and query analysis, produced maps paddy field Subak zoning (zoning map) Lestari, buffer (limited conversion). Figure 13 shows the of rice field that can be converted much less than rice field area in Figure 14, the result of numerical classifications are different according to the needs of land in the development. Classification of rice fields for the next 20 years, it was determined the total value of $\geq$ 100 (sustainable rice fields), the total value of 75-100 (buffer rice fields), and the total value <75 (rice fields can be converted). As for the 40 years to come, the total value of $\geq$ 125 (sustainable rice fields), the total value of 100-125 (rice fields buffer), and the total value of <100 (rice fields can be converted).

Zonation mapping rice fields for the next 20 years the fields available to most sustainable paddy fields except in coastal areas. Coastal areas are around the way Ida Bagus Mantra (national road) classified rice fields buffer. The existence of this buffer rice fields for the next 40 years of development needs changed into land that can be converted to non-agricultural development. Rice fields that can be converted are in the border city of Denpasar, and access roads Ida Bagus Spells as a national road. This area experienced many non-agricultural development. In other words, the results Rice fields zoning classification in accordance with the needs of development. Development needs will increase with time. For the total value of the classification to be improved, so get rice area that can be converted bigger and smaller sustainable rice fields.

Location rice fields sustainable for the next 40 years there is a need in the upper and eastern, which is bordered by Klungkung Regency. While the bordering city of Denpasar as buffer rice fields. Rice fields around the national road access can be converted. Based on the results of the zoning classification rice fields, both for the construction of the next 20 years or RTRW 1x and 2x RTRW can be used as a basis in the preparation or revision of the RTRW or Detailed Spatial Plan.

4. Conclusions

Word View satellite imagery can be used for mapping of rice field with 99% accuracy rate. paddy field conversion is caused by allocation of space for non-agricultural cultivated area and the pressure development in growth centers.

Land use change and food supply is inversely proportional to the rate of population growth and food needs. The population is increasing, but the area of rice field decline. Paddy fields decreased -34.61 ha yr^{-1} and food supplies decreased by -1.7 tonnes yr^{-1}. In contrast, the population increased by +12 560 people and increased food needs of +1,632 tons year $^{-1}$.

Classification zoning of rice field water control system using 10 parameters: (1) land use maps, (2) map suitability of land with a map RTRW (protected areas and agriculture), (3) map the morphology of Regions watershed, (4) the height of the map, (5) relief map, (6) the rainfall map, (7) irrigation folder, (8) agroekosistem land suitability maps, (9) a map of rice production, and (10) map distance from the city center, Then performed weighting of 1-10, and scoring 1-3, in accordance with its role in preserving the environment, generating group: sustainable rice field, rice field rice buffer and can be converted according to the needs of development in a period of 20 years and 40 years to come.

Rice fields need to be protected (10,973 ha), as a buffer (3,855 ha ha) and can be converted (311 ha) for the next 20 years of development needs. For the next 40 years of development, rice field that need to be protected (8,019 ha), as a buffer (5,855 ha), and which can be converted to non-agricultural construction area of 3,124 ha.

Remote sensing and GIS technology, as well as a numerical classification can be used for the establishment of Subak rice field area as LP2B within a certain period, and is able to inform the geographic locations (spatial) is appropriate, in accordance with the needs of development time.

Acknowledgement

We would like to thank the Director General of Strengthening Research and Development, Ministry of Research, Technology and Higher Education, Rector of Udayana University and Mr. Dean of the Faculty of Agriculture Udayana University who has mensuport funds and facilities. To Mr / Ms who have helped in this study we thank you

References

[1] Act No. 41 Year 2009 On Land Protection Sustainable Agriculture, Surf the internet http: // www google.com/
[2] Presidential Regulation No. 51 of 2014 on the Amendment of the presidential regulation No. 45 of 2011 on Spatial planning urban area of Denpasar, Badung, Gianyar and Tabanan
[3] Lanya Indayati, N. Netera Subadiyasa, Ketut Sardiana, Gst Putu Adi Ratna (2015) Land use control strategies to reduce the negative impact of tourism in Bali, *The final report MP3EI Research* (Udayana University, Bali Jimbaran Hill)
[4] UNESCO 2012 Subak System as Manifestation of the Tri Hita Karana Philosophy (St Petersburg: Rusia)
[5] Central Bureau of Statistics (CBS) 2013 Gianyar in Figures http://www .gianyarkab. bps.go.id
[6] Coordinating Agency for Surveys and Mapping (Bakosurtanal) (2002) Topographic Map scale 1: 25. 000 Sheets Gianyar and Bangli
[7] Gianyar Government 2012 Regional Regulation No. 16 In 2012 the Spatial Plan (RTRW) Gianyar regency from 2012 to 2032.

Urban Spatial Pattern and Interaction based on Analysis of Nighttime Remote Sensing Data and Geo-social Media Information

Nila Ratnasari[1*], Erika Dwi Candra[1], Defa Herdianta Saputra [1], Aji Putra Perdana[2]

[1]Undergraduate Program of Cartography and Remote Sensing, Department of Geographic Information Science, Faculty of Geography, Universitas Gadjah Mada, Yogyakarta 55281, Indonesia
[2]Geospatial Information Agency – Indonesia

Corresponding author: nila.ratnasari@mail.ugm.ac.id, erikadwic@gmail.com, defasaputra@gmail.com, perdana.ajiputra@gmail.com

Abstract. *Urban development in Indonesia significantly incerasing in line with rapid development of infrastructure, utility, and transportation network. Recently, people live depend on lights at night and social media and these two aspects can depicted urban spatial pattern and interaction. This research used nighttime remote sensing data with the VIIRS (*Visible Infrared Imaging Radiometer Suite*) day-night band detects lights, gas flares, auroras, and wildfires. Geo-social media information derived from twitter data gave big picture on spatial interaction from the geospatial footprint. Combined both data produced comprehensive urban spatial pattern and interaction in general for Indonesian territory. The result is shown as a preliminary study of integrating nighttime remote sensing data and geospatial footprint from twitter data.*

Keywords: *remote sensing, geo-social media, spatial pattern, spatial interaction, urban*

1. Introduction

Urban development in Indonesia increasing significantly caused by development in infrastructure, utilities, and transport networks. Urbanity activity can be characterized by the activity of people in longer time. The higher of the level urban area had the longer community activities even until night. Community activities at night need an electric light at public space or settlement. The use of light at night indicated community activity at urban area.

Recently, there has been a renewed interest in using nighttime images of Earth that show visible light emissions, providing a picture of urbanization through the detection of city lights. The high contrast provided by this type of image between lighted and unlighted lands and the sensor's moderate spatial resolution (45 m) make it a potent choice for identifying areas where significant human activity is being carried out. Dependence of communities on the light at night can be identified through remote sensing image is the Visible Infrared Imaging Radiometer Suite (VIIRS). VIIRS image has two bands and one of it is night band. VIIRS image of night gives an overview of spatial community activities through the lights at night. Satellite imaging of stable anthropogenic lights provides an accurate, economical and

IOP Conf. Series: Earth and Environmental Science **47** (2016) 012038 doi:10.1088/1755-1315/47/1/012038

straightforward way to map the global distribution of urban areas [4,5]. Urban areas account for a small fraction of Earth's surface area but exert a disproportionate influence on their surroundings in terms of mass, energy and resource fluxes.

Along with the development of technology, the use of social media was one of the needs of urban communities. With the increasing capabilities of social media, individuals leave behind footprints of their interaction with urban environments. In this context, cell phones have become one of the main sensors of human behavior, thanks, among others, to their growing penetration and wealth of social applications. The use of social media can be obtained geo-social media is social media data that has location attributes. Geo-social media twitter derived from these data that could give an idea of interaction of geospatial footprint. One of the widely used social media community is Twitter. Particularly georeferenced Twitter data is a promising opportunity to understand geographic processes inside online social networks. The enormous potential of interactive social media platforms like Twitter has been increasingly recognized by numerous research domains over the last years [2,6,7].

VIIRS imagery and geo-information provide an overview of social media and the pattern of urban spatial interaction in the area of Indonesia. In this paper, we propose to analyze the pattern and spatial interaction based on nighttime remote sensing data and geo-social media information The results of this study represent an early review the integration of nighttime remote sensing data and geo-social media information

2. Material and Methods

2.1 Study Area

The study area was the Java Island, Indonesia (Fig. 1). Until this year, the highest urban development happened on Java Island beside the others islands [10]. Its mean java has complexity spatial pattern and interaction. The urban development that used to appeal others islands is infrastructure, utility, and transportation network. Based on rapid development in java, people now live in the fast rhythm. Due to this habit people live to depend on lights at night and social media. These two factors can figure out the spatial pattern and interaction [1].

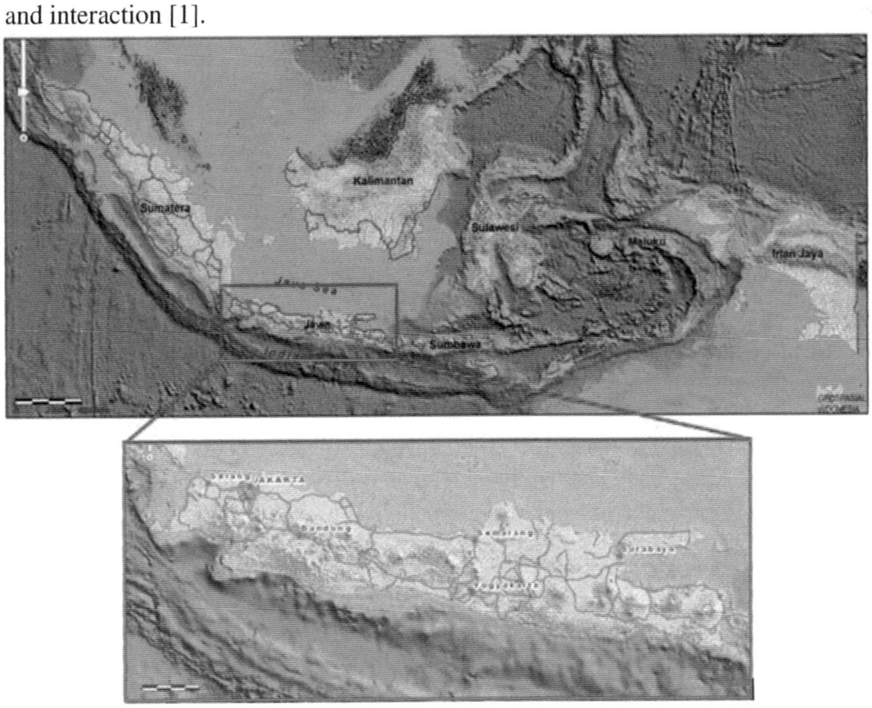

Figure 1. Java Island, Indonesia as study area
(source: http://tanahair.indonesia.go.id/)

2.2 Data Collection

a. Remote Sensing Night-time data

Remote Sensing Night-time data is a reliable source of information on light emissions only if the clear imagery is available. In addition, nightlight imagery is available at a low temporal resolution due to the uniqueness of the satellite collection and its orbit. Visible Infrared Imaging Radiometer Suite (VIIRS) data established from Earth observations Group (EOG) NOAA/NGDC with National Polar-orbiting Partnership (NPP) imagery that specific to detects lights, gas flares, auroras, and wildfires [9]. Remote sensing data that used to this research is VIIRS Day/Night Band (DNB) Cloud free composites data were acquired on July 2016. VIIRS DNB Cloud free composites are one of the provided data with scope cloud free and visible region based on average light on one month.

Human activity, especially in that night, depends on light. Remote sensing night-time data probably give some information for the area which has high activity in the night that shown by high reflectance of light. The other hands, Dark region in imagery indicate low activity of light. Figure 2 point out light reflectance in java island so it can absorb spatial distribution of light from the picture.

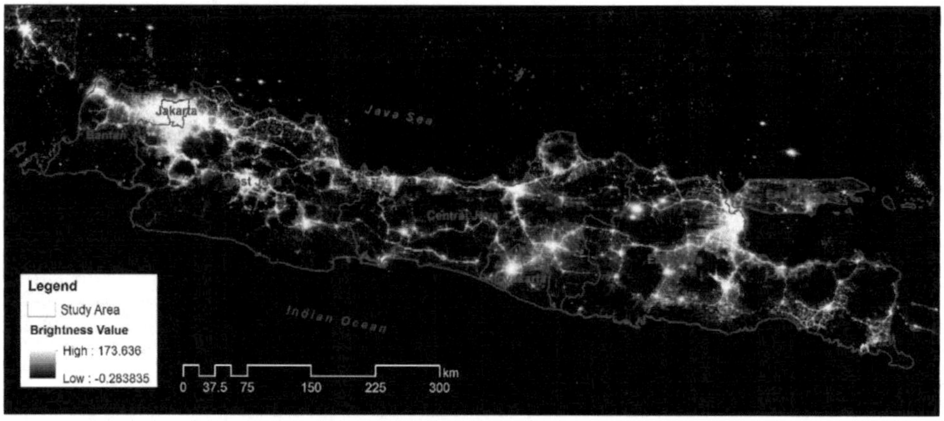

Figure 2. Java Island at night.
(*http://ngdc.noaa.gov/eog/viirs.html*)

b. Night Light Development Index (NLDI)

Development of night light information in the certain area be known by compare remote sensing night-time data with density inhabitant, it called Night Light Development Index (NLDI). The calculation of NLDI based on co-distribution light at night and inhabitant. NLDI data is development imagery data from NOAA/NGDC and the value based on state and province. Table 1 shows NLDI in each province of Java. NLDI is additional data to support how to know urban development in a certain area.

Table 1. NLDI of Java Island [9]

No.	Country	County	NLDI
1	Indonesia	East Java	0.629397
2	Indonesia	Jakarta	0.407950
3	Indonesia	West Java	0.652290
4	Indonesia	Central Java	0.656312
5	Indonesia	Yogyakarta	0.521561

c. Social Media Data (Twitter)

Real-time collection of social media is the most reliable method for gaining a sufficient quantity of streaming tweets during an event. Geo-social media information on this research is twitter data that has a large amount of data in the word. It collected by using Twitter API utility. Then API data visualized spatially using Mapbox application or website developed by Eric Fischer. Historical twitter data that had been collected able to visualize in dot map suit at a certain scale.

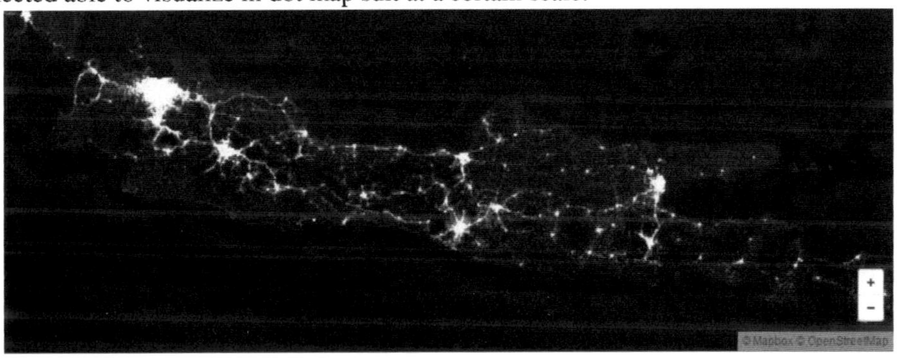

Figure 3. Twitter Spatial Data
(https://api.tiles.mapbox.com/)

2.3Analysing Data

Spatial analysis is a horde of methods to determine and depicted the pattern of the spatial phenomenon, so it can understand easily. There are various types approach to analyse spatial pattern and interaction, ranging from visual interpretation till applied statistic/mathematic [3]. This research is preliminary work, experiment design to catch information for spatial pattern and interaction from data above use visual interpretation both of the data (VIIRS data and Twitter data). Visual interpretation very useful to find and make clear pattern and interaction a few object and spatial phenomenon in the surface of the earth. If doing visual correctly, the complicated spatial pattern can detect quickly. Visual analyse of urban development divide into distributed of point object and spatial tessellation. The important things that're have to concern to detect spatial pattern are distance, interaction, and movement.

Spatial pattern and interaction urban development in java island determine based on [8]. According to Yunus [8] spatial pattern divides into nine elements, that is:

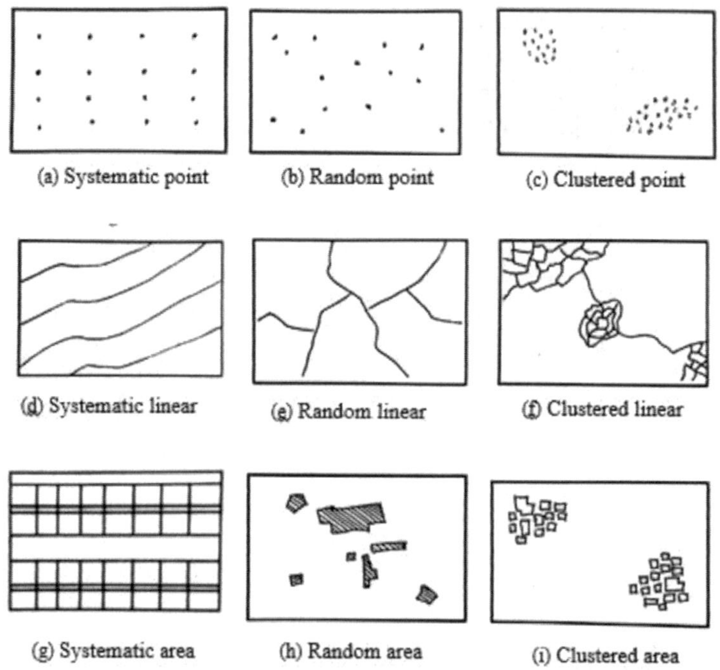

Figure 4. Types of spatial [8]

Each of the data either VIIRS data and twitter data should be analysed with the same way, which is used spatial pattern method using visual interpretation. Similarities and differences of the result are the points of view discussion. Additional data (NLDI) is support and prove how urban or rural area grown up and help to detect the spatial pattern and interaction information.

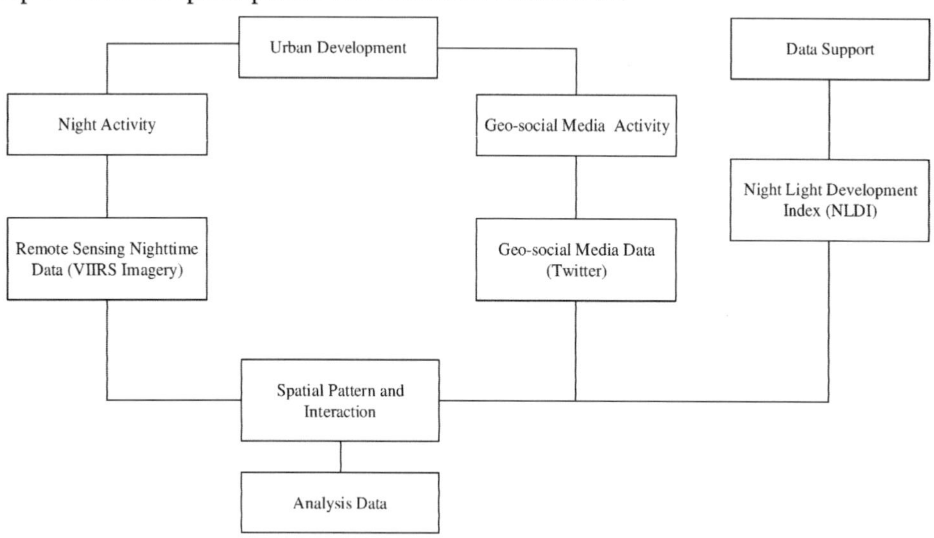

Figure 5. Workflow Design

Spatial interaction is a process influencing another object at the certain space [8]. There are two types of spatial interaction Yunus [8] that are stimuli-responses i.e. urban-rural interaction and push-pull i.e. one-way interaction and two ways interaction. This research used the spatial interaction above to identify in case study area.

3. Result

The result is shown as a preliminary study of integrating nighttime remote sensing data and geospatial footprint from twitter data. The integration can depict urban spatial pattern and interaction in Java Island. Based on the visual analysis we can see three types of spatial pattern: systematic, random, and clustered. Clustered area in Java Island can be seen from VIIRS image and twitter data visually, especially located in big cities such as Jakarta, Bandung, Semarang, Yogyakarta, Solo, and Surabaya. Geospatial footprint from twitter for the big cities has the same pattern with the image as can be seen in Figure 5. Clustered area has the most dominant light resulting in the geospatial footprint as the brighter area formed originally from points (Figure 6). Urban activities are higher than rural area perform the higher needs of light and more concentrated. The availability of facilities and infrastructures in urban areas gives more active the people live in urban with social media such as twitter.

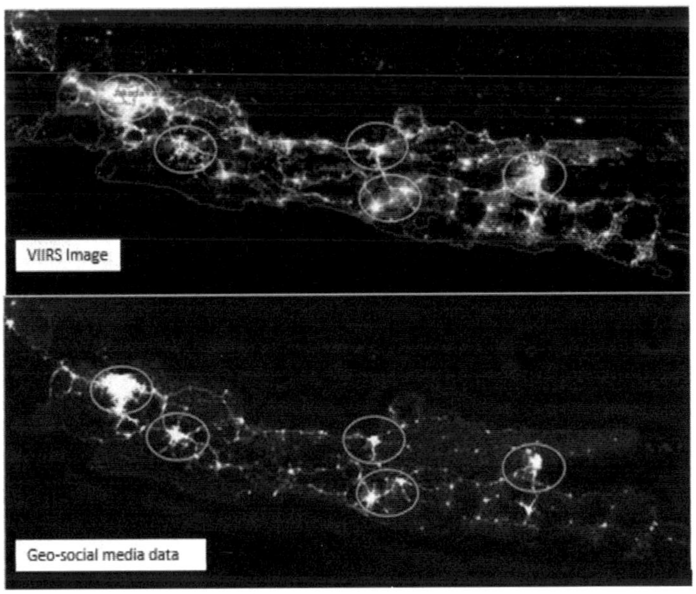

Figure 5. Clustered spatial of pattern from VIIRS image and Geo-social media data

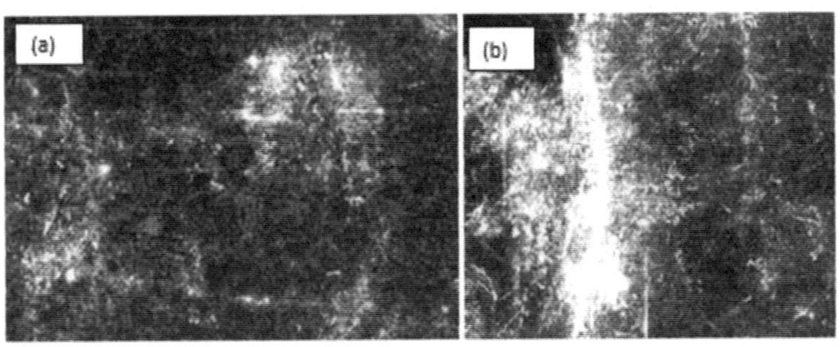

Figure 6. (a) Clustered Point of Spatial pattern in Gresik, (b) Clustered Area Pattern in Jakarta and big city

The spatial pattern may reflect the interaction between objects. As described by Yunus [8], spatial interaction as a process of mutual influence between the two terms has different types. Interactions that occur in urban areas included in the category of interaction balance. Balance interaction occurs because of the area adjacent or neighbouring affect each other, but each is independent of one another. The area adjacent to Jakarta indirectly be affected economically, socially, culturally and administration, such as

Tangerang, Bogor, Bekasi, and Serang. Land use of these areas mostly built-up areas which depend on the light becomes high.

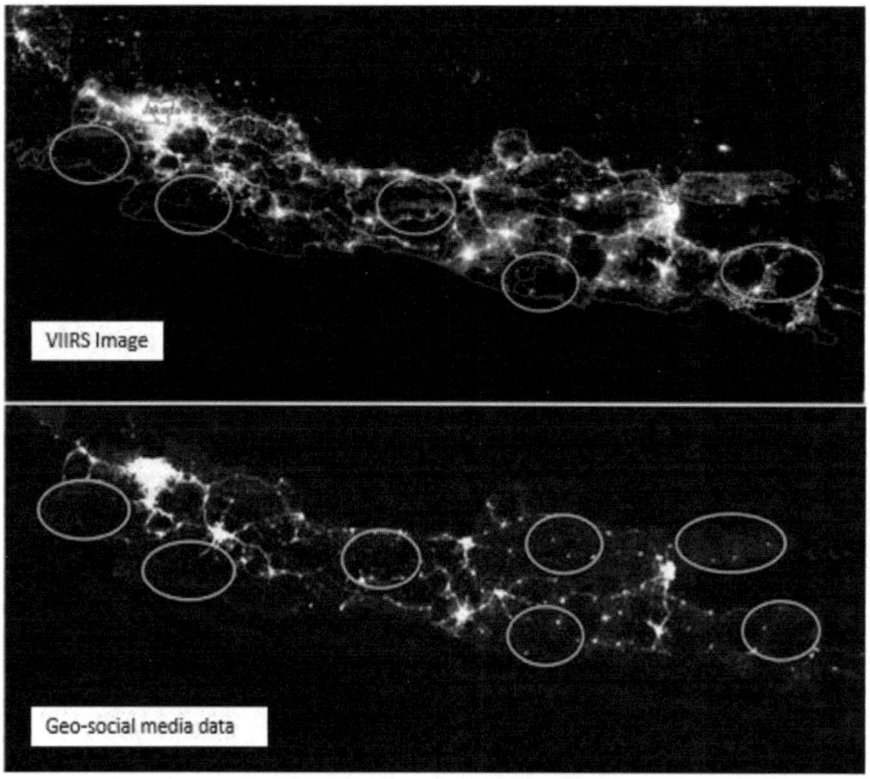

Figure 6. Random Spatial Pattern

The different can be seen in rural areas with a random spatial pattern as shown in Figure 6. The dependence of local people to light is not too high and likely to spread to adjust from the settlement, as well as social media activity is not dominant because of the limited facilities and infrastructure. Random pattern illustrates the distribution of data as point features. Areas with random pattern can be seen as spot areas in the southern regions of the province of Banten, West Java, and other districts in Java which has rural characteristics (Figure 7).

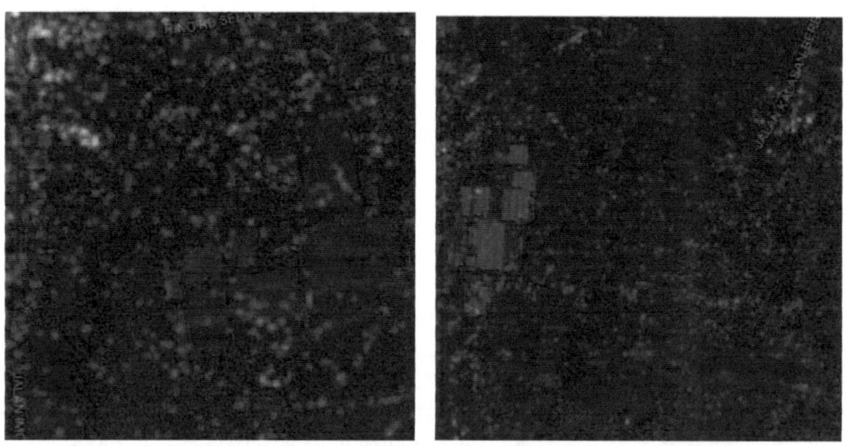

Figure 7. Random point spatial pattern in Bantul and another county

Rural areas also categorized as balance interaction as well, where the region immediately surrounding rural areas will also be affected, for example, land conversion. Land use in these areas still dominated by agriculture, social and cultural situation are still identic with rural activities. Built-up areas are still inclined slightly so that the light intensity is not dominant.

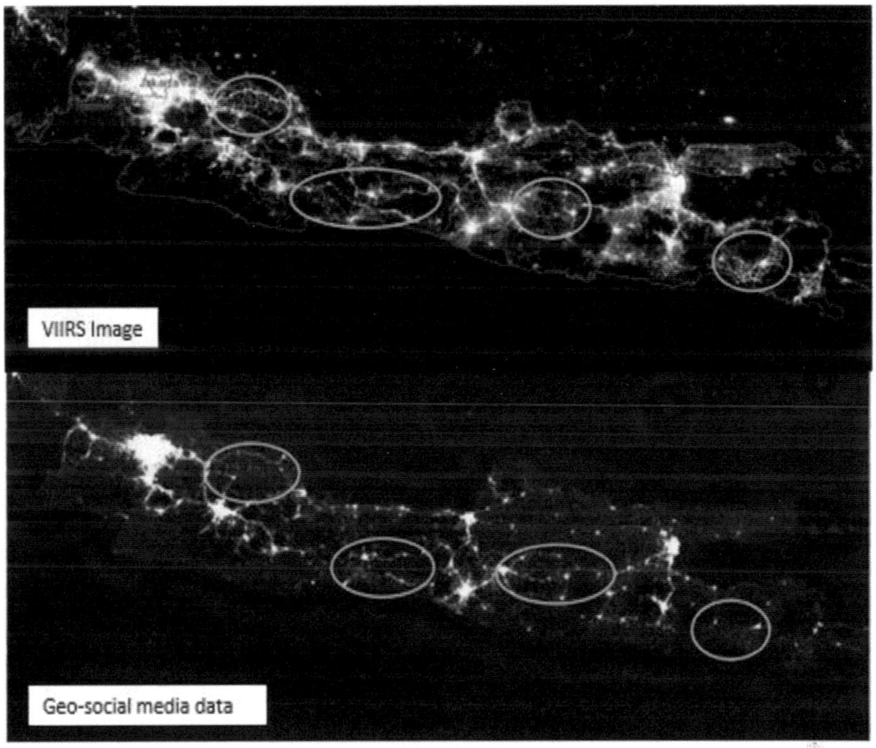

Figure 8. Regular spatial pattern

Regular spatial pattern has a tendency to follow a particular object in this example, the road pattern as can be shown in Figure 8. There is a spatial pattern of scattered points regularly follow the road so it can be seen forming into line feature, such as roads in northern of Java and the road connecting the town with other areas (figure 9). The pattern following the network of the roads. Figure 9 shows the distribution pattern of random lines formed in the city of Yogyakarta and Surabaya. Interactions that occur can be included in the category of interaction balance and solid interaction. Solid interaction which is defined as the interaction between the two regions are interdependent.

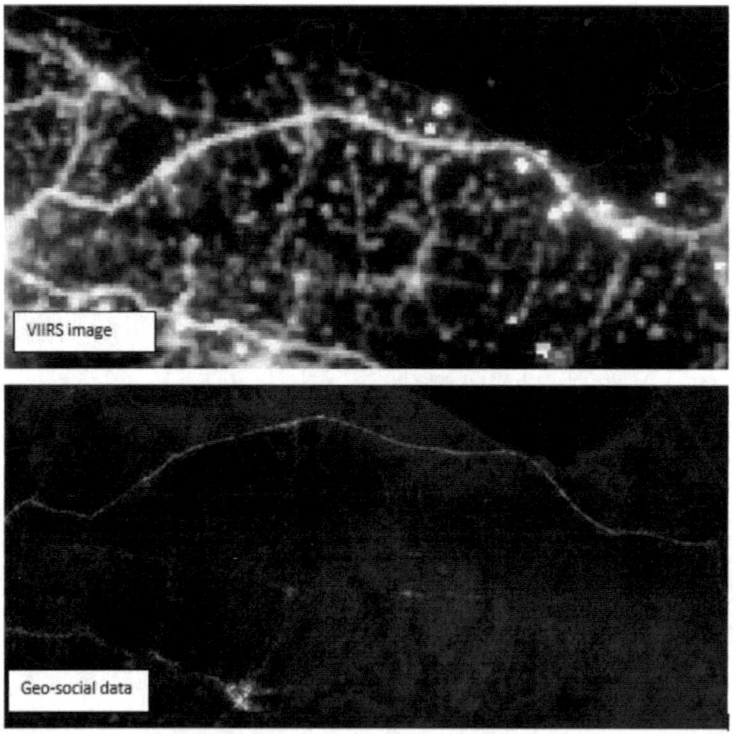

Figure 9. Systematic point of spatial pattern

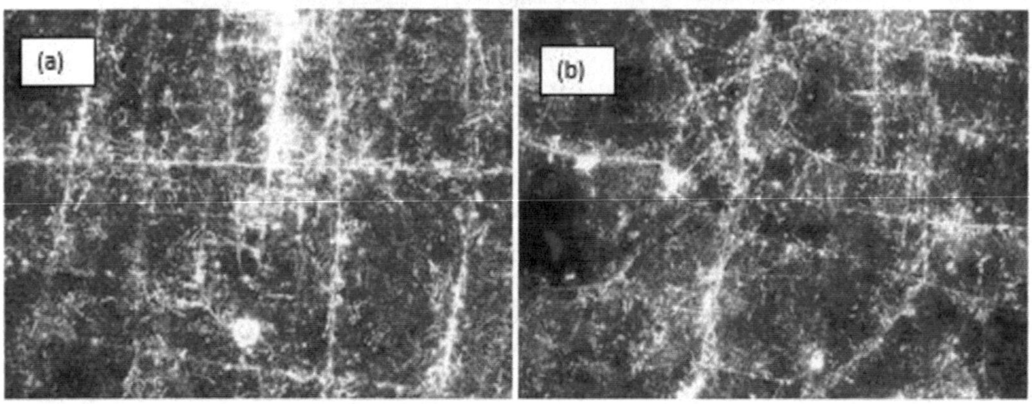

Figure 10. Random linear of spatial pattern in Yogyakarta City (a) and Surabaya City (b)

Spatial pattern from VIIRS image and geo-social media data has depict the same characteristic of urban spatial pattern. Mode detailed identification really depend on the level of analysis because different scale can give the different perform of spatial pattern to be analyse.

4. Conclusion

This research is still preliminary and using visual interpretation and analysis to recognized the urban spatial pattern. Future works will considering the developing semi-automatic or automatic spatial pattern recognition from these two kinds of dataset. We may conclude that integration of night time remote sensing data and geospatial footprint from geo-social media can help us in understanding urban development from spatial pattern and interaction.

References

[1] Hultquist C, Simpson M, and Cervone G 2015 *Using Nightlight Remote Sensing Imagery and Twitter Data to Study Power Outages* (Department of Geography: University of Wisconsin-Madison)

[2] Mitchell L, Frank M R, Harris K D, Dodds P S, Danforth C M 2013 The Geography of Happiness: Connecting Twitter Sentiment and Expression, Demographics, and Objective Characteristics of Place *Plos One* **8** 1-15

[3] Sadahiro Y 2010 Analysis of the spatial relations among point distributions on a discrete space, *International Journal of Geographical Information Science* **24** (7), 997-1014.

[4] Small C, Pozzi F, and Elvidge C D 2005 Spatial analysis of global urban extent from DMSP-OLS night lights *Remote Sensing of Environment* **96** 277-291

[5] Sutton P C 2003 A scale-adjusted measure of ''Urban sprawl'' using nighttime satellite imagery *Remote Sensing of Environment* **86** 353–369

[6] Steiger E, Westerholt R, Resch B, and Zipf A 2015 Twitter as an indicator for whereabouts of people? Correlating Twitter with UK census data *Computers, Environment and Urban Systems* **54** 255–265

[7] Wessel G, Ziemkiewicz, and Sauda E 2015 Revaluating urban space through tweets: An analysis of Twitter-based mobile food vendors and online communication *new media & society* (reprints DOI: 10.1177/1461444814567987 nms.sagepub.com)

[8] Yunus H S 2010 *Metodologi Penelitian Wilayah Kontemporer* (Pustaka Pelajar, Yogyakarta)

[9] National Centers For Environmental Information http://ngdc.noaa.gov/eog/viirs.html

[10] Ellis P D 2009 Indonesia Urban's Development Towards Inclusive and Sustainable Economic Growth http://siteresources.worldbank.org/INTURBANDEVELOPMENT/Resources/336387-1296405826983/Ellis.pdf

[11] Indonesia Goverment http://tanahair.indonesia.go.id/

[12] Mapbox https://api.tiles.mapbox.com/

Spatial temporal analysis of urban heat hazard in Tangerang City

Adi Wibowo[1], Kuswantoro[1], Ardiansyah[1], Andry Rustanto[1], Iqbal Putut Ash Shidiq[1]

[1]Department of Geography, Faculty of Mathematics and Natural Science, University of Indonesia

Corresponding author: adi.w@sci.ui.ac.id, iqbal.putut@yahoo.com

Abstract. Urban heat is a natural phenomenon which might caused by human activities. The human activities were represented by various types of land-use such as urban and non-urban area. The aim of this study is to identify the urban heat behavior in Tangerang City as it might threats the urban environment. This study used three types of remote sensing data namely, Landsat TM, Landsat ETM+ and Landsat OLI-TIRS, to capture the urban heat behavior and to analysis the urban heat signature of Tangerang City in 2001, 2012, 2013, 2014, 2015 and 2016. The result showed that urban heat signature change dynamically each month based on the sun radiation. The urban heat island covered only small part of Tangerang City in 2001, but it was significantly increased and reached 50% of the area in 2012. Based on the result on urban heat signature, the threshold for threatening condition is 30 °C which recognized from land surface temperature (LST). The effective temperature (ET) index explains that condition as warm, uncomfortable, increase stress due to sweating and blood flow and may causing cardiovascular disorder.

Keywords: land surface temperature, urban heat signature, spatial-temporal analysis, urban environment threat

1. Introduction

The IPCC 2007 report on the "Science of Climate Change" showed a small increase in temperature and rainfall for the Southeast Asia Region in the last 50 years or so. There is a general agreement amongst scientists that the changing behavior patterns of the el-Nino ENSO, Monsoons and to a certain extent the Indian Dipole Oscillation circulation systems are triggering extreme weather and variability to influence changing behavior patterns of hydro-meteorological and geomorphological events (floods, haze pollution and slope failures) in the region in general. The last half decade has showed that societies and their whole rubric of livelihood systems (that include low income economic systems and modern production systems) are becoming more threatened and more vulnerable to climate change induced hazards and their ability to adapt to these imposing conditions are becoming more demanding and would get worst in the near future. The significance of climate "variation" or "change" depends not only on the behavior of the change itself but also on the characteristics of society and systems exposed to the changes [1].

Asian cities are the most rapidly growing regions of the world nowadays and 16 of the world's 24 mega cities (cities with more than 10 million people) will be located in Asia by the year 2015 [2, 3]. Urbanization including residential, commercial and industrial developments initiated one of the most dramatic human-induced changed of a natural ecosystem: a natural landscape, often containing transpiring vegetation and a pervious surface, is converted to a built, largely impervious landscape made up of rigid, sharp-edged roughness elements [2]. Urban heating causes many problems for the inhabitants of some cities and areas, in particular those with a tropical environment. Urban heating could deteriorate our living environment, increase energy consumption, elevate ground-level ozone and even increase mortality rates. There are three billion people (48% of the world population) living in urban areas, they are directly exposed to urban heating problems and more people will be vulnerable to these problems as the number of people living in urban areas is expected to grow to five billion by 2030 [3].

With rapid urbanization, there has been a tremendous growth in population and buildings in cities. The rapid growth of a city has concluded that the urbanized areas had significantly higher day-time surface temp as compared to those of the surrounding rural with vegetated areas [2]. The accelerated rate of urban growth in tropical cities highlights the critical necessity of creating more outdoor spaces for leisure and recreation activities of citizens [4]. In Singapore, the primary root of heat island in cities is due to the absorption of solar radiation by mass building structures, roads, and other hard surfaces during daytime and it is doubtless that the UHI effect is aggravated mainly due to the loss of green areas in the urban environment as consequences of rapid population has led to demands for converting natural areas to public housing [5]. In Malaysia, number of factors contributes to the occurrence and intensity of heat island and these include weather, geographic location, time of day and season, city form and city functions [6]. The urban conurbation within urbanized in Selangor has a remarkable heat island despite the less urbanized area in 1988 compared to 1999. The urban land cover can be associated with buildings, road pavement, highways, green parks and also bare soil due to earthwork activities [7]. The urbanized areas had significantly higher day-time surface temperature as compared to those of the surrounding rural with relatively moist vegetated areas. In February 2002, the mean rural temperatures in Bangkok, Manila and Ho Chi Minh City were 29.5 °C, 26. 5 °C, and 30 °C respectively. Bangkok has the highest daytime surface UHI intensity of 8 °C, followed by Manila (7 °C) and Ho Chi Minh City (5 °C) in dry season [2].

The condition is similar to temperature trends in Indonesia. The rise of temperature is a phenomenon caused by human activities such as land-use change. It means that changes in land-use are in line with human growth. This research is to identify those circumstances. The study took place in Tangerang City, one of the cities in Banten Province with fast growth, as it is near with Jakarta.

2. Material and method

As it has been explained in the first part of this paper, the urban heat is very much related with land-use and land-cover condition. This research is focusing on analysis on land surface temperature (LST), determine urban heat island (UHI) and its distribution and observation on the impact to urban environment. Tangerang City is chosen as study area, where the rate of urbanization is very rapid, hence the dynamic change of land-use/land-cover were seen.

2.1. Image processing

This paper used indirect data collecting by employing satellite data form Landsat TM, Landsat ETM+ and Landsat OLI-TIRS. The list of Landsat imageries used in this study is showed in Table 1. Thermal band from each data were collected and processed to generate LST. The pixel size is determined at 100 x 100 meter, accordance with [8] who measured UHI in Japan. The grid size is also determined based on spatial representative of land-use and land-cover. Land surface temperature is representing the heat from the land which generated by radiated sun energy [5, 9, 10, 11].

Table 1. List of Landsat imageries in this study

Year	Path/row	Acquisition date	Remote sensing imagery
2001	122/64	15-07-2001	Landsat TM [15]
2012	122/64	27-06-2012	Landsat ETM+ [15]
2013	122/64	12-10-2013	Landsat OLI-TIRS
2014	122/64	15-10-2014	Landsat OLI-TIRS
2015	122/64	18-10-2015	Landsat OLI-TIRS
2016	122/64	13-05-2016	Landsat OLI-TIRS

LST is generated through several steps. The thermal band of Landsat imagery is the main "ingredient" of LST. First, digital number (DN) of the thermal band needs to be converted to spectral radiance. The formula is represented in equation 1 and 2 (Eq. 1 is applied for Landsat TM and ETM+, eq.2 is applied for Landsat OLI-TIRS) [12].

$$L\lambda = \left[\frac{(LMAX\lambda - LMIN\lambda)}{(QCALMAX - QCALMIN)}\right] \times (QCAL - QCALMIN) + LMIN\lambda \quad (1)$$

$$L\lambda = (M \times DN) + A \quad (2)$$

where, $L\lambda$ is spectral radiance ($wm^{-2}sr^{-1}\mu m^{-1}$), M is Multiplicative digital number value at thermal band, DN is digital number of Landsat's thermal band and A is additive value of spectral radiance at thermal band.

Second, the spectral radiance resulted from equation 1 is being converted to estimate land surface temperature. The formula is represented in equation 2 [10, 11].

$$T = K2/\ln((K1/L\lambda) + 1) \quad (2)$$

where *T* is the temperature at the satellite sensor (Kelvin), *K*1 is the calibration constant 1 for Landsat, *K2* is the calibration constant 2 for Landsat, and $L\lambda$ is the spectral radiance of band. The utilization of remote sensing imagery to collect urban heat signature have been performed in several studies [5, 11, 12, 13, 14, 15].

Finally, the estimate of LST is being transformed from Kelvin to Celsius by the following formula [11, 16].

$$LST\ (celcius) = T - 272.15 \quad (3)$$

Based on the equation, low value of LST will be seen in maximum vegetation cover, while minimum vegetation cover will be overlaid with high temperature value [9, 17-19].

2.2. Universal Thermal Climate Index (UTCI) and Effective Temperature (ET) Index

More people are affected by higher temperatures for longer period, higher temperature not only makes life uncomfortable for urban residents it is also adversely affecting people's health [14, 16]. The comfort thermal for tropical area ranges between 22 °C to 27 °C [20,21,22].

An index called Universal Thermal Climate Index (UTCI) was developed to closely monitor the impact climatic condition on human [16, 23]. There are five categories of heat stress based on the level of temperature, such as Extreme Heat Stress (> +46 °C), Very Strong Heat Stress (+38 to +46 °C), Strong Heat Stress (+32 to +38 °C), Moderate Heat Stress (+26 to +38 °C) and No Thermal Stress (+9 to +26 °C) and In this study, the UTCI is being combined with effective temperature (ET) index to explain the relation between temperature condition and the level of sensation (very hot, hot, warm, slightly warm and neutral), comfort (very uncomfortable, uncomfortable and comfortable), psychology (body heating, increasing stress due to sweating and blood flow, and normal regulation by sweating and vascular change), and health (circulatory collapse, increasing danger of heat stroke, cardiovascular embarassment and normal health) [16, 23]. Those indices will determine the urban heat hazard in our study area. Spatio-temporal analysis of urban hazard based on LST have been performed by previous studies [2, 9, 17, 19, 24].

3. Result and discussion.

3.1. Spatial analysis

The distributions of urban heat signature (UHS) were different each year. This study categorizes UHS with high, moderate and low classes based on the LST. The moderate temperature (signed by yellow color) covered 30 to 60 % of Tangerang City in 2013 and 2014. The distribution and density were different in between these two years. For the high temperature (red color), the coverage was higher in 2014 compare to 2013. However, according to LST the maximum temperature was decreased from 37 °C in 2013 to 30 °C in 2014 and becomes 26 °C in 2015, however it increased back to 30 °C in 2016. The area with high temperature in 2013 is mainly agglomerated in the center then it is spread to the west part of Tangerang City. It becomes more equally distributed in 2014 and spread from the center to east part in 2015 and 2016 (Figure 1).

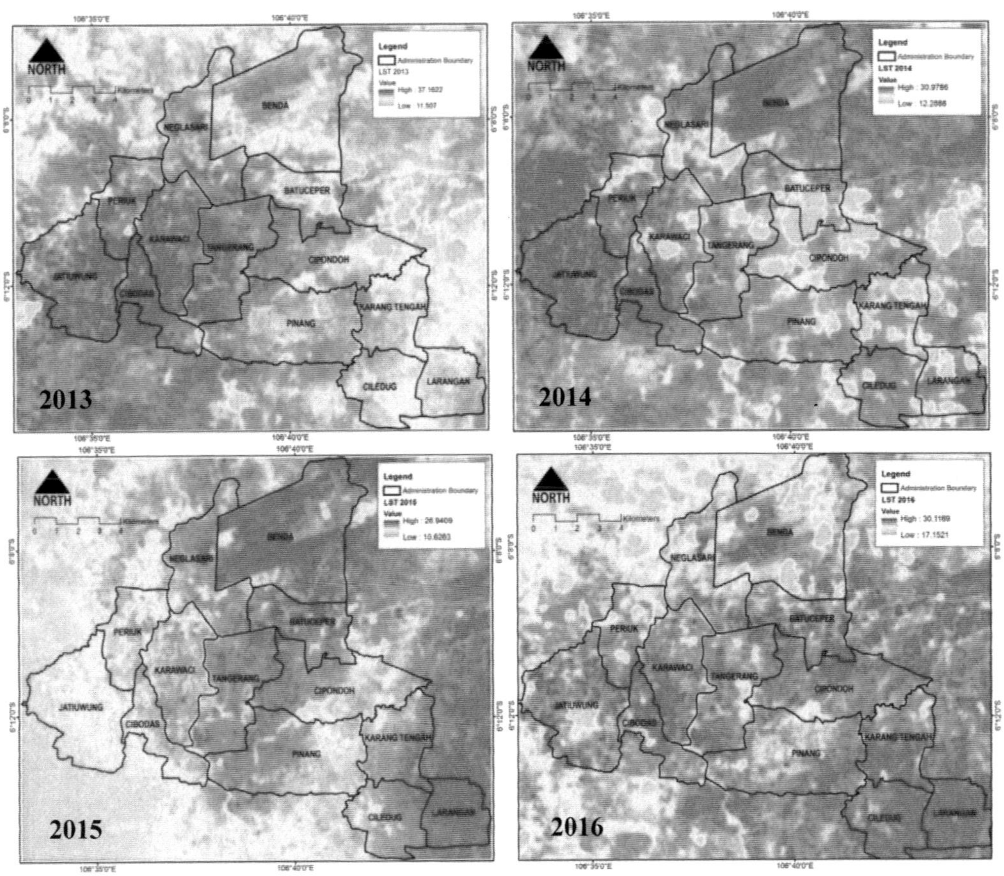

Figure 1. Land surface temperature of Tangerang City from 2013 to 2016

The +30 °C threshold is applied on LST to highlight the distribution of urban heat island (UHI) during 2013 and 2016 in Tangerang City. There is a slight adjustment for 2015, where the threshold is set to +28 °C, due to the maximum LST of 29 °C in 2015. The distribution of UHI in Tangerang City can be seen in figure 2 below. As we can see, the largest area of UHI was occurred in 2013, where it was mostly agglomerated in the southwest part of Tangerang City (Karawaci, Cibodas and Jatiuwung). It was significantly reduced in 2014, where only small area of UHI were gathered in north (Benda) and sparsely distributed in the southwest part of Tangerang City. The area of UHI were increased again in 2015 and mostly distributed in the north (Benda) and southeast part of Tangerang City (Larangan). The area of UHI were slightly reduced in 2016, where it is mostly found in southeast of Tangerang City (Larangan).

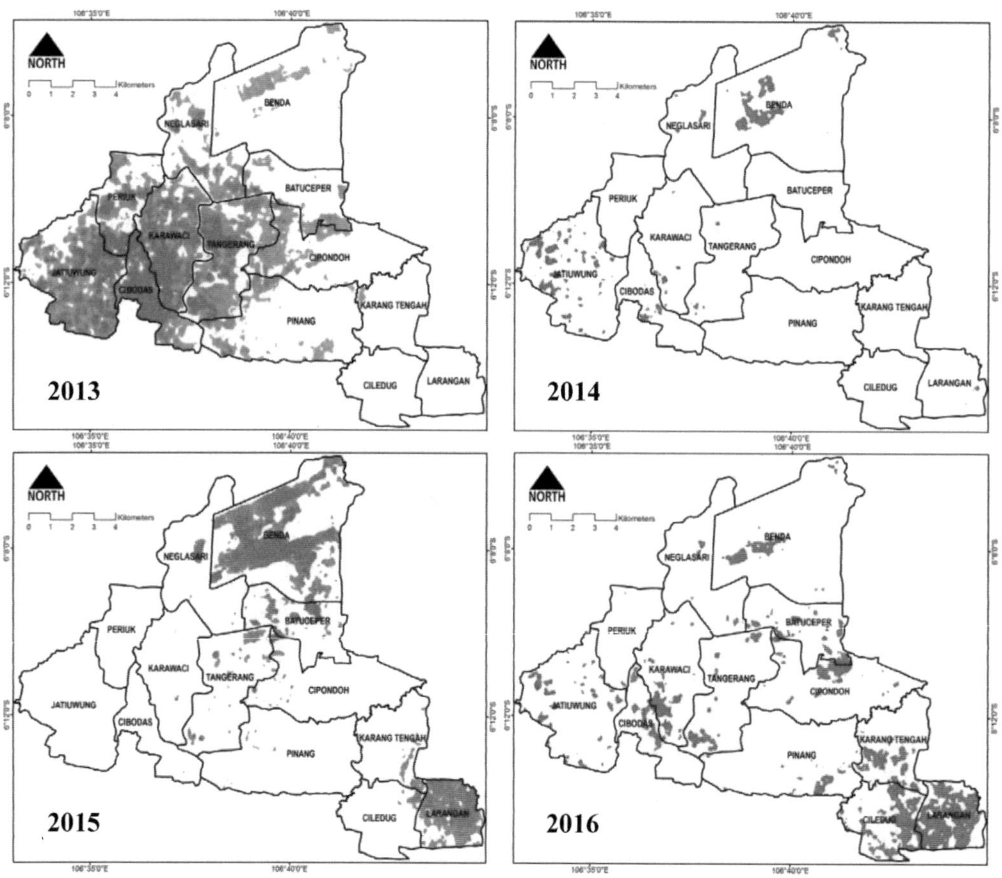

Figure 2. Urban heat island in Tangerang City from 2013 to 2016

Figure 3 shows the result from [15], who explained the distribution of UHI in Tangerang City in 2001 and 2012. Based on the result, the UHI were sparsely distributed in north and south part of Tangerang City. During 10 years period, it was greatly increased and found almost in every part of Tangerang City. We can assume that changes of land-use from vegetated area to built-up area as the main reason of these circumstances. If compared with the result from [15], it is known that the area of UHI were significantly reduced during 2012 and 2013.

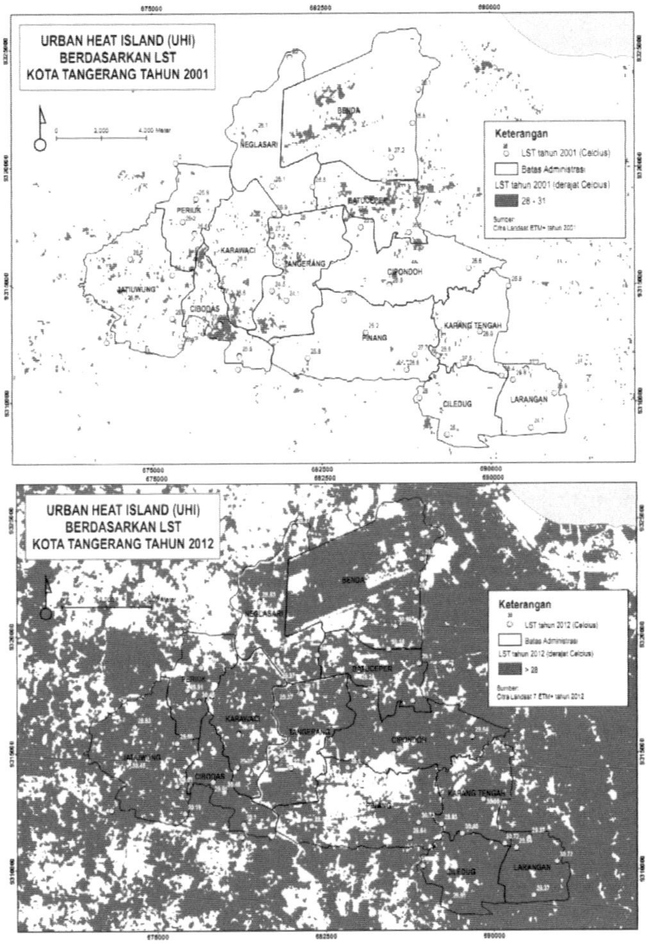

Figure 3. Urban heat island in Tangerang City in 2001 and 2012

3.2. Temporal analysis

Based on the LST, the trends show a declining in terms of surface temperature during 2013 and 2015 (Figure 4). The highest temperature was 38.3 °C and it was founded in 2012, while the lowest was 10.62 °C and it was founded in 2015. Based on the LST, we can safely say that 2012 is the warmest period in Tangerang City.

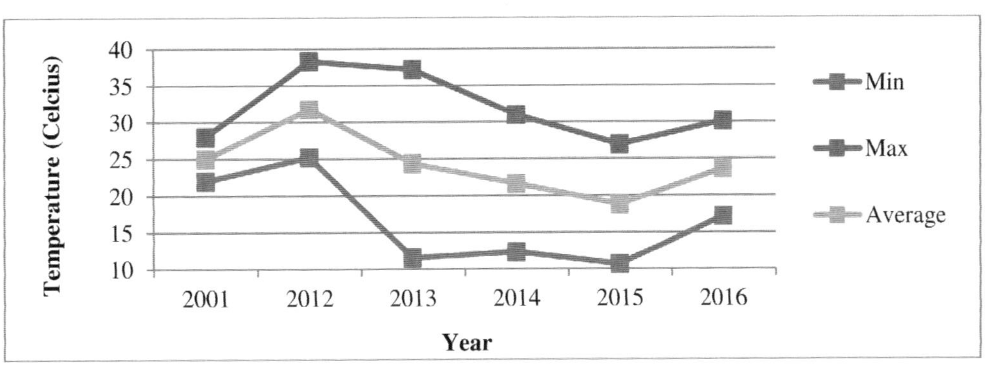

Figure 4. Temporal trend of urban heat signature in Tangerang City during 2001 until 2015

3.3. Urban heat hazard

The urban heat hazards were determined by relating the temperature value from LST with UTCI. Based on the index, the hazards were occurred in 2012 and 2013, with stress category from strong heat stress to very strong heat stress. Fortunately, there is no extreme heat stress situation available in this study. The relations between temperature and heat stress category are presented in Table 2 below.

Table 2. Universal Temperature Climate Index (UTCI) in Tangerang City

Range of maximum temperature	Stress category	Year
Above 46 ^{0}C	Extreme heat stress	-
+38 to +46 ^{0}C	Very strong heat stress	2012
+32 to +38 ^{0}C	Strong heat stress	2012, 2013
+26 to +32 ^{0}C	Moderate heat stress	2001, 2014, 2015, 2016
+9 to +26 ^{0}C	No thermal stress	-

To observe more about the impact of urban heat to human body, the effective temperature (ET) index was employed. Based on the ET index, the urban heat were very dangerous in 2012, where high temperature mostly occurred, causing very uncomfortable condition due to the increasing stress caused by sweating and blood flow and increase the probability of heat stroke (Table 3).

Table 3. Effective Temperature (ET) index in Tangerang City

Year	Temperature (^{0}C)	Sensation		Psychology	Health
		Temperature	Comfort		
				Body Heating, Failure of Regulation	Circulatory Collapse
	>40	Very Hot	Very Uncomfortable	Increasing Stress Caused by Sweating and Blood Flow	Increasing Danger of Heat Stroke
2012 and 2013	35 – 40	Hot			Cardiovascular Embarrassment
-	32 – 35	Warm	Uncomfortable		
2014	30 – 32	Slightly Warm		Normal Regulation by Sweating and Vascular Change	
2001, 2015 and 2016	25 - 30	Neutral	Comfortable		Normal Health

4. Conclusion

The UHS on Tangerang City with temperature more than 30 °C would become a threat for urban environment. Based on effective temperature index, such condition will bring warm sensation in temperature, which is uncomfortable and might cause increasing stress due to sweating and blood flow and may damage the cardiovascular organ. This condition is dangerous and can be categorized as urban heat hazard for everyone living in urban area. This result addresses the dangerousness of urban heat hazard to the community and urban environment.

Acknowledgements
The authors would like to thank Department Geography Faculty of Mathematics and Natural Sciences, University of Indonesia for the support of the research.

References

[1] Khairulmaini, O S and Ghaffar F A 2008 Vulnerability and adaptation to climate change threat: Issues and challenges for Malaysia In *Proceedings of the 3rd Regional Symposium on Environment and Natural Resources: Conservation for a green future* (Kuala Lumpur) pp. 1-22

[2] Tran H, Uchihama D, Ochi, S and Yasuoka, Y 2006 Assessment with satellite data of the urban heat island effects in Asian mega cities *International Journal of Applied Earth Observation and Geoinformation* **8**(1) 34-48

[3] Memon R A, Leung D Y, and Liu C H 2009 An investigation of urban heat island intensity (UHII) as an indicator of urban heating, *Atmospheric Research* ***94***(3) 491-500

[4] Makaremi N, Salleh E, Jaafar, M Z, and Ghaffarian H A (2012) Thermal comfort conditions of shaded outdoor spaces in hot and humid climate of Malaysia *Building and environment* ***48*** 7-14

[5] Wong N H and Yu C 2005 Study of green areas and urban heat island in a tropical city *Habitat International* **29**(3) 547-558

[6] Ishak A, Hassan Z N C, Edros N H, Zamberi M H, and Rahman M N A 2011 The Effect of Local Climate on Urban Heat Island Trend; A Case Study in Urban Areas of Ipoh and Kuantan *Malaysian Meteorological Department (MMD) Ministry of Science, Technology and Innovation* (Kuala Lumpur, Malaysia)

[7] Ahmad S, and Hashim N M 2007 Effects of soil moisture on urban heat island occurrences: case of Selangor *Malaysia Humanity & Social Sciences Journal* **2**(2) 132-138

[8] Suzuki C 2008 Improvements of heat island monitoring network in tokyo

[9] Srivanit M and Hokao K 2013 Evaluating the cooling effects of greening for improving the outdoor thermal environment at an institutional campus in the summer *Building and Environment* **66** 158-172

[10] Mirzaei P A and Haghighat F 2010 Approaches to study urban heat island–abilities and limitations *Building and Environment* **45**(10) 2192-2201

[11] Wibowo, A, Raditya, A, Harmantyo, D, and Semedi J M. (2015) Land Surface Temperature as Urban Hazard in Education Area (A Case Study: University of Indonesia) *Proceeding the First International Conference of Indonesian Society for Remote Sensing* (Surabaya, Indonesia)

[12] USGS 2014 "Landsat 5 History." Accessed on March 21, 2016. http://landsat.usgs.gov/about_landsat5.php.

[13] Tursilowati L, Tetuko Sri Sumantyo J, Kuze, H, and Adiningsih E S 2012 Relationship between urban heat island phenomenon and land use/land cover changes in Jakarta-Indonesia *Journal of Emerging Trends in Engineering and Applied Sciences* **3(4), 645-653**

[14] Ichinose T, Matsumoto F, and Kataoka K 2008 Urban thermal environment and its mitigation through urban planning process *Geographical Reports of Tokyo Metropolitan University* **43, p33-40.**

[15] Hernina R, Ismullah, I H, and Wikantika, K 2008 The Analysis of Urban Heat Island Using Satellite Image (Case Study Bekasi West Java) *Jurnal Geografi* **1**(2), 73-80 (Department of Geography, Faculty of Mathematics and Natural Science, University of Indonesia, Depok)

[16] Tursilowati L 2002 Urban heat island and their contribution on climate change and relationship with land use change, *Proceeding on National Seminar on Global Warming and Global Change: Fact, Mitigation and Adaptation* **ISBN: 978-979-17490-0-8.**

[17] Wibowo A, Rustanto A, and Shidiq I P A 2013 Spatial-temporal analysis of urban heat island in tangerang city, *The Indonesian Journal of Geography* *45*(2), 101-112

[18] Wong N H, Jusuf S K, La Win A A, Thu H K, Negara T S, and Xuchao W 2007 Environmental study of the impact of greenery in an institutional campus in the tropics *Building and environment* ***42***(8) 2949-2970
[19] Wong N H and Jusuf S K 2008 GIS-based greenery evaluation on campus master plan *Landscape and urban planning* **84**(2) 166-182
[20] Karyono T H 2001 Teori dan Acuan Kenyamanan Termis dalam Arsitektur *Catur, Libra Optima*
[21] Zoer'aini Djamal Irwan 1997 Tantangan lingkungan dan lansekap hutan kota CIDES
[22] Chen L and Ng E (2012) Outdoor thermal comfort and outdoor activities: A review of research in the past decade *Cities* **29**(2), 118-125
[23] Park S, Tuller S E, and Jo M 2014 Application of Universal Thermal Climate Index (UTCI) for microclimatic analysis in urban thermal environments *Landscape and Urban Planning* ***125*** 146-155
[24] Yeon-Hee K and Jong-Jin B 2005 Spatial and temporal structure of the urban heat island in Seoul *Journal of Applied Meteorology and Climatology* **44**(5) 591

Using Visible Infrared Imaging Radiometer Suite (VIIRS) Imagery to identify and analyze light pollution

Wahyu Nurbandi*, Febrina Ramadhani Yusuf, Ruwanda Prasetya, Mousafi Dimas Afrizal

Faculty of Geography, Universitas Gadjah Mada, Yogyakarta

Corresponden: wahyunurbandi.ipa130@gmail.com

Abstract. Light pollution is any adverse effect of artificial lighting including sky glow, glare, light trespass, light clutter, decreased visibility at night, and energy waste. Remote sensing is the acquisition of information about an object or phenomenon without making physical contact with the object. Remote sensing can be used for identification of light pollution. The purpose of this study is to identify and analyze the light pollution by using remote sensing imagery. This study uses VIIRS DNB Free Cloud Composites imagery to identify light pollution in Yogyakarta province and surrounding areas. VIIRS imagery which obtained is processed to get information of light pollution by classifying the information into several classes presented in a map. Selected few sample points as test sites to determine the actual condition. Field work conducted at theree location, they are Yogyakarta City, Depok Beach, and Gajah Mungkur reservoir. Night sky condition analysis conducted field tests as well as night time shooting the night sky conditions. Analysis of the night sky conditions are calculated qualitatively using Bortle Dark-Sky Scale with a value range of 1-9. Field test results show that Yogyakarta City has a value of 8, Depok has a value of 3, and Gajah Mungkur Reservoir has a value of 4. The conclusion of study is VIIRS imagery can be used for identification light pollution and calculation analysis of light pollution can use Bortle Dark-Sky Scale.

1. Introduction

This time the use of artificial light is highly increased, thus at night the light intensity of artificial light is also increased. The artificial light is used by settlement, industries, highways, and decorating lights. The excessive intensity of artificial light at night results in an increase of light pollution. Light pollution is as any adverse effect of artificial light including sky glow, glare, light trespass, light clutter, decreased visibility at night, and energy waste. Generally, light pollution thriving along with the change of environment, especially in urban areas. It is also along with the change of land use from farms to settlements or industries that also affect the increase of artificial light usage at night [1].

Light pollution is generally divided into two categories: astronomical and ecological. The astronomical definition of light pollution involves too much light blocking out the starry night sky, which seriously impede the studies of astronomers. Astronomical light pollution was the astronomers who turned their attention to light pollution and harmful effect of limitless lighting for the first time. Disturbing light form luminous sources made it almost impossible to accomplish astronomical research, since they were many fold more intensive than the natural background lighting. Researchers talk about light pollution when the artificial lighting reaches 10% of the natural background lighting. Ecological light pollution usually refers to the effect that too much light has on the environment, including humans [2]. Overnight darkness probably plays a significant role in the human body. Light pollution also basically determines the decisions, behavior and direction of migration of animals, even if they have to find the place of food sources or the area reproduction. The direction of the sea is defined by the starlight reflecting the surface of water or by the moonlight. Due to the impact of artificial lights the turtles start for the wrong direction not for the water but for the land, which leads to their death [3].

Light pollution is a relatively new phenomenon. The first light bulb was invented only 160 years ago. Since then light emissions have grown steadily, even exponentially in certain areas. Artificial lighting has become integral part of our culture. In conveys a feeling of safety and wealth. Nevertheless, scientist warn of detriment effects of excessive light emissions. Among the negative effects of light pollution are: 1) disturbance of biological rhythms, 2) psychological effects, and 3) environmental degradation [4,5]. Artificial light causes a lot of trouble for insect, too. Light sources lure night-flying insect species from their natural habitats and feeding place. Excessive light emissions do not only have negative effects on the natural environment. They are also enormous water resources. A sustainable use of light and energy pays off for the municipalities an d the individual household. Some countries have introduced laws to limit the light pollution [6].

The problem of light pollution needs to be considered in this era. The study of light pollution becomes important for government policy. Remote sensing technologies have been able to map the light pollution. Information of radiation values of artificial light on earth can be obtained from satellite imageries. Visible Infrared Imaging Radiometer Suites (VIIRS) DNB Free Cloud Composites is one of imagery that can be used to obtain information of the amount of light pollution on earth. This study aims to get light pollution information using VIIRS imagery and identifying the night sky condition in study area.

2. Study area and data

2.1 Study area

This study is performed in Daerah Istimewa Yogyakarta (DIY) and Central Java provinces. The area of DIY province is 3.185,80 km2 and astronomically located on 7033' – 8012' South Latitude and 110000' - 110050' East Longitude. DIY is located in Java Island which is bounded by Indian Ocean on south. Climate circumstances of DIY province are classified as tropical climate with annual rainfall average of 3.000 mm/year and average of 25-310C. Central Java province is astronomically located on 5040' - 8030' South Latitude dan 108030' - 111030' East Longitude which has tropical climate with temperature average of 24,80 – 31,80 C and rainfall average of 2.618 mm/year. The region which has low rainfall and frequently droughts is located in Blora, Rembang, half of Grobogan and surroundings, and southern Wonogiri [7]. **Figure 1** shows the map of study area i.e. DIY and Central Java provinces

2.2 Remote sensing data

Visible Infrared Imaging Radiometer Suites (VIIRS) DNB Free Cloud Composites imagery is one of remote sensing data that can provide information of light pollution. This imagery is produced by Suomi National Polar Partnership (SNPP) which is collaboration between National Oceanic Atmospheric Administration (NOAA) dan United Nation Aeronautics and Space Administration (NASA). VIIRS imagery has 3000 km wide swath and wavelength spectrum between 0.5 – 0.9 μm. Radiometric resolution of the imagery is up to 14 bits and capable of detecting light limit up to `~2E $\times 10^{-9} \text{watt}/\text{cm}^2 * \text{sr}$. VIIRS has the ground pixel footprint at nadir of 724x724 m (0.55 km^2). Some VIIRS data sets are available and can be downloaded from the National Oceanic and Atmospheric Administration. DIY and Central Java are included in the recorded area. The imagery used for this study is the result of recording in September 2015. The chosen imagery on that time is based on assumption that September is driest month in Indonesia that has minimum cloud cover. If the cloud cover is minimum artificial light on earth which cause the light pollution can be recorded optimally.

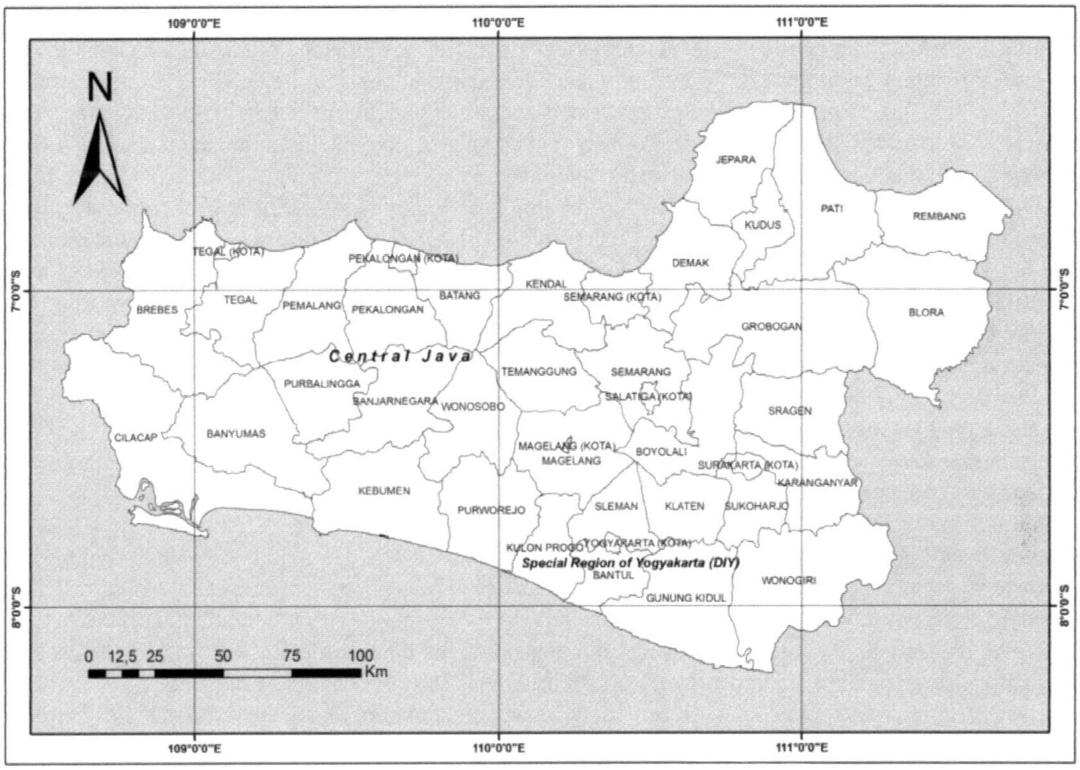

Figure 1. Map of study area (DIY and Central Java provinces).

3. Methodology

3.1 Image Processing

VIIRS DNB Free Cloud Composites imagery which the result of Suomi National Polar Partnership (SNPP) satellite is used to detect faint radiant emissions. The use of VIIRS Imagery that catch the radiance value of earth surface also better to detect dimmer lightning than any imagery, like DMSP. The radiance information is processed to get light pollution information. Baugh (2015) stated that, quantitatively, VIIRS DNB Free Cloud Composites had been calibrated, both radiometric and geometric by NOAA and NASA [8]. So that in image processing cutting/masking has been performed according to study area by using ENVI 5.0. The available data in official website of NOAA has been divided into 6 files of (00N/060W).

3.2 Fieldwork

Extraction of information of light pollution is not just obtained by imagery interpretation but also performed by fieldwork to validate the actual circumstances whether it is appropriate with the imagery or not. The result of performed fieldwork is used to identify and transformed into Bortle Dark-Sky Scale so that the light pollution can be classified. The adapted fieldwork is based on the imagery, that is low and high radiance value. Fieldwork was also performed by observing astronomic objects. Sky photography was also conducted at 00.00- 03.00 p.m.-the most effective time for astronomical observation-. The camera which used was Canon EOS 600D with f-stop of f/3.5; exposure time 30 sec; ISO speed ISO-1600; Focal length of 10 mm; and partial metering mode. The checklist which was used contained information of coordinates, area description, land cover description, circumstances description, astronomic objects that can be identified, and location sketch were used for area analysis.

3.3 Bortle Dark-Sky Scale interpretation

The Bortle Dark-Sky Scale is a nine-level numerical scale that measures the night sky's brightness of a particular locations. It quantities the astronomical observably of celestial objects and the interference caused by light pollution. John E. Bortle created the scale and published it in the February 2001 edition of sky & telescope magazine to help amateur astronomers evaluate the darkness of an observing site, and secondarily, to compare the darkness of observing sites. The scale range from Class 1, the darkest skies available on Earth, through Class 9, inner-city skies. It gives several criteria for each level beyond naked-eye limiting magnitude (NELM). The accuracy and utility of the scale have been questioned in recent research. Naked eye in astronomy is the visibility of astronomic objects is strongly affected by light pollution. Limiting magnitude is the faintest apparent magnitude of a celestial body that is detectable or detected by a given instrument. In some cases, limiting magnitude refers to the utter threshold of detection [9].

Class 1: Excellent dark-sky site. The zodiacal light-gegenschein and zodiacal band (S&T: October 200, page 116) are all visible – the zodiacal light to a striking degree, and the zodiacal band spanning the entire sky. Even with direct vision, the galaxy M33 is an obvious naked-eye object. The Scorpios and Sagittarius region of the Milky Way cast obvious diffuse shadows on the ground. To the unaided eye the limiting magnitude is 7.6 to 8.0 (with effort); the presence of Jupiter or Venus i the sky seem to degrade dark adaptation. Airglow (a very faint, naturally occurring glow most evident within about 15 of the horizon) is readily apparent. With 32-centimeter (12.5 – inch) scope, stars to magnitude 17.5 can be detected with effort, while a 50-cm (20-inch) instrument used with moderate magnification will reach 19th magnitude. If you are observing on a grass-covered field bordered by tress, your telescope, companions, and vehicle are almost totally invisible. This is an observer's Nirvana.

Class 2: Typical truly dark site. Airglow may be weakly apparent along the horizon. M33 is rather easily seen with direct vision. The summer Milky Way is highly structured to the unaided eye, and its brightest parts look like veined marble when viewed with ordinary binoculars. The zodiacal light is still bright enough to cast weak shadows just before dawn and refer dusk, and its color can be seen as distinctly yellowish when compared with the blue-white of the Milky Way. Any clouds in the sky are visible only as dark holes or voids on the starry background. You can see your telescope and surroundings only vaguely, except where the project against the sky. Many of the Messier globular clusters are distinct naked-eye objects. The limiting naked-eye magnitude is as faints as 7.1 to 7.5 while a 32-cm telescope reaches to magnitude 16 or 17.

Class 3: Rural sky. Some indication of light pollution is evident along the horizon. Clouds may appear faintly illuminated in the brightest parts of the sky near the horizon but are dark overhead. The Milky Way still appears complex, and globular clusters such as M4, M5, M15, and M22 are all distinct naked-eye objects. M33 is easy to see with averted vision. The zodiacal light is striking in spring and autumn (when is extends 60 above the horizon after dusk and before dawn) and its color is at least weakly indicated. Your telescope is vaguely apparent at a distance of 20 or 30 feet. The naked-eye limiting magnitude is 6.6 to 7.0 and a 32-cm reflector will reach to 16th magnitude.

Class 4: Rural/suburban transition. Fairly obvious light-pollution domes are apparent over population centers in several directions. The zodiacal light is clearly evident but doesn't even extend halfway to the zenith at the beginning or end of twilight. The Milky Way well above the horizon is still impressive but lacks all but the most obvious structure. M33 is a difficult averted-vision object and is detectable only when at an altitude higher than 50. Clouds in the direction of light-pollution sources are illuminated but only slightly so, and are still dark overhead. You can make out your telescope rather clearly at a distance. The maximum naked-eye limiting magnitude is 6.1 to 6.5, and a 32-cm reflector used with moderate magnification will reveal stars of magnitude 15.5.

Class 5: Suburban sky. Only hints of the zodiacal light are seen on the best spring and autumn nights. The Milky Way is very weak or invisible near the horizon and looks rather washed out overhead. Light sources are evident in most if not all directions. Over most or all of the sky, clouds are quite noticeably brighter than the sky itself. The naked-eye limit is around 5.6 to 6.0, and a 32-cm reflector will reach about magnitude 14.5 to 15.

Class 6: Bright Suburban sky. No trace of the zodiacal light can be seen, even on the best nights. Any indications of the Milky Way are apparent only towards the zenith. The sky with 35 of the horizon glows grayish white. Clouds anywhere in the sky appear fairly bright. You have no trouble seeing eyepieces and telescope accessories on an observing table. M33 is impossible to see without binoculars, and M31 is only modestly apparent to the unaided eye. The naked-eye limit is about 5.5, and a 32-cm telescope used at moderate powers will show stars at magnitude 14.0 to 14.5.

Class 7: Sub-urban/urban transition. The entire sky background has a vague, grayish white hue. Strong light sources are evident in all directions. The Milky Way is totally invisible or nearly so. M44 or M31 may be glimpsed with the unaided eye but are very indistinct. Clouds are brilliantly lit. Even in moderate-size telescopes, the brightest Messier objects are pale ghost of their true selves. The naked-eye limiting magnitude is 5.0 if you really try, and a 32-cm reflector will barely reach 14th magnitude.

Class 8: City sky. The sky glows whitish gray or orangish, and you can read newspaper headlines without difficulty. M31 and M44 may be barely glimpsed by an experienced observer on good nights, and only the bright Messier objects are detectable with a modest-size telescope. Some of the stars making up the familiar constellation patterns are difficult to see or are absent entirely. The naked eye can pick out stars down to magnitude 4.5 at best, if you know just where to look, and the stellar limit for a 32-cm reflector is little better than magnitude 13.

Class 9: Inner-city sky. The entire sky is brightly lit, even at the zenith. Many stars making up familiar constellation figures are invisible, and dim constellations such as Cancer and Pisces are not seen at all. Aside from perhaps the Pleiades, Messier objects are visible to the unaided eye. The only celestial objects that really provide pleasing telescopic views are the Moon, the planets, and a few of the brightest star clusters (if you can find them). The naked-eye limiting magnitude is 4.0 or less.

4. Results and discussions

4.1 VIIRS DNB Free Cloud Composites imagery information

The result of information extraction of VIIRS DNB Free Cloud Composite consists of two main information, which are cloud cover and light pollution (recorded artificial usage on earth). The exposure on earth at night were recorded by the satellite sensor can be directly identified using VIIRS DNB Free Cloud Composite imagery. The radiance value of the imagery is correlated with the amount of recorded lights from earth surface, the higher radiance value, the brighter the pixel hue, and vice versa. The high brightness of the pixel value represents the earth surface condition which is brighter due to amount of artificial lights. In contrast, the regions which look darker are light pollution minimum. The distribution of location with quite plenty use of lights is shown in **Figure 2.** The area which have bright hue indicates an excessive use of lights on Northern coastal of study area i.e Northern coastal of Java and on the Southern, Yogyakarta and Surakarta. On the other hand, most of study area has minimum lighting condition. This surely can affect the light pollution level that occurs on earth surface.

The cloud cover information which obtained from extraction of DNB Free Cloud Composite shows the amount of clear sky of an area which is not covered by clouds. **Figure 3** represents clouds cover information. The lowest radiance value of the imagery of 0 while the highest value of 17 . The high pixel value shows that the area is rarely covered by clouds and vice versa. The area which often covered by clouds in the imagery is visualized with darker hue, compared with the area which rarely covered by clouds. Spatial distribution of region which rarely covered by clouds is on Northern of study area, which are Wonogiri, Blora, Gunungkidul, Rembang, and Grobogan regions. This recorded clouds coverage information is very useful as one one of factors of light pollution level.

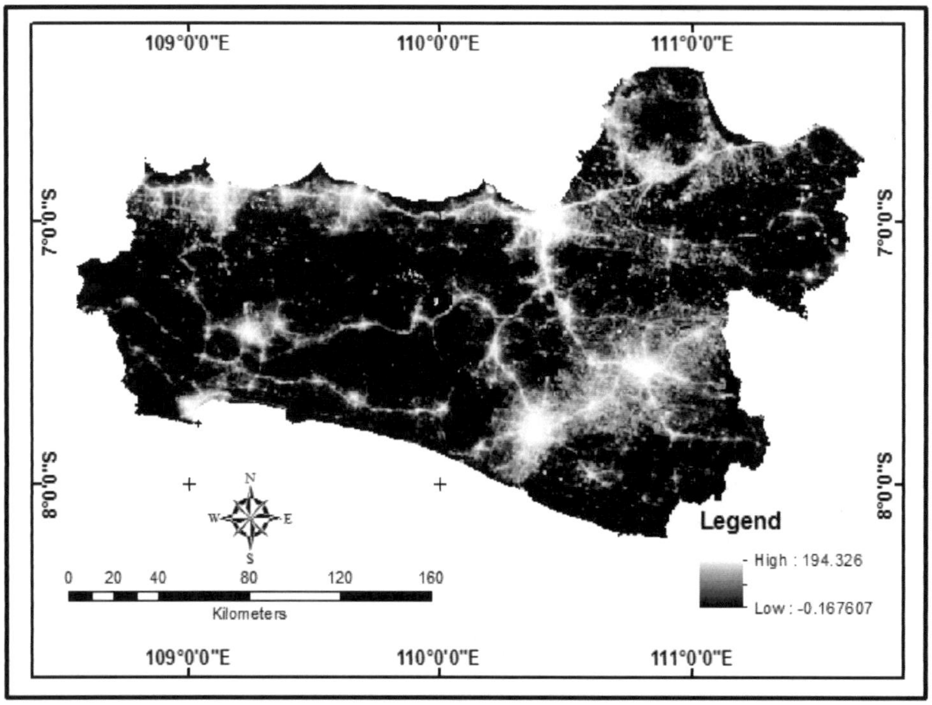

Figure 2. Light radiance information of VIIRS DNB Free Cloud Composites Imagery.

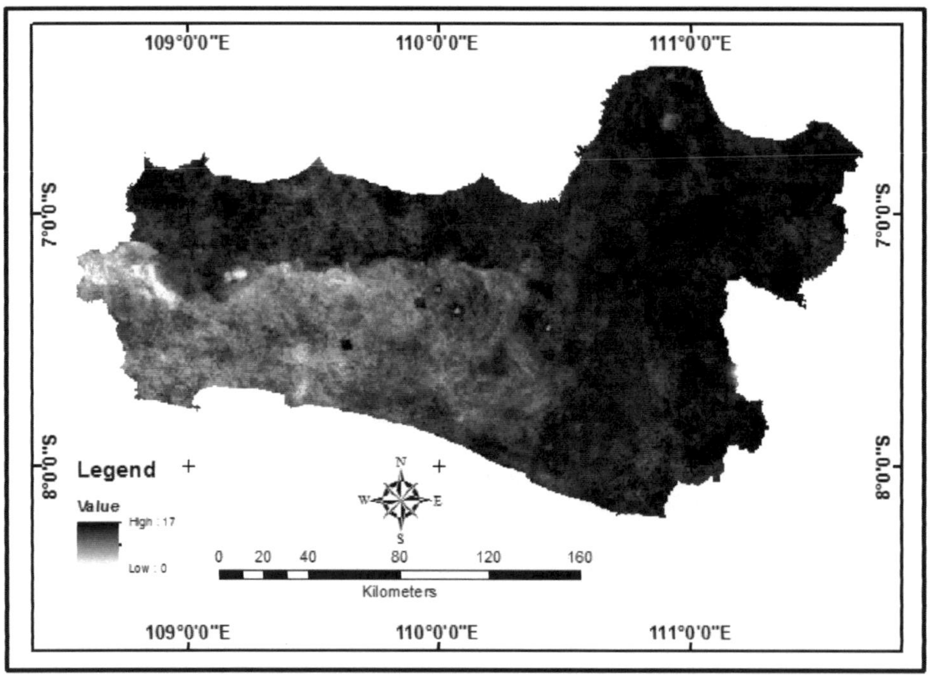

Figure 3. Information of cloud coverage of VIIRS DNB Free Cloud Composites.

4.2 Fieldwork

The night sky condition of Gajah Mungkur reservoir is showed in **Figure 4**. The radiance value of the area is in range of 0.207987 to 0.69997 x10^{-9} watt/cm^2*sr and in range of 11 to 13 for cloud coverage information. **Figure 4** shows the night sky condition at Gajah Mungkur reservoir. Fieldwork result shows that night sky condition at Gajah Mungkur reservoir is free from disruption of artificial lights, or in other word is light pollution minimum. For astronomical observation, the Gajah Mungkur reservoir is observable without optical aids such as Telescope. Many of astronomic objects such as planet (Mars), space dust, Milky Way, and other objects can be observed clearly, including constellations. The result of Bortle Dark-Sky Scale interpretation shows that the area is classified as **Class 4**, which is Rural/Suburban transition. This is marked by galaxy of Milky Way which could be seen clearly under horizon. Generally, other astronomic objects also can be identified. The result of performed fieldwork gives information that light pollution of Gajah Mungkur reservoir and surroundings came from local activities such as fishermen, industries, and etc.

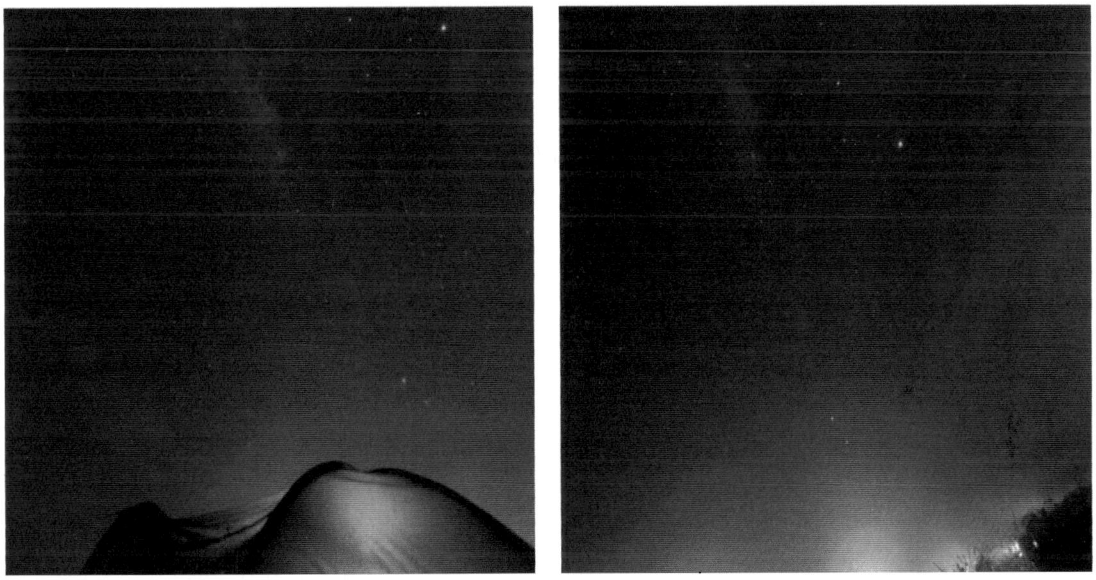

Figure 4. Night sky condition in Gajah Mungkur reservoir.

Figure 5 shows the night sky condition at Depok Beach. The night sky condition in Depok Beach and surroundings is very low of light pollution. Compared with night sky at Gajah Mungkur reservoir, this area is much better because the appearance of astronomic objects is very clear. Based on the imagery, the radiance value of the location of 0.406948 x10^{-9} watt/cm^2*sr to 0.554071 x10^{-9} watt/cm^2*sr and for cloud coverage information is up to 13, which means that there were 13 times of clear sky in a month (September). The result of Bortle Dark-Sky Scale interpretation for this location shows that is classified as **Class 3** or Rural Sky. Rural sky condition is indicated by very minimum of light pollution. Many of astronomic objects can be identified easily, including planets (Mars and Saturn) and constellations such as Scorpion, Indian, Peacock, Altar, Sea Goat, and etc. **Figure 6** shows some of astronomic objects that can be identified.

IOP Conf. Series: Earth and Environmental Science 47 (2016) 012040 doi:10.1088/1755-1315/47/1/012040

Figure 5. Night sky condition in Depok Beach.

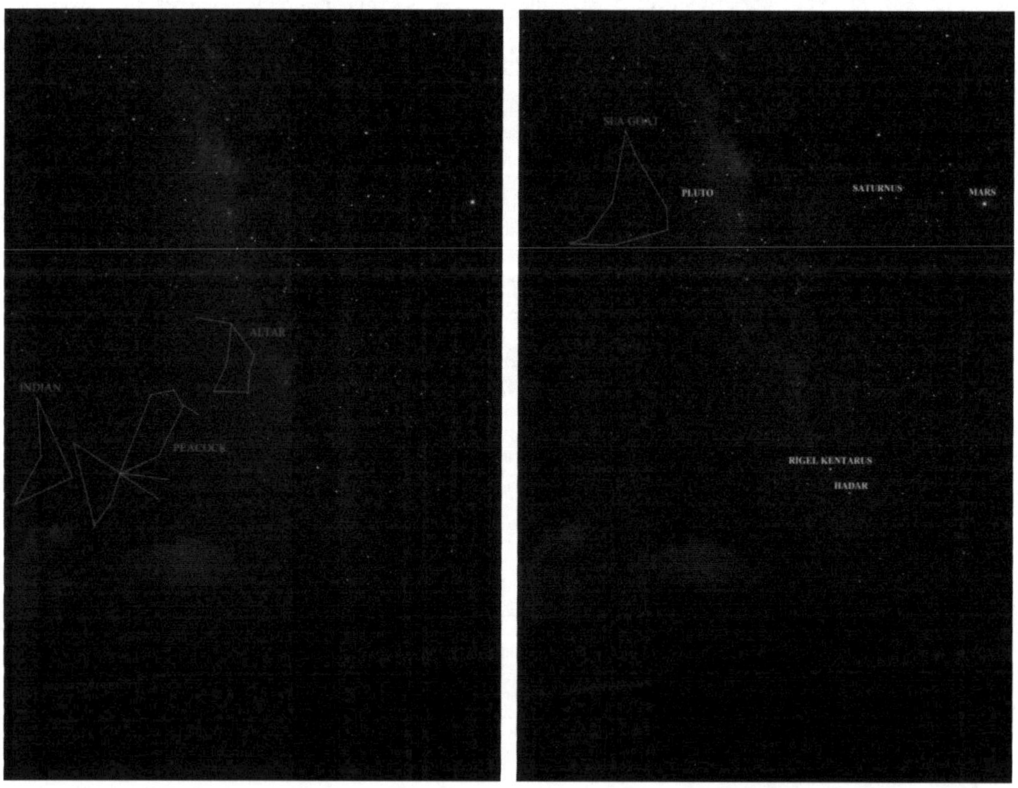

Figure 6. Identification of astronomic object at Night Sky in Depok Beach (Constellations of India, Peacock, Altar, and Seagot; Pluto, Saturnus, Mars, Rigel Kentarus, and Hadar).

Yogyakarta City is one of the location for performing fieldwork that has high lighting condition. **Figure** 7 shows the night sky condition at Yogyakarta, which has high illuminations of reflectance from earth surface. In the VIIRS imagery, this location has quite high radiance value of 23.814596 watt/cm^2*sr to 44.069031 watt/cm^2*sr and cloud coverage information of 9 to 13. Based on Bortle Dark-Sky Scale, this location is classified as Class 8, which is City Sky. The sky even looks orange that indicates the amount of artificial lights usage, or in other word, this location had been disrupted by light pollution. It is very difficult to identify astronomic objects here, compared with the other location (Gajah Mungkur reservoir and Depok Beach). The high light pollution level is a result of artificial lights usage, which most of them is caused by settlements. There are some reasons why the location had a high light pollution level: high settlements density that result a high electricity consumption, the use of decorating lights and infrastructure, and traffic activities.

Figure 7. Condition night sky in Yogyakarta City.

5. Conclusions

VIIRS DNB Free Cloud Composites Imagery can be used for identification and analysis of light pollution. The imagery contains information of light pollution (recorded artificial lights usage on earth) and cloud coverage. Bortle Dark-Sky Scale is one of several methods that can be used to determine the level of light pollution by considering the radiance value of earth surface. Based on the VIIRS imagery, the radiance value of Gajah Mungkur reservoir is in range of 0.207987 to 0.69997 x10^{-9} watt/cm^2*sr, which is classified as Rural/suburban transition (Class 4) in Bortle Dark-Sky Scale. The radiance value in Depok Beach of 0.406948 to 0.554071 x 10-9 watt/cm2 *s, which is classified as Rural (Class 3), and the radiance value of Yogyakarta City is up to 23.814596 watt/cm^2*sr to 44.069031 watt/cm^2*sr, which is classified as City Sky (Class 8).

For future research, it is better to use more sample sites so the distribution of night sky condition can be identified accurately. It is also crucial to conduct measurement of night sky by using Sky Quality Meter to get the information of light pollution level quantitatively. Temporarily, this research can be used for identifying night sky dynamics in study region.

Acknowledgements
This research is funded by the Ministry of Research, Technology, and Higher Education of Indonesia through Program Kreativitas Mahasiswa 2015/2016. VIIRS DNB Free Cloud Composites imageries are obtained for free from official website of NOAA. We would like to express our greatest gratitude to Muhammad Kamal, Ph.D. who has been willing to be mentor of this research. And also to all member of research who have been involved and friends who have supported this research.

References

[1] International Dark-Sky Association 2009 *IDA Mission and Goals*, retrived June 02, 2016 from http://www.darksky.org/mc/page.do?sitePageId=56411&orgId=idsa
[2] Longcore T and Rich C 2004 *Ecological Light Pollution Frontiers in Ecology and The Environment* **2**(4) 191-198
[3] Nagy Pinterne 2012 Light Pollution and its Effect on the Living World *International Scientific Conference on Sustainable Development & Ecological Footprint*, Marc 26-27 2012 Sopron, Hungary.
[4] Shaflik C 1997 Technical paper prepared at University of British Columbia, Dept. of Civil Engineering **p. 9.**
[5] Borg V (1996) *Geogr. Mag.* 68, 56
[6] Rabaza O, Galadi-Enriquez D, Estrella A, Espin Dols, F Aznar 2010 All-Sky Brightness Monitoring of Light Pollution with Astronomical Methods, *Journal of Environmental Management* **91** 1278-1287
[7] Pemda Yogyakarta (2010) *Kondisi Geografis Yogyakarta* Retrived 05 Juni 2016 from jogjaprov.go.id/pemerintahan/situs-tautan/view/kondisi-geografis
[8] Baugh K 2015 *VIIRS Day-Night Band Cloud-free Composites* (Colorado, USA)
[9] Bortle John E 2001 The Brotle Dark-Sky Scale Sky & telescope, Sky Publishing Corporation: Retrieved 2013-02020

Impact of human interventions on mangrove ecosystem in spatial perspective

Abd. Rasyid [1], M. Akbar AS [2], Nurjannah Nurdin [1], 2, Ilham Jaya [1], Ibrahim [2]

[1] Marine Science and Fisheries Faculty, Hasanuddin University, Jl. Perintis Kemerdekaan 10, Makassar, 90245. Indonesia.

[2] Center for Regional Development and Spatial Information, Hasanuddin University, Jl. Perintis Kemerdekaan 10, Makassar, 90245. Indonesia

Email: fayufi@yahoo.com

Abstract : Climate change components that affect mangroves include changes in sea level, high water events, storms, precipitation, temperature, and oceanic circulation. Cumulative impacts of these factors have a distinct synergy with respect to an accelerated rate of mangrove degradation. development of coastal agricultural land and shrimp farming ponds in intertidal areas are considered as the major factors behind mangrove delineation. There is no agreement for the extensive of mangrove forest in Indonesia, but invarious forums it is usually used the number of 4.25 million ha for that. At approximately 9 years ago, the extensive vast of mangrove forest in Indonesia was about 4.13 million ha but now it is only 2.49 million ha (60%). The methodology of this study consisted of two main steps: (1) image analysis, and (2) post classification. The aim of this study is to analysis the human intervention base on spatial dynamic of mangrove. The study site was selected Sagara island in Spermonde Archipelago. The results of analyze shows that decreasing of mangrove is caused human intervention.

Keywords: mangrove, fishpond, landsat, Spermonde, spatial

1. Introduction

1.1 Background

Mangroves are considered to behave like a natural barrier against ocean dynamics along the shoreline. Their ability to protect shoreline and inland areas from natural hazards (hurricanes, cyclones, tsunamis) was recently discussed [2]. They can break the force of waves and help to prevent coastal-erosion processes [5]. Mangrove ecosystems support aquatic food chains and form habitats for marine fauna, such as juvenile crabs, prawns, offshore fish, reef fish, and larvae [7].

South Sulawesi is facing massive mangrove forest destruction, for over the past 30 years deforestation and pollution have taken their toll and damaged almost 90 percent of the total original areas. Before 1980s, the mangrove forest in South Sulawesi was over 214,000 hectares. The conversion of the mangrove forest into fishpond reduced the area into 23,000 hectares in the early 1991, or a decrease of 61 percent. The conversion of mangrove forests into fish farms was a result of a government policy in the early 1980s that instructed the offices of local fisheries and maritime affairs to carry out intensification of shrimp cultivation, which was then a major export commodity and earned the state a huge amount of money.

Deforestation led by increasing demand for land and climate change events such as the rise of sea level and reduction in freshwater flow are considered as major players behind the continuous annihilation of mangrove; however, climate change events may have an impact of only 10-15% reduction of mangrove habitats in the distant future whereas the immediate threat comes from uncontrolled exploitation and deforestation [1]. This has resulted in serious concerns among conservationists and has exposed the coastal communities to a further increasing threat of climate change and hydro meteorological disasters.

Remote sensing is an indispensable tool for assessing and monitoring mangrove forests primarily because many mangrove swamps are inaccessible or difficult to reach for a field survey. Remote sensing data providing synoptic coverage and those of historical satellites dating back to the 1960s are available. Global mapping initiatives have failed to map the extent and rate of deforestation with sufficient details. For example, only extensive mangrove areas were mapped as part of the Global Land Cover of 2000 survey [10]. The quantitative information achieved from interpretation image might be useful for forest managers to devise better plans for long-term management of mangrove ecosystems in Sagara island.

1.2 Objectives

The aim of this study is to analysis the human intervention base on spatial dynamic of mangrove.

2. Methodology

2.1 Study area

The study area is located in Sagara island, Pangkajene District, South Sulawesi, Indonesia. The geographical boundary of the study area between 119^{0}127'18"East and 04^{0}42'9" South (Figure 1).

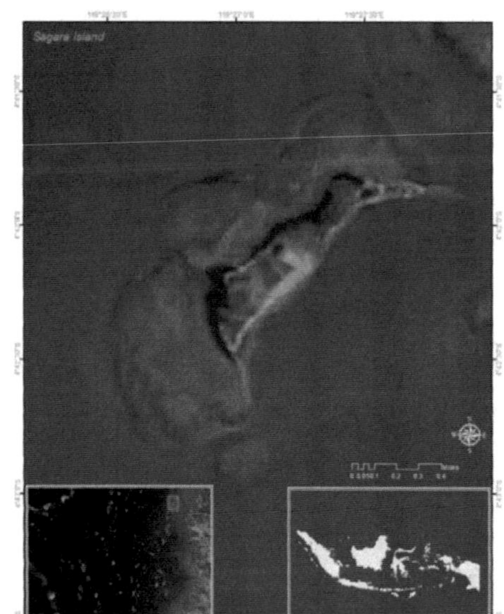

Figure 1. Sagara island is one of the small islands in Spermonde Archipelago, Indonesia as study area

2.2 Data collection

We used Landsat GeoCover and recently acquired a Landsat Thematic Mapper (TM) in 1990 and Operational Land Imager (OLI) data in 2015, made available through the US Geological Survey (USGS) Center for Earth Resources Observation and Science (EROS). Landsat TM data have seven spectral bands, with a spatial resolution of 30 m for bands 1–5 and 7. The Landsat 8 OLI-TIRS data have nine spectral bands with a spatial resolution of 30 m for bands 1-7 and 9.

2.3 Image Processing

Landsat satellite imagery t=1990 and t=2015 were being corrected in radiometric aspect and geometric aspect by rectify/resampling pixel position in image, and geo-referencing the image into the selected projection system. Optical band of imageries had 30 m of resolution. There were several characteristics in interpretation such as tone, pattern, texture, shape, size, shadow and association. In this study, the process was done by visual classification method to obtain land use for two time series. The combination bands of 453 is commonly used in the interpretation of satellite image. This band combination is used because in the visual interpretation, the appearance of natural vegetation and planted areas can be seen clearly. Accuracy assessment was being conducted by Kappa coefficient (κ) for accuracy assessment which relies on image training area. Image validation was counted based on the above table such as *Overall Accuracy* (OA), *Producer's Accuracy* (PA), and *User's Accuracy* (UA). OA is a percentage of sample units that were classified accurately. A classification scheme was developed to obtain a broad level of classification, to derive various land use classes, i.e., mangrove, fishfond, bare land and settlement. A post-classification change matrix function was applied between 1990 to 2015.

3. Result And Discussion

3.1 Spatial Distribution of Mangrove

Textural and spectral characteristics of the canopy and leaves are the main features used to distinguish among mangrove communities [8]. Their structural appearance, partially more homogeneous or heterogeneous, depends on several factors, such as species composition, distribution pattern, growth form, density growth, and stand height. The spectral variations of the canopy reflectance as a function of several optical properties, such as leaf area index (LAI), background reflectance, and leaf inclination [6]. Identification of mangrove in Sagara island using Landsat image selected Landsat TM and Landsat 8 to the years 1990 and 2015. The classification result for the 1990 and 2015 can be seen in Figure 2.

Table 1. Areal and percentages of land use/cover changes

Classes	Mangrove area (ha)		Mangrove change (ha) 1990 to 2015	
	1990	2015	Ha	%
Fishpond	26,71	39,88	(+)13,17	33,03
Mangrove	21,02	10,94	(-)10,08	47,97
Setlement	4,03	6,54	(+)2,51	38,39
BareLand	6,96	2,17	(-)4,79	68,77

Monitoring deforestation at a regional scale using moderate resolution satellite images over a long period of time requires the processing of large volumes of data. By classification of image in 1990 and 2015, the resulst of mangrove for two time periods are presented in Table 1. Time-series analysis of

TM and OLI-TIRS data has revealed a net loss of 10,08 ha (47.97%) of mangrove in the region from 1990 to 2015 and the area of fish fond and setlement had increased by 13.17 ha and 2.51 ha.

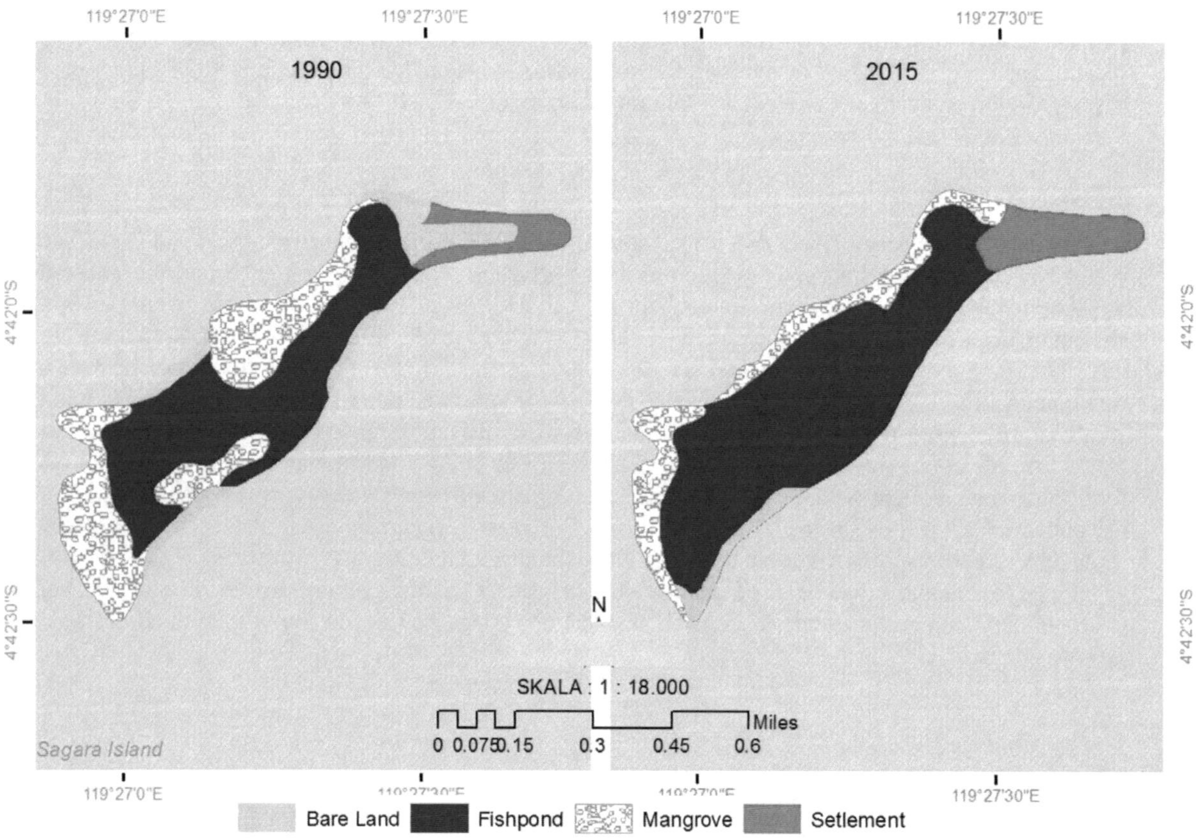

Figure 2. Mangrove classification maps in 1990 and 2015

While the measured net loss of mangrove forest is not that high, the change matrix (Table 2) shows that turnover was much greater than net change. mangrove changed to fishpond and bare land by 2016. The largest category of mangrove change was loss to fishfond by 8,05 ha.

Table 2. Areal change of mangrove during 25 years

Years in 1990 to 2015	Areal change of mangrove (ha)				
	Fishpond	Mangrove	Setlement	Bare land	Sea
Fishpond	26.71	0	0	0	0
Mangrove	12.07	8.05	0	0.9	0
Setlement	0	0	4.03	0	0
Bare land	1.11	2.07	2.51	1.27	0
Sea	0	0.81	0	0	0

3.2 Human Intervention of Mangrove

Monitoring deforestation at a regional scale using moderate resolution satellite images over a long period of time requires the processing of large volumes of data. We used simple but such estimates on the sustainability of mangrove resources may contribute to the evaluation of impacts of mangrove degradation on socio-economic systems [3]. According to hierarchy theory [4], processes at a particular organization level can be explained by constraints at higher levels along with mechanisms at lower levels of organization. Thus, it is essential to evaluate the climatic and landform characteristics of coastal regions which result in local and often gradual environmental gradients that represent top-down constraints of mangrove forest development. At the same time, tree performance, growth response, and interactions among trees affect bottom-up patterns of forest development [9]. Clearing forests for farms on a large scale by investors from the main land of Pangkajene district in cooperation with the local community in Sagara island resulted in loss of mangrove. Factors influencing loss of mangrove in Sagara island from the human intervention are conversion of mangrove for fishpond on a large scale and utilization of mangrove trees for domestic use. Other abandoned fishpond areas are very difficult to rehabilitate or regenerate, mainly because these sites are highly degraded by pollution and pesticides. Delineations of degraded mangrove forests are needed to promote regrowth and enrichment planting.

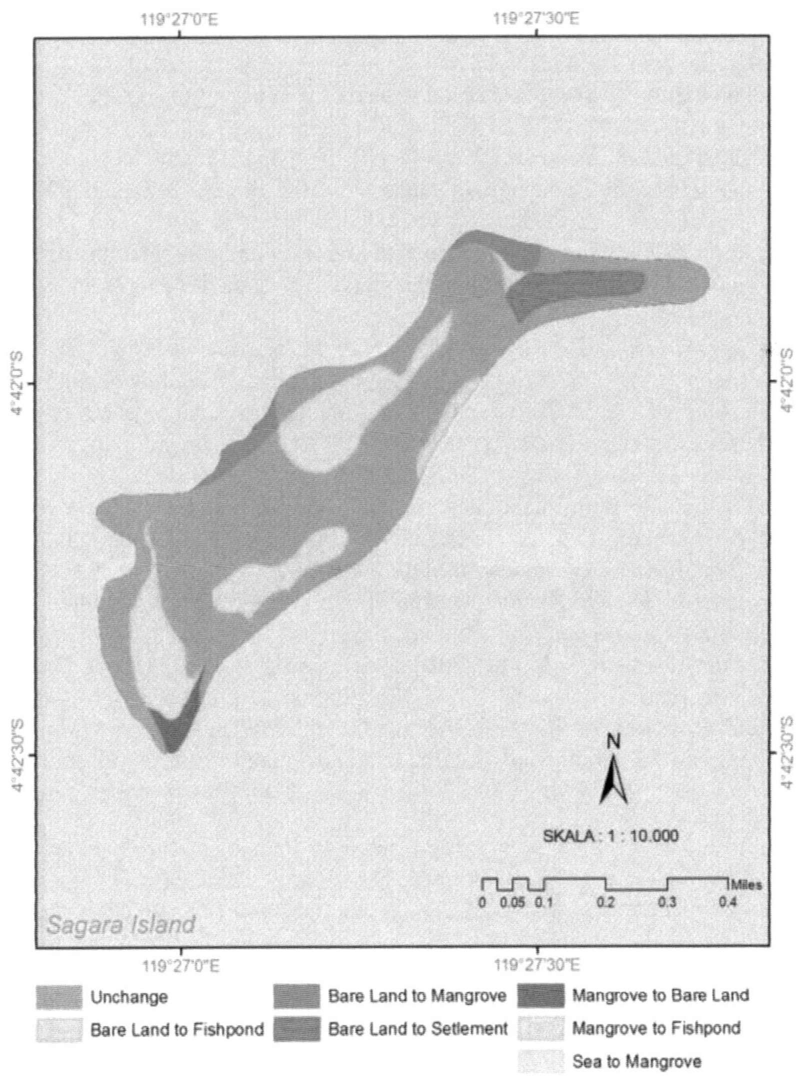

Fifure 3. Post classification of mangrove in Sagara island of 1990 to 2015

4. Conclusion

Our measure of extent for the Sundarbans mangrove forest shows large change in net area (approximately 33.03% loss) in the last 25 years. Factors influencing loss of mangrove in Sagara island from the human intervention are conversion of mangrove for fishpond, clearing of mangrove trees for charcoal production on a large scale, and utilization of mangrove trees for domestic use.

Acknowledgement

We would like to thank to Ministry of Research Technology and Higher Education, Indonesia, for funding sponsor. We gratefully acknowledge Regional Development and Spatial Information, Hasanuddin University for the field survey team support.

REFERENCES

[1] Alongi D M 2008 Mangrove forests Resilience; protection from tsunamis and responses to global climate change *Estuar Coast.Shelf Sci* **76** 1-13
[2] Barbier E B 2006 Natural barriers to natural disasters Replanting mangroves after tsunami *Front Ecol Environ* **4** 124-131.
[3] Davis SM Childers DL, Lorenz JJ Wanless HR Hopkins TE 2005 A conceptual model of ecological interactions in the mangrove estuaries of the Florida Everglades Wetlands **25** 832–842
[4] Holker F Breckling B 2002 Concepts of scales, hierarchies and emergent properties in ecological models In Holker F (Ed.) Scales Hierarchies and Emergent Properties in Ecological Models Peter Lang Frankfurt pp 7–27
[5] Mazda Y, Magi M, Nanao H, Kogo M, Toyohiko M, Kanazawa N, Kobashi D 2002 Coastal erosion due to long-term human impact on mangrove forests *Wetlands Ecol Manage* **10** 1-9.
[6] Meza D B Blackburn G A 2003 Remote sensing of mangrove biophysical properties Evidence from a laboratory simulation of the possible effects of background variation on spectral vegetation indices *Int. J Remote Sens 24* 53-73
[7] Nagelkerken I, Blaber S J, Bouillon S, Green P, Haywood M, Kirton L G, Meynecke J O, Pawlik J, Penrose H M, Sasekumar A, Somerfield P J. 2008 The habit function of mangroves for terrestrial and marine fauna A review *Aquat Bot* **89** 155-185
[8] Ramsey E W, Jensen J R 1996 Remote sensing of mangrove wetlands Relating canopy spectra to site-specific data *Photogramm. Eng Remote Sensing.* **62** 939-948
[9] Smith. 1992. Forest structure. In: Robertson, A.J., Alongi, D.M. (Eds.), Tropical Mangrove Ecosystems. American Geophysical Union, Washington, DC, pp. 101–136
[10] Stibig, Belward, Roy, Rosalina, Wasrin, Agrawal, Joshi, Hildanus B, Fritz, Mubareka, Giri 2007 A land-cover map for South and Southeast Asia derived from SPOT-VEGETATION data *Journal of Biogeography* **34** 625–637

Comparison Effectiveness of Pixel Based Classification and Object Based Classification Using High Resolution Image In Floristic Composition Mapping (Study Case: Gunung Tidar Magelang City)

Prama Ardha Aryaguna[1], Projo Danoedoro[2]

[1] Master of Remote Sensing Program, Faculty of Geography, University of Gadjah Mada, Yogyakarta, Indonesia
[2] Lecture Faculty of Geography, University of Gadjah Mada, Yogyakarta, Indonesia

[1] pramaardhaaryaguna23@gmail.com, [2] projo.danoedoro@geo.ugm.ac.id

Abstract. Developments of analysis remote sensing have same way with development of technology especially in sensor and plane. Now, a lot of image have high spatial and radiometric resolution, that's why a lot information. Vegetation object analysis such floristic composition got a lot advantage of that development. Floristic composition can be interpreted using a lot of method such pixel based classification and object based classification. The problems for pixel based method on high spatial resolution image are salt and paper who appear in result of classification. The purpose of this research are compare effectiveness between pixel based classification and object based classification for composition vegetation mapping on high resolution image Worldview-2. The results show that pixel based classification using majority 5x5 kernel windows give the highest accuracy between another classifications. The highest accuracy is 73.32% from image Worldview-2 are being radiometric corrected level surface reflectance, but for overall accuracy in every class, object based are the best between another methods. Reviewed from effectiveness aspect, pixel based are more effective then object based for vegetation composition mapping in Tidar forest.

Keywords: Remote sensing, Object based classification, Pixel based classification, Worldview-2, Floristic composition, effectiveness

1. Introduction

Multispectral classification mostly being used for Landover and land use mapping. Forest study itself more emphasis on extensive mapping, monitoring, biophysical and structure. The classification for more detailed object such as floristic composition mapping still rarely performed. Floristic composition mapping is seen from the structure of the leaf and canopy appearance difference is very thin, between vegetation, so it's needed remote sensing imagery with high spatial resolution and spectral resolution. Type of forest in tropical Indonesia, mostly heterogeneous with irregularity pattern grows in the forest. This heterogeneous manifold causes its own difficulties in mapping floristic composition using digital analysis.

Classification using high resolution image very good for visual interpretation because the object looks clear, but using digital classification such pixel based classification and object based classification

IOP Conf. Series: Earth and Environmental Science **47** (2016) 012042 doi:10.1088/1755-1315/47/1/012042

be trend for today. Every method have advantage and weakness in classification result. Visual interpretation will give accuracy but need a lot of time and effort especially if the large area of interest. Pixel based classification often being used than object based classification because pixel based classification only consider spectral value or one aspect for boundary class and that's make this classification more easy and fast. The problems occur when pixel size and spatial resolution very detail so salt and paper will occur, which shadow effect or sun angel could make same object will be different because spectral value different.

Along with development of satellite technologies, occur new method classification that is object based classification. This classification not depend on one aspect in every pixel but this classification considering another aspect such texture, scale and color and also this method using cluster pixel or segment to classing information. Different characteristic between pixel based classification and object classification makes it possible produce different accuracy in classification result. Object based classification itself is also not perfect [5], object based classification still have problems when processing very large data. Even if the object based classification is more effective than pixel-based classification, segmentation process performed on the multispectral image pixel-based classification is a process that is too heavy. The processing time and different difficulty levels become considered in choosing the method of classification, which accuracy result will more significant or not, compare to time processing and difficulty level and can be declared as effective method for composition vegetation mapping.

2. Material and Study Area

2.1. Study area

Location of Tidar forest in middle of Magelang city with geographic location in 110°12'49.95"–110°13'22.99" eastern and 7°29'34.09"- 7°30'7.85" southern (Figure 1). Tidar forest have purpose for forest city. Tidar forest was located in two district which are Magersari district and Jurangombo district. In 2008, Tidar forest have 701,673 m^2, but in 2013, but in 2013, area of Tidar forest got reduction to 690,526 m^2 because land conversion into golf area and settlement. Tidar forest are heterogeneous type forest because it consist of several different types vegetation just like in Table 1.

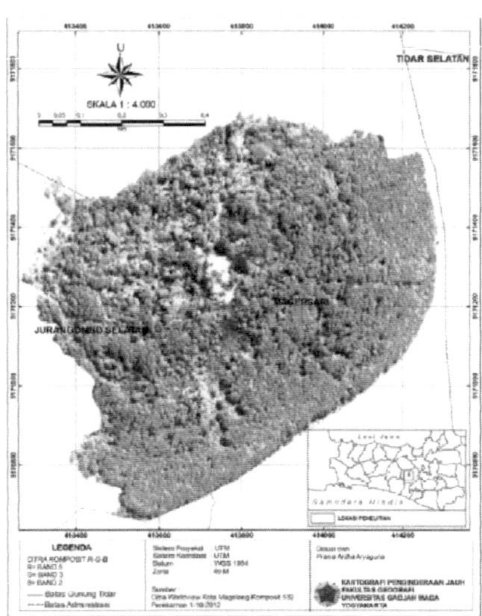

Figure 1. Study area Tidar forest

Tabel 1. Tidar forest vegetation variation

No.	Vegetation latin name	Nama Daerah Tumbuhan
1	*Delonix regia*	Flamboyan
2	*Swietenia macrophylla*	Mahoni
3	*Albitzia lopantha*	Sengon jowo
4	*Albitzia paraserianthes*	Sengon laut
5	*Leucaena glauca*	Lamtoro, manding
6	*pteris enchyformis*	Pakis
7	*Glyseridia sp*	Kleresida
8	*Ficus benyamina*	Beringin
9	*Tamarindus indica*	Asem Jawa
10	*Pinus mercusii*	Pinus
11	*Bischofia javanica*	Gadog (Meranti Putih)
12	*Hibiscus similis Bl*	Waru gombong
13	*Calliandra calothyrsus*	Kaliandra
14	*Spathodea campanulata*	Sepatu Dia

Source: Field survey, June 2013

3.1. Preparing data

Preparing primary data such Worldview-2 image and secondary data such statistic data and administration boundary. Before using Worldview-2 image, first step to do was doing geometric and radiometric correction and make a boundary for area interest. Geometric correction needed because research location have hill topography so the position not like a real condition.

Table 2. Floristic composition

Class Floristic Composition	Vegetation Species Description (Local name)									
	Pinus	Mahoni	Beringin	Sengon	Gombong	Aren	Kaliandra	Gadog	Sepatu Dea	Klereside
A			√							
B								√		
C				√						
D						√	√			√
E	√					√				
F	√									
G									√	
H				√	√					
I					√					
J		√								

3.2. Pre-survey data

This step is make a tentative map for mapping unit field survey and class floristic composition (Table 2). The method for make tentative map are visual interpretation because when use unsupervised classification the result will came salt and paper just like Figure 2.

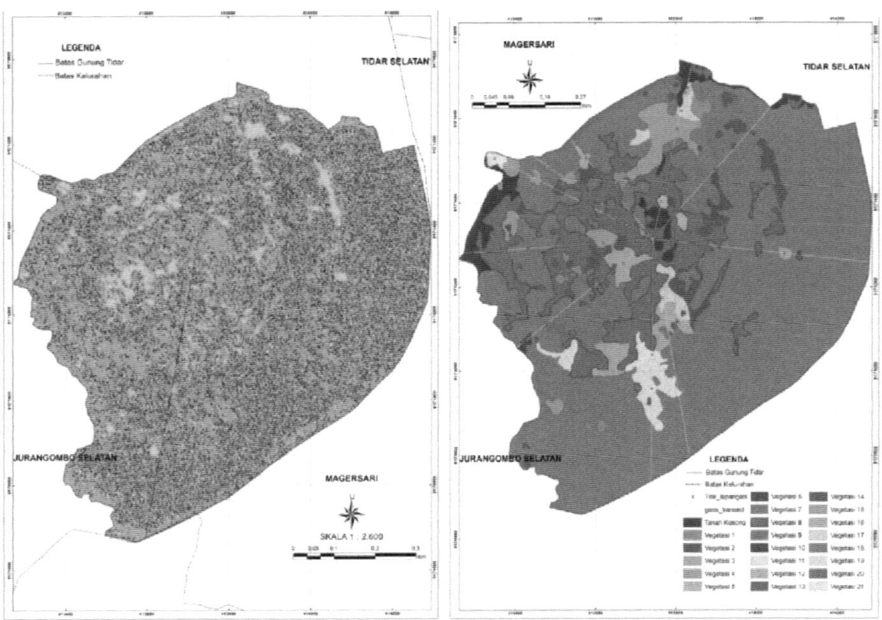

(a) (b)

Figure 2. Pre-Survey maps (a) Using Unsupervised (ISODATA) method, and (b) Using Visual Interpretation

3.3. Field Survey

The field survey aim is to verification floristic composition in field for input data classification and accuracy assessment. Sampling method using stratified random sampling (Figure 3). Information which take from field survey is:

1. Coordinate sample using GPS
2. Floristic composition sample such leaf, density pattern and dominant vegetation.
3. Sketch of Composition vegetation sample location
4. Field photo

Figure 3. Location of samples

3.4. Pixel Based Classification

The algorithm for pixel based classification using MLC (Maximum likelihood). This method are supervised classification which needed sample for classification. Input sample got from field survey. Step of pixel based classification is :

a. Training area : This step is to prepare key interpretation
b. Classification : Every pixel compare with sample for every category
c. Output : Layouting for result classification

3.5. Object Based Classification

Object based classification method not only see from spectral aspect but also from shape and texture. Object based classification using segmentation method. This method divide spectral to be segment-segment. Step of object based classification are :

a. Make Segment : This step is to divide pixel to cluster pixel
b. Training segment : Segmentation process depend on network level-level from parameter scale.
c. Classification : This function is classing the segment depend on classification result

3.6. Assessment Accuracy

Accuracy assessment using confusion matrix table to count accuracy result (producer accuracy), user accuracy and overall accuracy.

4. Result and Discussion

Class of floristic composition got reduction from 21 class be 13 class with 10 class of floristic composition and 3 class of non-floristic composition. Input image of both classification from 3 kind of radiometric correction. Before do the classification, sample must be do separability with every image. Class B and C have the smallest separability 0.8 because the similarity of both composition canopy. Pixel based classification using non correction data have 63.56% for overall accuracy. Class J have small accuracy for individual class floristic composition accuracy that is 2.22%. Pixel based using TOA reflectance image have better overall accuracy than non-correction 65,09% and pixel based classification using surface reflectance image have the best accuracy 66.22% but for class J reduce to be 1.80%. From 3 different input radiometric correction, the result show the best accuracy from surface reflectance but that's still not improve accuracy for a few class such class J and I. Class J with species Mahoni have small accuracy because this vegetation was not to dominant in this location so input sample very few and Mahoni canopy looks same like (Figure 4)

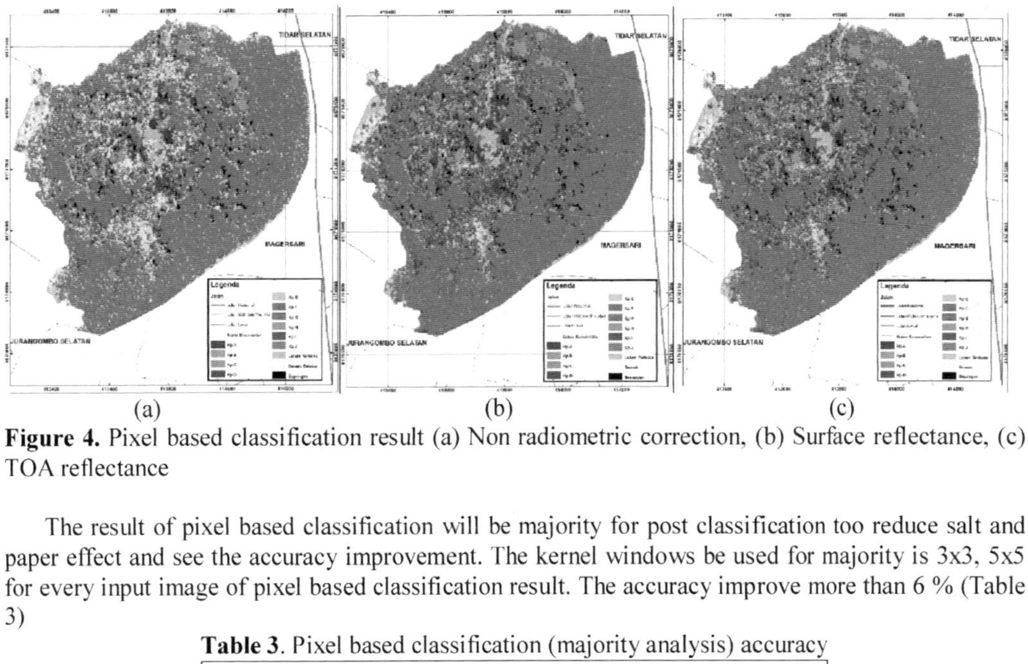

(a) (b) (c)

Figure 4. Pixel based classification result (a) Non radiometric correction, (b) Surface reflectance, (c) TOA reflectance

The result of pixel based classification will be majority for post classification too reduce salt and paper effect and see the accuracy improvement. The kernel windows be used for majority is 3x3, 5x5 for every input image of pixel based classification result. The accuracy improve more than 6 % (Table 3)

Table 3. Pixel based classification (majority analysis) accuracy

Pixel Based Classification (Majority Analysis)					
Original		Toa Reflectance		Surface Reflectance	
3x3	5x5	3x3	5x5	3x3	5x5
70,75%	72,75%	69,94%	69,94%	71,16%	73,32%

Second classification is object based classification. This classification have 2 step, first is segmentation process and second is classification process. Every wavelength have different respond for every object, for vegetation green, yellow, red edge, NIR and NIR2 have sensibility for vegetation. In segmentation every spectral column can be given different weight.

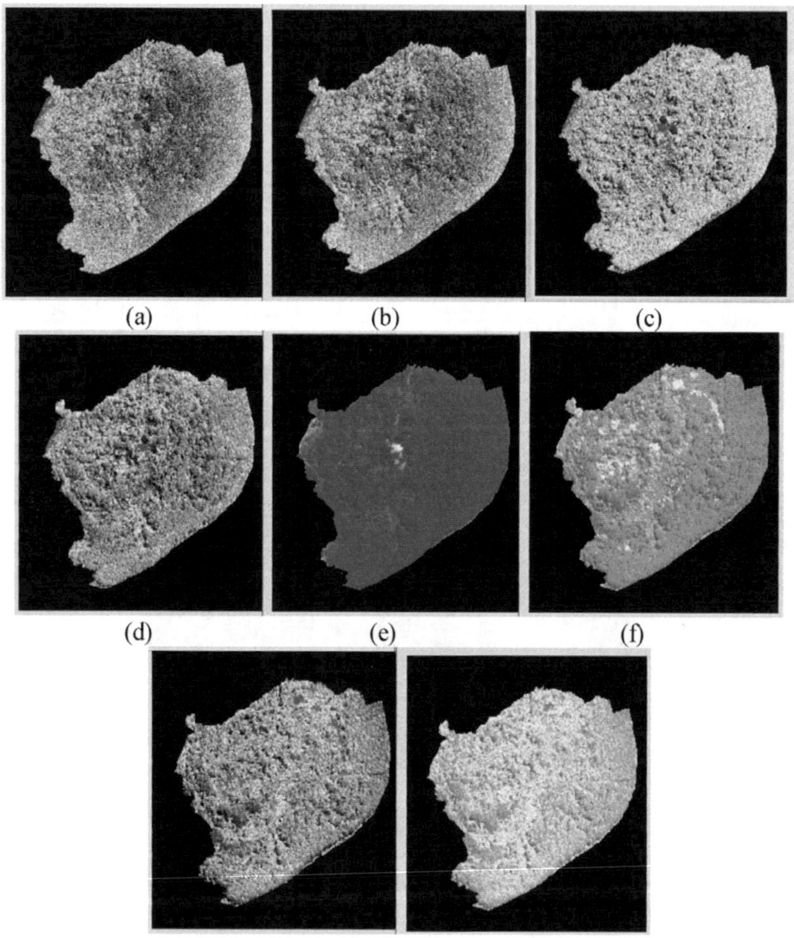

(a) (b) (c) (d) (e) (f) (g) (h)

Figure 5. Image every wavelength for vegetation (a) Cyan, (b) Blue, (c) Green, (d) Yellow, (e) Red, (f) Red Edge, (g) NIR1, (h) NIR2

The weight every wavelength can be subjective depend on user from try and error. In this research weight for segmentation can be seen in Table 4. Mean and variance weighted use comparison 2:8, because object of interpretation very detailed.

Table 4. Weighted Wavelength

Band	Cyan	Blue	Green	Yellow	Red	Red Edge	NIR1	NIR2
Weighted	0,005	0,005	0,005	0,245	0,005	0,245	0,245	0,245

Parameter for segmentation not only from weighted every wavelength but also similarity tolerance, mean and variance. Value of similarity tolerance will make different result of segment. Smaller the value of similarity tolerance, more detailed the segment result. A few value of parameter segment was be tried to found the best segment for represented floristic composition (Figure 6)

(a) (b)

(c) (d) (e)

Figure 6. Segmentation using different similarity tolerance (a).value 0 (b).value 9 (c).value 10 (d).value 15 (e).value 20

Segmentation is unit analysis for object based classification, just like pixel for pixel based classification. The best segment value is 9 because not to general n the border of segment proportional. Post classification for object based classification using maximum-likelihood. Sample for classification using same sample like pixel based classification so it can be compared (Figure 7)

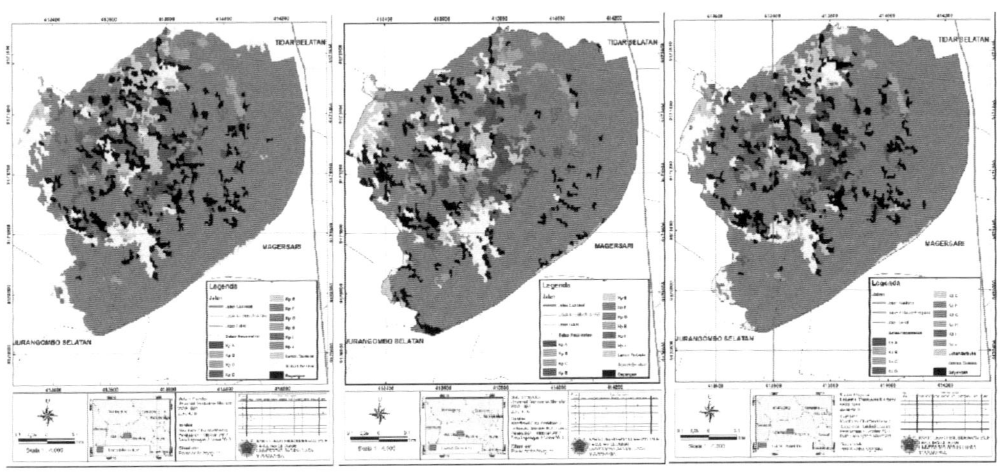

Figure 4. Object based classification result (a) Non radiometric correction, (b) Surface reflectance, (c) TOA reflectance

There is different result for object based classification. The best accuracy from image without radiometric correction 70.24%. This show that radiometric correction not always give better accuracy for classification because sometime if not accurately correction will make pixel change (Table 5).

IOP Conf. Series: Earth and Environmental Science **47** (2016) 012042 doi:10.1088/1755-1315/47/1/012042

Table 5. Accuracy assessment object based classification

Object Based Classification		
Non Correction	TOA Reflectance	Surface Reflectance
70,24%	69,63%	67,72%

Effectivity of classification can be seen from 3 parameter is, accuracy assessment, difficulty and processing time. Even object based can give better accuracy assessment but it's need more processing time and more difficult than pixel based classification. (Table 6)

Table 6. Effectiveness classification method from (a) accuracy assessment, (b) time processing and (c) difficulty level

(a)	Pixel Based			Object Based		
	Non Correction	Toa Reflectance	Surface Reflectance	Non Correction	Toa Reflectance	Surface Reflectance
Overall Accuracy	63,56%	65,09%	66,22%	70,24%	69,63%	67,72%
Value	1			2		

Pixel Based (Majority Analysis)					
Non Correction		Toa Reflectance		Surface Reflectance	
3X3	5X5	3X3	5X5	3X3	5X5
70,75%	72,75%	69,94%	69,94%	71,16%	73,32%
2					

(b)	Time cost		
	Qualitative	Quantitative	Value
Pixel Based	Fast	10 Days	2
Pixel Based (Majority Analysis)	Fast	12 Days	2
Object Based	Long	20 Days	1

(c)	Difficulty	
	Qualitative	Value
Pixel Based	Easy	3
Pixel Based (Majority Analysis)	Medium	2
Object Based	Hard	1

5. Conclusion

The best overall accuracy assessment for floristic composition is 73.32% using pixel based with majority 5x5. Total class of floristic composition is 10 class and 3 class for non-floristic composition. The dominant vegetation is *Pinus mercusii* about 62% in Tidar forest. Pixel based classification with majority 5x5 more effective then object based. Salt and paper problems from pixel based classification can be minimalized using majority.

References

[1] Chubey, M.S., Franklin, S.E., Wulder, M.A., 2006 *Object-based Analysis of Ikonos-2Imagery for Extraction of Forest Inventory Parameter*, Photogrammetric Engineering & Remote Sensing. Vol. 72, No.4, April 2006, pp.383-394. Available at http://www.asprs.org/ publications /pers/ 2006 journal/ april/ 2006_apr_383-394.pdf, accessed on March, 16th 2008.

[2] Danoedoro, Projo.1996. *Pengolahan Citra Digital: Teori dan Aplikasinya dalam Bidang Penginderaan Jauh*. Yogyakarta: Faculty of Geography, Universitas Gadjah Mada

[3] Definiens Imaging, 2004. *Whitepaper-eCognition Professional 4.0*, Definiens Imaging,GmbH, Munich, Germany. Available on http://www.gis.unbc.ca/ help/ software/ecognition4/ ELuserguide.pdf, access on Desember, 5th 2006.

[4] Fraser, RH., A Abuelgasim, and R Latifovic. 2005. A method for detecting large-scale forest cover change using coarse spatial resolution imagery. *Remote Sensing of Environment* 95:414-27.

[5] Jyothi, B.N., G.R. Babu, and I.V.M. Krishna. 2008. Object oriented and multi-scale image analysis: strengths, weaknesses, opportunities, and threats – a review. *J. of Computer Science* 4(9): 706-712.

[6] Lindgren, D.T. 1985. *Land Use Planning and Remote Sensing*, Martinus Nijhoff Publishers, Doldrecht.

[7] Misra R. 1973. *Ecology Work Book*. New Delhi: Oxford & IBH Publishing Co.

[8] Mueller-Dombois and Ellenberg H. 1974. *Aims and Methods of Vegetation Ecology*. New York; John Wiley & Sons.

[9] Noritoshi Kamagata, dkk. 2005. *Comparison of pixel-based and object-based classifications of high resolution satellite data in urban fringe areas*. Japan: Tokyo University

[10] Richards, JA. 1993. Remote *Sensing Digital Image Analysis: An Introduction*. Berlin: Sringer-Verlag

[11] Richard PW. 1966. The *Tropical Rain Forest an Ecological Study*. Cambridge an The University Press. London.

[12] Rusdi, Muhammad. 2005. *Perbandingan Klasifikasi Maximum Likelihood dan Object Oriented Pada Pemetaan Penutup/Penggunaan Lahan Studi Kasus Kabupaten Gayo Lues, NAD HTI PT Wirakarya Sakti Jambi dan Taman Nasional Lore Lindu Sulawesi Tengah.Bogor*. Institut Pertanian Bogor.

[13] Soerianegara I dan A. Indrawan. 2005. *Ekosistem Hutan Indonesia*. Bogor: Forestry Ecological Laboratory, Faculty of Forestry, IPB.

[14] Sutanto. 1987.*Penginderaan Jauh Jilid 2*. Yogyakarta: Gadjah Mada University Press.

[15] Wicaksono, P. (2010). *Integrated Model of Water Column Correction Technique for Improving Satellite-Based Benthic Habitat Mapping*. Yogyakarta: Gadjah Mada University.

[16] Wikantika, Ketut. (2008). *Sekilas Tentang Sistem Klasifikasi Land Use/Land Cover*. Access on Desember, 24th 2012 from http://wikantika.wordpress.com.

[17] Yan Gao, dkk. 2005. Comparison of pixel-based and object-oriented image classification approaches—a case study in a coal fire area, Wuda, Inner Mongolia, China. *International Journal of Remote Sensing*.

[18] Zan, Q. 2003. A Hierarchical Object-Based Approach for Urban Land-Use Classification from Remote Sensing Data, *ITC Dissertation* No.103.

The Utilization of Landsat 8 Multitemporal Imagery and Forest Canopy Density (FCD) Model for Forest Reclamation Priority of Natural Disaster Areas at Kelud Mountain, East Java

S Himayah[1], Hartono[2], P Danoedoro[3]

[1] Remote Sensing, Geography Faculty, Gadjah Mada University, Yogyakarta, Indonesia 55281
[2] Cartography and Remote Sensing, Geography Faculty, Gadjah Mada University, Yogyakarta 55281
[3] PUSPICS, Yogyakarta, Indonesia 55281

[1]shafirahimayah@gmail.com;[2]hartono_ge@ugm.ac.id;
[3]projo.danoedoro@geo.ugm.ac.id

Abstract. Remote sensing has the advantage in terms of temporal resolution that can be used to examine changes of the forest canopy density as occurred in Kelud Mountain after the eruption of 2014. Canopy density changes then used as a consideration for forest reclamation priority. This study aims to assess the ability of Landsat 8 multitemporal imagery and Forest Canopy Density (FCD) modeling for canopy density changes at Kelud forest before and after the eruption, as well as take advantage of the canopy density changes from FCD and biophysical condition of forest to make a forest reclamation priority. This research using a Landsat 8 imagery (26 June 2013 and 4 September 2015). The method that used is FCD modeling to obtain canopy density. Forest reclamation priority is determined based on the canopy density change after the eruption and biophysical factors such as slope, soil fertility and native vegetation. Landsat 8 can used to determine the forest canopy density of Kelud before and after eruption with an accuracy of 83.73% and 81.14%. Kelud forest reclamation priorities are divided into nine classes based on priority level. The most prioritized class is 1a with an area of 865 ha and class 1b with an area of 2.085 ha. Then class 1c (0 ha), 1d (413 ha), and 1e that most dominate (5.454 ha). Beside that, there is class 2a (1.900 ha) and 2b (243 ha), and the last is class 3a (467 ha) and 3b (1.172 Ha).ntroduction

1. Introduction

Temporal resolution of remote sensing imagery can be used to monitor phenomena or objects changes within a certain time. Vegetation includes as an easy objects to identify from remote sensing imagery. Infrared band that owned by Landsat 8 system is a band that is most sensitive to vegetation. The vegetation reflection at infrared band are often used to calculate the vegetation index. Vegetation index is then developed to expand the remote sensing studies of vegetation.

The advantages of multitemporal remote sensing imagery and reflection of vegetation in the near infrared band makes it possible to do research on forest canopy density changes as occurred in the forest of Mount Kelud after the eruption. Forest Canopy Density (FCD) is a biophysical analysis model that can be used to analyze a forest canopy density changes after the eruption with involving four factors:

the vegetation, bare soil, shadow, and temperature. Data of percentage per pixel on four factors obtained from advance vegetation index (AVI), bare soil index (BI), shadow index (SI), and thermal index (TI), [11].

Forest degradation should be addressed to restore its functionality. One attempt for recovery of forest function is with forest reclamation. FCD Model through Landsat 8 multitemporal can facilitated the implementation of the reclamation. Analysis of changes in canopy density through FCD Model and forest biophysical conditions then used to make a priority of forest reclamation.

1.1 Study area

This study area located at a forest of Mount Kelud, administratively belonged to Kediri, Blitar, and Malang, East Java, Indonesia. The peak of Mount Kelud is located at an altitude of 1731 meters above sea level, and the forest of Mount Kelud has an area of 13.381 hectares. Mount Kelud included in stratovolcano type with explosive eruptions characteristics, and last erupted on February 13-14, 2014. Forest of Mount Kelud is a tropical rain forest located in 112°14'00"-112°22'20"E and 7°51'40"-8°0°40"L. According to the height of the sea surface, the forests of Mount Kelud divided into the low rain forest zone (0-1000 meters above sea level) and the middle rain forest zone (1000-3300 meters above sea level). Forests of Mount Kelud selected as the study area because it suffered damage caused by the eruption.

1.2 Data

This study is using Landsat 8 image acquired on 23rd June 2013 and 4th September 2015. Landsat 8 Operational Land Imager (OLI) and Thermal Infrared Sensor (TIRS) images consist spectral bands with a spatial resolution of 30 meters for Bands 1 to 7 and 9. The resolution for Band 8 (panchromatic) is 15 meters and Thermal bands 10 and 11 are collected at 100 meters. Landsat 8 Operational Land Imager (OLI) and Thermal Infrared Sensor (TIRS) images consist of eleven bands : coastal aeorosol (0.43 - 0.45 μm), blue (0.450 - 0.51 μm), green (0.53 - 0.59 μm), red (0.64 - 0.67 μm), near-infrared (0.85 - 0.88 μm), SWIR 1(1.57 - 1.65 μm), SWIR 2 (2.11 - 2.29 μm), panchromatic (PAN) (0.50 - 0.68 μm), cirrus (1.36 - 1.38 μm), TIRS 1 (10.6 - 11.19 μm) and TIRS 2 (11.5 - 12.51 μm). Beside that, this study also using ASTER GDEM scene S08E112 and S09E112, image acquired on 17th October 2011.

2. Methods

2.1 Forest Canopy Density Model

Forest Canopy Density (FCD) is a method for estimating forest canopy density by considering vegetation, bare soil, temperature, and shadow. FCD using biophysical phenomena modeling such as Advanced Vegetation Index (AVI), Bare Soil Index (BI), Shadow Index (SI) and Thermal Index (TI) (Rikimaru, 2002). These indices is using band 2 (blue), band 3 (green), band 4 (red), band 5 (near infrared), band 6 (SWIR 1), band 7 (SWIR 2), dan band 10 (TIR1) of Landsat 8 imagery.

Advanced Vegetation Index (AVI) is one of the indices that examine the characteristics of the vegetation chlorophyll. The index uses the equation:

$$AVI = ((Band5 + 1)\ (256 - Band4)\ (Band5 - Band4))1/3 \quad \ldots\ldots\ldots\ldots\ldots\ldots\ldots\ (2.1)$$

Bare Soil Index (BI) formulated using SWIR 1 band, NIR band, red band, and blue band. This index is using equation :

$$(((B6+B4)-(B5+B2))/(((B6+B4)+(B5+B2)))\times 100+100 \quad \ldots\ldots\ (2.2)$$

Shadow Index (SI) test a shadow characteristic by utilize a spectral information from forest shadow and thermal informatin, with an equation :

$$SI = ((256 - Band2) \times (256 - Band3) \times (256 - Band4))1/3 \quad \ldots\ldots\ (2.3)$$

Thermal Index (TI) is based on two kind of factor that affect cold relativity condition inside the forest. The first one is shielding effect from forest canopy that blocking and absorbing energy from the sun.

The second one is leaf area surface which reduces heating (Rikimaru, 2002). Temperature data then used to separating a soil and a non-tree shadow. Equation that use for *Thermal Index* (TI) is :

$$T= K\_2/(\ln(K\_1/L\_\lambda +1)) \quad \ldots\ldots (2.4)$$

Explanation of FCD four indices combination is :

Table 3.1 Four Indice Combination Characteristic

Index	High- FCD	Low-FCD	Grass land	Bare land
Advanced Vegetation Index (AVI)	High	Middle	High	Low
Bare Soil Index (BI)	Low	Low	Low	High
Shadow Index (SI)	High	Middle	Low	Low
Thermal Index (TI)	Low	Middle	Middle	High

Source : Rikimaru, 2002

If the value of AVI higher then most likely a value of FCD was high too. Low AVI value and high BI value means bare land area, meanwhile low BI value means a vegetation area. SI have a low value for objects such as bushes and open ground area. SI value indicates impairment along with a value of FCD, meanwhile TI is increased with a reduction of vegetation and have a high value in bare land areas. Vegetation Density (VD) is a proportion of soil surface that enclosed by plant crown projection. VD result from Principal Component Analysis (PCA) method because AVI and BI has a negative correlation. [11];[4].

SSI is the histogram value of SI that has been adapted by TI to distinguish the forest and vegetation ground such as shadows and surface temperature. Areas that have zero value of SSI indicates a forest with a lowest shadows and a 100 SSI value indicates forests that is likely to have the highest shadow value. With the development of the SSI, then will be able to distinguish between vegetation canopy and vegetation in the soil. [7]; [11] Integration of VD and SSI produces the value of forest canopy density. Forest canopy density index value is calculated using the equation:

$$FCD = (VDxSSI+1)1/2-1 \quad \ldots\ldots (2.5)$$

The integration of these models results the canopy density of data per pixel. FCD percentage value from 0% to 100%. 0% indicates a lack of coverage of the canopy and the area is dominated by bare land. Along with the greater percentage of the value of the FCD, the pixel area more covered by canopy.
Forest canopy density modeling before and after the eruption then tested using the Standard Error of Estimation (SEE). SEE is a test method that considering the difference between the accuracy of canopy density from FCD model with actual density from field measurement. Field measurements to obtain the actual canopy density was conducted in March 2016, with the total number of sample are 41 plots. FCD modeling accuracy test results that have is further used to analyze changes of canopy density after the eruption of 2014.

2.2 Forest Canopy Density Changes

Canopy density modeling map results from FCD in 2013 and 2015 are used to detect canopy density changes with a combine methods in ArcGIS 10.2. Combine is a raster overlay method, in which each pixel in the same geographical position with two different time (or in this case may also be called cell) analyzed and then given a new identity.

2.3 Forest Reclamation Priority

Forest reclamation is an attempt to repair or restore a damaged land and vegetation in order to function to its optimum. (Permenhut 48, 2013 on Guidelines for the Forest Reclamation of Natural Disasters Area). Priority of reclamation created to facilitate the implementation of the reclamation. Priority refers to the inventory based on changes in the forest canopy density from FCD result, slope derived from

DEM data, as well as the fertility of the soil and vegetation conditions as the results of field measurements.

Utilization of remote sensing in the protected forest reclamation priority of Mount Kelud did not involve biophysical factors such as land use, forest functions, and climate. This is because the land use in the this study is focused only on protected forest areas which is has a tropical climate.

2.3.1Topographic Form

Slope classification obtained from the DEM data by using the 3D Analyst tools in ArcGIS software. Topographic are qualitatively termed as landscape, while quantitatively expressed in grade slope (%), namely: Class I (0-8%), Class II (9-15%), Class III (Kinda Steep 16-25%), Class IV (26-45%), Class V (> 45%). Topographic considered as a biophysical parameter in the inventory of forest reclamation due to its relevance in terms of accessibility of forest reclamation. Flat areas can facilitate the implementation of the reclamation because its easier to reach, and conversely, areas with steep slopes will be difficult for reclamation because it is difficult to reach. In addition, the flat slopes and ramps will be easier for field measurements.

2.3.2Soil Fertility

Soil fertility is obtained based on field measurements were conducted in March 2016. The parameters used to determine the soil fertility is color, texture, and pH.

2.3.3Native Vegetation

Vegetation condition before the occurrence of a natural disaster can be represented through the canopy density. While the native vegetation can be expressed by a species of vegetation that are dominant in the study area. Plant species in the forests of Mount Kelud which includes as an area of lowland forest (<1,000 meters above sea level) consists of *Dendrocalamus asper, Croton argyratus, Ficus infectoria, Laportea stimulant, Villebrunea rubescens, Ostodes paniculata, Altingia, Bischofia, Castanopsis, Ficus, Gossampinus* and plants of the *Leguminosae* family [10].

Lower montane forests or middle rainforests (> 1000 meters above sea level) generally have the type of dominant vegetation such as *Lithocarpus, Castanopsis, Altingia excelsa, Dacrycarpus imbricartus, Engelhardia spicata, Schima wallichii, Cyathea spp, and Weinmannia blumei* [12; [5];[8].

Native vegetation type compared to other types of vegetation after the eruption. So then it can obtained species that survives, or whether there is a change of the dominant species, or perhaps there are species disappearing.

2.3.4Determining the Priority Class of Forest Reclamation

Map of canopy density changes after the eruption is overlayed with a slopes map, also with the information that obtained from the field measurements. Factors to be examined first is the forest canopy, and then topography, soil fertility, and vegetation type. The results of these analysis will be used to make a priority concept of forest reclamation. Variables such as canopy density changes obtained from the FCD modeling, slope obtained from ASTER GDEM image processing, while the fertility of the soil and vegetation changes obtained from field work.

Class of reclamation priority is divided into three, namely 1,2, and 3. The division of this class is based on canopy density changes : reduced, stable, and increased. Areas which is canopy density reduced is the most prioritized area. All three classes are then subdivided into several sub-class with considering a slope (topography), the fertility of the soil and vegetation conditions.

3. Results and Discussion

3.1 Forest Canopy Density Model

Advanced Vegetation Index (AVI), *Bare Soil Index* (BI), *Shadow Index* (SI) and *Thermal Index* (TI) processed using FCD Mapper software to obtain a canopy density in percentage (%). AVI and BI calculation result a Vegetation Density (VD), whereas SI and TI calculation result a Scaled Shadow Index (SSI). VD and SSI then analyzed to make a FCD estimation in percentage (%) which represent a canopy density for each pixel. FCD in 2013 (before eruption) and in 2015 (after eruption) are reclassified into 10 classes to make a change detection process easier.

Canopy Density of Mount Kelud Map is using color symbol adjusted to FCD Mapper results, namely blue, cyan, magenta, red, orange, yellow, and gradation of green. The appearance of canopy density before the eruption dominated by the density of 72-80% with an area of 4698.90 ha. This density classes represent the condition of forest with a canopy cover 71-80% of the area of pixels with different strata of stands clearly visible as well as the heterogeneity of types is increasing. Meanwhile, the most visible appearance is the red color (21-30%) which is shaped elongated resembling mountain slopes.

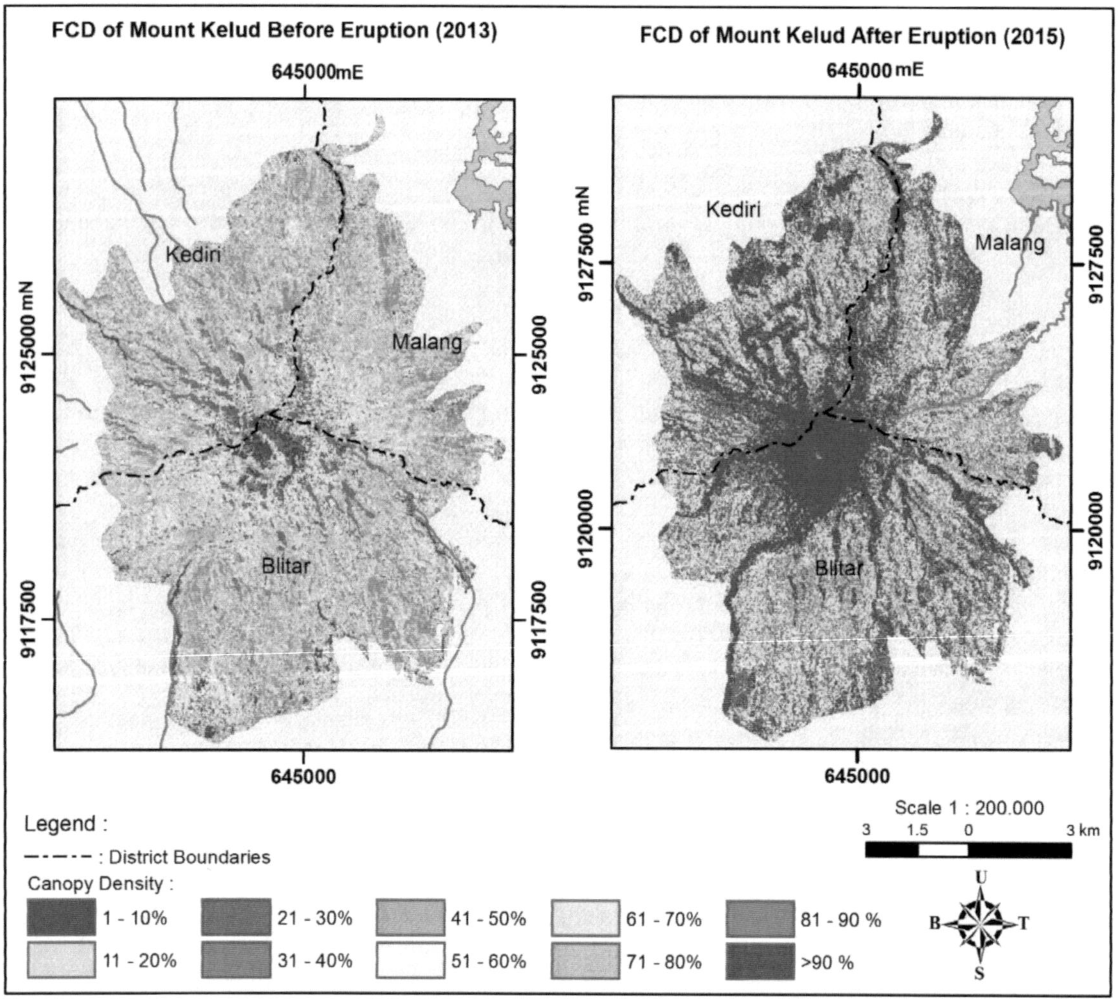

Figure 4.1 Forest Canopy Density of Mount Kelud before and after the eruption.

After the eruption, the percentage 1-10% of canopy density is wider than before the eruption. The canopy density of 1-10% is located on the top of the mountain and distribute to different directions. The area density of 1-10% is an area of 2878.8 hectares.

3.2 Forest Canopy Density Changes

Map of forest canopy density changes of Mount Kelud after eruption in 2014 contain an information about the canopy density change for each pixels. The it used to calculate a quantity of a changing pixels (reduced, stable, or increased). The decreasing canopy density forest is 8833.95 hectares, while the stable canopy density is 2149.38 ha, and increased forests canopy density is 1643,31Ha.

Canopy density changes on a map shown with a contrasting color for an interactive informatiom display. Decreasing canopy density displayed in red, stable canopy density displayed in yellow, and increased canopy density displayed in green. Based on Canopy Density Changes Map, the most dominating is a reduced canopy density (red). Eruption caused a largely decreased of Kelud forest canopy density, the decrease was spread both in Kediri, Malang, Blitar. Stable canopy density and increased canopy density were also seen spread, but with a much narrower area than the reduced canopy density.

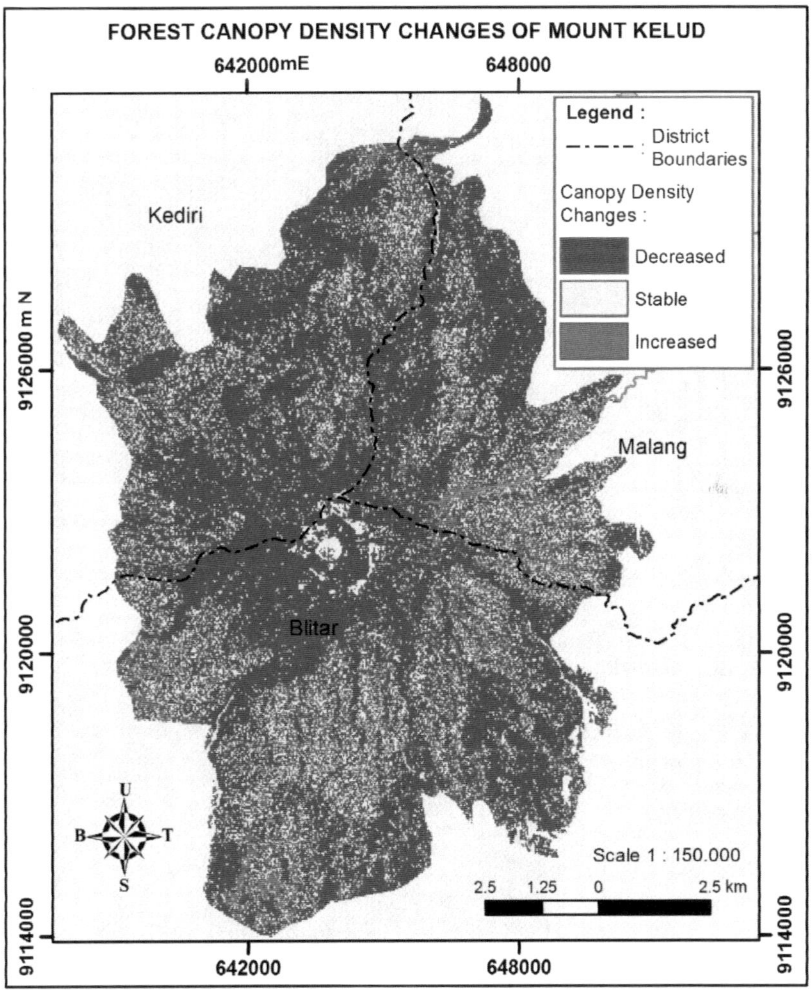

Figure 4.2 Forest Canopy Density Changes of Mount Kelud before and after the eruption.

3.3 Priority class of Forest Reclamation

Slope classification obtained by processing GDEM ASTER data into a degree of slope. Slope classified to 5 type: class 1 (flat), class 2 (sloping), class 3 (somewhat steep), class 4 (steep) and class 5 (very steep) with shades of brown. Increasingly steep slope then shown with a darker color.

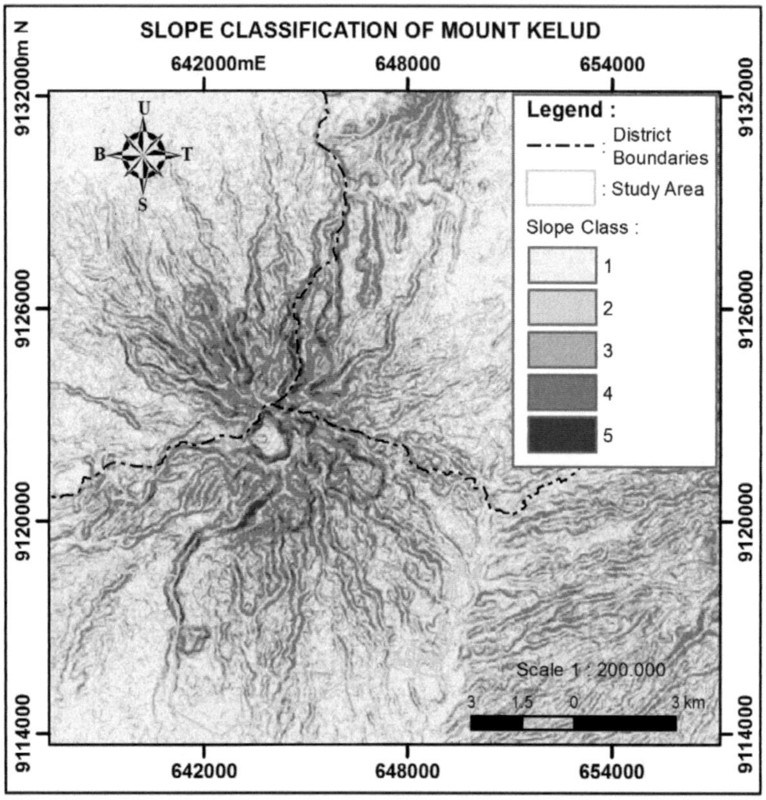

Figure 4.3 Slope classification of Mount Kelud.

Scene S08E112 has a range of 0 to 85.36 degrees, while scene S09E112 has a range from 0 to 77.7 degrees.

Soil fertility is determined by observing the current condition field measurement of pH, color and soil friability. The results of field measurements show that the area of forest under 1300 meters above sea level has fertile soil, while the area above 1300 meters above sea level has a less fertile soil. It was concluded that the soil is less fertile because its affected by the eruption and has not recovered. This is confirmed by the absence of vegetation at sample plots with less fertile soil (Figure 4.4)

Figure 4.4 Bare land area with less fertile soil at 1305 m above sea level.
Location : Ngancar, Kediri Coordinat : x = 642624.57 y = 9122875.39

Vegetation type based on field observations divided by altitude. The type of vegetation in sample plots at an altitude of 525-586 meters dominated by agroforestry crops. There are four sample plots at an altitude of 605-692 m, two of them were dominated by grasses and shrubs and the other two are dominated by agroforestry crop. The type of vegetation in sample plots with an altitude of 873-972 m also dominated by grasses and shrubs.

Sample plots were located at an altitude above 1000 m is largely dominated by shrubs and red calliandra also there were dead trees. Sample plots that have no vegetation except for a bit of grass is a plot at an altitude above 1300 m or closest to the crater of Mount Kelud. This is likely due to the result of the eruption that occurred in 2014, and the region has not recovered so there is no vegetation.

Percentage of FCD results are used to test the accuracy. FCD before the eruption (in 2013) compared with the canopy density visual interpretation hrough Google Earth's imagery, while the FCD modeling results after the eruption (2015) compared with the field measurements. The analysis used to test the accuracy is the standard error of estimation and confidence level by looking at the standard deviation from the predicted data and the actual data. Based on calculation accuracy test, a score of accuracy for FCD in 2013 amounted to 83.73%. While FCD 2015 has an accuracy of 81.14%.

Forest reclamation priority class is divided into class 1a, 1b, 1c, 1d, 1e, 2a, 2b, 3a, and 3b. The results showed a lack of class priority 1c, it indicates that the research area is not encountered an area with decreased canopy density, the flat-sloping areas with infertile soil conditions, as well as native vegetation is reduced or even disappear.

Forest reclamation class of Mount Kelud most preferred class 1a and 1b. Because in these two classes canopy density decreased, and native vegetation is reduced or lost but a soil conditions is fertile. Based on the calculation and processing, reclamation class 1a has an area of 865 hectares and class 1b has an area of 2085 Ha. Both classes are symbolized by shades of red, where the class 1a as the most prioritized is colored darker red than the class 1b (Figure 4.7). Although reclamation class 1a is the most prioritized, apparently the most dominant class is a class 1e.

Tabel 4.3. Priority class of Forest Reclamation

Class of Reclamation	Condition	Area (Ha)
1a	Decreased canopy density, flat-sloping slope, fertile soil, native vegetation is decreased or dissapear.	865
1b	Decreased canopy density, somewhat steep-steep slope, fertile soil, native vegetation is decreased or dissapear.	2085
1d	Decreased canopy density, somewhat steep-steep slope, less fertile soil, native vegetation is decreased or dissapear.	413
1e	Decreased canopy density and native vegetation is stable or increased	5454
2a	Stable canopy density and native vegetation is decreased or dissapear	1900
2b	Stable canopy density and native vegetation is stable or increased	243
3a	Increased canopy density and native vegetation decreased or dissapear	467
3b	Increased canopy density and native vegetation is stable or increased	1172

Based on measurements and field work, the area that goes on class 1a and 1b lies at an altitude of more than >900m. Altitude> 900 m is dominated by red calliandra (*Calliandra calothyrsus*). Calliandra can be categorized as a pioneer plant with a high life probability. Therefore, it can be concluded that original vegetation of these area are reduced or disappear.

Areas included in the class of reclamation 1d are above 1300 meters with a total area of 413 hectares. These area is dominated by sand or bare land. The most dominant priority class is 1e (Decreased canopy density and native vegetation stable or increased) with area of 5454 Ha. This may be due to the prevalent agroforestry areas or land use that combines trees with crops. Agricultural crops can be affected or even damaged due to the impact of the eruption, but a field check on the sample plot shown its already functioning back to normal. So it was concluded that in these sample plots the native vegetation is stable or increases. The types of agricultural crops which common in the sample plot of agroforestry is coffee and chili.

Figure. 4.5. Red Calliandra
Location : Ngancar, Kediri Coordinat : x = 641507.98 y = 9122634.33

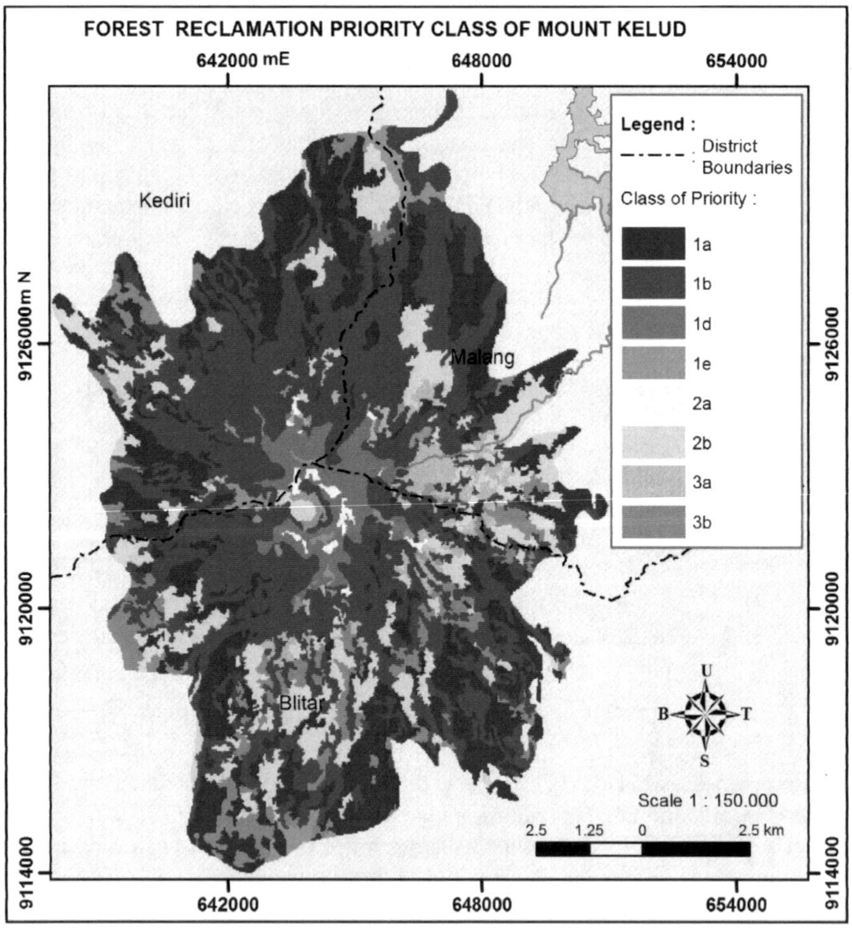

Figure 4.6 Forest reclamation priority classes of Mount Kelud.

Reclamation class 2 has a stable percentage of canopy density, and are divided into two sub classes : 2a is the area that the native vegetation is gone and class 2b is areas of a remains or increases native vegetation. Both classes are symbolized by the color gradation of greenish yellow. Class 2a has a larger area than the class 2b. Class 2a has 1900 Ha whereas classes 2b is 243 hectares.

The last class of canopy density is class 3, which is a canopy density increased after the eruption. Reclamation class 3 divided by the original vegetation and symbolized by shades of green. Class 3a has an area of 467 hectares, is located at altitudes above 900 meters above sea level. Type of native vegetation is lost and then replaced by a red calliandra (*Calliandra calothyrsus*) and *Chromolaena odorata* (lower plants). While class 3b has an area of 1172 ha, located at an altitude below 900 meters above sea level and dominated by crop plants *Neolamarckia cadamba*.

Priority classes is generalized to make an easier implementation in a real world. Single pixel can obstruct the implemetation of reclamation in real world, so it must generalized and combined with a majority pixel around it.

Based on Priority of Reclamation at Mount Kelud Forest Map, it can be seen that the area around crater of Mount Kelud (near a boundary of Kediri, Malang, and Blitar) is classified as 2a (yellow) and 1a (red). And there is class 1d (bright red) which has a less fertile soil and there is no vegetation. Class of 2a, 2b, 3a, and 3b is distributed.

4. Conclusion

Based on the processing of Landsat 8 results canopy density changes: reduced canopy density with area of 8833.95 ha, canopy density that does not change is 2149.38 hectares and increased canopy density is 1643.31 ha. Accuracy test result of Forest Canopy Density (FCD) modeling is amounting to 83.73% for the FCD of 2013 and 81.14% for the FCD of 2015. So it was decided that the FCD model results can be used to make a priority of forest reclamation.

Reclamation priority of protected forest area of Mount Kelud is divided into nine classes in the order based on its priority level : class 1a, 1b, 1c, 1d, 1e, 2a, 2b, 3a, and 3b. However, on a protected forest area of Mount Kelud there are no class 1c. The most prioritized reclamation is class 1a with an area of 865 hectares, while the most dominating class is 1e with an area of 5454 Ha.

References

[1] Azizi, Z. Najafi A & Sohrabi H 2008 Forest Canopy Density Estimating, Using Satellite Image The International Archive of the Photogrammetry *Remote Sensing and Spatial Information Science* XXXVII B8 Beijing 2008

[2] Chuvieco, E. & Huete, A 2010 *Fundamentals of Satellite Remote Sensing* (CRC Press United States of America)

[3] Coppin, P. R. & Bauer M. E 1996 *Change Detection in Forest Ecosystems with Remote Sensing Digital Imagery* Remote Sensing Reviews 13:207-234.

[4] Danoedoro P 2012 *Pengantar Penginderaan Jauh Digital* Penerbit ANDI Yogyakarta.

[5] Indriyanto 2012 *Ekologi hutan* (Bumi Aksara Jakarta)

[6] ITTO/JOFCA 2003 *FCD-Mapper Version.2 User Guide Semi-Expert Remote Sensing System for Canopy Density Mapping*.

[7] Jamalabad, M. S. Akbar A. A 2000 *Forest Canopy Density Monitoring Using Satellite Images* ISPRS Commission **VII** 17(Holland).

[8] Kartawinata Kuswata. 2013 *Diversitas Ekosistem Alami Indonesia* (LIPI Press & Yayasan Pustaka Obor Jakarta)

[9] Kementrian Kehutanan 2013 *Peraturan Menteri Kehutanan Republik Indonesia Nomor : P.48/Menhut-II/2013 tentang Pedoman Reklamasi Hutan pada Areal Bencana Alam* (Jakarta)

[10] Larashati, I 2004 Keanekaragaman Tumbuhan dan Populasinya di Gunung Kelud Jawa Timur. *Biodiversitas* **5** (2) : 71-76.

[11] Rikimaru, A. Roy P S. & Miyatake S 2002 *Tropical Forest Density Mapping* Tropical Ecology **43** 1 : 39-47.

[12] Whitten, T. Soeriaatmaji R.E. Afiff, S.A 1999 *Ekologi Jawa dan Bali* (Prenhallindo Jakarta)

[13] Yulianto, S 2010 *Kajian Kemampuan Model Forect Cover Density (FCD) Berbasis Citra Landsat ETM+ Multitemporal untuk Memantau Suksesi Hutan Rawa Gambut di Kalimantan Tengah* (Fakultas Geografi Universitas Gadjah Mada Yogyakarta)

Species Distribution Model using Hyperspectral Data Application in Peatland, Central Kalimantan

Muhammad Dayuf Jusuf

Earth Science Systems Technology (Geostech), BPPT.
PUSPIPTEK Region - Serpong, Tangerang - Banten. Indonesia.

dayuf.yusuf@bppt.go.id, dayuf.jusuf@mail.ugm.ac.id, dayuf2012@gmail.com

Abstract. Tumih / Parepat (*Combretocarpus-rotundatus* Mig. *Dencer*) of the family Anisophylleaceae and Meranti (*Shorea Belangerang, ShoreaTeysmanniana Dyer ex. Brandis*) of Dipterocarpaceae are groups of species of vegetation that is modeled for the designation of its spread. Pioneer species are predicted as an indicator of ecosystem restoration succession of wet tropical peatland characteristic (*unique*) and very susceptible (*fragile*) in an endemic Sundaland hotspot. Projected climate change and conservation planning a topic of heated discussion, the analysis of alternative approaches and the development of a combination of search algorithm projection models within the fabric of the diffusion of the species to attain this through a geospatial information system technology. Model approach used to work out problems in research on this vegetation species level based machine learning is a method (hybrid), which is a combination of wavelet and neural network. Often the approach to field data into general use in the management of natural resources and biodiversity assessment. Expression of the hybrid models provides encouraging results, to decide the classification of the species Tumih / Parepat and Meranti through mapping of vegetation species that are backed up by the pattern of the spectral curve of the field spectrometer.

Keywords: species mapping, algorithms, hybrid models.

1. Introduction

Healthy peatland forest store huge amount of soil and biomass carbon. The composition florists peat land forests of Central Kalimantan, has a variety of flora species that come down into group family botanical endemic hotspot Sundaland, is a species Tumih / Parepat (*Combretocarpus-rotundatus* Mig. *Dancer*) of the family *Anisophylleaceae* and groups of species Meranti (*Shorea white, red and yellow, Shorea Belangerang, Shorea Teysmanniana Dyer ex. Brandis*) of *Dipterocarpaceae*. Both families are considered can be used as an indicator of the presence of peat burned and antrophogenic cause, with the approach of geospatial information through the vegetation cover.

Flora mapped resultant distribution enabled visualization of vegetation homogeneous, heterogeneous mix of dissimilar species of plants, without the demand for a unique spectral signature or characteristics of the data collection. Forest degradation in peatland forest is directed to correlate to dissolved organic carbon from artificial canals, poor growth vegetation, and forest disturbance.

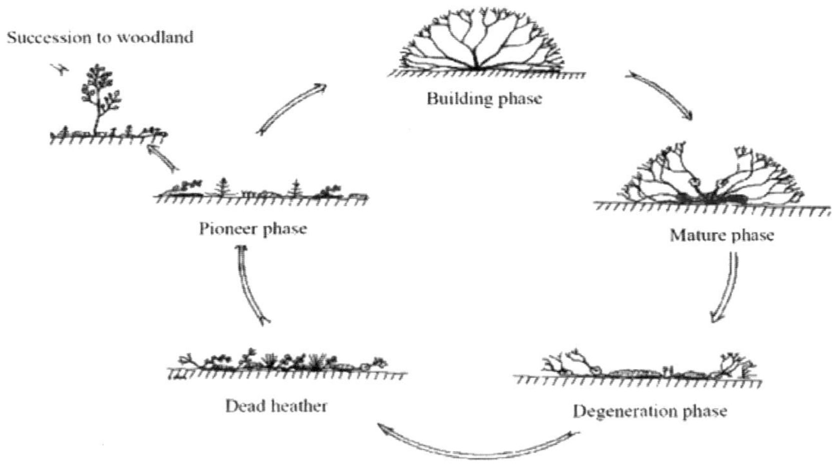

Figure 1. Phase difference vegetation species growing in peat land(Delalieux, *et al.*, 2012)

Groups of plant species and vegetation of peatlands are a pioneer species that develops when it is over peat fires. Floristic composition of wet tropical Peatland has a wealth of flora and biodiversity species to provide environmental services functions as a natural habitat with the ability of absorbing and storing carbon (*carbon sink*) from the atmosphere and is not found in other ecosystems [1].

These plants are beneficial to plants recovery peat soil adaptability to environmental conditions, because it contains fungus *Mikoriza Arbuskula*, which acts as an important factor for the ecosystem such as (*composition of the soil, plants, energy absorber*), and the effect is better than other types of plants [3]. Equally we recognize that the plan of attack to acquire geospatial information that has been carried out using a mixture of technologies multispectral imagery, but previous approaches did not offer detailed information classification in the instance of certain plant species level peatland forests.

In this inquiry, how potential hyperspectral data to assess and characterize forest degradation types, discrimination of plant/tree species (through biophysical parameteres), detection from spectral signature features extraction, soil types, etc. To development of algorithms to accept vantage of the availability of hyperspectral image (HyMap) and field spectral data collections (spectral library) for troubleshooting assessment of natural resource inventories. Our principal contribution of this inquiry is to map the vegetation species of peat based methods of machine learning the application of a hybrid model; wavelet and artificial neural networks, assess the relevance and accuracy of the classification of regular maps with maps hybrid design results. Calibration is performed by referring to a map correction measurement data library of spectral reflectance field to HyMap image correct.

2. Method

Technical information WANN methods below, is the integration of both the image data and Spectrometer HyMap field to compare the output with the regular distribution map distribution map design results in a hybrid model WANN (*Wavelet + Artificial Neural Networks*). The purpose of this mannequin was employed to examine the appropriateness and suitability of the input data set that is designed as an input data set for the distribution of 3 (three) classification of the vegetation species peatland type of dataset of species, namely: Tumih / Parepat (*Combretocarpus-rotundatus*), Meranti (*Shorea: White, Red, Yellow*), and the categorization is not peatlands.

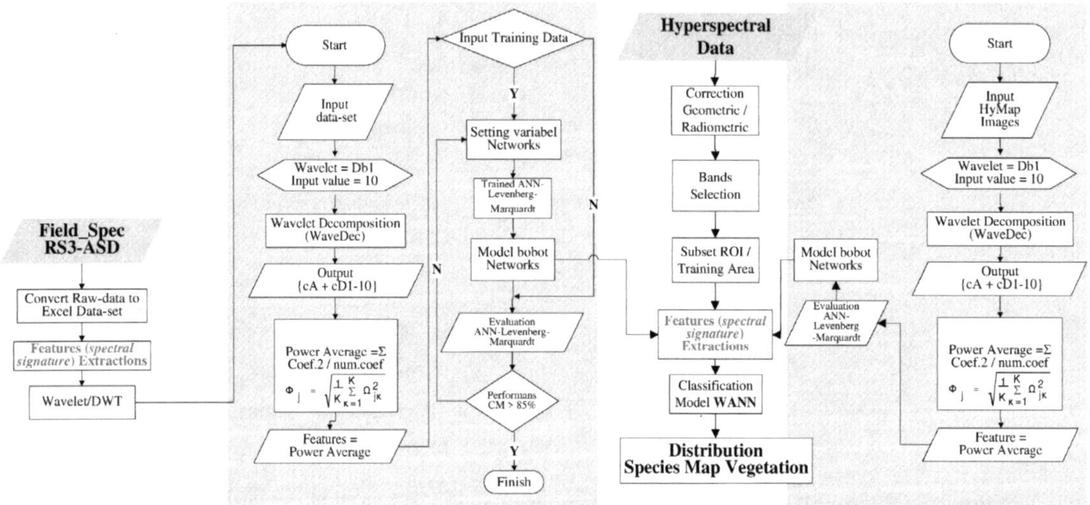

Figure 2. Integration classification WANN model for distribution map of vegetation species.

3. Results and Discussions

The effects of model validation design, distribution map design results in a hybrid model WANN (Wavelet + Artificial Neural Networks) with a regular distribution map output, as follows:

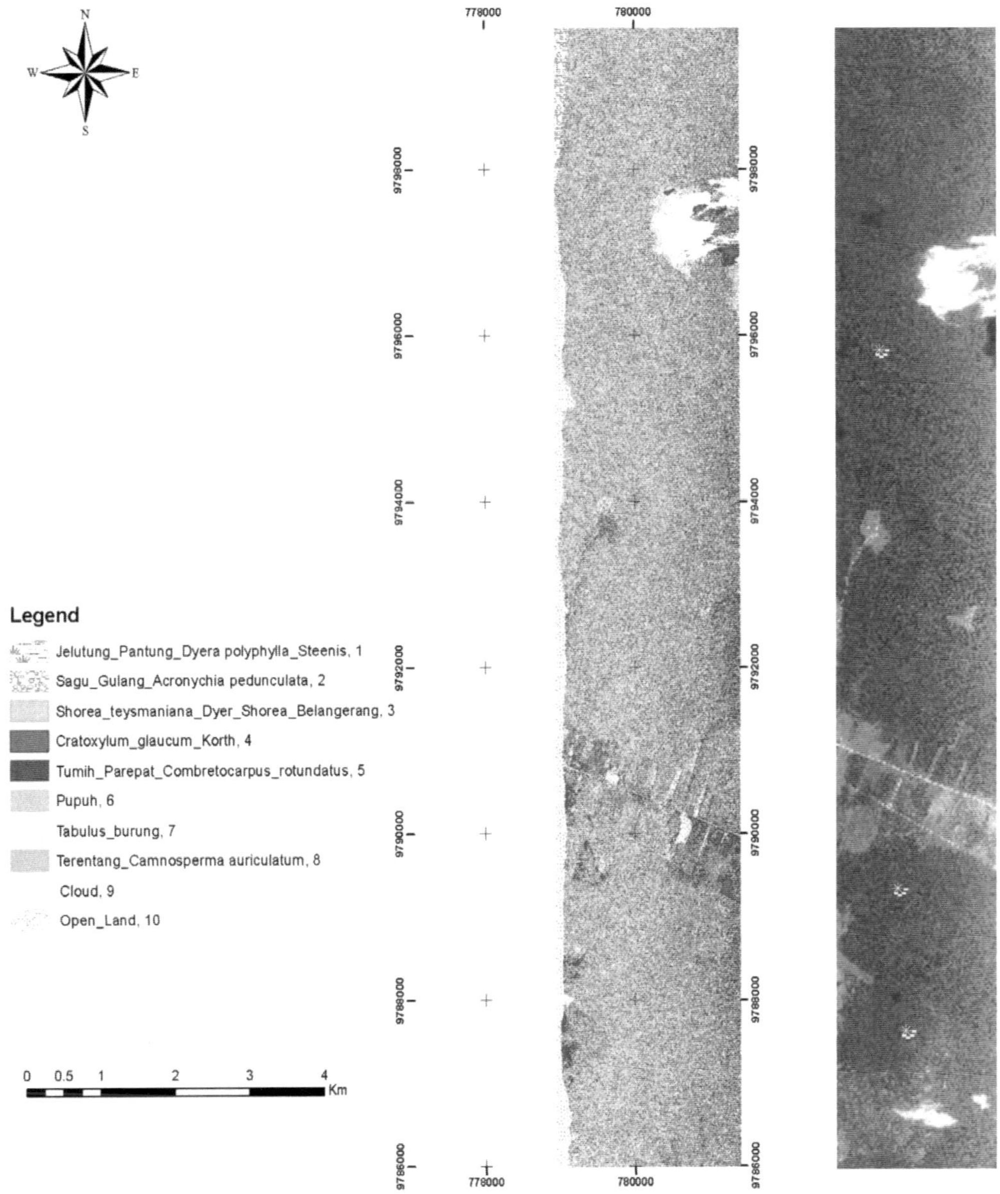

Figure 3. Distribution map of peatland species vegetation (left) and validation results using model WANN of (*red dots: Sagu gulang and yellow dots: Tumih/Parepat*) on Strip_HU01_04, HyMap Imagery (right).

Nevertheless, the successful mapping of the make-up of those species and plant in this model is dependent on how Peatland plant functional types are defined. The classification of species into existing well-recognized may not always be the most appropriate grouping from a remote sensing perspective and furthermore, the same definition of of those species and plant may not be useful for elucidating all interaction effect lansdcape to the atmosphere cycling processes. But at least we will get some ideas in advance from this research and consequently, further work should focus on understanding how the Peatland vegetation function and ecological theory can be more effectively linked to optical properties behaviour of plant [2].

4. Conclusion and Recommendation

WANN hybrid method showed quite good results of calculations which show the value of MSE (*minimum square error*) is smaller than the ANN training data, which totaled performance MSE = 0.0837 and 0.0547. Method development of hyperspectral with programming in software MatLab-based hybrid method WANN (Wavelet + Artificial Neural Networks) at the location of sample point positions vegetation species Tumih / Parepat (*Combretocarpus rotundatus* Mig. *Dancer*) and Meranti (*Shorea white, red and yellow, Shorea Belangerang, Shorea Teysmanniana Dyer ex. Brandis*) in HyMap imagery target of this study, showing results quite accurate (precision) right after cross-marked by the placement of GPS Spectrometer sampling locations in the area. The validation level attained by the above computation results reaches an accuracy rate of 90% so that the computation can be trusted design algorithm development (*source code*) programming can be applied on other images that are either multispectral, hyperspectral imaging and also recording results drone.

5. Acknowledgements

I would like to thanks are conveyed on behalf of deceased our brother Mr. Muhammad Evri (Agency for the Assessment and Application of Technology, BPPT) who delivered previous talks and adviced me to this research and Mr. Hendrik Segah from the Faculty of Forestry, University of Palangkaraya (UNPAR) - Central Kalimantan, which has been willing to provide a means of handheld spectrometer measurements in the field. Which supply me of hiperspectral data (HyMap) images from cooperation projects between BPPT- UNPAR, Indonesia and JST-JICA Japan.

References

[1] Arunarwati, B. 2015 Pendekatan Special Lahan Gambut [Presentasi Power Point] Retrieved from www.cifor.org/ipn-toolbox/
[2] Harris, A., Charnock, R., & Lucas, R. M. 2015 Hyperspectral remote sensing of peatland floristic gradients*Remote Sensing of Environment* **162** 99–111http://doi.org/10.1016/j.rse.2015.01.029
[3] Suryadi, U. E. 2013 *Dampak Pembendungan Kanal Terhadap Watak Fisika dan Kimia Gambut Ombrogen dalam Usaha Pemulihan Lahan Terdegradasi* (Gadjah Mada Yogyakarta)

Learning Methods of Remote Sensing In the 2013 Curriculum of Secondary School

Lili Somantri, Nandi

Department of Geography Education, Universitas Pendidikan Indonesia
Jl. Dr. Setiabudi No. 229 Bandung, 40154, Indonesia

lilisomantri@upi.edu

Abstract. The new remote sensing material included in the subjects of geography in the curriculum of 1994. For geography teachers generation of 90s and over who in college do not get the material remote sensing, for teaching is a tough matter. Most teachers only give a theoretical matter, and do not carry out practical reasons in the lack of facilities and infrastructure of computer laboratories. Therefore, in this paper studies the importance about the method or manner of teaching remote sensing material in schools. The purpose of this paper is 1) to explain the position of remote sensing material in the study of geography, 2) analyze the Geography Curriculum 2013 Subjects related to remote sensing material, 3) describes a method of teaching remote sensing material in schools. The method used in this paper is a descriptive analytical study supported by the literature. The conclusion of this paper that the position of remote sensing in the study of geography is a method or a way to obtain spatial data earth's surface. In the 2013 curriculum remote sensing material has been applied to the study of land use and transportation. Remote sensing methods of teaching must go through a practicum, which starts from the introduction of the theory of remote sensing, data extraction phase of remote sensing imagery to produce maps, both visually and digitally, field surveys, interpretation of test accuracy, and improved maps.

Keywords: 2013 Curriculum, Secondary School, Learning Methods, Remote Sensing.

1. Introduction

The new remote sensing material included in the subjects geography curriculum in Indonesia since 1994. For geography teachers generation of 90s and over who never get a course of remote sensing, remote sensing for teaching is a difficulty. It can be seen on every visit to the school or the teacher training activities, they always complain the difficulty for teaching remote sensing. Most teachers just give the material form of the theory, they also complained of the lack of facilities and infrastructure of computer laboratories for teaching remote sensing.

Whereas remote sensing material is very important in the study of geography is as one method to obtain data on the earth's surface that can save time, money and energy. The fact today is many teachers who aren't able to distinguish the position of remote sensing and GIS in the study of geography. Geographers in Indonesia views remote sensing are in the same field of study and have a very similar approach. It is though the remote sensing has fundamental differences in the study of geography [1].

Learning remote sensing must be done effectively and efficiently. Effective learning is incompatibility between the lessons and learning objectives formulated in core competencies (KI) and basic competence (KD) in the curriculum. The study is the accuracy of the election efficient learning method in accordance with the characteristics of the material being taught. So that learning can be done effectively and efficiently then the teacher must improve the competence, either deepening mastery of material and teaching methods. Therefore, the need for innovative learning methods for Remote Sensing is essential.

The purpose of the writing is 1) to explain the position of the material in the field of Geography Remote Sensing, 2) analyze the Geography Curriculum 2013 Subjects relating to the material Remote Sensing, 3) describes a method of remote sensing of learning materials in schools. The method in this paper is a descriptive analytical study was supported by the literature.

2. Study of Remote Sensing in Geography

Position remote sensing in the study of geography is as a method to solve the problems that are spatial. Remote sensing plays a role in obtaining the data using the vehicle is not in direct contact with the object. Results of data from remote sensing along with other data sources such as from secondary data and field survey data were processed through GIS analysis to produce new information in decision making.

The position of remote sensing in the study of geography is merely a tool (tools) [1]. At the end of the 1980s along Remote Sensing Geographic Information Systems entered the new field, the geographic information science. In orthodox geography, the position of remote sensing techniques to get in on a clump of geography, while the integrated geography get in on the analysis of spatial approaches (spatial). The core of the spatial analysis is one theory of spatial information. Remote sensing in the study of geography is used to analyze the data objects depicted in the image will be displayed in the form of a map. Remote sensing approaches in the form of spatial, spectral, and temporal used in analyzing geographic phenomena, such as the distribution of socio-economic conditions of urban society, and other types of land use.

Furthermore remote sensing in the study of geography takes a holistic approach is to map the object in the earth's surface through units (unit) mapping analysis, for example, the classification of the landscape, land use, and land units settlements. In the analysis of spatial problem solving, remote sensing always be integrated so that a more thorough study [1].

3. Remote Sensing Materials in the 2013 Curriculum of Secondary School

Remote sensing material has been included in the subjects geography curriculum since 1994. In 2013 the curriculum load sensing material core competencies (KI), basic competence (KD), and the subject matter. The core competence related to the ability to live and practice the religion professed (spiritual attitude), develop behavior (social attitudes), understand and apply knowledge (cognitive), and develop skills (psychomotor).

Basic competence is a translation of core competencies. Spiritual attitudes in remote sensing material must be associated with an increased sense of faith learners, for example, grateful for the gift of God for the ability of the human brain in a remote sensing technique that can make satellite in order to produce data of the earth surface is fast and accurate. This is because the limited human physical capability to map areas that are difficult to reach in order to save time, cost, and energy.

Competence social attitudes or behavior on learning remote sensing, among others capable of cooperation between learners, respect the opinions of others from starting the process of image interpretation to presentation of results interpretation. In addition, the material of remote sensing in the 2013 curriculum should develop the attitude of learners among others honest, disciplined, responsible, caring, and polite.

Competence of knowledge on learning material remote sensing seen from the material on the syllabus, which is the subject matter in the form of remote sensing for land use and transportation, with sub subject the concept of remote sensing (type, aspect of interpretation, the benefits and advantages of remote sensing), remote sensing for land use, remote sensing for transport networks, and governance and remote sensing institutes in Indonesia [2].

Competency skills in remote sensing materials for learners to create land use maps of visual and digital interpretation of remote sensing imagery or aerial photography.

n the 2013 curriculum, learning remote sensing should use the scientific approach which consists of 5M: to observe, ask, collect data, associate and communicate.

In remote sensing material, the skill can be observed land use observe objects depicted on aerial photographs or digital images. Skills ask such learners can ask the seven elements of visual interpretation on aerial photographs or digital images of high spatial resolution. Skills to collect data in the form of learners can delineate the boundaries of land use on aerial photographs or remote sensing image based on the elements of interpretation. Skills can associate the form of an analysis distribution of land uses associated with the transport network. As for the skills to communicate in the form of presenting the results of visual interpretation of the data of land which is used in relation to the transport network.

Remote sensing material is given in schools to create awareness of students about the Earth and its environment by introducing methods of observation from space [3]. The education system should open the students' understanding of the importance of remote sensing and satellite image processing program. Then create student interest in scientific innovation to create a system tool receiver remote sensing image data. The hope is students will continue to develop space technology for multiple uses remote sensing satellites in the future.

4. Materials and Learning Methods of Remote Sensing

Learning is a teaching and learning activity so that learners gain experience, knowledge and competence appropriate to the learning objectives. If learners do not acquire new competencies is that of the learning process is not optimal and should see the issues contained in the learning component.

Learning component consists of teachers, students, teaching methods, teaching materials, media and learning tools, and evaluation tools. Teachers have a very important role in learning because as a driver of the learning process. Professional teachers are teachers who have competence in planning the learning process (RPP), preparing teaching materials, preparing instructional media, teaching methods, and make the evaluation. Therefore, teachers must be competent at each stage of learning, which includes planning, learning activities, and assessment (evaluation / reflection) [4].

The learning method is an effective way used by teachers in presenting the content of learning to achieve a certain competence in accordance with the purpose of learning. There are other types of learning methods, there are lectures, question and answer, discussion, demonstration, practical learning, field trips, projects, and tutorials [5].

Lecture method has the characteristics of teacher orally delivered material. This method is suitable to explain the cognitive abilities such as learning materials such as definitions and concepts. Question and answer method of conveying material through a question and answer both between teacher-student and student-student. A fundamental question posed covers 5 WH (what, why, when, where, who, how). Excess methods become active student in asking questions, thinking critically about the subject matter.

Discussion method has the characteristics of information or knowledge sharing between students, students actively discuss, and teacher as facilitators are on a theme or issue to be discussed. Methods of demonstration or demonstration, characterized displays detailed steps of a process. This method will give a more concrete understanding of the process so students easily master certain skills.

Practical learning methods, typically to acquire certain skills. The location can be in the laboratory and in the field. Practice in the laboratory to train basic competence with some simulations and in the field to apply the basic competence of the laboratory directly in the field. As for the field trips, the characteristics of students take to the field to observe the state of the object or the real issues on the ground.

Method is an integrated learning project on specific competencies to produce a specific product or work. Competence, for example in the form of a map-making skills, knowledge of the mapped region.

The method tutorial, is characterized learners to learn independently through modules. Teachers provide guidance if learners have difficulty understanding and applying these modules.

Geography learning method consists of the method in the room (indoor study) and methods of outdoor (outdoor study). Methods in the room include a lecture, question and answer, discussion, and sociodramas. While outdoors duties and trips [6]. Each method has advantages and disadvantages of learning, so the selection of methods should consider the basic competencies, learning objectives, instructional material characteristics, the characteristics of the student and teacher competence in using the model [4].

Generally appropriate learning methods for remote sensing material that is through the method of practicum (practice). Before using practical methods, remote sensing teaching material must be preceded by a lecture, question and answer, discussion, and demonstration. The detailed study remote sensing should use methods that vary depending on the learning objectives, learning materials and characteristics of learners.

The learning method of remote sensing based material, namely 1) the concept of remote sensing should use the lecture method, question and answer, and discussion. 2) The material interpretation, the benefits and advantages of remote sensing by using question and answer, discussion, demonstration, and practice. 3) The material of remote sensing for land usage can be applied using question and answer, discussion, tutorial, practice, field trips and projects. 4) The material of remote sensing to the transport network, can use the method of discussion, tutorial, practice, field trips and projects. 5) The material governance and remote sensing institutions in Indonesia with lecture, question and answer, discussion, field trips, and tasks.

Methods of laboratory work in remote sensing material, in the laboratory, among others interpreting aerial photographs or imagery visually delineate the boundaries of land use, and make the land use map. Practicum in that field surveys to match the suitability of land use between the interpretation and real situation in the field.

Practical method of remote sensing is done by training and mastery of the operating modules for a wide range of remote sensing applications [3]. Even students are expected to operate a system of remote sensing data receiving in school. Furthermore, the lab system that interactive learning remote sensing through online modules and utilize internet facilities, can enhance the activity and enthusiasm of the students so as to provide a pleasant experience for the students [7].

Teaching material not only in the form of remote sensing theory and practical work in the computer lab, but students must be brought to experiment in the field. Students are invited to obtain data on the earth's surface directly in the field through the camera. One model of learning for the benefit of students is Inquiry Based Learning. Through this model, students do a hypothesis about a theory and predict an event, students will focus on learning, develop the capabilities and skills to perform the investigation of an event [8].

5. Conclusion and Recommendation

The conclusion of this paper that 1) the position of remote sensing in the study of geography is a method or a way to obtain spatial data earth's surface, 2) In the curriculum 2013 matter remote sensing has been applied to the study of land use and transport, 3) Methods of teaching remote sensing to go through a practicum , which starts from the introduction of the theory of remote sensing, phase data extraction from remote sensing imagery in the form of image interpretation both visual and digital, surveying the field to check the situation on the ground, accuracy test to see correspondence between the interpretation and the results of field surveys, repair information on a map and displays a map corresponding cartographic rules.

References

[1] Danoedoro P 2008 Posisi Penginderaan Jauh dalam Perkembangan Ilmu Geografi. Proceeding Filsafat Sains Geografi Yogyakarta 12 Juli 2008 Program Studi Pembangunan Wilayah Fakultas Geografi Universitas Gadjah Mada, 182-189

[2] Somantri, L and Huda, N 2015 Buku Siswa Aktif dan Kreatif Geografi untuk kelas XII Sekolah Menengah Atas/Madrasah Aliyah Peminatan Ilmu-Ilmu Sosial (Grafindo Media Pratama. Bandung)

[3] Yusoff, N.M., Shafri, H.Z.M., Mohamed, A.R 2008 Remote Sensing Educational Ground Receiving System for interest creation in space science and technology in education *International Journal of Education and Development using Information and Communication Technology (IJEDICT)n*2008 Vol. **4** Issue 4 pp 171-182

[4] Ningrum, E 2009 Pendekatan, Model dan Metode Pembelajaran Geografi Makalah Disampaikan pada Kegiatan Pelatihan Induksi Lesson Study dan Team Teaching bagi Guru Geografi SMA se Kabupaten Bandung Tanggal 29 Juni 2009 Jurusan Pendidikan Geografi Fakultas Pendidikan Ilmu Pengetahuan Sosial, Universitas Pendidikan Indonesia

[5] Gintings, A 2008 Esensi Praktis Belajar & Pembelajaran: Disiapkan untuk Pendidikan profesi dan Sertifikasi Guru-Dosen (Humaniora Bandung)

[6] Sumaatmadja, N 1996 Metodologi Pengajaran Geografi (Bumi Aksara Bandung)

[7] Joyce, K.E. dan White, B 2015 Remote Sensing Tertiary Education Meets High Intensity Interval Training *The International Archives of the Photogrammetry, Remote Sensing and Spatial Information Sciences* Volume **XL-7/W3** 2015 36th International Symposium on Remote Sensing of Environment, 11–15 May 2015 Berlin Germany

[8] Mountrakis, G dan Triantakonstantis, D 2012 Inquiry-Based Learning in Remote Sensing: A Space Balloon Educational Experiment *Journal of Geography in Higher Education* Vol. **00**, No 0 1–17 2012

The Use of Hotspot Spatial Clustering and Multitemporal Satellite Imagery to Facilitate Peat Land Degradation in West Kalimantan, Indonesia (Case Study in Mensiku Miniwatershed of Kapuas River)

I Yanuarsyah[1], Y Suwarno[2], S. Hudjimartsu[3]

[1]Geoinformatic of Informatics Engineering Department, Faculty of Engineering, Ibn Khaldun University, Jalan KH Soleh Iskandar KM 2, Tanah Sereal, Bogor, West Jawa, Indonesia
[2]Geospatial Information Agency (BIG), Jalan Raya Jakarta – Bogor KM 46, Cibinong West Jawa, Indonesia
[3]Post Graduate Student, Computer Science Department, Bogor Agricultural University (IPB), Jalan Meranti Wing 20 Level V Kampus Dramaga IPB Bogor, West Jawa Indonesia

iksal.80@gmail.com, yatinsuwarno@yahoo.com, shudjimartsu@gmail.com

Abstract. Peat land in Indonesia is currently a matter of interest to economic activity. In addition to having the uniqueness of the ecosystem which is reserve a huge of biodiversity and carbon storage, peat land is grow an alternative expansion of agriculture and plantation. Mensiku miniwatershed is a subset of Kapuas Watershed with the domination of the peat soil type. It located in the upstream from the Kapuas River and supporting for the continuation of the river ecosystem. The research objective is to facilitate peat land degradation by using hotspot spatial clustering and multitemporal satellite imagery. There have three main processes which are image processing, geoprocessing and statistical process using DBSCAN to determine hotspot clustering. The trend of LUC changes for 14 years (2002 to 2016) shows that the downward occurred in secondary peat forest (0.9% per year) and swampy shrub (0.6% per year). The upward occurred in mixed farms (0.6% per year) and plantations (0.8% per year). degradation rate of peat land over 14 years about 4.6 km^2 per year. Hotspot predominantly occurrence in secondary peat forest with 200-250 centimeter depth and Saprists type. DBSCAN clustering obtain 2 clusters in 2002, obtain 4 clusters in 2009 and obtain 1 clusters in 2016. Regarding LUC platform, average density value over 14 years about 0.063 hotspot per km2. DBSCAN is common used to examine the cluster and perform the distribution and density with spatial analysis

Keyword: peat land degradation, LUC change, clustering, hotspot clustering

1. Introduction

Peat land in Indonesia is currently a matter of interest to economic activity. In addition to having the uniqueness of the ecosystem which is reserve a huge of biodiversity and carbon storage, peat land is grow an alternative expansion of agriculture and plantation. It intensified after the depletion of mineral land reserves for intensively agricultural and plantation. This activity led to the spread of peatlands in Indonesia, particularly in Sumatra and Kalimantan islands increasingly degraded both located in the forest concession area and non-forest concession area.

Nevertheless, the peatlands of Sumatra and Kalimantan continue to have numerous ecological, hydrological and biochemical functions and societal values [1] as much at local as at regional level. In recent years, the remaining peatland forests have also become an increasingly important refuge for endangered animal species [2], due to shrinking area of lowland forests on mineral soils [3].

Based on Center for Agricultural Land Resources Research and Development, Ministry of Agriculture in 2014, total peat land about14.9 Mha. In 2000, peat forest covered 57% of peat land, and decreased 4% until 2005 and so become 49% covers in 2010 [4].

Peat is undecomposed plant material that has accumulated over thousands of year. Due to the absence of oxygen in water-saturated environments, the decomposition of this material has been halted. Through drainage, the dry peat is in contact with the air and starts oxidating, decomposing, and emitting carbon dioxide. This process happens very rapidly in the tropics, and is often accelerated by wildfires. These fires can last for weeks, sometimes even months, burning thick layers of peat over large areas. The peat fires in South-east Asia can burn millions of hectares in one dry season and can last for weeks, sometimes even months, burning thick layers of peat and covering the region in thick smoke. Of the 21 million hectares of peatland in Indonesia 9 million hectares are drained and are decomposing or even burning. [5]

Degraded peatlands is a source of environmental problems as they are a source of greenhouse gas (GHG) emissions and fire prone. These land are unproductive, while the need for the expansion of agricultural land is very high. Research of the Indonesian Climate Change Trust Fund II was carried out to delineate the degraded peatlands, developed methods of management, evaluate the management impacts on crop yields, profitability and GHG emission reduction as well as analyzing the cost (opportunity cost) of emission reduction. The experiment was conducted from September 2012 to August 2013 in the provinces of Riau, Jambi, West Kalimantan, Central Kalimantan and Papua. The results showed that there are about 3.7 million ha (26% of total area of 14.9 million ha) of degraded peatland covered by shrubs plus about 0.6 million ha of bareland from open mining. Most of these lands have the potential for expansion of agriculture. In addition to increasing the economic value, the rehabilitation of degraded peatlands to agriculture tends to reduce GHG emissions and simultaneously improve the condition of the land. Degraded peatland are distributed in various areas of land allotment including non-forest production area (APL), protection forest (HL), convertible production forest (HPK) and forest production (HP) areas. It takes an allotment area rationalization and repositioning policies that degraded land inside the forest area can be rehabilitated for agriculture and forest land in the APL can be conserved to maintain the environmental function [6].

Hotspot is one of forest and land fire indicators that is widely used for developing fire early warning system in the purpose of preventing, controlling and monitoring of wildfire in the area that has high fire risk such as peat land [7].

Initially the hotspot is associated with the point of fire, but in fact not all hotspots indicates a fire points. The term of hotspot is more accurately when it synonymous with the spot of a hot point. A hotspot with high surface temperature value can be indicated of fire existence. This term is widely used in various countries to detect and monitor forest fires and land fires from satellite observation. Hotspot only provide little information and should be supported by further analysis and advanced interpretation. Groups of hotspot and come up continuously is an indicator for the fire occurrences. The determination of the fire-prone areas can be accessed through to landuse and cover change approach from open access of logged forest stand to alternation agriculture and plantations with a massive burned land clearing system.

Mensiku miniwatershed is a subset of Kapuas Watershed with the domination of the peat soil type. It located in the upstream from the Kapuas River and supporting for the continuation of the river ecosystem. Based on the landuse history, most part of forest stand in Mensiku miniwaterhsed had exploited as forest concessions business for 30 years. Secondary peat forest area has degraded into shrub, swamp and open land. This condition may lead the emergence of hotspot that can encourage dangerous condition of fire.

The research objective is to facilitate peat land degradation by using hotspot spatial clustering and multitemporal satellite imagery of Mensiku miniwatershed. This research is useful to examine the pattern and density of the hotspot emergence due to peat landuse change (LUC).

2. Method

Provision of the early vulnerability information of forest fires and land fires is one of the important efforts that need to know the threat of fire and tool of fire controlling. Hotspot spreading data detected by satellite remote sensing are indicators of forest fires and land fires so that analysis of the data needs to be done. Some recent research has been conducted to integrate sophisticated techniques remote sensing, spatial data mining, Geographical Information System, and smoothing algorithms for the new processing and analyzing spatial data such as hotspot data and other supporting data so that it can be obtained by spatial model, map detection and prediction in order to achieve level of forest fires and land fires vulnerability. Some related research has been implementing clustering approach on data mining to perform clustering on a large spatial hotspot data [7].

This research applies some analysis which is hotspot clusters analysis and peat LUC analysis. Hotspot cluster analysis using the DBSCAN algorithm for determining density and spatial pattern. Meanwhile, LUC analysis obtained with comparing landuse cover the period of 7 years of 2002, 2009 and 2016. The tendency focus on the rate of national peatland degradation.

2.1 Location

This research is located at 111°32'35'' E and 0°13'42'' N in Mensiku miniwatershed of Kapuas Waterhsed of 638.64 km2, in administrative of Sintang Regency (West Kalimantan Province), consists of Kelam Permai Sub-districts (10% of Mandiri Jaya Village), Binjai Hulu Sub-districts (10% of Binjai Hulu Village, 12% of Mensiku Village, and 9% of Ampar Bedang Village), Ketungau Hilir Sub-districts (11% of Tanjung Baung Village, 20% of Nanga Ketungau Village, and 27% of Setungkup Village).

2.2 Spatial Dataset

This research required spatial data such as i) Landsat images path 120 Row 60 (1 scene of Landsat 7 acquition in 2002, 1 scene of Landsat 5 acquition in 2009, and 1 scene of Landsat 8 acquition in 2016) with free download at USGS Global Visualization Viewer (http://glovis.usgs.gov/); ii) West Kalimantan administrative boundary with scale 1:250000 produced by BIG in 2010; iii) National peat land deliniation with scale 1:250000 produced by Ministry of Agriculture in 2011; iv) Hotspot data point (acquition in 2002, 2009, and 2016) produced by Ministry of Environment and Forestry and combined with Terra/Aqua MODIS of NASA (https://earthdata.nasa.gov/data/near-real-time-data/firms; v) Peat survey with 40 log point (depth, type). Data processing required particular software such as Quantum GIS 2.12.3 Lyon and R Statistic.

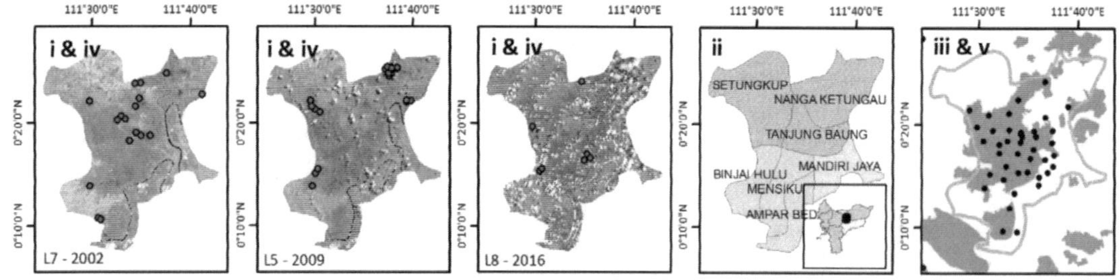

Figure 1. Spatial dataset

2.3 Process Flow

The process flow of this research described in Figure 2, where all the data will be confined by study location (AOI) thus data only be focused on the national peat boundary as actually AOI (AOI_peat). Commonly there has three main processes which are image processing, geoprocessing and statistical process.

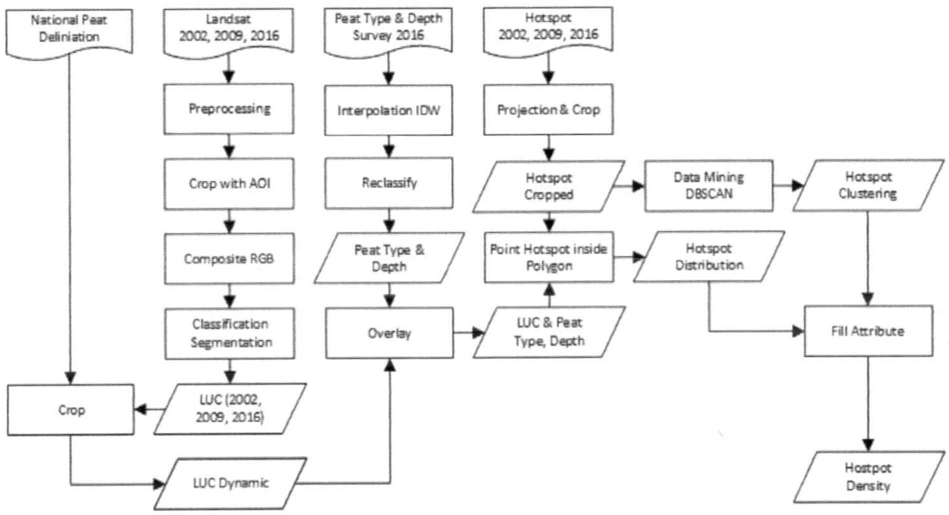

Figure 2. Data processing flow

Performing first step is preprocessing of Landsat imagery which are acquisition 2002 (Landsat 7/L7), 2009 (Landsat 5/L5), and 2016 (Landsat 8/L8). Preprocessing consists corrections (radiometric and geometric) and enhancement (contrast adjustment). The, cropping images in accordance with AOI (Mensiku miniwatershed) and RGB composition (RGB 542 to L7 2002, RGB 542 to L5, 2009 and RGB 1702 to L8 2016) required for unsupervised classification. The classification is based on object segmentation and refer to the Indonesia landuse cover standard (SNI Number 7645 year 2010). As the result, LUC (2002, 2009, 2016) geospatial data will match with national peat delineation (AOI_Peat) to adjust the dynamic rate of landuse cover change in peatland.

In the other side, performing interpolation by using IDW method is needed to map peat survey data (type and depth). Regarding the national interpretation of peat ripeness is grouped as Fibrists (immoldy), Hemists (half-moldy) and Saprists (moldy). In this study, peat type and depth is combined into several group such Peat Group One (< 150 centimeter, D2, Fibrists); Peat Group Two (150-200 centimeter, D2, Hemists); Peat Group Three (200-250 centimeter, D3, Saprists); Peat Group Four (250-300 centimeter, D3, Saprists); Peat Group Five (300-350 centimeter, D4, Saprists); Peat Group Six (350-400 centimeter, D4, Hemists); and Peat Group Seven (> 400 centimeter, D4, Hemists).

Overlay geospatial data is the next stage which are peat type-depth data, LUC data and hotspot data in order to gain the amount of hotspot in each landuse type and period. Statistical clustering approach applied by DBSCAN method in R software to obtain number of hotspot cluster in properly condition such as low of noise, little cluster, well-balanced of epsilon (eps), little minimum points in cluster (MinPts) and small number of error (SSE). The last stage is overlay hotspot cluster and mixed geospatial data to perform hotspot density in each peat group and landuse type.

Clustering algorithms are attractive for the task of class identification in spatial database. However, the well-known algorithms suffer severe drawbacks when applied to large spatial database. The main reason of recognizing the clusters is to promote typical density of points within each cluster which is considerably higher than outside of the cluster. The density within the areas of noise is lower than the density in any of the clusters [8].

Density-Based Spatial Clustering Algorithm with Noise (DBSCAN) is a clustering algorithm that is based on data density. For example, if the amount of data within an epsilon radius is more than or equal to the MinPts (the minimum amount of data within a radius Eps) then the data is included in the category of desired density, where the amount of data within the radius including the data itself. These algorithms perform well on the high density of data and it can find well for any form of groups, so it can separate between the high data density and the low data density of points as well as resistant to noise [9].

3. Result and Discussion

3.1 Landuse cover classification

LUC classification in Mensiku miniwatershed generated 8 classes in 2002, 8 classes in 2009, and 9 classes in 2016, which are Secondary Peat Forest (HRS), Mixed Farms (KBC), Plantation (PKB), Settlement (PMK), Mining (PTB), Agriculture (PTN), Swamp (RWA), Swampy Shrub (SBR), and Water Body (TBA). LUC in 2002 was dominated by HRS about 39.7%, followed by SBR about 26.2% and by 16.5% of agricultural land. LUC in 2009 was dominated by HRS about 37.4%, followed by SBR about 26.7%, agriculture about 16.8% and by 10.1% of swamp area. Meanwhile, LUC in 2016 was dominated by HRS about 27%, followed by SBR about 18.5%, agriculture about 17.1%, mixed farms about 12.2% and plantations about 10.8%.

The trend of LUC changes for 14 years (Figure 3) shows that the downward trend occurred in secondary peat forest about 0.9% per year and swampy shrub about 0.6% per year. The upward trend occurred in mixed farms about 0.6% per year and plantations over 0.8% per year. This condition showed in Mensiku miniwatershed, peat forests has been degraded into plantations and mixed farms.

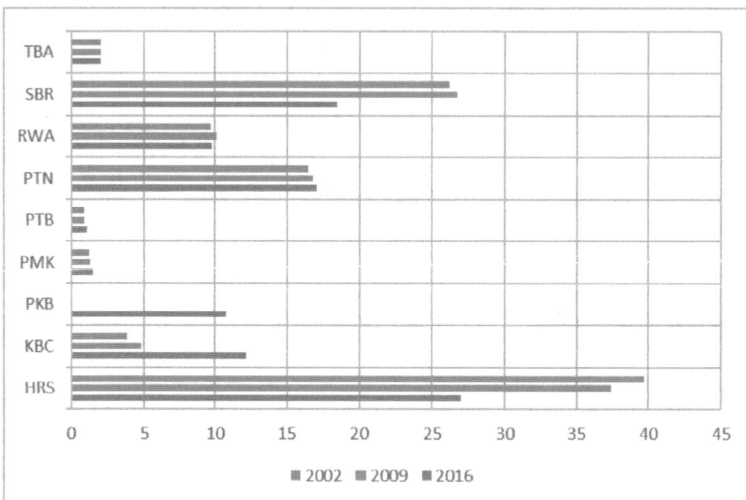

Figure 3. Percentage of LUC 2002, 2009, 2016

3.2 Change detection

More detailed explanation of land cover change detection for particular secondary peat land can be obtained by comparing among periods. During the period 2002 to 2009, secondary peat forests has been converted into other types of LUC such mixed farms about 21.34 ha, swamp area over 134.01 ha and swampy shrub over 488.36 ha. Degradation rate of peat land in period 2002 to 2009 over 91.96 ha per year or 0.92 km^2 per year (Table 1).

Table 1. LUC comparison period 2002 and 2009

		LUC 2009 (Ha)								
		HRS	**KBC**	**PMK**	**PTB**	**PTN**	**RWA**	**SBR**	**TBA**	**Total**
LUC 2002 (Ha)	**HRS**	19520.05	21.34				134.01	488.36		20163.76
	KBC		183.82							183.82
	PMK			35.76						35.76
	PTB				3.02					3.02
	PTN					716.71				716.71
	RWA						1264.92			1264.92
	SBR		13.66				67.18	2109.99		2190.83
	TBA								0.46	0.46
	Total	19520.05	218.82	35.76	3.02	716.71	1466.11	2598.35	0.46	24559.26

During the period 2009 to 2016, secondary peat forest has been converted to other types of LUC in the form of mixed farms about 223.64 ha, plantation over 5079.85 ha, settlement about 13.37 ha, agriculture about 16.44 ha, amount to 61.76 ha of swamp area and swampy shrub about 396.41 ha. HRS degradation rate of peat land in the period 2009 to 2016 amounted to 827.35 ha per year or 8.3 km^2 per year (Table 2). Definitely the degradation rate of peat land over 14 years (period 2002 to 2016) about 4.6 km^2 per year.

Rapid population growth in Indonesia leads a peat land converted into agricultural land in order to support food security, paper industry raw materials, plantations and bioenergy development. Most of peat lands utilization is the conversion for agricultural land, pulp plantations and oil palm plantations. One of the main trigger is the government program to look for an alternative energy by utilizing palm oil as renewable alternative energy. Human activities in the peat ecosystem has increased rapidly in the last two decades so that many areas of peat land become degraded an easier to burn [11].

Table 2. LUC comparison period 2002 and 2009

		LUC 2016 (Ha)									
		HRS	**KBC**	**PKB**	**PMK**	**PTB**	**PTN**	**RWA**	**SBR**	**TBA**	**Total**
LUC 2009 (Ha)	**HRS**	13728.59	223.64	5079.85	13.37		16.44	61.76	396.41		19520.05
	KBC		206.76		12.06						218.82
	PMK				35.76						35.76
	PTB					3.02					3.02
	PTN			0.42			716.29				716.71
	RWA		20.53	50.52				1395.06			1466.11
	SBR		331.96	106.63		10.25	123.66	13.24	2012.61		2598.35
	TBA									0.46	0.46
	Total	13728.59	782.88	5237.42	61.19	13.27	856.39	1470.06	2409.02	0.46	24559.26

3.2 Peat detection

A simple spatial analysis or geoprocessing is used to overlay LUC and peat type and depth in each period (2002, 2009 and 2016) in order to obtained a description of peatland conversion trends associated with peat group. Figure 4 presents the information of each peat group that had been developed. It shows the conversion occurrence over 14 years (2002 to 2009) of secondary peat forest. Large quantities occurred in Group Three (200-250 centimeters, D3, Saprists) and Group Five (300-350 centimeters, D4, Saprists).

IOP Conf. Series: Earth and Environmental Science **47** (2016) 012046 doi:10.1088/1755-1315/47/1/012046

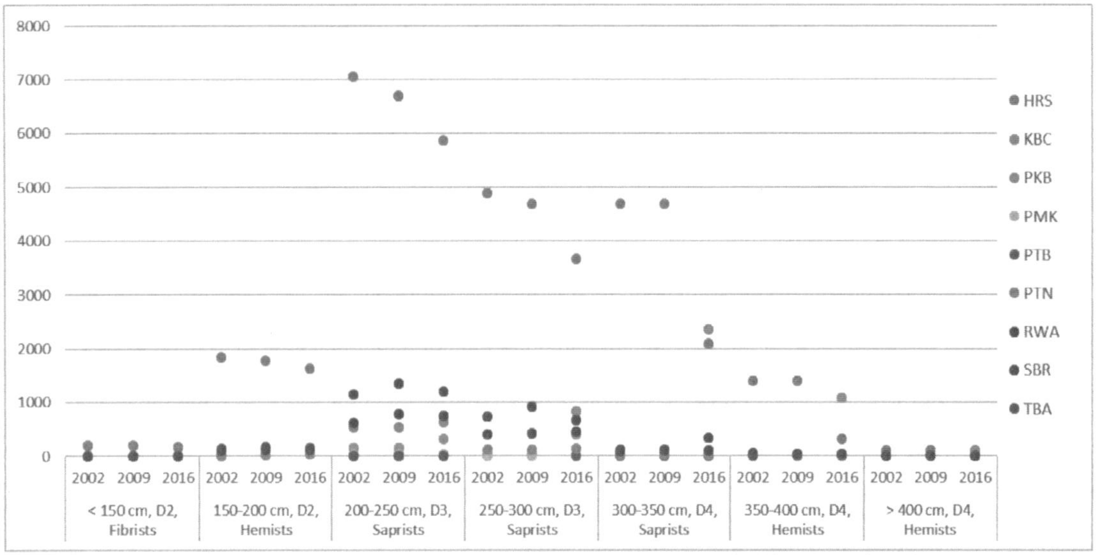

Figure 4. Peat type-depth and LUC comparison in 2002, 2009, 2016

3.3 Hostpot distribution

Hotspot data acquisition actually had recorded in daytime satellite orbit during 02-18 UTC. Regarding national peat boundary in Mensiku miniwatershed, there has 18 hotspot data occurred in 2002 and 25 hotspot data occurred in 2009 and 7 hotspot data occurred in 2016. Figure 5 presents the information of hotspot distribution in each LUC and peat group. It shows the predominantly occurrence along 3 periods (2002, 2009, and 2016) in secondary peat forest with depth over 200-250 centimeter and Saprists type.

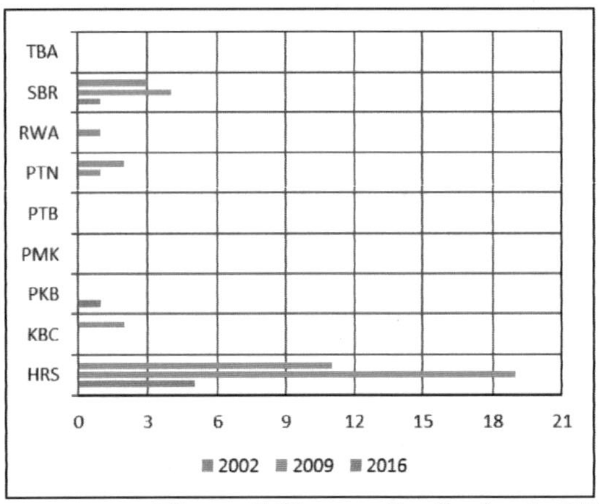

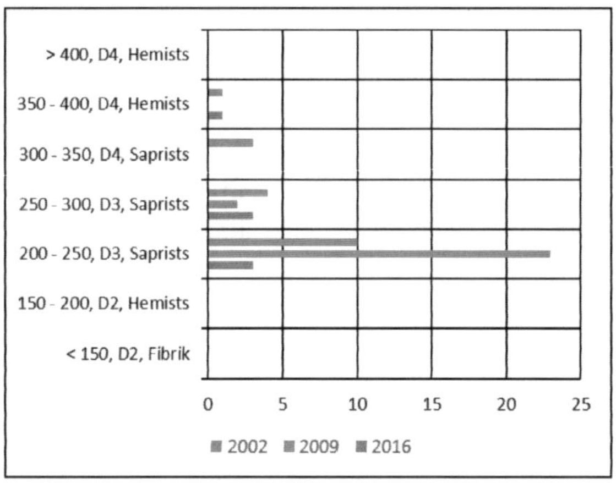

Figure 5. Hotspot distribution in LUC and peat type-depth (2002, 2009, 2016)

The highest density of hotspots often occurs in the farms, the closest distance to the river, the longest distance of the downtown and the closest distance to the road. The fact in Indonesia, any natural forest hardly of forest fires occurrence due to tropical climate with heavy rainfall and high humidity boost the hotspot and another fuel difficult to arise [10]. Secondary peat forest in Mensiku miniwatershed particularly in national peat delineation inside vegetation conditions is uncompact. It accompanied by a number of irregular open access may cause unscrupulous easily enter and exploited forest stand.

Many People assume that the weather or climate can cause forest and land fires. Basically, humans are the main driving factors n fire incidents. When doing burning activities, humans tend to not care about the surrounding environmental condition, so that often found the fire that hard to extinguish or quickly spread to the other land and become uncontrollable. These events have been severely aggravated by improper burning time. People who burns may think low rainfall a high temperature as a help so that burning activities can be done faster and easier. But on the other side, it also makes the effect worse [11].

3.4 Hotspot clustering

Some text Hotspot clustering by using DBSCAN algorithm with appropriate condition obtain compromised result due to determine parameters. DBSCAN clustering for 18 hotspots in 2002 obtain 2 clusters and 0 noise points with 0.07 Eps (epsilon), 2 minPts (minimum points in cluster) and 0.043 SSE (sum of squared errors). DBSCAN clustering for 25 hotspots in 2009 obtain 4 clusters and 0 noise points with 0.025 Eps, 2 minPts and 0.002 SSE. DBSCAN clustering for 7 hotspots in 2016 obtain 1 clusters and 0 noise points with 0.08 Eps, 2 minPts and 0.014 SSE.

3.5 Hotspot density

Hotspot density is composed to adjust the distribution and density of hotspot with LUC and peat type-depth platform. Regarding LUC platform particularly of secondary peat forest, the density value in 2002 obtained of 0.055 hotspot per km^2 (occupied by 201.638 km^2 peat area with 11 hotspot objects) and the density value in 2009 obtained of 0.097 hotspot per km^2 (occupied by 195.20 km^2 peat area with 19 hotspot objects), and the density value in 2016 obtained of 0.036 hotspot per km^2 (occupied by 137.286 km^2 peat area with 5 hotspot objects). The average density value over 14 years (2002 to 2016) about 0.063 hotspot per km^2.

Regarding peat type-depth platform particularly of 200-250 centimeters deep and Saprists peat type, the density value in 2002 obtained of 0.105 hotspot per km^2 (occupied by 95.39 km^2 peat area with 10 hotspot objects) and the density value in 2009 obtained of 0.0241 hotspot per km^2 (occupied by 95.39 km^2 peat area with 23 hotspot objects), and the density value in 2016 obtained of 0.031 hotspot per km^2 (occupied by 95.39 km^2 peat area with 3 hotspot objects). The average density value over 14 years (2002 to 2016) about 0.126 hotspot per km^2.

The study revealed that in 2002, the highest density of hotspot found in Riau province with the density value of 0.226 km2 and South Sumatra province with the density value of 0.201 km2. Distribution of hotspot in Riau province were mostly found in the area that has peat types of Hemists/Saprists (60/40) with very deep peat, and Hemists/Saprists (60/40) with moderate depth and spread evenly on the area with a moderate depth (100-200 cm), in the deep peat (200-400 cm), and very deep peat (> 400 cm) consecutively. Hotspot distribution in South Sumatera province were mostly found in peat land area that has peat types of Hemists/mineral (90/10) and mostly found in areas with moderate depth (100-200 cm). Land covers of peat land in both regions are dominated by peat swamp forest. In the year of 2013, the highest density of hotspots on peat land area of Sumatra was found in Riau province with the density value of 0.056 km2, where most of the hotspot was on peat types of Hemists/Saprists (60/40), and very deep peat land. Land covers of peat land are dominated by peat swamp forest. It seems that there were changes in the pattern of distribution of hotspot and land cover of peat land from 2002 to 2013, where in 2002 the distribution of hotspot are mostly found in moderate depth (100-200 cm) peat land, whereas in 2013, the distribution of hotspot were mostly found in very deep (> 400 cm) peat land area [12].

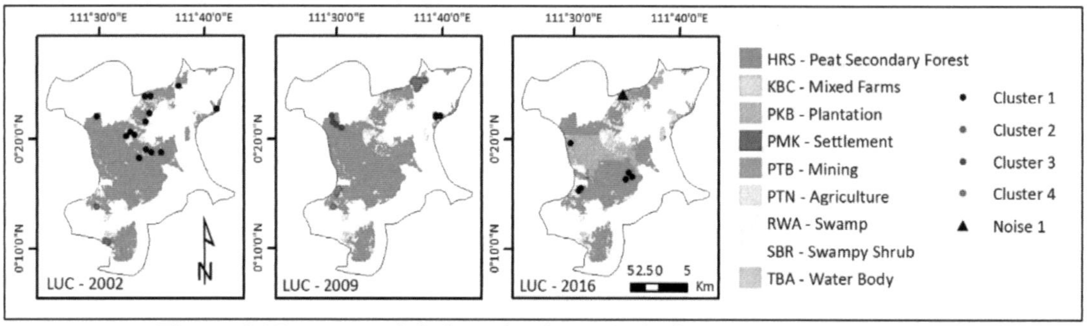

Figure 6. Hotspot spatial clustering in LUC platform (2002, 2009, 2016)

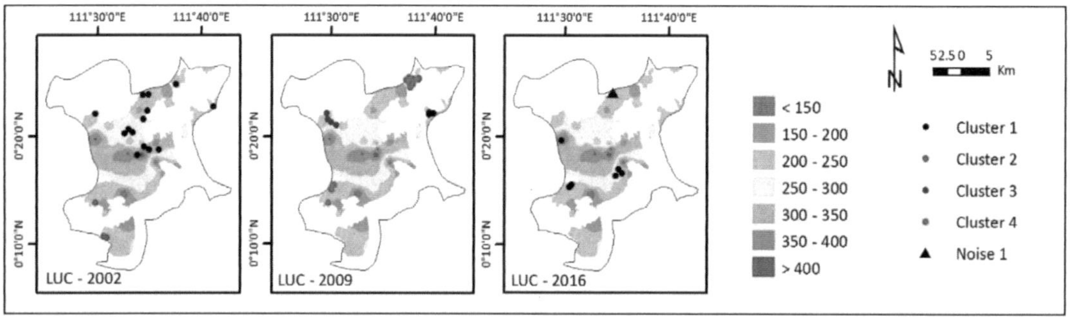

Figure 7. Hotspot spatial clustering in peat depth platform (2002, 2009, 2016)

4. Conclusion

This study demonstrates the enormous potential hotspot clustering and multi temporal imagery to facilitate peat land degradation. DBSCAN is common used to examine the cluster and perform the distribution and density with spatial analysis. Future study could be held to investigate hotspot spatial clustering considering degradation level for achieving clearly result of peat degradation.

References

[1] Rieley J O and Page S E eds 2005 *Wise use of tropical peatlands: Focus of Southeast Asia* Wageningen The Netherlands ALTERRA Wageningen University and Research Centre and the EU INCO STRAPEAT and RESTORPEAT Partnership

[2] Morrogh-Bernard H Husson S Page S E Rieley J O Population status of the Bornean orangutan (*Pongo pygmaeus*) in the Sebangau peat swamp forest Central Kalimantan Indonesia *Biological Conservation* 2003 **110**:141–152

[3] Miettinen J and Liew S C 2010 *Status of Peatland Degradation and Development in Sumatra and Kalimantan* Royal Swedish Academy of Sciences *Ambio* 2010;39:394-401 p 1 2

[4] Agus F 2015 Degradation and Sustainable Management of Peat Soils in Indonesia *MARCO Symposium Tsukuba International Congress Center Japan p 7*

[5] Wetland 2006 *Peatland Degradation Fuels Climate Change* Wetland International

[6] Agus F Wahyunto Sosiawan H Subiksa I G M Setyanto P Dariah A Maswar Nurida N L Nurida Mamat H S Las I 2014 *Sustainable Management Of Degraded Peatland : Trade-Offbetween Economic And Environmental Benefits* Prosiding Seminar Nasional Pengelolaan Berkelanjutan Lahan Gambut Terdegradasi untuk Mitigasi GRK dan Peningkatan Nilai Ekonomi Badan Penelitian dan Pengembangan Pertanian Kementerian Pertanian Jakarta ISBN 978-602-8977-83-8 p 19

[7] Usman M 2014 *Spatial Clustering* Berbasis Densitas Untuk Persebaran Titik Panas Sebagai Indikator Kebakaran Hutan Dan Lahan Gambut Di Sumatera Muhammad Usman Sekolah

Pascasarjana Institut Pertanian Bogor Bogor p 5 16

[8] Ester M Kriegel H Sander J Xu X 1996 A Density-based algorithm for discovering cluster in large spatial database with noise *Proceeding 2nd Int Conf. on Knowledge Discovery and Data Mining*. https://www.aaai.org/Papers/KDD/1996/KDD96-037 p 2 6

[9] Patil Y S Vaidya M B *A technical survey on cluster analysis in data mining* International Journal of Emerging Technology and Advance Engineering 2012; 2 (9): pp 503-13.

[10] Andini N Anita Z Samsuri 2014 Model Spasial Tingkat Kerawanan Kebakaran Hutan dan Lahan di Kabupaten Rokan Hilir *Jurnal Universitas Sumatera Utara* www.jurnal.usu.ac.id index.php/PFSJ/article/download/14151/6267 p 9

[11] Nurdiana A Setiawan Y Pawitan H Permatasari Prasetyo L B Ayu P 2015 Land changes monitoring using MODIS time-series imagery in peat lands areas, Muaro Jambi, Jambi Province, *Indonesia The 2nd International Symposium on LAPAN-IPB Satellite for Food Security and Environmental Monitoring LISAT-FSEM p 2 6*

[12] Usman M Sitanggang I S Syaufina L 2015 Hotspot Distribution Analyses Based on Peat Characteristics Using Density-based Spatial Clustering *The 1st International Symposium on LAPAN-IPB Satellite (LISAT) for Food Security and Environmental Monitoring p 9*

Land Cover Mapping using GEOBIA to Estimate Loss of *Salacca zalacca* Trees in Landslide Area of Clapar, Madukara District of Banjarnegara

Anggi Permata, Anwar Juniansah, Eka Nurcahyati, Mousafi Dimas Afrizal, Muhammad Adnan Shafry Untoro, Na'ima Arifatha, Raden Ramadhani Yudha Adiwijaya, Nur Mohammad Farda

Department of Geographic Information Science Faculty of Geography Gadjah Mada University, Indonesia

muhammad.adnan.s@mail.ugm.ac.id

Abstract. Landslide is an unpredictable natural disaster which commonly happens in highslope area. Aerial photography in small format is one of acquisition method that can reach and obtain high resolution spatial data faster than other methods, and provide data such as orthomosaic and Digital Surface Model (DSM). The study area contained landslide area in Clapar, Madukara District of Banjarnegara. Aerial photographs of landslide area provided advantage in objects visibility. Object's characters such as shape, size, and texture were clearly seen, therefore GEOBIA (Geography Object Based Image Analysis) was compatible as method for classifying land cover in study area. Dissimilar with PPA (PerPixel Analyst) method that used spectral information as base object detection, GEOBIA could use spatial elements as classification basis to establish a land cover map with better accuracy. GEOBIA method used classification hierarchy to divide post disaster land cover into three main objects: vegetation, landslide/soil, and building. Those three were required to obtain more detailed information that can be used in estimating loss caused by landslide and establishing land cover map in landslide area. Estimating loss in landslide area related to damage in Salak (*Salacca zalacca*) plantations. This estimation towards quantity of Salak tree that were drifted away by landslide was calculated in assumption that every tree damaged by landslide had same age and production class with other tree that weren't damaged. Loss calculation was done by approximating quantity of damaged trees in landslide area with data of trees around area that were acquired from GEOBIA classification method.

Keywords: GEOBIA, Land Cover Mapping, Aerial Photography, Landslide, Loss Estimation

1. Introduction

Landslide is described as a mass movement of materials, including rocks, soil or debris that happens in slope area and is affected by gravitation force. Landslide mainly occurs in area with steep slope, significant rainfall, developed soils, and hilly topography. This area can be found commonly in Banjanegara region. On March 25th 2016, an event of landslide occurred and caused loss in infrastructure yet no human injury. The landslide area extends from north-west to south east of Clapar area. It became a necessity to collect and measure data of landslide area and land cover affected by the event.

OBIA (Object-Based Image Analysis) is a classification method based on object which can be applied in imagery data. This method is divided into two parts of processes, which are image segmentations and classification on each segment [5]. This method classifies object recorded by aerial photography or satellite imagery based on three parameters which are object scale, object shape, and object compactness.

This method has more advantages related to object classification, resulting more accurate and higher precision of classification. This method is also easier and more time-efficient in image processing and object classification than pixel-based classification [1]. Furthermore, this classification considers not only object's spectral aspect, but also its spatial aspect [4].

This paper presents how OBIA classification detecting differences between each land cover that surrounds over landslide area. The segmentation process results three types of image segmentation based on three types of land covers, which are landslide area itself, vegetation, and building/local settlements. The result was used to generate a land cover map, and also to measure the extent of each type of land covers. These measurements can be used to calculate on how much loss in each type of land cover area, especially in salak plantations. This paper shows the manifestation of using OBIA into identifying and measuring loss caused by local disasters in small scale area by detecting and differentiating between damaged land cover and undamaged land cover.

2. Study Site

Clapar village is located in an area with a high-level intensity of landslide hazard. This level can be seen from how the landslide is occurred frequently in this region, Banjarnegara. Data provided by Badan Nasional Penanggulangan Bencana (BNPB) shows that there were 25 landslide locations had been recorded, including Clapar Village, in range from 11-12 December 2014. The largest landslide in Clapar Village itself was occurred in March 25, 2016, swallowing almost half of the village's region. The hazard happened due the spring that should be flowing down form the river but because the lack of space, the water seeped onto the ground, making the soil very soft, and causing the ground to flow as gravitation pull it down. The physical characteristics of the site itself also had quite-impacting factors to the landslide, such as low-stability type of soil and steep slope, making the motion to move faster.

Topology area with steep slope makes the condition of soil moving faster. Rainfall reaching in Clapar is calculated for about 301 to 400 millimeters per month. Clapar has a specific rock formation called Rambatan Formation. Crown Clapar landslide in the village is located at an elevation of 837 meters above sea level with a slope of 23%. While the tip of the tongue landslide Clapar located at an elevation of 705 meters above sea level with a slope of 12%. The crown part of the landslide was known to be used in plantation of *Salacca zalacca* or Salak. Clapar possible landslides village lied in fault zones or fault of the north-south direction.

IOP Conf. Series: Earth and Environmental Science **47** (2016) 012047 doi:10.1088/1755-1315/47/1/012047

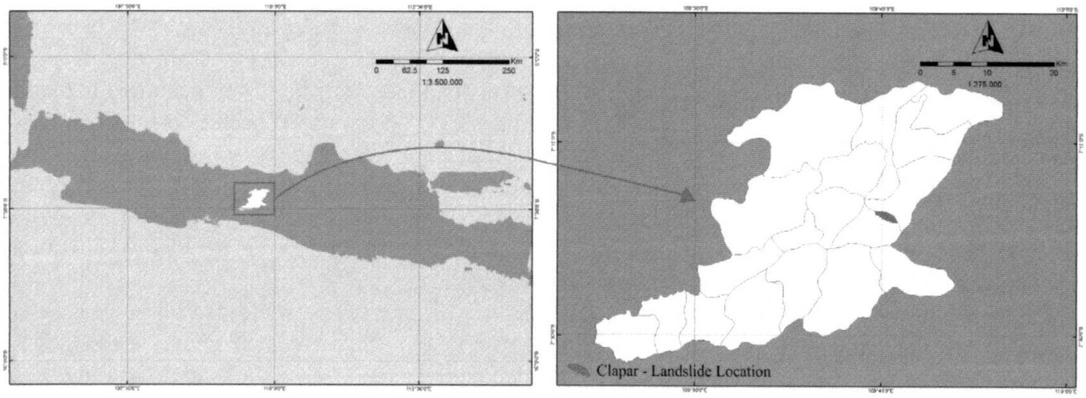

Figure 1. Location of landlides on Clapar, Madukara District of Banjarnegara.

3. Methods

Land cover mapping used GEOBIA to estimate loss of Salak trees consist of 5 steps: 1) processing aerial photographs segmentation using the example based feature extraction method edge in ENVI 5.2; 2) sampling for each object defined as landslides, vegetation, and settlement; 3) classifying aerial photographs segmentation results based on the sample that has been taken; 4) masking landslide areas with consideration of landslide potential area; and 5) estimating loss of Salak using the data of landslide areas and data of plantation space of Salak tree.

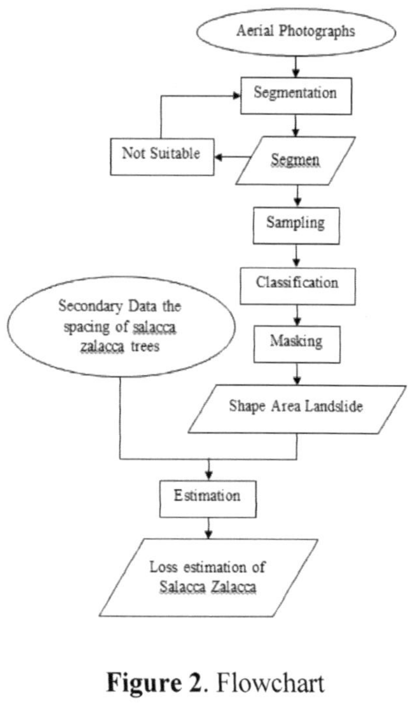

Figure 2. Flowchart

Example based feature extraction is a method of segmentation in ENVI 5.2 utilizing the sample as a basis for classifying objects. This method is a simple OBIA method that uses two algorithms, which are intensity algorithms and edge algorithms. Intensity is suitable method for data that has a continuous spectral value. The algorithm works by converting pixel values into intensity values on

Object Creation Panel. Edge is suitable method for objects that have clear boundaries and internal parts consisting identical pixel value. The land cover objects existing in aerial photographs shows clear boundaries between each type of land cover, so that edge algorithms for segmentation is selected. To combine identical segments, merge function is used. There are two algorithms on the merge function, which are Full Lamda Schedule and Fast Lamda. Fast Lamda combines segments pixels with similiar values and limits. Values merge level selected number 13.5 with Lamda Fast algorithm that produced the segment is sized to fit.

Segmentation result needs to be classified as types of required land covers. Samples are required for each required type so that the system can recognize the characteristics of each segments. The required amount of sample was taken for each class adjusting the level of diversity. For landslide, 3029 segments were taken as sample, 565 segments were taken as sample for local settlement, and 5414 segments were as sample for vegetation. Samples were taken equally on each spectral range for each class. Classification was processed using gained samples into three required types of existing land use.

Loss of damaged Salak tree was estimated in landslide area, therefore, measurement of landslide area was needed. Estimating landslide area was limited in area formerly identified as Salak plantations, which was determined by association between each land cover. The results determining potential landslide area was calculated by dividing Salak area with plantation space between each Salak tree.

$$L = \frac{SA}{ST}$$

L : Loss value of *Salacca Zalacca*
SA : Area of landslide (m^2)
ST : Spacing distance between each Salak tree/Salak canopy (m^2)

Plantation space of salak tree was used to model the area of one salak canopy. Seven samples of plantation space was measured in the field. The means of the samples is 197,85 cm. Sample's mean was used to build salak canopy model to estimated thea area of one salak tree. Area of one salak tree canopy is 3.9145 m^2.

Table 1. Plantation space between samples in the filed

Sample	Space (cm)
1	243
2	222
3	123
4	175
5	212
6	166
7	244
Mean	197.85

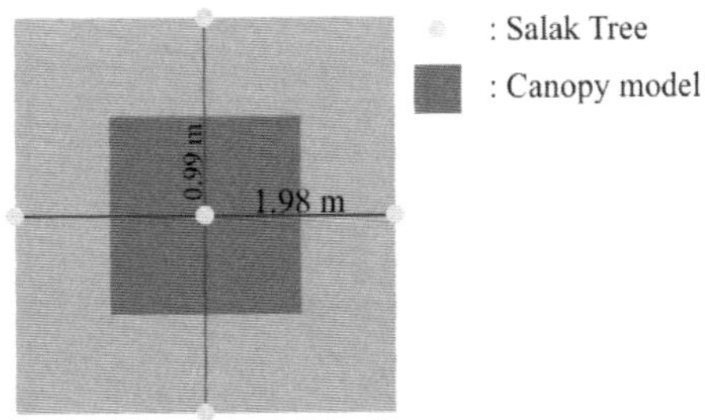

Figure 3. Canopy Model

4. Results

4.1. Land cover map

Land cover map generated from GEOBIA classification consisted of three types of land cover, which were landslide area/soil, vegetation, and building/local settlement. This simple classification of land covers was intended to gain clear form of boundaries between landslide area and other land covers.

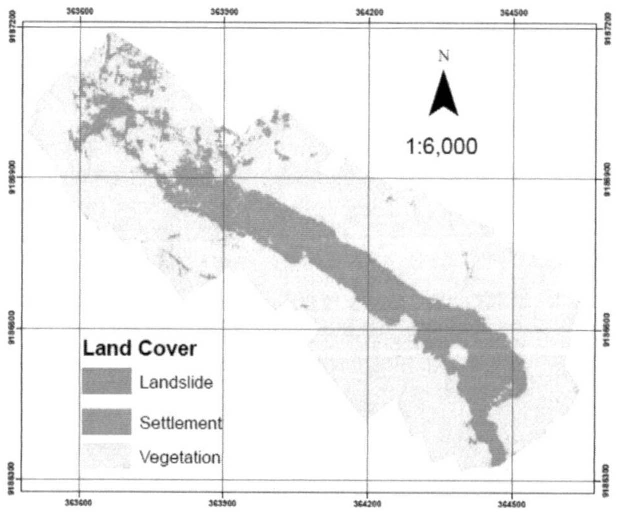

Figure 4. Land cover map generated by GEOBIA method

Accuracy assessment was performed to check the accuracy of the land cover map. 35 samples generated using stratified random sampling method. Unfortunately, only 20 samples checked because of the hard terrain and bad weather. All samples were classified correctly so the accuracy of this output map is 100%. Main factor why this land cover map accuracy is 100% because the classification quite simple. It's only consists of three main object, there there are landslide/soil, vegetation, and building/local settlement.

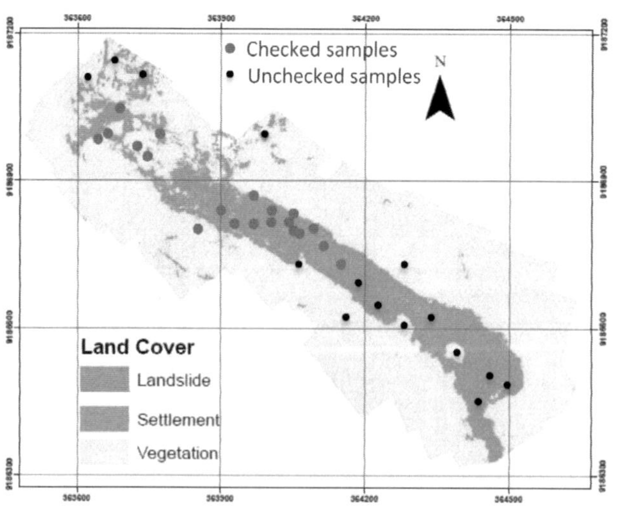

Figure 5. Distribution of samples

4.2. Estimation result

About 16.585 Salak trees were lost due to landslides and 64,923.9197 m2 of land area damaged by landslide were estimated from GEOBIA method. The result was obtained from the assumption of the extent of tree canopy calculated as 3.9145 m^2.

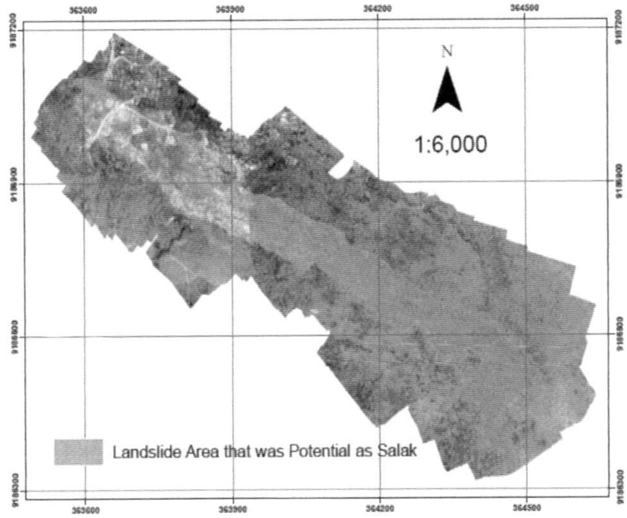

Figure 6. Landslide area formerly identified as Salak plantations map

5. Discussion

Segmentation process with OBIA method generated three types of land cover. Segmentation were processed by edge method which differentiated each segments by considering angular differences of each object. This method was suitable with aerial imageries that shown clear differences between objects and spectral gradation. In this segmentation method, scale factor value dan merge level value were needed and adjusted by aerial imageries condition with various type of land cover. The greater the scale and merge level, the classification will be generalized. Hence, scale factor around 55 and merge level around 30.1 were selected so that the segmentation result would be suitable with aerial imageries's appearance.

The output of segmentation was classified by example-based method. This method was using example samples based on each type of required classes. The distribution of sample was considering characteristics of each object. The amount of sample can be reviewed in figure 5, with amount of sample of landslide material was 3,029 segments, amount of sample of local settlement was 565segments, and amount of sample of vegetation was 5,414 segments.

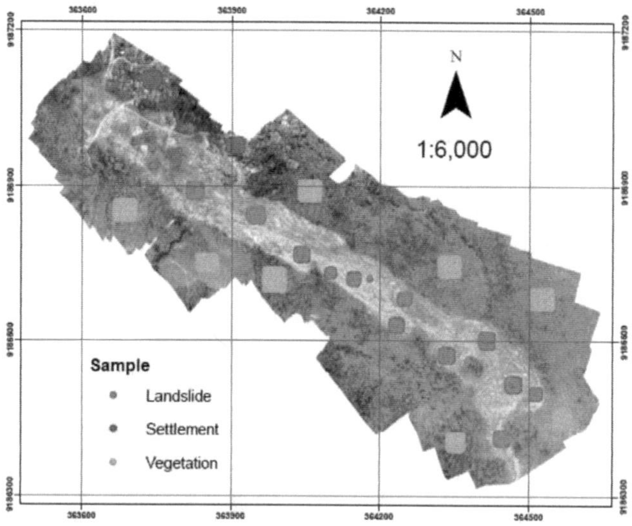

Figure 7. Samples distribution map

The result of classification showed three different types of land cover around landslide area, which was vegetation, local settlement, and materials of landslide itself. This result can be used to measure area of each land cover type, with the result showed that vegetation has largest area in the study area. By comparing between Salak area affected or not affected by landslide, we can estimate the damaged area of Salak plantation. Determining the extent of Salak area damaged by landslide was done with masking method. Masking method was processed by intersecting area formerly identified as Salak area. After that, interpretation manual were processed to determine the extent of masked area. The result showed that the masked area, also identified as damaged Salak area, extended to 64,923.9197 m2. This result showed that there were 16.585 canopies of Salak were damaged. This result also showed that OBIA method was suitable to estimate the amount of canopies which were damaged by landslide. Total canopies damaged by landslide were calculated by area of damaged Salak divided with area of each canopy.

6. Conclusion

GEOBIA segmentation can be used to land cover mapping and estimate loss of Salak in Clapar, Banjarnegara. The result of classification showed land use's appearance which was assumingly suitable with real object condition. The size of landslide area that was formerly potential as Salak area was calculated to 64,923.9197 m2 with 16,585 damaged Salak canopies were estimated. It is considered quite reasonable because the size of the landslide area and the density of canopies or spaces between each Salak tree. The recommendation for further research is to use this method in other area with different land cover to check its capability.

References

[1] Noviar H Carolita I and Cahyono J S 2012 Uji Akurasi Training Sampel Berbasis Objek Citra Landsat Di Kawasan Hutan Provinsi Kalimantan Tengah *Pusat Pemanfaatan Penginderaan Jauh, LAPAN Jurnal Ilmiah Geomatika* **18** No 2 Desember 2012 page 134

[2] Uddin K Gurung K R Giriraj A and Shrestha B 2013 Application of Remote Sensing and GIS for Flood Hazard Management A Case Study from Sindh Province, Pakistan *American Journal of Geographic Information System* 2163-1131

[3] Selby M J 1985 *Earth's Changing Surface an Introduction to Geomorphology* (Clarendon Press *Oxfor*)

[4] Wibowo TS and Suharyadi *Aplikasi Object-Based Image Analysis (OBIA) untuk Deteksi Perubahan Penggunaan Lahan Menggunakan Citra ALOS AVNIR-2* page 131

[5] Xiaoxia S Jixian Z and Zhengjun L 2004 *A Comparison of Object-Oriented and Pixel-Based Classification Approachs Using Quickbird Imagery* (Chinese Academy of Surveiing and Mapping Beijing China)

Institute of Physics
Dirac House, Temple Back
Bristol BS1 6BE UK

ISSN: 1755-1307
ISBN 978-1-5108-3521-4

9 781510 835214

CURRAN ASSOCIATES INC.
proceedings.com